MONOGRAPHS ON THE PHYSICS AND CHEMISTRY OF MATERIALS

General Editors
H. FRÖHLICH, P. B. HIRSCH,
N. F. MOTT

Nonequilibrium Thermodynamics and its Statistical Foundations

H. J. KREUZER

University of Alberta

CLARENDON PRESS · OXFORD

Oxford University Press, Walton Street, Oxford OX2 6DP
London Glasgow New York Toronto
Delhi Bombay Calcutta Madras Karachi
Kuala Lumpur Singapore Hong Kong Tokyo
Nairobi Dar es Salaam Cape Town
Melbourne Auckland
and associates in
Beirut Berlin Ibadan Mexico City Nicosia

Oxford is a trade mark of Oxford University Press

Published in the United States
by Oxford University Press, New York
© H.J. Kreuzer 1981

All rights reserved. No part of this publication may be reproduced,
stored in a retrieval system, or transmitted, in any form or by any means,
electronic, mechanical, photocopying, recording, or otherwise, without
the prior permission of Oxford University Press

This book is sold subject to the condition that it shall not, by way
of trade or otherwise, be lent, re-sold, hired out or otherwise circulated
without the publisher's prior consent in any form of binding or cover
other than that in which it is published and without a similar condition
including this condition being imposed on the subsequent purchaser

First published 1981
First published in paperback 1983

Kreuzer, H.J.
Nonequilibrium thermodynamics and its statistical
foundations.—(Monographs on the physics and
chemistry of materials)
1. Statistical thermodynamics
I. Title   II. Series
536'.7   QC311.5

ISBN 0-19-851375-5

Printed in the United States of America

Für
meine
Eltern

Preface

Nonequilibrium thermodynamics is a vast field of scientific endeavour with roots in physics and chemistry. It has applications in all branches of the physical sciences and engineering, and more recently in a number of interdisciplinary fields, including environmental research and, most notably, the biological sciences. To cover all of this ground thoroughly in one volume is obviously impossible. Rather than simply touch upon many of the very interesting aspects and applications of the theory, therefore, I have restricted myself to covering a few essential topics in depth.

The first half of the book presents the phenomenological theory of nonequilibrium thermodynamics. I briefly review the established fields with some emphasis on the historical background, and give, where possible, applications of the theory. All of chapter 6 on Bénard convection is to be understood as an application of the theory and of ideas of previous chapters to a system where both theory and experiment have advanced to a very high level of sophistication.

The second half of the book presents the statistical foundations of nonequilibrium thermodynamics and includes both a concise exposition of kinetic theories and a derivation of the hydrodynamic and thermodynamic balance equations. The task facing nonequilibrium statistical mechanics—namely the derivation and justification of kinetic equations from the equations of motion of an N-body problem, the derivation of macroscopic balance equations of nonequilibrium thermodynamics, and the calculation of transport coefficients appearing in the constitutive laws—has so far been completed only for a classical dilute gas. I follow this program in chapter 7, which may be viewed as the centerpiece of the second part of the book. I also present a thorough discussion of irreversibility, a century-old problem for which one can now give some answers in a careful analysis of the objectives of a physical theory. Where active areas of research are touched on, I have tried to present a critical assessment of what has been established and of what is still at the speculative stage. Where battle is still raging among different schools of thought, for example over the derivation and role of master equations, an admittedly biased report of the various strategies is presented.

Some interesting topics had to be omitted entirely from the book. These choices have been difficult, and only slightly eased by the fact that excellent books and review articles on some of the topics related to but not covered in this book have recently appeared. For example the exciting field of turbulence has not been touched at all, but the reader is referred here to two long books (with a combined total of 1633 pages!) on statistical fluid mechanics: mechanics of turbulence by A.S. Monin and A.M. Yaglom (MIT Press, Cambridge, 1971, 1975). Another relevant and very active area of nonequilibrium thermodynamics, relativistic thermodynamics and relativistic kinetic theory, had to be left out in its entirety. For an introductory survey and assessment of the field I refer the reader to an article by W. Israel and J.M. Stewart (1979; see bibliography). To supplement the discussion in chapter 4 on stability and in chapter 5 on chemical oscillations I would like to suggest a very thorough article by B.L. Clarke on stability of complex reaction networks in *Advances in Chemical Physics*, Vol. 43, Eds. I. Prigogine and S.A. Rice, Wiley, New York, 1980. Ergodic theory, also not covered in this book, has seen a revival in the last ten to fifteen years; I refer to a review by J. Ford on the transition from analytic dynamics to statistical mechanics (*Advances in Chemical Physics*, Vol. 24 (1973) 155–185) and to a review by O. Penrose on foundations in statistical mechanics in *Reports on Progress in Physics*, Vol. 42 (1979) 1937–2006. In the last five to ten years the theory of chaos has attracted the attention of mathematicians, chemists, and physicists alike. Its impact on statistical physics, e.g. in ergodic theory and for far from equilibrium phenomena, cannot be grasped yet. This field is evolving so rapidly that I can only advise the reader to watch for conference proceedings on the topic.

To keep this book to a manageable size I have refrained from adding the customary chapters on prerequisite topics like quantum mechanics, thermodynamics, and equilibrium statistical mechanics. I feel that such chapters are, of necessity, too short to be useful and, moreover, very seldom read. Excellent treatments of these topics include the textbooks by L.I. Schiff or A. Messiah on quantum mechanics, H.B. Callen on thermodynamics, and J. Kestin and E.D. Dorfman on equilibrium statistical mechanics. I should mention that some chapters (like 8, 10, 12) are at a somewhat higher level of mathematical sophistication than others. I have tried to develop the necessary techniques (like field quantization) as I go along. I feel that the statistical foundations of nonequilibrium thermodynamics demand mathematical rigor, which can be achieved most easily in the theoretical framework that is most adapted to the description of many-body systems, namely second quantization.

This book grew out of graduate lectures on nonequilibrium statistical physics which I have given at the University of Alberta since 1974. A set of lecture notes on Topics in Nonequilibrium Physics was prepared by J. Beamish. While working on the manuscript I had the chance of presenting the material in courses in 1977–78 at the University of Regensburg, the Technion-Israel Institute of Technology in Haifa, and the Universities of Adelaide and New South Wales in Australia. I wish to take this opportunity to thank Professors G.M. Obermair and U. Krey of Regensburg; C.G. Kuper and S. Eckstein of Haifa; H.S. Green and P. Szekeres of Adelaide; and J. Oitmaa and D.N. Lowy of Sydney for many stimulating and clarifying discussions during and after my lectures. In particular, I wish to thank P. Szekeres for reading and commenting on chapters 2 and 3, and Professor H.S. Green for his comments on chapter 7. I also feel deeply indebted to my colleague M. Razavy for patiently listening to my lectures, for innumerable long discussions on nonequilibrium statistical physics, for reading the entire manuscript, and for help in preparing the index. I would like to thank Z.W. Gortel for help in proofreading. The first draft of the manuscript was typed by my wife Uta; the final version was produced by Mary Yiu. Many thanks to both of them!

Edmonton H.J.K.
July 1980

Contents

List of Symbols, xv

1. **Introduction, 1**
2. **Balance equations of irreversible thermodynamics, 18**
   2.1 Balance equations for mechanical quantities, 18
   2.2 Balance equation for the viscoelastic stress tensor, 28
   2.3 Entropy balance and the second law of thermodynamics, 32
   2.4 Balance equations for solids, 36
3. **Linear phenomenological laws, 40**
   3.1 Choosing forces and fluxes, 40
   3.2 The Onsager reciprocity relations, 44
   3.3 Transformation properties of Onsager relations, 56
   3.4 A first example: thermodiffusion, 60
   3.5 A second example: thermoelectricity, 63
   3.6 Minimum entropy production, 67
4. **Stability and fluctuations, 71**
   4.1 Stability theory: an outline, 71
   4.2 Stability of equilibrium states, 76
   4.3 Stability of nonequilibrium states, 82
   4.4 The general evolution criterion (Glansdorff and Prigogine), 85
5. **Chemical reactions, 91**
   5.1 Equations of motion and stability, 91
   5.2 A chemical reaction model with a nonequilibrium phase transition, 95

5.3 Volterra-Lotka model, 98
5.4 Chemical oscillations with limit cycle behavior, 107

6. Bénard convection, 111
6.1 Statement of the problem, 111
6.2 Linearized excess balance equations and boundary conditions, 116
6.3 Normal mode analysis in the linear theory, 118
6.4 Convection cell patterns from the linear theory, 123
6.5 Experimental results, 128
6.6 Nonlinear theory, 135
6.7 Variational principles and stability criterion of Glansdorff and Prigogine, 143

7. Classical statistical mechanics and kinetic theory, 149
7.1 Introduction, 149
7.2 The BBGKY hierarchy, 155
7.3 Microscopic derivation of balance equations, 158
7.3.1 Mass balance, 159
7.3.2 Momentum balance, 160
7.3.3 Energy balance, 164
7.3.4 Entropy balance, 168
7.4 Derivation of constitutive laws, 169
7.4.1 Vicscosity, 170
7.4.2 Thermal conduction, 174
7.5 Simple kinetic equations: Vlasov and Boltzmann, 177
7.5.1 Preliminaries and a derivation of Vlasov's equation, 177
7.5.2 Derivation of Boltzmann's equation, 179
7.5.3 Discussion of the Boltzmann equation, 186
7.6 Balance equations from the Boltzmann equation, 190
7.7 Constitutive laws and transport coefficients from the Boltzmann equation, 196
7.8 Derivation of Ohm's law from the Boltzmann equation, 206

8. Microscopic derivation of balance equations: quantum-mechanical theory, 209
8.1 Outline of approach and introduction of reduced density matrices, 209

- 8.2 Fröhlich's derivation of the equations of hydrodynamics in the reduced density formalism, 214
- 8.3 Operator balance equations, 221
- 8.4 Macroscopic balance equations from the operator hierarchy, 226

9. Linear response theory, 235

- 9.1 The formalism of linear response theory, 235
- 9.1.1 Introductory remarks, 235
- 9.1.2 Classical response theory, 239
- 9.1.3 Quantum-mechanical response theory, 243
- 9.2 General properties of response functions, 247
- 9.2.1 Symmetries, analyticity, and dispersion relations, 247
- 9.2.2 Sum rules and fluctuation-dissipation theorem, 251
- 9.2.3 Current response and Onsager reciprocity relations, 258
- 9.2.4 Density-density correlation functions from scattering experiments, 261
- 9.3 Hydrodynamic fluctuations and transport coefficients, 269
- 9.3.1 Linearized hydrodynamics, 269
- 9.3.2 Connection with linear response, 277
- 9.4. Practical results, 280
- 9.4.1 The ideal gas, 280
- 9.4.2 Electrical conductivity of metals, 286

10. Master equations, 296

- 10.1 Introduction, 296
- 10.2 Pauli's master equation, 297
- 10.3 Van Hove's master equation, 306
- 10.4 Prigogine's approach to nonequilibrium statistical mechanics, 312

11. Irreversibility and the approach to equilibrium, 322

- 11.1 Defining the problem, 322
- 11.2 Irreversibility in the Boltzmann equation, 327
- 11.2.1 The $\mathcal{H}$ theorem, 327
- 11.2.2 The Loschmidt paradox: time reversal, 330
- 11.2.3 The Zermélo paradox: Poincaré recurrence, 336

11.3   Irreversibility in systems of coupled oscillators, 341
11.3.1 The thermodynamic limit in the classical harmonic chain, 341
11.3.2 Time evolution of a reacting two-component gas, 348
11.4   Irreversibility in an ensemble, 358
11.5   Generalized $\mathcal{H}$ theorems, 364

12. Transient effects in the time evolution of an ideal gas in an external potential, 370

12.1   Formulation of the problem, 270
12.2   Time evolution of operators $a_k(t)$, 373
12.3   Density evolution, 376
12.4   Weak potentials, 379
12.5   Boundstates, 386
12.6   Resonances, 390
12.7   Appendix: a single particle in a separable potential, 396

Bibliography, 403
Author Index, 425
Subject Index, 429

List of Symbols

(Number in parentheses indicates page where symbol is defined or first introduced.)

LATIN SYMBOLS

- a dimensionless wave number in Bénard convection (119)
- $a_\mathbf{k}$ annihilation operator of a particle in momentum state $\mathbf{k}$ (281)
- $a_\mathbf{k}^\dagger$ creation operator of a particle in momentum state $\mathbf{k}$ (281)
- A_δ affinity of component δ (92)
- B Bulk modulus (30)
- $\mathbf{B}$ magnetic field (45)
- c number of components in multi-component system (19)
- c_γ mass fraction (77)
- c_p specific heat per unit mass at constant pressure (113)
- c_v specific heat per unit mass at constant volume (78)
- C_p specific heat per mole at constant pressure (9)
- C_V specific heat per mole at constant volume (8)
- d thickness of fluid layer in Bénard convection (112); diameter of hard sphere molecule (188)
- $\dfrac{D}{Dt}$ barycentric derivative (20)

	e	total energy density (23); electronic charge (64)
	E	total energy density (255)
	$\mathbf{E}$	electric field (64)
$f_N(\boldsymbol{\xi}_1,\ldots,\boldsymbol{\xi}_N,t)$		N-particle distribution function (155)
	$\mathbf{F}_\gamma$	external force per unit mass (21)
	$\mathbf{g}$	gravity (112)
	G	Gibbs free energy (46); shear modulus (30)
	$\hbar$	Planck's constant
	$h(\mathbf{r},t)$	local $\mathscr{H}$ quantity (192)
H_N, H, H_0		Hamiltonian (152, 239)
$\hat{H}, \hat{H}_0$		Hamiltonian operator (209, 212)
	i,j	labels
	$\mathbf{I}$	unit tensor (22)
	$\mathbf{j}_F$	current of F (19)
	$\mathbf{j}_\gamma^{\text{diff}}$	diffusion current of component γ (20)
	$\hat{\mathbf{j}}(\mathbf{x})$	particle current operator in quantum mechanics (215)
	J_δ	rate for reaction δ (19)
	k	label
	$\mathbf{k}$	particle momentum (281); wave number (118)
	k_B	Boltzmann's constant
	l	mean free path (3, 13); label
	L_{ik}	transport coefficients in linear phenomenological laws (42)
	m	label; particle mass
	n	label; mole number (7)
	n_γ	mole number of component γ (7)
	N	number of particles (4)
	N_A	Avogadro's number (13)
	Nu	Nusselt number (130)
	p	hydrostatic pressure (3)
	$\mathbf{p}$	momentum

LIST OF SYMBOLS

	P	pressure tensor (21)
	Pr	Prandtl number (119)
	$P[F]$	production of F (18)
	q	heat per unit mass (24)
	Q	heat (33)
	r	position vector with cartesian components r_α
	r_{12}	distance between **r** and $\mathbf{r}_1$
	R	gas constant (13); Rayleigh number (115)
	R_c	critical Rayleigh number
	$s(\mathbf{r}, t)$	entropy density per unit mass (33)
	S	entropy (3)
	$\hat{S}$	instantaneous (fluctuating) entropy (7)
	$S(|\mathbf{k}|)$	structure factor (267)
	$S_{mn}(\mathbf{r}, t; \mathbf{r}_1, t_1)$	correlation function (253)
	t	time
	T	temperature
	Tr	Trace of a matrix or operator
	u	internal energy per unit mass (23)
	u	displacement vector with cartesian components u_α (29)
	U	internal energy (3)
	v	specific volume per unit mass (77); speed
	v_s	speed of sound (4)
	v	velocity with cartesian components v_α
	V	volume
	$V_{ij} = V(\mathbf{r}_i, \mathbf{r}_j)$	two-body interaction potential (156)
	W	probability function (7, 48)
	X_i	thermodynamic force (40)
	Y_i	thermodynamic flux (40)

GREEK AND SCRIPT SYMBOLS

	α	thermal expansion coefficient (9); label
	α_i	thermodynamic variable (41)

	β	inverse temperature; label
	γ	thermodynamic phase space spanned by thermodynamic variables (52); 6-dimensional single-particle phase space; label
	Γ	6 N-dimensional phase space (52)
	$\delta\mathcal{P}$	fluctuation in $\mathcal{P}$ (6)
	$\Delta\mathcal{P}$	finite increment in $\mathcal{P}$ (6)
$\boldsymbol{\nabla}_{r_i} = \partial/\partial \mathbf{r}_i$		gradient operator
	$\delta^2 s$	second order differential of s (excess s) (77)
	$\boldsymbol{\varepsilon}$	strain tensor with cartesian components $\varepsilon_{\alpha\beta}$ (29)
$\varepsilon^{(K)}(\mathbf{r}, t)$		kinetic energy density (158)
$\varepsilon^{(V)}(\mathbf{r}, t)$		interparticle potential energy density (158)
	ζ	bulk viscosity (26, 30)
	η	shear viscosity (26, 30)
	η_{rot}	rotational viscosity (26)
	θ	angle; temperature fluctuation (116)
	κ	thermal diffusivity (115)
	κ_T	isothermal compressibility (9)
	κ_{ji}, κ_3	rate constants (45, 95)
	λ	thermal conductivity (14)
	λ_1	intrinsic (kinetic) length scale (e.g., mean free path, or size of local equilibrium cell) (5)
	Λ	rate-of-strain tensor with cartesian coordinates $\Lambda_{\alpha\beta}$ (29)
	μ	chemical potential per mole (3)
	μ_γ	chemical potential per mole of component γ (3)
	ν	kinematic viscosity (115)
	ν_{coll}	collision frequency (12)
	$v_{\gamma\delta}$	stoichiometric coefficient (19)
$\boldsymbol{\xi} = (\mathbf{r}, \mathbf{p})$		combination of position and momenta vector in phase space (155)
	Π	Peltier coefficient (65)

	Π	viscous (dissipative) part of pressure tensor (24, 26)
$\rho(\mathbf{r}, t)$		mass density (18)
ρ_γ		mass density of component γ (19)
ρ		ensemble density in classical statistical mechanics (153)
$\hat{\rho}$		density matrix (statistical operator) in quantum statistical mechanics (211)
$\hat{\rho}_0$		equilibrium density matrix (212)
$\hat{\rho}(\mathbf{x}', \mathbf{x}''; t)$		nonlocal particle density operator (222)
$\rho_n(\mathbf{x}'_1, \ldots, \mathbf{x}'_n; \mathbf{x}''_1, \ldots, \mathbf{x}''_n; t)$		reduced nth order density matrix (212)
	σ	molecular diameter (14); electrical conductivity (65)
$\sigma(\chi, |\mathbf{p} - \mathbf{p}_1|)$		scattering cross section (185)
	σ_F	source density of F (19)
	$\boldsymbol{\sigma}$	stress tensor with cartesian components $\sigma_{\alpha\beta}$ (29)
	$d\boldsymbol{\Sigma}$	surface element (19)
	τ_0	microscopic (dynamic) time scale (e.g., duration of a binary collision) (150)
	τ_1	intrinsic (kinetic) time scale (e.g., for regression of fluctuations or τ_{coll}) (5)
	τ_{coll}	collision time (151)
	τ_{ev}	time scale of macroscopic evolution (5)
	$\phi[F]$	flow of F (18)
$\phi_{mn}(\mathbf{r}, \mathbf{r}_1; t - t_1)$		linear response function (246)
	$\Phi(\mathbf{r}_i)$	external potential (156)
$\Phi_{mn}(\mathbf{r}, \mathbf{r}_1; t)$		linear relaxation function (246)
	χ	angle
$\chi_{mn}(\mathbf{k}, z)$		Fourier transformed linear response function (247)
	ψ	potential energy per unit mass (22)
	ψ_γ	potential energy per unit mass of component γ (22)

$\psi(\mathbf{x}, t)$	field operator in second quantization (209)
$\boldsymbol{\omega}$	vorticity (116); angular velocity tensor (25)
$\mathscr{F}_k$	intensive thermodynamic variable (7)
$\mathscr{H}(t)$	Boltzmann's $\mathscr{H}$ quantity (192)
$\mathscr{P}$	some thermodynamic property (6)
$\mathscr{X}_k$	extensive thermodynamic variable (7)
$\hat{\mathscr{X}}_k$	instantaneous (fluctuating) thermodynamic variable (7)

1
Introduction

NONEQUILIBRIUM thermodynamics‡ is a phenomenological macroscopic field theory concerned with states and processes in systems out of equilibrium. Primarily, the theory gives a unified treatment of steady-state and transport phenomena in continuous media, but it also deals with the approach of systems toward steady states and their stability, and it examines relaxation phenomena during the approach to equilibrium. The scope of the theory thus goes far beyond equilibrium thermodynamics where, according to Planck, only 'reversible' processes are considered. In nonequilibrium thermodynamics we are then concerned with finite and irreversible processes which are, in general, induced by opening a system to an external supply of energy or matter. The fact that the system has been taken away from equilibrium reflects itself in a finite, i.e. macroscopic, change in some thermodynamic variables that may now also become space- and time-dependent. The resulting processes, whether they are stationary or time-dependent, will always evolve with a positive entropy production.

‡ A scholastic dispute has been going on among the experts on what the appropriate name should be for this discipline of physics. It is argued that equilibrium thermodynamics should be renamed thermostatics for obvious reasons, whereas nonequilibrium thermodynamics should be simply referred to as thermodynamics because only in this theory are dynamical changes in macroscopic systems dealt with. In this book one name is used for the whole discipline—thermodynamics. The attributes of equilibrium or nonequilibrium are added to delineate our particular field of endeavor. The latter branch will also be referred to as thermodynamics of irreversible processes.

In nonequilibrium situations, systems in a steady state play an equally important role as thermally isolated systems in equilibrium. Just as the latter are characterized by a maximum of the entropy, we find for some of the former, namely those that are close to equilibrium in some sense, a minimum in the entropy production. However, this principle is not universally true for systems in a steady state. In particular, it does not hold for the very interesting situations far from equilibrium that to some extent can even depend on the preparation, i.e. the history, of the system.

Historically,‡ the roots of thermodynamics of irreversible processes are found in the phenomenological laws of viscous flow (Newton, 1687), heat conduction (Fourier, 1822), diffusion (Fick, 1855), and electrical conduction (Ohm, 1826). The modern unified approach seems to have started with Bertrand (1887) who first pointed out the central role of entropy production in nonequilibrium systems. The general theory, particularly for systems in which two or more linear transport phenomena are coupled together—such as the electrokinetic effect (Reuss, 1809) and the thermoelectric effect (Thomson, or Lord Kelvin, 1854)—was greatly advanced by Onsager (1931). He introduced certain symmetry relations and made the basic connections with microscopic physics. Thermodynamics of irreversible processes was developed to its present form, following pioneering work by Eckart (1940) in the Aachen–Amsterdam–Brussels triangle (Meixner, de Groot, Prigogine, and coworkers) in the years since with the establishment of a set of general balance equations governing thermodynamic systems out of equilibrium. The theory currently encompasses detailed studies of the stability of systems far from equilibrium, including oscillating systems. In this context, the notion of nonequilibrium phase transitions is gaining in importance as a unifying theoretical concept. (See Haken, 1975.)

The thermodynamic theory of irreversible processes restricts itself to large systems that can be treated as continuous media and can be assumed to be in local equilibrium. That is to say, we assume that we can divide the system under study into cells small

‡ A small collection of historical notes on thermodynamics and statistical mechanics can be found in *Handbuch der Physik* III/2 (Springer, Berlin, 1959). See, in particular, the articles by E. A. Guggenheim and by J. Meixner and H. G. Reik.

enough so that the thermodynamic properties of the system vary little over each cell but large enough so that the cells can be treated as macroscopic thermodynamic subsystems in contact with their surrounding. By this, we wish to imply that in each cell we can define thermodynamic variables and functions such as pressure, density, temperature, internal energy, and entropy as constants in each cell. However, we also wish to vary them from cell to cell in such a way that the thermodynamic variables can be described as continuous space- and time-dependent fields such as mass density $\rho(\mathbf{r}, t)$, velocity $\mathbf{v}(\mathbf{r}, t)$, internal energy $u(\mathbf{r}, t)$, temperature $T(\mathbf{r}, t)$, and entropy $s(\mathbf{r}, t)$. It will be our first task to establish differential balance equations for these field variables, giving us a local description of the thermodynamics of the system.

It is the assumption of local equilibrium that makes it possible to meaningfully define a local entropy $s(\mathbf{r}, t)$ that is the same function of the local thermodynamic variables as the equilibrium entropy is of the equilibrium thermodynamic parameters. This implies that the fundamental differential form

$$dS = \frac{1}{T} dU + \frac{p}{T} dV - \sum_{\gamma=1}^{c} \frac{\mu_\gamma}{T} dn_\gamma \tag{1.1}$$

is valid locally and that the local equation of state is independent of field gradients. Here S denotes the entropy, U the internal energy, T the temperature, p the pressure, V the volume, c the number of chemical components, μ_γ the chemical potential per mole of the γth component, and n_γ the corresponding mole number.

To explicitly state the conditions under which the assumption of local equilibrium is valid, the methods of nonequilibrium statistical mechanics are required. So far, this has been done rigorously and explicitly only for a dilute gas. In this case (Meixner, 1941, 1943) one studies the Chapman–Enskog solution of the Boltzmann equation (Enskog, 1929) and finds that local equilibrium can be assumed to hold in systems in which temperature variations ΔT over a mean-free path l are much smaller than the average temperature in a cell of that size, i.e.

$$\frac{\Delta T}{T} \approx \frac{l\, |\boldsymbol{\nabla} T|}{T} \ll 1 \tag{1.2}$$

Here $\boldsymbol{\nabla}$ denotes the spatial gradient. Similarly, one would demand for pressure variations that

$$\frac{\Delta p}{p} \approx \frac{l \, |\boldsymbol{\nabla} p|}{p} \ll 1 \tag{1.3}$$

Moreover, macroscopic velocities $\mathbf{v}(\mathbf{r}, t)$ should vary little over a mean-free path l as compared to the sound velocity v_s, i.e.

$$\frac{l \, |\boldsymbol{\nabla} \cdot \mathbf{v}|}{v_s} \ll 1 \tag{1.4}$$

One would, of course, like to have criteria like eqns (1.2)–(1.4) in any system. We will now show that general criteria can, indeed, be gotten without reference to specific kinetic models.

To develop our argument, we first consider a large, but finite system in thermodynamic equilibrium in which a thermodynamic parameter, e.g. the volume V of the system, is changed by a 'small' fraction dV resulting in a 'small' change, e.g. in the internal energy U, namely dU. If this change occurs at a vanishingly small rate and through a continuous sequence of equilibrium states, it is said to have occurred via a quasistatic process. If the latter does not increase the entropy of the system, we call it a reversible process. We should ask ourselves at this point by what mechanism the system can actually adjust itself in this reversible manner.

The answer has to come from the fact that any system studied in thermodynamics is a large, but *finite* many-body system of N particles, and it will exhibit relative fluctuations in its macroscopic thermodynamic variables that are typically of order $N^{-\frac{1}{2}}$. These statistical fluctuations are, of course, time-reversal invariant and are, indeed, the only universal reversible phenomena in nature. The response of a system to a 'small' change in some externally controlled parameters is then invoked by these omnipresent fluctuations and a new equilibrium will be achieved on the characteristic time scale of these fluctuations. To ensure that a process induced in the system is evolving reversibly, i.e., that it proceeds through a sequence of equilibrium states, we must therefore insist that the external control variables of the system are changed slowly on the time scale of fluctuations and that at each step the changes remain within the limits of these fluctuations. A 'small' relative change in a thermodynamic parameter therefore has to be of the order $N^{-\frac{1}{2}}$ during a reversible, or more generally, a quasistatic process.

To suppress fluctuations in equilibrium thermodynamics one limits oneself to infinitely large systems. But this necessitates that reversible (quasistatic) processes must be thought of as proceeding by infinitesimally small increments at an infinitesimally slow rate. In a real, i.e. finite, physical system, 'infinitesimally small' means within the fluctuations of the system and 'infinitesimally slow' means slow on the time scale of fluctuations.‡

Let us now extend these ideas to nonequilibrium systems in local equilibrium. We suppose that such systems can be subdivided into cells small enough that any thermodynamic properties—such as mass density ρ, pressure p, and temperature T—which in nonequilibrium situations can be functions of space and time, vary slightly over one cell. On the other hand, these cells must be large on a microscopic scale—large enough so that they can still be treated as thermodynamic subsystems in contact with their surroundings. In particular, the number N_j of constituent particles in cell j must be large enough that statistical mechanics can be done on them. In other words, N_j must be large enough that thermodynamic variables can be defined through a partition function Ξ_j in each cell j. This necessitates that all fluctuations stay within bounds. In particular, for the number fluctuations in cell j we must have $\delta N_j/N_j \ll 1$, where an upper tolerance on the size of such fluctuations must be either postulated theoretically or provided experimentally.

The subdivision of a nonequilibrium system into local equilibrium cells is not merely a theoretical trick but must be inherent in the system. There must, first of all, exist at least one intrinsic length scale λ_1 of the system such that the number of constituent particles in a cell of size λ_1, $N_j = (N/V)\lambda_1^3$, where (N/V) is the average number density of particles, must show small fluctuations $\delta N_j/N_j \ll 1$. Next, there must also exist an intrinsic time scale τ_1 in the system such that a macroscopic fluctuation dies away within one cell much faster than the overall time evolution of the system, which we characterize by a time scale τ_{ev}. This implies that over time intervals Δt, such that $\tau_1 \ll \Delta t \ll \tau_{ev}$, local equilib-

‡ We note that close to a phase transition fluctuations in extensive parameters diverge, growing to macroscopic size, and can cause finite reversible changes, i.e. discontinuities, in thermodynamic functions.

rium is maintained in each cell. But this implies that (up to fluctuations) a partition function can be constructed from which we can define such macroscopic thermodynamic variables as pressure p, internal energy u, and temperature T as well as derive a local equation of state. In a system in macroscopic motion we can further introduce a local velocity $\mathbf{v}$ as the center of mass velocity of each cell. For processes that vary slowly in the above coarse-grained space-time manifold we can next perform a continuum limit on the (so far discrete) thermodynamic variables (e.g., by formally letting τ_1 and λ_1 approach zero in the appropriate way), thus introducing continuous space- and time-dependent fields like $T(\mathbf{r}, t)$, $u(\mathbf{r}, t)$, $\rho(\mathbf{r}, t)$, and $\mathbf{v}(\mathbf{r}, t)$.

As implied above, the local equilibrium cells must be open for energy and/or mass transport in order to account for the overall macroscopic time evolution of the system. Resulting changes in the local properties of the system must, however, be compatible with the set of general balance equations derived in Chapter 2. The question then arises: How big can externally generated field gradients and fluxes, such as pressure gradients, temperature gradients, heat fluxes, etc., be so that local equilibrium can still be maintained? To answer this we must realize that changes in local thermodynamic properties of a given cell produced by the exchange of matter and/or energy with its neighbors can only come about through local fluctuations in that cell. We note in particular that in systems in local equilibrium, spatially neighboring cells have thermodynamic states that are also close to each other in thermodynamic phase space. This implies, according to the arguments about equilibrium systems made above, that relative changes in the thermodynamic parameters of neighboring cells must be of the order of the fluctuations in each cell.

In order that local equilibrium is not destroyed, we must obviously insist that field gradients must be such that induced relative changes in thermodynamic variables over a cell, i.e. over distances λ_1, must be less than or of the order of equilibrium fluctuations in that cell. So if a local thermodynamic property $\mathcal{P}$ has fluctuations $\delta \mathcal{P}$ in a cell of size λ_1, and if an external gradient $\nabla \mathcal{P}$ produces a change $\Delta \mathcal{P} = \lambda_1 |\nabla \mathcal{P}|$ over a distance λ_1, then in order for local equilibrium to be maintained in the system we must demand that

$$\frac{\Delta \mathcal{P}}{\mathcal{P}} \lesssim \frac{\delta \mathcal{P}}{\mathcal{P}} \ll 1 \qquad (1.5)$$

This is our basic criterion for the validity of the local equilibrium assumption in nonequilibrium systems. It must, of course, be supplemented with a condition on the rate with which macroscopic fields can change in time, namely

$$\tau_{ev} \gg \tau_1 \tag{1.6}$$

The criterion (1.5) can only be evaluated if the intrinsic length scale λ_1 and the relevant equilibrium fluctuations $\delta\mathcal{P}/\mathcal{P}$ for the particular system under study are known explicitly. We therefore collect here the basic results from the theory of equilibrium fluctuations.

Let us assume that the equilibrium states of a system with c different components are characterized by a set $\{\mathcal{X}_i\}$ of extensive variables like internal energy $\mathcal{X}_0 = U$, volume $\mathcal{X}_1 = V$, and mole numbers $\mathcal{X}_2 = n_1$ to $\mathcal{X}_{c+1} = n_c$. If the system is in contact with appropriate reservoirs, these extensive variables will not be constant but will fluctuate randomly around their equilibrium values. In Tisza's postulational approach to thermodynamics (Callen, 1960) one would then specify (1) that there exists an instantaneous entropy

$$\hat{S} = \hat{S}(\hat{\mathcal{X}}_0, \hat{\mathcal{X}}_1, \ldots) \tag{1.7}$$

which is a function of the instantaneous extensive variables $\{\hat{\mathcal{X}}_i\}$, and (2) that the probability $W\,d\hat{\mathcal{X}}_0, \ldots, d\hat{\mathcal{X}}_{c+1}$ that the instantaneous extensive variables are in a range $d\hat{\mathcal{X}}_i$ around $\hat{\mathcal{X}}_i$ is given by

$$W(\hat{\mathcal{X}}_0, \ldots, \hat{\mathcal{X}}_{c+1}) = W_0 \exp\left\{\frac{1}{k_B}\left[\hat{S} - \sum_{k=0}^{c+1} \mathcal{F}_k \hat{\mathcal{X}}_k - S(\mathcal{F}_0, \ldots, \mathcal{F}_{c+1})\right]\right\} \tag{1.8}$$

where $k_B = 1.38 \times 10^{-23}\,J/K$ is Boltzmann's constant, $\mathcal{F}_k$ are the equilibrium values of the intensive parameters in the entropy representation, i.e.

$$\begin{aligned}
\mathcal{F}_0 &= \left(\frac{\partial S}{\partial U}\right)_{V, n_1, \ldots, n_c} = \frac{1}{T} \\
\mathcal{F}_1 &= \left(\frac{\partial S}{\partial V}\right)_{U, n_1, \ldots, n_c} = \frac{p}{T} \\
\mathcal{F}_k &= \left(\frac{\partial S}{\partial n_k}\right)_{U, V, n_1, \ldots, n_c} = -\frac{\mu_k}{T} \quad \text{for } k = 2, \ldots, c+1
\end{aligned} \tag{1.9}$$

characterizing the reservoirs. Finally,
$$S(\mathscr{F}_0, \ldots, \mathscr{F}_{c+1}) = S - \sum_k \mathscr{F}_k \mathscr{X}_k \qquad (1.10)$$
is the Legendre transform of the equilibrium entropy with $\mathscr{X}_k$ being the equilibrium values of the extensive parameters, i.e.
$$\mathscr{X}_k = \langle \hat{\mathscr{X}}_k \rangle = \int W \hat{\mathscr{X}}_k \, d\hat{\mathscr{X}}_0, \ldots, d\hat{\mathscr{X}}_{c+1} \qquad (1.11)$$
The normalization constant is fixed by demanding that
$$\int W \, d\hat{\mathscr{X}}_0, \ldots, d\hat{\mathscr{X}}_{c+1} = 1 \qquad (1.12)$$
Concentrating on the second-order fluctuation moments for the deviations
$$\delta \hat{\mathscr{X}}_i = \hat{\mathscr{X}}_i - \mathscr{X}_i \qquad (1.13)$$
one can show from eqn (1.8) that
$$\langle \delta \hat{\mathscr{X}}_i \, \delta \hat{\mathscr{X}}_j \rangle = -k_B \left(\frac{\partial \mathscr{X}_i}{\partial \mathscr{F}_j} \right)_{\{\mathscr{F}_l, \mathscr{X}_k\}} = -k_B \left(\frac{\partial \mathscr{X}_j}{\partial \mathscr{F}_i} \right)_{\{\mathscr{F}_l, \mathscr{X}_k\}} \qquad (1.14)$$
where the subscripts $\{\mathscr{F}_l, \mathscr{X}_k\}$ indicate that all variables except $\mathscr{X}_i$ and $\mathscr{F}_j$ are held constant. In particular, if a single component system is in contact with a heat reservoir, with all other extensive variables kept constant, we find for the internal energy fluctuations
$$(\delta U)^2 = \langle (\delta \hat{U})^2 \rangle$$
$$= -k_B \left(\frac{\partial U}{\partial (1/T)} \right)_{V,n} = k_B T^2 n C_V \qquad (1.15)$$
where
$$C_V = \frac{T}{n} \left(\frac{\partial S}{\partial T} \right)_{V,n} \qquad (1.16)$$
is the specific heat per mole at constant volume. If such a system is simultaneously in contact with heat and volume reservoirs, then both U and V can fluctuate and we get
$$\langle (\delta \hat{U})^2 \rangle = -k_B \left(\frac{\partial U}{\partial (1/T)} \right)_{p/T,n}$$
$$= k_B T^2 n C_p - k_B T^2 p V \alpha + k_B T p^2 V \kappa_T \qquad (1.17)$$
$$\langle \delta \hat{U} \, \delta \hat{V} \rangle = k_B T^2 V \alpha - k_B T p V \kappa_T$$
$$\langle (\delta \hat{V})^2 \rangle = k_B T V \kappa_T$$

where
$$C_p = \frac{T}{n}\left(\frac{\partial S}{\partial T}\right)_p \quad (1.18)$$
is the specific heat at constant pressure,
$$\alpha = \frac{1}{V}\left(\frac{\partial V}{\partial T}\right)_p \quad (1.19)$$
is the coefficient of thermal expansion, and
$$\kappa_T = -\frac{1}{V}\left(\frac{\partial V}{\partial p}\right)_T \quad (1.20)$$
is the isothermal compressibility. From eqn (1.17) we find for the relative fluctuations in volume,
$$\frac{\delta V}{V} = \frac{\sqrt{\langle(\delta\hat{V})^2\rangle}}{V} = \left(\frac{k_B T \kappa_T}{V}\right)^{\frac{1}{2}} \quad (1.21)$$

Observe that $\delta V/V \to 0$ as $V \to \infty$, i.e., only finite systems exhibit fluctuations, implying that the reservoirs themselves, being infinite, do not exhibit fluctuations.

Let us point out here that the second-order fluctuation moments (but no higher moments!) can be calculated exactly from an approximate gaussian distribution
$$W = W_0 \exp\left(\frac{1}{2k_B}\sum_{i,j} S_{ij}\,\delta\hat{\mathscr{X}}_i\,\delta\hat{\mathscr{X}}_j\right) \quad (1.22)$$
sometimes referred to as Einstein's fluctuation formula (Einstein, 1910). Here, S_{ij} are the second-order coefficients in the expansion of the instantaneous entropy $\hat{S}$ around its equilibrium value S, i.e.
$$\hat{S} = S + \sum_k \mathscr{F}_k\,\delta\hat{\mathscr{X}}_k + \frac{1}{2}\sum_{i,j} S_{ij}\,\delta\hat{\mathscr{X}}_i\,\delta\hat{\mathscr{X}}_j + \cdots \quad (1.23)$$

Equation (1.22) reads explicitly for a one-component system with constant mole number n
$$W = W_0 \exp\left\{-\left[T^{-1}(\delta\hat{U})^2 + \left(\frac{p^2}{T} + \frac{C_p n}{\kappa_T V}\right)(\delta\hat{V})^2 \right.\right.$$
$$\left.\left. + 2\left(\frac{\alpha}{\kappa_T} - \frac{p}{T}\right)\delta\hat{U}\,\delta\hat{V}\right]\bigg/(2nC_V k_B T)\right\} \quad (1.24)$$

For the mole number fluctuations in a one-component system, this gives

$$\langle(\delta n)^2\rangle = k_B T\left(\frac{\partial n}{\partial \mu}\right)_{T,V} \qquad (1.25)$$

and

$$\frac{\delta n}{n} = \left(\frac{k_B T \kappa_T}{V}\right)^{\frac{1}{2}} \qquad (1.26)$$

So much for the fluctuations of extensive variables. It is the purpose of this review of the theory of equilibrium fluctuations to provide the tools to evaluate the criterion (1.5) for the validity of local equilibrium. Typically, what we want to establish then is not how much an extensive parameter can change over a local equilibrium cell but how big externally controlled gradients in intensive parameters, i.e. temperature and pressure gradients, can be. We must therefore consider fluctuations in intensive parameters next. As this is a fairly controversial subject with frequent misunderstandings, we will proceed rather carefully.

First, recall that intensive parameters are introduced in a postulational approach to thermodynamics as the derivatives of the equilibrium entropy (in the entropy representation) or of the equilibrium internal energy (in the energy representation) with respect to the equilibrium values of the extensive variables. [See eqn (1.9).] From the standpoint of statistical mechanics, some of the intensive variables like temperature and chemical potentials are introduced in the ensemble partition function as Lagrange multipliers conjugate to those extensive variables for which the system is in contact, i.e. in exchange, with reservoirs. By their very definition, Lagrange multipliers are constants of the distribution and therefore cannot fluctuate. Ensemble theory therefore says (Münster, 1959; Kittel, 1973) that of any two conjugate variables only one can fluctuate, namely the one that is in contact with a reservoir. For example, if a system is in contact with a heat reservoir but is otherwise closed, its internal energy will fluctuate according to eqn (1.15) but its temperature as the Lagrange multiplier in the canonical partition function will be constant and equal to the temperature of the infinite heat reservoir, i.e. the fictitious canonical ensemble. On the other hand, if it were possible to prepare an open system at a fixed energy, such a system would then exhibit temperature fluctuations (Guggenheim, 1939).‡

Before we subscribe to these conclusions we should note the physical limitations of ensemble theories (Schrödinger, 1960). A logically satisfactory definition of an ensemble is its identification with a large (possibly infinitely large) collection of independent identical replica of the system under study. It is only for such an abstract construction that the above conclusions are rigorously valid. In a more physical approach (Landau and Lifshitz, 1958) we can identify the individual members of an ensemble as parts (subsystems) of the system under study in contact with their surroundings (rather than abstract reservoirs), interacting weakly through their mutual interfaces. In this situation the mathematical basis of ensemble theory, namely the statistical independence of the individual members, is only valid approximately and the above statement that only one of a pair of conjugate thermodynamic variables can fluctuate does not hold in all circumstances. We can see this point quite clearly in the following example. Inserting a sufficiently sensitive and small thermometer into a system we will no doubt observe its readings to fluctuate as a function of time. This is due to the fact that a thermometer primarily measures the instantaneous energy of its immediate surroundings (von Laue, 1917; McFee, 1973). We can therefore calculate fluctuations in intensive parameters (temperature) as a measure of and resulting from fluctuations in extensive parameters (internal energy). We define

$$\delta\hat{\mathscr{F}}_i = \hat{\mathscr{F}}_i - \mathscr{F}_i = \sum_j S_{ji}\,\delta\hat{\mathscr{X}}_i \tag{1.27}$$

and, using Einstein's fluctuation formula (1.22), get immediately

$$\langle \delta\hat{\mathscr{F}}_i\,\delta\hat{\mathscr{F}}_j \rangle = \sum_{k,l} S_{ki}S_{lj}\langle \delta\hat{\mathscr{X}}_k\,\delta\hat{\mathscr{X}}_l \rangle \tag{1.28}$$

‡ Observe that from eqns (1.15) and (1.32) we get $\delta U\,\delta T = k_B T^2$, independent of the material properties of the system. This and similar relations have been interpreted by Bohr (1932) and Rosenfeld (1962) as implying some sort of complementarity between the thermodynamic and the statistical description of macroscopic systems. Thus, in order to assign to the system a definite temperature, it is necessary to allow it to exchange energy with a heat reservoir, and it is impossible to assign to its energy any definite value. Conversely, in order to keep its energy constant, one must isolate the system, and we cannot assign a temperature to it. Also observe that $\delta U\,\delta T = RT^2/N_A$ is of order N_A^{-1}.

For an explicit calculation we eliminate $\delta\hat{U}$ from eqn (1.24) and get (Kestin and Dorfman, 1971)

$$W = W_0 \exp\left[\frac{-\left(\delta\hat{T}\,\delta\hat{S} - \delta\hat{p}\,\delta\hat{V} + \sum_{\gamma=1}^{c}\delta\hat{n}_\gamma\,\delta\hat{\mu}_\gamma\right)}{2k_B T}\right] \quad (1.29)$$

For a one-component system at constant mole number (Landau and Lifshitz, 1958), we obtain

$$W = W_0 \exp\left[-\frac{nC_V}{2k_B T^2}(\delta\hat{T})^2 + \frac{1}{2k_B T}\left(\frac{\partial p}{\partial V}\right)_T (\delta\hat{V})^2\right] \quad (1.30)$$

or

$$W = W_0 \exp\left[\frac{1}{2T}\left(\frac{\partial V}{\partial p}\right)_s (\delta\hat{p})^2 - \frac{1}{2nC_p}(\delta\hat{S})^2\right] \quad (1.31)$$

We then find for the temperature fluctuations

$$(\delta T)^2 = \langle(\delta\hat{T})^2\rangle = \frac{k_B T^2}{nC_V} \quad (1.32)$$

or

$$\frac{\delta T}{T} = \left(\frac{k_B}{nC_V}\right)^{\frac{1}{2}} \quad (1.33)$$

and for pressure fluctuations‡

$$(\delta p)^2 = \langle(\delta\hat{p})^2\rangle = -k_B T\left(\frac{\partial p}{\partial V}\right)_s \quad (1.34)$$

or

$$\frac{\delta p}{p} = \left(\frac{k_B T C_p}{V p^2 C_V \kappa_T}\right)^{\frac{1}{2}} \quad (1.35)$$

Let us proceed to evaluate the local equilibrium criterion (1.5) in a few typical systems and situations. We start with gases. The intrinsic time scale is given by the inverse collision frequency— the time a gas particle is more or less in free flight before it undergoes another effective collision—

$$\tau_1 = \nu_{\text{coll}}^{-1} \quad (1.36)$$

‡ Pressure fluctuations within the framework of ensemble theories have been discussed by Münster (1959) with reference to earlier work. As pointed out above, we do not share his criticism of von Laue's and Landau's approach.

Table 1.1. Collision times and mean free paths for various gases at $T = 393$ K and $p = 1$ atm

Gas	Collision time $10^{10} \cdot \nu_{\text{coll}}^{-1}$ (sec)	Mean-free path $10^6 \cdot l$ (cm)
He	2.2	27.45
A	2.5	9.88
CO_2	1.6	6.15
H_2	1.0	17.44
N_2	2.0	9.29
O_2	2.2	9.93

Source: Handbook of Chemistry and Physics, 53rd edition. Ed. R. C. Weast, Chemical Rubber Co., Cleveland, Ohio, 1972.

Macroscopic changes in a gas must be slow on this time scale. A few typical numbers are given in Table 1.1. The intrinsic lengthscale in a gas is correspondingly the mean-free path $\lambda_1 = l$. In a cell of size $V_j = l^3$, we have $N_j = (\rho/m)l^3$ particles.

In argon gas at room temperature and atmospheric pressure, we have $\rho/m = 2.7 \times 10^{19}$ cm^{-3} and (see Table 1.1) $l = 9.88 \times 10^{-6}$ cm, so that $N_j = (\rho/m)l^3 = 2.7 \times 10^4$. Because this gas satisfies the ideal gas law very well,

$$pV = nRT \qquad (1.37)$$

where $R = N_A k_B$ is the gas constant and $N_A = 6.027 \times 10^{23}$ is Avogadro's number, we immediately obtain from eqn (1.26) $\delta N_j/N_j \approx 6 \times 10^{-3}$, justifying the use of statistical mechanics in such a cell. In an ideal gas, all (relative) fluctuations are proportional to $N_j^{-\frac{1}{2}}$. To evaluate the maximum permissible gradient of any one of the macroscopic fields $T(\mathbf{r}, t)$, $p(\mathbf{r}, t)$, $u(\mathbf{r}, t)$, etc., we can therefore concentrate on just one of them, e.g. temperature, and find for (1.5), using (1.33), that

$$\frac{\Delta T}{T} = \frac{l|\boldsymbol{\nabla} T|}{T} \lesssim \frac{\delta T}{T} = \left(\frac{2}{3N_j}\right)^{\frac{1}{2}} \approx 5 \times 10^{-3} \qquad (1.38)$$

That is, the relative change in temperature over a mean-free path must be small. This is a criterion that one also finds in a detailed study of the Chapman–Enskog solution to the Boltzmann equation, as we will see in detail in Section 7.7. This example demonstrates very nicely the dual nature of our criterion (1.5). On the one hand, recall that we calculate equilibrium fluctuations in an ideal, i.e. noninteracting, gas. These, on the other hand, are used to estimate acceptable gradients in the temperature field. But a temperature gradient gives rise to heat conduction, i.e. energy dissipation, according to Fourier's law

$$\mathbf{j}_q = -\lambda \nabla T \tag{1.39}$$

where λ is the thermal conductivity. Elementary kinetic theory‡ gives the connection between l and λ as

$$\lambda = \tfrac{1}{2}\rho \bar{v}_m l \left(\frac{C_V}{m}\right) \tag{1.40}$$

where $\bar{v}_m$ is a typical molecular velocity and m is the molar mass. We can then rewrite eqn (1.38) as

$$\frac{\Delta T}{T} = 2 \frac{|\mathbf{j}_q|}{|\rho u \bar{v}_m|} \lesssim \frac{\delta T}{T} \tag{1.41}$$

where $u = C_V T/m$ is the internal energy per unit mass in an ideal gas. Hence we see that $\Delta T/T$ is twice the ratio of the conductive (dissipative) over the convective (reversible) internal energy transport carried at a typical molecular speed, say, the speed of sound. Our criterion then says that this ratio should be of the order of the local equilibrium fluctuations.§

Let us finally note that elementary kinetic theory calculates the mean-free path to be

$$l = \frac{V}{\sqrt{2}\pi N\sigma^2} \tag{1.42}$$

where σ is the molecular diameter. We therefore get from eqn (1.38)

$$\frac{\Delta T}{T} = \frac{1}{\sqrt{2}\pi\sigma^2} \frac{1}{N} \frac{|\nabla T|}{T} \lesssim \frac{\delta T}{T} \ll 1 \tag{1.43}$$

This implies that for a very dilute gas (i.e. density small) under

‡ See, for example, Kestin and Dorfman (1971).

§ In liquids and solids, the lengthscale $\lambda m/\rho C_V \bar{v}$, which in gases is, by construction, equal to the mean-free path, is of the order of the interatomic separation and therefore cannot be a measure for the size of the local equilibrium cells.

a fixed temperature gradient, local equilibrium cannot be assumed because the mean-free path and thus the cell size go to infinity. Physically, this is so because there are simply not enough collision partners in a given cell to achieve equipartition of energy. A similar comment applies to a low-temperature Fermi-Dirac gas where the exclusion principle freezes out the interaction and thus the collisions.

We turn next to a solid, and estimate an upper bound on temperature gradients for which local equilibrium could be assumed in a heat conduction experiment. We recall that in a simple nonmetallic solid heat is conducted via the phonon system and, ignoring boundary effects, is limited by, among other things, phonon-phonon collisions. The phonon mean-free path is therefore a relevant intrinsic lengthscale. In a NaCl crystal it is about 23 Å for Umklapp processes at a temperature $T = 273$ K. The number density of (free) phonons at that temperature is given by

$$n_{\text{ph}} = 6N_u \int_0^{\omega_{\max}} \frac{g(\omega)}{e^{\beta\hbar\omega}-1} d\omega \qquad (1.44)$$

where $N_u = 0.228 \times 10^{23}$ cm^{-1} is the number of unit cells per cm^3 and $g(\omega)$ is the phonon density of states. In a local equilibrium cell all phonons have to be equilibrated sufficiently fast. With eqn (1.43) we find that $N_j = 1.7 \times 10^3$ phonons are in a cell of the size of the mean-free path for Umklapp processes. Thus the inequality $\delta N_j/N_j \sim N_j^{-\frac{1}{2}} \ll 1$ is satisfied reasonably well, and inequality (1.38) imposes a bound on allowable temperature gradients.

Turning finally to a liquid, we wish to find a bound on the maximum temperature gradients that can be maintained with the liquid still locally in equilibrium. From eqn (1.34) we have to demand that

$$\frac{\Delta T}{T} \lesssim \frac{\delta T}{T} = \left(\frac{k_B}{N_j/N_A C_V}\right)^{\frac{1}{2}} \ll 1 \qquad (1.45)$$

where N_j is the number of constituent particles in a local equilibrium cell. To determine the size of the latter, we observe that the concept of a mean-free path is meaningless in a liquid. The only intrinsic lengthscales are the interatomic spacing and the two-particle correlation length, both of which are of the same order and too short to qualify as the dimension of a local equilibrium cell. We therefore invoke eqn (1.26) to determine the volume V_j

of a local equilibrium cell,

$$V_j = \frac{k_B T \kappa_T}{(\delta N_j/N_j)^2} \tag{1.46}$$

and demand that $\delta N_j/N_j \sim 10^{-2}$ or some similar small number. As an example, consider a simple liquid like argon. At temperature $T = 100$ K and at vapor pressure $p = 3.21$ atm, we have (Stephenson, 1975) $\kappa_T = 3.3 \times 10^{-4}$ atm^{-1} (1 atm = 1.013×10^5 N/m^2 = 1.013×10^5 Pa). From eqn (1.46) we then get with $(\delta N_j/N_j) \sim 10^{-2}$ that $V_j \sim 4.6 \times 10^{-20}$ cm^3 and for the characteristic cell size $\lambda_1 = V_j^{\frac{1}{3}} = 36$ Å. At the corresponding density $N/V = 2 \times 10^{22}$ cm^{-3} we find $N_j = 900$ particles in a local equilibrium cell.‡ With the specific heat at constant volume $C_V = 18.7$ JK^{-1} mol^{-1} we then get for eqn (1.45)

$$\frac{\Delta T}{T} = \frac{\lambda_1 |\nabla T|}{T} \lesssim \frac{\delta T}{T} \approx 0.022 \tag{1.47}$$

giving a bound on the maximally allowed temperature gradients. Whereas in a gas pressure fluctuations in a local equilibrium cell with linear dimension of a mean free path are of the order of number and temperature fluctuations, it turns out that in a liquid pressure fluctuations in cells with number fluctuations of the order $\delta N_j/N_j \approx 10^{-2}$ are extraordinarily large due to the smallness of κ_T. If a pressure gradient is maintained in a liquid, we must therefore find local equilibrium cells in which $\delta P/P$ is small. For liquid argon at $T = 100$ K and at the vapour pressure $p = 3.21$ atm, for example, one finds $\lambda_1 \approx 3500$ Å with some 9×10^8 atoms in a cell. The maximum pressure gradients that can be maintained in such a liquid then turn out to be about the same as in a gas at room temperature and atmospheric pressure. Observe finally that as the critical point is approached, $\lambda_1 = V^{-\frac{1}{3}}$ from eqn (1.46) tends to infinity, because κ_T grows indefinitely, and we find that $\delta T/T$ approaches zero, implying that local equilibrium cannot be maintained for a fixed temperature gradient, because fluctuations in extensive variables grow without bounds.

‡ To see the typical range of these numbers in a liquid, look at a point on the coexistence curve close to the critical point. Namely, at $T = 140$ K, $p = 31.3$ atm, $N/V = 1.4 \times 10^{22}$ cm^{-3}, we have $\kappa_T = 3.7 \times 10^{-3}$ atm^{-1} and obtain $\lambda_1 = 90$ Å and $N_j = 10^4$. On the other hand, on the fusion curve at $T = 140$ K we have $p = 2642.7$ atm, $N/V = 2.4 \times 10^{22}$ cm^{-3}, $\kappa_T = 5.6 \times 10^{-5}$ and obtain $\lambda_1 = 22$ Å and $N_j = 260$.

In the first half of this book we will restrict the discussion to macroscopic systems in local equilibrium. It will be our first task to establish local differential balance equations for the relevant macroscopic fields linking the (thermodynamically open) local equilibrium cells to their respective local surroundings. In Chapters 7–9 we will derive these balance equations microscopically and establish criteria for local equilibrium for particular kinetic models using the methods of nonequilibrium statistical mechanics.

2
Balance Equations of Irreversible Thermodynamics

2.1. Balance Equations for Mechanical Quantities‡

CONSIDER a macroscopic system in a volume V, bounded by a closed surface Σ. In order to derive equations describing the time evolution of various extensive thermodynamic functions in V and fluxes through the surface Σ, it is useful to write an arbitrary extensive quantity $F(t)$ as

$$F(t) = \int_V \rho(\mathbf{r}, t) f(\mathbf{r}, t) \, dV \tag{2.1}$$

where $f(\mathbf{r}, t)$ represents the density of $F(t)$ per unit mass (specific variable) and $\rho(\mathbf{r}, t)$ is the mass per unit volume as a function of position and time. Since the quantity $F(t)$ is not necessarily conserved, the most general equation for $F(t)$ is of the form

$$\frac{dF(t)}{dt} = P[F] + \phi[F] \tag{2.2}$$

where $P[F]$ is a source term describing production of F in the volume V and $\phi[F]$ is a flow term describing the exchange of F between V and its surroundings through the surface Σ. These

‡ Some general references on thermodynamics of irreversible processes are: J. Meixner and H. G. Reik, *Handbuch der Physik* III/2 (Springer, Berlin, 1955); S. R. de Groot and P. Mazur, *Non-Equilibrium Thermodynamics* (North-Holland, Amsterdam, 1969); R. Haase, *Thermodynamics of Irreversible Processes* (Addison-Wesley, Reading, Mass., 1969); I. Prigogine, *Introduction to Thermodynamics of Irreversible Processes*, 3rd ed. (Interscience, New York, 1969); P. Glansdorff and I. Prigogine, *Thermodynamic Theory of Structure, Stability and Fluctuations* (Wiley-Interscience, New York, 1971); L. C. Woods, *The Thermodynamics of Fluid Systems* (Clarendon, Oxford, 1975).

BALANCE EQUATIONS OF IRREVERSIBLE THERMODYNAMICS

source and flow terms may be written as

$$P[F] = \int_V \sigma_F \, dV$$

$$\phi[F] = -\int_\Sigma \mathbf{j}_F \cdot d\mathbf{\Sigma} \tag{2.3}$$

where σ_F is the source or sink density in V and $\mathbf{j}_F$ is the density of the current flow through Σ. The vector $d\mathbf{\Sigma}$ of the surface element is pointing out of the volume V.

To cast this into a local form, we use the Euler coordinate system, which is fixed in space as opposed to the Lagrange or center-of-mass system which moves with the material. We can write

$$\frac{dF(t)}{dt} = \frac{d}{dt} \int_V \rho(\mathbf{r}, t) f(\mathbf{r}, t) \, dV = \int_V \frac{\partial}{\partial t}(\rho f) \, dV$$

$$= \int_V \sigma_F \, dV - \int_\Sigma \mathbf{j}_F \cdot d\mathbf{\Sigma} \tag{2.4}$$

Applying the divergence theorem to the continuous function $\mathbf{j}_F$ gives us the differential equation for the local balance of f:

$$\frac{\partial(\rho f)}{\partial t} + \boldsymbol{\nabla} \cdot \mathbf{j}_F = \sigma_F \tag{2.5}$$

In order to do physics, we must next specify a set of extensive variables F relevant to the particular thermodynamic system under study and construct the corresponding currents $\mathbf{j}_F$ and source densities σ_F. We will do this for the balance equations for mass, momentum, energy, angular momentum, and entropy.

For mass transport, we set $f = 1$ and, since mass is conserved, $\sigma = 0$. The mass current is $\rho \mathbf{v}$ and, for a one-component system, eqn (2.5) reduces to the continuity equation

$$\frac{\partial \rho}{\partial t} + \boldsymbol{\nabla} \cdot (\rho \mathbf{v}) = 0 \tag{2.6}$$

More generally, if there are c different components, labeled by γ and with densities ρ_γ, they satisfy the equation

$$\frac{\partial \rho_\gamma}{\partial t} + \boldsymbol{\nabla} \cdot (\rho_\gamma \mathbf{v}_\gamma) = \sum_{\delta=1}^{r} v_{\gamma\delta} J_\delta \tag{2.7}$$

where $v_{\gamma\delta}J_\delta$ is the rate of production of component γ in the δth of r chemical reactions; J_δ is the reaction rate for reaction δ, representing the total mass transformed per unit volume and unit time; and $v_{\delta\gamma}$ is proportional to the stoichiometric coefficient with which component γ appears in the chemical equation for reaction δ. It is negative (positive) if it appears on the left- (right-) hand side of the reaction equation.

Defining the total density

$$\rho \equiv \sum_{\gamma=1}^{c} \rho_\gamma$$

center of mass velocity,

$$\mathbf{v} \equiv \sum_{\gamma=1}^{c} \frac{\rho_\gamma \mathbf{v}_\gamma}{\rho}$$

hydrodynamic (barycentric, material, or substantial) derivative

$$\frac{D}{Dt} \equiv \frac{\partial}{\partial t} + \mathbf{v} \cdot \boldsymbol{\nabla}$$

and the diffusion flows

$$\mathbf{j}_\gamma^{\text{diff}} \equiv \rho_\gamma(\mathbf{v}_\gamma - \mathbf{v})$$

the continuity equations become

$$\frac{D\rho_\gamma}{Dt} + \rho_\gamma \boldsymbol{\nabla} \cdot \mathbf{v} + \boldsymbol{\nabla} \cdot \mathbf{j}_\gamma^{\text{diff}} = \sum_{\delta=1}^{r} v_{\delta\gamma} J_\delta \qquad (2.8)$$

and

$$\frac{D\rho}{Dt} + \rho \boldsymbol{\nabla} \cdot \mathbf{v} = 0 \qquad (2.9)$$

Before we proceed to balance equations for variables other than mass, let us rewrite eqn (2.5) as

$$\frac{\partial(f\rho)}{\partial t} = \rho \frac{\partial f}{\partial t} - f \boldsymbol{\nabla} \cdot (\rho \mathbf{v})$$

$$= \rho \frac{\partial f}{\partial t} + \rho \mathbf{v} \cdot \boldsymbol{\nabla} f - \boldsymbol{\nabla} \cdot (\rho f \mathbf{v})$$

$$= \sigma_F - \boldsymbol{\nabla} \cdot \mathbf{j}_F \qquad (2.10)$$

which yields balance equations in the center of mass frame

$$\rho \frac{Df}{Dt} = \sigma_F - \nabla \cdot (\mathbf{j}_F - \rho f \mathbf{v})$$

$$= \frac{\partial (f\rho)}{\partial t} + \nabla \cdot (f\rho \mathbf{v}) \qquad (2.11)$$

Thus the current $\mathbf{j}_F$ observed at a fixed point in space is reduced for an observer moving along with the fluid by an amount $\rho f \mathbf{v}$ attached as a convective current to a moving fluid element.

The balance equation for the momentum of a unit mass in the system is nothing but the equation of motion of continuum mechanics, and the source terms are known. They arise from external potentials and from internal forces that appear as a pressure. Using the material derivative, we obtain

$$\rho \frac{D\mathbf{v}}{Dt} = \sum_{\gamma=1}^{c} \rho_\gamma \mathbf{F}_\gamma - \nabla \cdot \mathbf{P} \qquad (2.12)$$

where $\mathbf{F}_\gamma$ is the external force acting on component γ and $\mathbf{P}$ is the pressure tensor due to short-range internal forces. Equation (2.12) can be written in the form of a local balance equation as

$$\frac{\partial (\rho \mathbf{v})}{\partial t} + \nabla \cdot (\mathbf{P} + \rho \mathbf{v}\mathbf{v}) = \sum_{\gamma=1}^{c} \rho_\gamma \mathbf{F}_\gamma \qquad (2.13)$$

where $\mathbf{v}\mathbf{v}$ is a tensor product (dyadic).‡

The term $\nabla \cdot \mathbf{P}$ describes the transport of momentum out of a volume due to conduction processes (momentum transport through internal forces) while the term $\nabla \cdot (\rho \mathbf{v}\mathbf{v})$ describes 'convection' of momentum out of the volume due to the macroscopic motion of the fluid. If there are no external forces acting on the system, the source terms on the right-hand side of eqn (2.13) vanish and we recover conservation of momentum,

$$\frac{\partial (\rho \mathbf{v})}{\partial t} + \nabla \cdot (\mathbf{P} + \rho \mathbf{v}\mathbf{v}) = 0 \qquad (2.14)$$

In the center of mass (Lagrange) frame, there is no convection

‡ We use the notations $(\mathbf{ab})_{ij} = a_i b_j$ and

$$(\nabla \cdot \mathbf{P})_i = \sum_{k=1}^{3} \frac{\partial}{\partial x_k} P_{ik}$$

term present and the pressure term gives the total momentum flow. In most cases, we can split

$$\mathbf{P} = \mathbf{P}^e + \mathbf{P}^{\text{diss}} \tag{2.15}$$

where $\mathbf{P}^e$ is the elastic part of the pressure tensor and includes a term $p\mathbf{I}$ ($\mathbf{I}$ is the unit tensor, $I_{ij} = \delta_{ij}$), which is simply the hydrostatic pressure; $\mathbf{P}^{\text{diss}}$ is the dissipative contribution arising from viscosity. For fluids in equilibrium, only the elastic term is present, i.e. $\mathbf{P} = \mathbf{P}^e = p\mathbf{I}$.

To set up the energy balance, we multiply eqn (2.12) by $\mathbf{v}$ to yield

$$\rho \mathbf{v} \cdot \frac{D\mathbf{v}}{Dt} = \rho \frac{D(\tfrac{1}{2}v^2)}{Dt} = -\mathbf{v} \cdot (\boldsymbol{\nabla} \cdot \mathbf{P}) + \sum_{\gamma=1}^{c} \rho_\gamma \mathbf{F}_\gamma \cdot \mathbf{v}$$

$$= -\boldsymbol{\nabla} \cdot (\mathbf{P} \cdot \mathbf{v}) + \mathbf{P} : (\boldsymbol{\nabla}\mathbf{v}) + \sum_{\gamma=1}^{c} \rho_\gamma \mathbf{F}_\gamma \cdot \mathbf{v} \tag{2.16}$$

where

$$\mathbf{P} : (\boldsymbol{\nabla}\mathbf{v}) \equiv \sum_{\alpha,\beta=1}^{3} P_{\alpha\beta} \frac{\partial v_\alpha}{\partial x_\beta} \tag{2.17}$$

As a local balance equation, this becomes

$$\frac{\partial(\tfrac{1}{2}\rho v^2)}{\partial t} + \boldsymbol{\nabla} \cdot (\tfrac{1}{2}\rho v^2 \mathbf{v} + \mathbf{P} \cdot \mathbf{v}) = \mathbf{P} : \boldsymbol{\nabla}\mathbf{v} + \sum_{\gamma=1}^{c} \rho_\gamma \mathbf{F}_\gamma \cdot \mathbf{v} \tag{2.18}$$

where $\tfrac{1}{2}\rho v^2 \mathbf{v}$ is a kinetic energy convection term and $\mathbf{P} \cdot \mathbf{v}$ is a kinetic energy conduction term. The sources of kinetic energy involve the power, i.e. work done per unit time, by the external forces in the term $\sum_{\gamma=1}^{c} \rho_\gamma \mathbf{F}_\gamma \cdot \mathbf{v}$ and the power of compression due to the pressure tensor in the term $\mathbf{P} : \boldsymbol{\nabla}\mathbf{v}$.

To arrive at the potential energy balance, we assume that the external forces are due to some potential, i.e.

$$\mathbf{F}_\gamma = -\boldsymbol{\nabla}\psi_\gamma \tag{2.19}$$

and set $\partial \psi_\gamma / \partial t = 0$ for time-independent forces. Using eqn (2.7) and the definition $\mathbf{j}_\gamma^{\text{diff}} = \rho_\gamma(\mathbf{v}_\gamma - \mathbf{v})$, where $\mathbf{v}$ is the barycentric velocity, we get a local balance equation for the total potential energy

$$\rho\psi \equiv \sum_{\gamma=1}^{c} \rho_\gamma \psi_\gamma \tag{2.20}$$

namely

$$\frac{\partial(\rho\psi)}{\partial t}+\nabla\cdot\left(\rho\psi\mathbf{v}+\sum_{\gamma=1}^{c}\psi_\gamma\mathbf{j}_\gamma^{\text{diff}}\right)=-\sum_{\gamma=1}^{c}\rho_\gamma\mathbf{F}_\gamma\cdot\mathbf{v}$$
$$-\sum_{\gamma=1}^{c}\mathbf{j}_\gamma^{\text{diff}}\cdot\mathbf{F}_\gamma+\sum_{\gamma=1}^{c}\sum_{\delta=1}^{r}\psi_\gamma v_{\gamma\delta}J_\delta \quad (2.21)$$

where $\rho\psi\mathbf{v}$ represents a convection current of potential energy; $\sum_{\gamma=1}^{c}\psi_\gamma\mathbf{j}_\gamma^{\text{diff}}$ represents the transport of potential energy due to diffusion; $-\sum_{\gamma=1}^{c}\rho_\gamma\mathbf{F}_\gamma\cdot\mathbf{v}$ represents a sink due to conversion of potential energy to kinetic energy (an equal but opposite term appears in the kinetic energy equation); $-\sum_{\gamma=1}^{c}\mathbf{j}_\gamma^{\text{diff}}\cdot\mathbf{F}_\gamma$ represents the conversion of potential energy to internal energy by diffusion; and $\sum_{\gamma=1}^{c}\sum_{\delta=1}^{r}\psi_\gamma v_{\gamma\delta}J_\delta$ is a source due to change in potential energy as a result of chemical reactions.

In most cases, the last term will be zero, since the property of the particles responsible for the potential interaction usually remains unchanged in a chemical reaction (e.g., mass in a gravitational potential). In this case, the equation for the total mechanical energy density $(\rho\psi+\tfrac{1}{2}\rho v^2)$ becomes

$$\frac{\partial(\rho\psi+\tfrac{1}{2}\rho v^2)}{\partial t}+\nabla\cdot\left\{(\rho\psi+\tfrac{1}{2}\rho v^2)\mathbf{v}+\mathbf{P}\cdot\mathbf{v}+\sum_{\gamma=1}^{c}\psi_\gamma\mathbf{j}_\gamma^{\text{diff}}\right\}$$
$$=\mathbf{P}:\nabla\mathbf{v}-\sum_{\gamma=1}^{c}\mathbf{j}_\gamma^{\text{diff}}\cdot\mathbf{F}_\gamma \quad (2.22)$$

The presence of two source terms indicates the fact that the internal energy must be included in order to have energy conservation. In this equation, $\mathbf{P}:\nabla\mathbf{v}$ is a source term arising from the conversion of internal energy to kinetic energy through compression (or vice versa) and $-\sum_{\gamma=1}^{c}\mathbf{j}_\gamma^{\text{diff}}\cdot\mathbf{F}_\gamma$ is a source due to conversion of internal energy to potential energy through diffusion processes.

The total energy e may be written as

$$\rho e = \tfrac{1}{2}\rho v^2 + \rho\psi + \rho u \quad (2.23)$$

where the internal energy density u includes the energies of thermal agitation and short-range molecular interactions.

Energy conservation

$$\frac{\partial(\rho e)}{\partial t}+\nabla\cdot\mathbf{j}_e = 0 \quad (2.24)$$

implies that the source term for the internal energy u is

$$\sigma_u = -\mathbf{P}:\nabla\mathbf{v} + \sum_{\gamma=1}^{c} \mathbf{j}_\gamma^{\text{diff}} \cdot \mathbf{F}_\gamma \qquad (2.25)$$

Analogous to the current for mechanical energy, the current for total energy has a convective term $\rho e \mathbf{v}$, the previously discussed mechanical and potential flux terms, $\mathbf{P}\cdot\mathbf{v}$ and $\sum_{\gamma=1}^{c} \psi_\gamma \mathbf{j}_\gamma^{\text{diff}}$, and a new internal energy flux $\mathbf{j}_q$. Thus, the balance equation for total energy is

$$\frac{\partial(\rho e)}{\partial t} + \nabla \cdot \left(\rho e \mathbf{v} + \mathbf{P}\cdot\mathbf{v} + \sum_{\gamma=1}^{c} \psi_\gamma \mathbf{j}_\gamma^{\text{diff}} + \mathbf{j}_q \right) = 0 \qquad (2.26)$$

and the balance equation for internal energy is

$$\frac{\partial(\rho u)}{\partial t} + \nabla \cdot (\rho u \mathbf{v} + \mathbf{j}_q) = -\mathbf{P}:\nabla\mathbf{v} + \sum_{\gamma=1}^{c} \mathbf{j}_\gamma^{\text{diff}} \cdot \mathbf{F}_\gamma \qquad (2.27)$$

These equations define $\mathbf{j}_q$, commonly called the heat flux.

Equation (2.27) is simply the first law of thermodynamics. It can be put in more familiar form as

$$\frac{Du}{Dt} = \frac{Dq}{Dt} - p\frac{D\rho^{-1}}{Dt} - \rho^{-1}\mathbf{\Pi}:\nabla\mathbf{v} + \rho^{-1}\sum_{\gamma=1}^{c} \mathbf{j}_\gamma^{\text{diff}} \cdot \mathbf{F}_\gamma \qquad (2.28)$$

where p is the scalar hydrostatic pressure; $\mathbf{\Pi} \equiv \mathbf{P} - p\mathbf{I}$ is the pressure tensor without the hydrostatic part; and q is the heat per unit mass, defined by $\rho(Dq/Dt) + \nabla \cdot \mathbf{j}_q = 0$.

The final mechanical density which will be considered is the angular momentum density $\mathbf{J}$, which can also be expressed in terms of an angular momentum density tensor $\mathbf{J}$ with components

$$J_{\alpha\beta} = \sum_{\gamma=1}^{3} \varepsilon_{\alpha\beta\gamma}(\mathbf{J})_\gamma \qquad (2.29)$$

where $\varepsilon_{\alpha\beta\gamma}$ is the antisymmetric Levi-Civita tensor.

In the Lagrange frame of reference moving with the fluid, the angular momentum conservation equation is, in the absence of external forces,

$$\rho\frac{DJ_{\alpha\beta}}{Dt} = -\sum_{\gamma=1}^{3} \frac{\partial}{\partial r_\gamma}(r_\alpha P_{\gamma\beta} - r_\beta P_{\gamma\alpha}) \qquad (2.30)$$

The right-hand side represents the flow of angular momentum

due to the torque exerted on a mass element by the pressure tensor.

The angular momentum density tensor **J** is now split into two parts:

$$\mathbf{J} = \mathbf{L} + \mathbf{S} \tag{2.31}$$

where $L_{\alpha\beta}(=r_\alpha v_\beta - r_\beta v_\alpha)$ is the usual angular momentum density due to the circulation of the material. In classical hydrodynamics, the material has no microstructure and **L** is the total angular momentum. More generally, there can be another contribution **S**, the internal angular momentum of the material. It arises from the fact that the molecules making up the material can have angular momentum (spin) without having a macroscopic fluid velocity. **S** can be written as

$$\mathbf{S} = \Theta\boldsymbol{\omega} \tag{2.32}$$

where $\boldsymbol{\omega}$ is the antisymmetric rotation tensor corresponding to $\boldsymbol{\omega}$, the angular velocity, and Θ is the average moment of inertia per unit mass.

Taking the equation of motion

$$\frac{D\mathbf{v}}{Dt} = -\boldsymbol{\nabla} \cdot \mathbf{P} \tag{2.33}$$

multiplying by **r** and subtracting a transposed term gives

$$r_\alpha \frac{Dv_\beta}{Dt} - r_\beta \frac{Dv_\alpha}{Dt} = \frac{DL_{\alpha\beta}}{Dt} \tag{2.34}$$

and therefore,

$$\frac{DL_{\alpha\beta}}{Dt} = -\sum_{\gamma=1}^{3} \frac{\partial}{\partial r_\gamma}(r_\alpha P_{\gamma\beta} - r_\beta P_{\gamma\alpha}) + (P_{\alpha\beta} - P_{\beta\alpha}) \tag{2.35}$$

Hence, from angular momentum conservation, we get

$$\frac{D\mathbf{S}}{Dt} = -2\mathbf{P}^{(a)} \tag{2.36}$$

where $\mathbf{P}^{(a)}$ is the antisymmetric part of **P**.‡

---

‡ A discussion of nonsymmetric tensors in ferroelectrics and other polar elastic materials has been given by Huntington (1958, p. 230) and Truesdell and Noll (1965, p. 389).

If $\mathbf{S} = 0$, the constituent particles have no angular momentum and $P_{\alpha\beta} = P_{\beta\alpha}$, as is usually assumed in continuum mechanics. The antisymmetric part of the pressure tensor is due to the internal body torques which are proportional to volume. These can arise if the constituent molecules are not spherically symmetric. If the molecules are spherical or if they can be regarded as structureless (as in a dilute gas), then $P_{\alpha\beta} = P_{\beta\alpha}$ and $\mathbf{S} = $ const. This means that internal and external angular momenta are separately conserved. In viscous fluids, however, the two angular momenta would not be expected to be independent. Instead, viscous effects would result in transfer between internal and external angular momentum.

To investigate the effect of $\mathbf{P}^{(a)}$ on the motion of the fluid, it is assumed that the pressure tensor is a generalization of that for a newtonian fluid. That is, the pressure tensor components are linear functions of the appropriate velocity gradients. For example, if the viscous part of $\mathbf{P}$ is separated out and elasticity is neglected, we have

$$\mathbf{P} = p\mathbf{I} + \mathbf{\Pi}$$

where $\mathbf{\Pi}$ is the viscous pressure tensor, written as

$$\mathbf{\Pi} = \Pi_0 \mathbf{I} + \mathbf{\Pi}_{tr=0}^{(s)} + \mathbf{\Pi}^{(a)} \qquad (2.37)$$

where the first term is the trace of $\mathbf{\Pi}$, the second term is the traceless symmetric part, and the last one accounts for the antisymmetric part. After linearization, we have

$$\Pi_0 = -\zeta(\mathbf{\nabla} \cdot \mathbf{v})$$

$$\mathbf{\Pi}_{tr=0}^{(s)} = -2\eta(\mathbf{\nabla}\mathbf{v})_{tr=0}^{(s)}$$

that is,

$$\Pi_{tr=0_{\alpha\beta}}^{(s)} = -\eta\left(\frac{\partial v_\alpha}{\partial r_\beta} + \frac{\partial v_\beta}{\partial r_\alpha} - \frac{2}{3}\mathbf{\nabla} \cdot \mathbf{v}\, \delta_{\alpha\beta}\right)$$

$$\mathbf{\Pi}^{(a)} = -\eta_{\text{rot}}(\mathbf{\nabla} \times \mathbf{v} - 2\boldsymbol{\omega}) \qquad (2.38)$$

($\mathbf{\nabla} \times \mathbf{v}$ is regarded as an antisymmetric tensor).

Three phenomenological coefficients have been introduced:

1. ζ, the bulk viscosity
2. η, the shear viscosity
3. η_{rot}, the rotational viscosity

If these viscosity coefficients are independent of position, then the (generalized) Navier-Stokes equation reads

$$\rho \frac{D\mathbf{v}}{Dt} = -\nabla p + \eta \nabla^2 \mathbf{v} + (\tfrac{1}{3}\eta + \zeta)\nabla(\nabla \cdot \mathbf{v}) + \eta_{\text{rot}} \nabla \times (2\boldsymbol{\omega} - \nabla \times \mathbf{v}) \quad (2.39)$$

In order to see the physical significance of the new term $\eta_{\text{rot}} \nabla \times (2\boldsymbol{\omega} - \nabla \times \mathbf{v})$, two special cases are studied.

First, pure expansion (no shears or rotations of the fluid) is considered for which

$$\mathbf{v} = \alpha \mathbf{r}$$
$$(\nabla \mathbf{v})^{(s)}_{tr=0} = 0$$
$$\nabla \times \mathbf{v} = 0$$
$$\Pi_0 = -\zeta(\nabla \cdot \mathbf{v}) = -3\alpha\zeta \neq 0 \quad (2.40)$$

Thus, bulk viscosity ζ appears in the equation of motion but shear viscosity does not. If the molecules have spin initially, the rotational viscosity will transmit the angular momentum and cause $\nabla \times \mathbf{v}$ to become nonzero, i.e., cause the fluid motion to depart from pure expansion.

Let us next consider rigid body rotation, where $\mathbf{v} = \mathbf{b} \times \mathbf{r}$ with $\mathbf{b}$ a constant vector. We then have $(\nabla \mathbf{v})^{(s)}_{tr=0} = 0$, $\nabla \cdot \mathbf{v} = 0$, and the only nonzero term in the viscous pressure tensor is $\mathbf{\Pi}^{(a)} = -\eta_{\text{rot}}(\nabla \times \mathbf{v} - 2\boldsymbol{\omega})$. The equation of motion for $\mathbf{S}$ is

$$\rho \frac{D\mathbf{S}}{Dt} = -2\mathbf{\Pi}^{(a)} = 2\eta_{\text{rot}}(\nabla \times \mathbf{v} - 2\boldsymbol{\omega}) \quad (2.41)$$

If we consider a case where the rotation of the fluid is constant ($\nabla \times \mathbf{v} = 2\mathbf{b}$, where $(\partial \mathbf{b}/\partial t) = 0$), we can set $\mathbf{S} = \Theta \boldsymbol{\omega}$ and find

$$\frac{d\boldsymbol{\omega}}{dt} = -\frac{4\eta_{\text{rot}}}{\rho \Theta}(\boldsymbol{\omega} - \mathbf{b}) \quad (2.42)$$

If initially $\boldsymbol{\omega} = 0$, i.e., the molecules have no initial spin, we obtain

$$\boldsymbol{\omega} = \mathbf{b}(1 - e^{-t/\tau}) \quad (2.43)$$

where $\tau = (\rho \Theta / 4\eta_{\text{rot}})$ is a relaxation time.

Thus, after a time $t \gg \tau$, $2\boldsymbol{\omega}$ is essentially equal to $\nabla \times \mathbf{v}$; the rotational viscous effects, particularly the antisymmetric part of the pressure tensor, become negligible and the internal and external angular momenta are in equilibrium.

2.2. Balance Equation for the Viscoelastic Stress Tensor

In the previous section we set up differential balance equations for such mechanical quantities as density, momentum, energy, and angular momentum. To complete a thermodynamic theory of irreversible processes, two operations remain to be done. First we have to include thermodynamics into the theory by setting up the balance equation for the local entropy density. This will be done in the next section. Second, in the following chapter, we will use experiments to arrive at a set of constitutive equations that describe in a phenomenological way the properties of the particular system under study—e.g., the local equation of state and linear transport equations relating currents to externally imposed constraints. Before we proceed we will consider in this section the pressure or stress tensor in a viscoelastic material, a quantity for which both a balance equation can be derived and constitutive relations have to be assumed.

In many situations, it is possible to consider either elastic or viscous effects separately. For example, in hydrodynamics, it is assumed that the fluid has no elasticity. Also, in solid mechanics, it is often assumed that there can be no flow or permanent deformation and hence the stress tensor is purely elastic. However, there are situations where both viscous and elastic phenomena are important. Examples are plastic substances that flow under sustained stress yet retain many of the elastic properties of other solids. There are also some liquids with appreciable elastic properties commonly referred to as *gels*.‡ Early work on the problem of viscoelasticity was summarized and unified by Boltzmann (1874) who presented a linear field theory of isotropic viscoelasticity. Interest in the subject was revived in past decades due to great advances in polymer rheology and related fields.§

Let us recall that a perfectly elastic material can be characterized by its ability to store mechanical energy of deformation

---

‡ There are certain gels which combine fluidity (low shear viscosity) with elasticity in such a way that immersing a pendulum in the gel actually decreases its period of oscillation. See Frenkel (1946), p. 218.

§ For an introduction into the field, see R. M. Christensen, *Theory of Viscoelasticity, An Introduction* (Academic, New York, 1971), or A. Lodge, *Elastic Liquids* (Academic, London, 1964).

without any dissipation. In contrast, a newtonian viscous fluid has the capacity of dissipating energy. A viscoelastic material then is one in which not all the work deforming it can be recovered.

We can also characterize these materials by their response to a suddenly applied surface traction. An elastic solid will respond with an instantaneous deformation, a viscous fluid with a slow flow, and a viscoelastic material with a sudden deformation followed by a flow or creep. The creep characteristics will in general change in time depending on the past history of the material, indicating that memory effects can be very important, particularly in the nonlinear regime. Two classes of materials constitute the simple ends of the spectrum of viscoelasticity. They can again be classified according to their response when subjected to a fixed shear state of deformation. A viscoelastic fluid will produce a stress that will eventually decay to zero, whereas at the other end of the spectrum, a viscoelastic solid will remain in a state of nonzero stress.

In purely elastic deformations, the quantities of interest are the stress tensor

$$\boldsymbol{\sigma} = -\mathbf{P} = -p\mathbf{I} - \mathbf{\Pi} \tag{2.44}$$

and the linear strain tensor

$$\varepsilon_{\alpha\beta} = \frac{1}{2}\left(\frac{\partial u_\alpha(\mathbf{r}, t)}{\partial r_\beta} + \frac{\partial u_\beta(\mathbf{r}, t)}{\partial r_\alpha}\right) \tag{2.45}$$

where $u_\alpha(\mathbf{r}, t)$ is the α component of the vector $\mathbf{u}$ which connects the equilibrium position $\mathbf{r}$ inside a material with its position after deformation.

In a linear elasticity theory, the material obeys a generalized Hooke's law

$$\sigma_{\alpha\beta} = C_{\alpha\beta\gamma\delta}\varepsilon_{\gamma\delta} \tag{2.46}$$

where the $C_{\alpha\beta\gamma\delta}$ are elastic constants.

We now want to consider a viscoelastic fluid in which small elastic effects can be added linearly. In this case the rate of strain tensor

$$\Lambda_{\alpha\beta} = \frac{1}{2}\left(\frac{\partial v_\alpha}{\partial r_\beta} + \frac{\partial v_\beta}{\partial r_\alpha}\right) \tag{2.47}$$

must be considered. It is conveniently split into an elastic part

$\Lambda^{(e)}$ and a viscous part $\Lambda^{(v)}$ as $\Lambda = \Lambda^{(e)} + \Lambda^{(v)}$, where

$$\Lambda^{(e)}_{\alpha\beta} = \frac{\partial \varepsilon_{\alpha\beta}}{\partial t} = \frac{1}{2}\left(\frac{\partial v^{(e)}_{\alpha}}{\partial r_{\beta}} + \frac{\partial v^{(e)}_{\beta}}{\partial r_{\alpha}}\right) \tag{2.48}$$

with $\mathbf{v}^{(e)} = \partial \mathbf{u}/\partial t$. In a newtonian fluid the viscous part of the rate of strain tensor is similarly related to the viscous part of the stress tensor by

$$(\boldsymbol{\sigma} + p\mathbf{I}) = -\boldsymbol{\Pi}_{\text{viscous}} = 2\eta[\Lambda^{(v)} - \tfrac{1}{3}Tr(\Lambda^{(v)})\mathbf{I}] + \zeta Tr(\Lambda^{(v)})\mathbf{I} \tag{2.49}$$

The viscous terms have been regrouped to separate effects due to changes in shape (first term) from those arising from changes in volume (second term).

Standard elasticity theory for an isotropic medium gives us the elastic contribution to the stress tensor

$$(\sigma_{\alpha\beta} + pI_{\alpha\beta})_{\text{elastic}} = G\left[\frac{\partial u_{\alpha}}{\partial r_{\beta}} + \frac{\partial u_{\beta}}{\partial r_{\alpha}} - \frac{2}{3}(\boldsymbol{\nabla}\cdot\mathbf{u})\,\delta_{\alpha\beta}\right] + B(\boldsymbol{\nabla}\cdot\mathbf{u})\,\delta_{\alpha\beta} \tag{2.50}$$

where, as before, the separation is into shape- and volume-dependent terms. Here, G is the shear modulus and B is the bulk modulus.

Taking the time derivative of eqn (2.50) and using the definition $(\partial \mathbf{u}/\partial t) = \mathbf{v}^{(e)}$ gives

$$\frac{1}{G}\frac{\partial}{\partial t}(\boldsymbol{\sigma} + p\mathbf{I})_{\text{elastic}} = 2[\Lambda^{(e)} - \tfrac{1}{3}Tr(\Lambda^{(e)})\mathbf{I}] + \frac{B}{G}(\boldsymbol{\nabla}\cdot\mathbf{v}^{(e)})\mathbf{I} \tag{2.51}$$

Splitting the total velocity field into an elastic contribution $\mathbf{v}^{(e)}$ and a viscous one $\mathbf{v}^{(v)}$, i.e. $\mathbf{v} = \mathbf{v}^{(e)} + \mathbf{v}^{(v)}$, where $\mathbf{v}^{(v)}$ is related to $\Lambda^{(v)}$ via an equation like (2.47), we can add eqns (2.49) and (2.51) to get‡

$$\left(\frac{1}{\eta} + \frac{1}{G}\frac{\partial}{\partial t}\right)(\boldsymbol{\sigma} + p\mathbf{I}) = 2[\Lambda - \tfrac{1}{3}Tr(\Lambda)\mathbf{I}] + A(\boldsymbol{\nabla}\cdot\mathbf{v})\mathbf{I} \tag{2.52}$$

This equation, valid for an isotropic, newtonian viscoelastic fluid, also assumes equal relaxation times $A = \xi/\eta = B/G$, which is justified in certain simple substances.

‡ The balance eqn (2.52) for the viscoelastic stress tensor is derived in Egelstaff (1967).

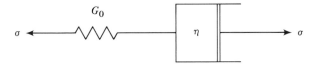

Fig. 2.1. One-dimensional Maxwell model for a viscoelastic fluid.

It should be emphasized that eqn (2.52) is only one possible model and is not a general equation for all effects involving viscous and elastic properties. The essential assumption (other than that of linear constitutive equations) was that the effects of viscous and elastic properties can be considered separately and that the net effect is simply additive. This corresponds to the Maxwell model for viscoelastic fluids, most simply illustrated by the one-dimensional apparatus in Fig. 2.1.

The spring (elastic effect) and the dashpot (viscous effect) are assumed to be linear, i.e. $\sigma = G_0 \varepsilon$ for the spring alone and $\sigma = \eta \, \partial \varepsilon / \partial t$ for the dashpot alone, where ε is the displacement, σ is the stress, G_0 is the spring constant, and η is the dashpot viscosity. The equation of motion for the system is

$$\left(\frac{1}{\eta} + \frac{1}{G_0} \frac{\partial}{\partial t} \right) \sigma = \frac{\partial \varepsilon}{\partial t} \qquad (2.53)$$

in obvious analogy to the tensor equation for an isotropic viscoelastic fluid [eqn (2.52)].

However, one should keep in mind that the Maxwell model is inadequate to explain viscoelastic effects in solids for which the Kelvin model, illustrated in Fig. 2.2, is more realistic.

The equation of motion for this one-dimensional model is

$$\sigma = G_0 \varepsilon + \eta \frac{\partial e}{\partial t} \qquad (2.54)$$

which obviously has a different structure from eqn (2.53).‡

A general theory of viscoelastic effects would have to start with a constitutive balance equation of the form

$$\sigma_{ij}(t) = G_{ijkl}(t) \varepsilon_{kl}(0) + \int_0^t G_{ijkl}(t - \tau) \frac{d\varepsilon_{kl}(\tau)}{d\tau} \, d\tau \qquad (2.55)$$

‡ Further models for viscoelastic materials are listed by Flügge (1967).

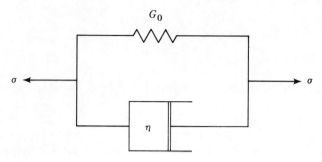

Fig. 2.2. One-dimensional Kelvin model for viscoelastic solids.

where the time integration obviously takes care of memory effects. It has been argued that in a linearized theory the present state of the material depends more on the recent than on the distant history, and a fading memory hypothesis‡ can be postulated by imposing the condition

$$\left|\frac{dG_{ijkl}(t)}{dt}\right|_{t=t_1} \leq \left|\frac{dG_{ijkl}(t)}{dt}\right|_{t=t_2} \tag{2.56}$$

for $t_1 > t_2 > 0$.

In an isotropic material, a simple exponential

$$G(t) = G_0 e^{-t/t_1} \tag{2.57}$$

can be assumed in the discussion of the Maxwell model with the additional identification that $\eta = G_0 t_1$.

2.3. Entropy Balance and the Second Law of Thermodynamics

After considering the balance equations for the various mechanical quantities, we now must make the connection to thermodynamics by writing a balance equation for the entropy. Let us briefly recall from equilibrium thermodynamics three main features of entropy.

First, it is additive, i.e., the entropy of the sum of two systems is the sum of their entropies. This implies that a differential

‡ For a discussion of this concept, see Truesdell and Noll (1965), pp. 101–117.

change of entropy dS can be written as

$$dS = dS_e + dS_i \tag{2.58}$$

Here, dS_e arises from the exchange of entropy with the system's surroundings and dS_i comes from internal production or destruction of entropy.

Secondly, dS_e can be positive, negative, or zero depending upon the system's interaction with its surroundings, but dS_i is always larger than or equal to zero, the latter occurring for reversible changes. Thus, for an isolated system, $dS_e = 0$ since there is no interaction with the surroundings, and hence $dS \geq 0$, which is nothing but the second law of thermodynamics.

Finally, if the system is free to receive heat from a reservoir at temperature T, but is otherwise isolated, then $dS_e = (dQ/T)$. Hence, $dS \geq (dQ/T)$, which is an alternate form of the second law for isothermal boundary conditions known as the Clausius inequality.

Let us turn now to a thermodynamic system out of equilibrium. The assumption of local equilibrium allows us to introduce a local entropy of density s, an internal entropy source density σ_s, and an entropy flow density $\mathbf{j}_s$, which includes both internal and external entropy flows.‡ Then, we can write

$$S = \int_V \rho s \, dV$$

$$\frac{dS_e}{dt} = -\int_\Sigma \mathbf{j}_s \cdot d\mathbf{\Sigma}$$

and

$$\frac{dS_i}{dt} = \int_V \sigma_s \, dV \tag{2.59}$$

Since, in local equilibrium situations, $s = s(u, v, c_\gamma)$ is a continuous function of the internal energy density u, the specific volume $v \equiv \rho^{-1}$, and the mass fraction $c_\gamma \equiv \rho_\gamma/\rho$, the divergence theorem

---

‡ Meixner (1970) has discussed the possibility of nonequilibrium thermodynamics without a local entropy. His theory starts from an integral form of the Clausius inequality and supposedly only involves the exact equilibrium thermodynamic functions. See also the discussion by Landauer (1975) and Hofelich (1969).

can be applied to yield

$$\frac{dS}{dt} = \int_V \frac{\partial(\rho s)}{\partial t}\, dV = \frac{dS_e}{dt} + \frac{dS_i}{dt}$$

$$= -\int_V (\boldsymbol{\nabla} \cdot \mathbf{j}_s)\, dV + \int_V \sigma_s\, dV \qquad (2.60)$$

Hence, the local balance equation for entropy reads

$$\frac{\partial(\rho s)}{\partial t} + \boldsymbol{\nabla} \cdot \mathbf{j}_s = \sigma_s \qquad (2.61)$$

where $\sigma_s \geq 0$, or alternatively

$$\rho \frac{Ds}{Dt} + \boldsymbol{\nabla} \cdot (\mathbf{j}_s - \rho s \mathbf{v}) = \sigma_s \qquad (2.62)$$

To find explicit expressions for $\mathbf{j}_s$ and σ_s, we write

$$ds(u, v, c_\gamma) = \frac{\partial s}{\partial u}\, du + \frac{\partial s}{\partial v}\, dv + \sum_{\gamma=1}^{c} \frac{\partial s}{\partial c_\gamma}\, dc_\gamma \qquad (2.63)$$

and use the definitions

$$\left(\frac{\partial s}{\partial u}\right)_{v,c_\gamma} = T^{-1} \qquad \left(\frac{\partial s}{\partial v}\right)_{u,c_\gamma} = pT^{-1}$$

$$\left(\frac{\partial s}{\partial c_\gamma}\right)_{v,u} = -\mu_\gamma T^{-1} \qquad (2.64)$$

where p is the equilibrium pressure in a fluid and μ_γ is the chemical potential per unit mass for component γ. This yields the fundamental relation of equilibrium thermodynamics:

$$T\, ds = du + p\, dv - \sum_{\gamma=1}^{c} \mu_\gamma\, dc_\gamma \qquad (2.65)$$

This, by virtue of the local equilibrium condition, is also valid, in a system out of equilibrium, for a small volume element in which the thermodynamic variables and functions can be assumed to be constant. In such a situation it is, however, imperative to formulate the Gibbs relation in the center of mass frame for a system in macroscopic motion because equilibrium thermodynamics cannot deal with convective phenomena. Moreover, it must be realized that in a system in local equilibrium, processes between neighboring cells are thermodynamically slow. This allows us to take the hydrodynamic derivative of the Gibbs equality yielding

$$T\frac{Ds}{Dt} = \frac{Du}{Dt} + p\frac{Dv}{Dt} - \sum_{\gamma=1}^{c} \mu_\gamma \frac{Dc_\gamma}{Dt} \quad (2.66)$$

Substituting from eqns (2.8) and (2.28) gives

$$\rho\frac{Ds}{Dt} = -\frac{1}{T}\boldsymbol{\nabla}\cdot\mathbf{j}_q - \frac{1}{T}\boldsymbol{\Pi}:\boldsymbol{\nabla}\mathbf{v} + \frac{1}{T}\sum_{\gamma=1}^{c}\mathbf{j}_\gamma^{\text{diff}}\cdot\mathbf{F}_\gamma$$

$$+ \frac{1}{T}\sum_{\gamma=1}^{c}\mu_\gamma\boldsymbol{\nabla}\cdot\mathbf{j}_\gamma^{\text{diff}} - \frac{1}{T}\sum_{\delta=1}^{r} J_\delta A_\delta \quad (2.67)$$

where $A_\delta = \sum_{\gamma=1}^{c} v_{\gamma\delta}\mu_\gamma$ is the chemical affinity. The terms in this equation can also be rearranged into a local balance equation

$$\frac{\partial(\rho s)}{\partial t} + \boldsymbol{\nabla}\cdot\mathbf{j}_s = \sigma_s \quad (2.68)$$

where the entropy current is

$$\mathbf{j}_s = \rho s\mathbf{v} + \frac{\mathbf{j}_q}{T} - \sum_{\gamma=1}^{c}\mu_\gamma\frac{\mathbf{j}_\gamma^{\text{diff}}}{T} \quad (2.69)$$

and the entropy production is given by

$$\sigma_s = \mathbf{j}_q\cdot\boldsymbol{\nabla}\left(\frac{1}{T}\right) + \frac{1}{T}\sum_{\gamma=1}^{c}\mathbf{j}_\gamma^{\text{diff}}\cdot\mathbf{F}_\gamma$$

$$- \sum_{\gamma=1}^{c}\mathbf{j}_\gamma^{\text{diff}}\cdot\boldsymbol{\nabla}\left(\frac{\mu_\gamma}{T}\right) - \frac{1}{T}\boldsymbol{\Pi}:\boldsymbol{\nabla}\mathbf{v} - \frac{1}{T}\sum_{\delta=1}^{r} J_\delta A_\delta \quad (2.70)$$

There is obviously some arbitrariness in separating the right-hand side of eqn (2.67) into a source term σ_s and an entropy current $\mathbf{j}_s$. Our choice was motivated (Meixner and Reik, 1955) by the fact that σ_s should not contain a divergence of a vector field which could easily have either positive or negative sign in different space-time points leading possibly to a trivial violation of the condition $\sigma_s \geq 0$. Let us next observe that for a system enclosed by impermeable, diathermal walls, we can write

$$\frac{DS}{Dt} = -\int_\Sigma \mathbf{j}_s\cdot d\Sigma + \int_V \sigma_s \, dV$$

$$\geq -\int_\Sigma \mathbf{j}_s\cdot d\Sigma \quad (2.71)$$

and have to demand that $\mathbf{v} = 0$, and $\mathbf{j}_\gamma^{\text{diff}} = 0$ at the surface. Thus, we recover the Clausius–Carnot theorem, namely,

$$\frac{DS}{Dt} \geq -\int_\sigma \frac{\mathbf{j}_q \cdot d\Sigma}{T} \tag{2.72}$$

The entropy production term, eqn (2.70), is a sum of products of thermodynamic fluxes and generalized thermodynamic forces. The term $\mathbf{j}_q \cdot \mathbf{\nabla}(1/T)$ is a heat conduction term. The flux is the heat current $\mathbf{j}_q$ and the force is $\mathbf{\nabla}(1/T)$ related to the temperature gradient. Next, $(1/T)\sum_{\gamma=1}^c \mathbf{j}_\gamma^{\text{diff}} \cdot \mathbf{F}_\gamma$ describes diffusion in an external field. The flux is $\mathbf{j}_\gamma^{\text{diff}}$ and the force is $(1/T)\mathbf{F}_\gamma$, the external force. The diffusion term $-\sum_{\gamma=1}^c \mathbf{j}_\gamma^{\text{diff}} \cdot \mathbf{\nabla}(\mu_\gamma/T)$ is due to chemical potential gradients. The flux is again $\mathbf{j}_\gamma^{\text{diff}}$ but the force is $-\mathbf{\nabla}(\mu_\gamma/T)$, related to the inhomogeneity of the chemical potential. In the viscous pressure term, $-(1/T)\mathbf{\Pi}:\mathbf{\nabla v}$, the flux is $\mathbf{\Pi}$, related to momentum flow, and the force is $-(1/T)\mathbf{\nabla v}$, the velocity gradient. Finally, $-(1/T)\sum_{\delta=1}^r J_\delta A_\delta$ is a chemical reaction term. The flux is J_δ, the reaction rate, and the force driving the reaction is the affinity A_δ. In general, the entropy production σ_s may be written as

$$\sigma_s = \sum_i Y_i X_i \tag{2.73}$$

where the Y_i are the thermodynamic fluxes ($\mathbf{j}_\gamma^{\text{diff}}$, $\mathbf{j}_q$, $\mathbf{\Pi}$, etc.) which are 'driven' by the conjugate thermodynamic forces [$\mathbf{\nabla}(\mu_\gamma/T)$, $\mathbf{F}_\gamma$, $\mathbf{\nabla}(1/T)$, $(1/T)\mathbf{\nabla v}$, etc.]. These entropy source terms will be studied in detail in Chapter 3.

2.4. Balance Equations for Solids

The balance eqns (2.6), (2.13), and (2.26) for the local mass density ρ, the local momentum density $\rho\mathbf{v}$, and the local energy density (ρe) are valid in any system. As we will see in Section 7.3, they are simply a consequence of the five microscopic conservation laws for mass, momentum, and energy. The entropy balance, on the other hand, can only be written for systems in local equilibrium, i.e., for which the fundamental relation (2.65) holds locally. This, in particular, implies that all macroscopic changes described by this set of balance equations must be slow on the time scale over which local equilibrium is established. Whereas the above balance equations are sufficient for a normal fluid, apart from the need of constitutive relations to get closure, they are, in general, incomplete for systems that exhibit broken continuous symmetries. In such a case, additional macroscopic variables will emerge that can exhibit slow thermodynamic change. For example, the staggered magnetization in an isotropic antiferromagnet or the director in a nematic liquid crystal can point in any direction, breaking the continuous rotational symmetry of the system. In such a case, a slow, continuous variation in the direction (a spin wave in the antiferromagnet) will require very little energy for its excitation and, indeed, less energy than to produce a deviation without allowing the system to readjust. The time rate of change of such a new macroscopic variable, the magnetization or the director, must therefore be small on the time scale of establishing local equilibrium, and additional macroscopic balance equations are needed—e.g. two more for nematic liquid crystals (see Forster *et al.*, 1971).

In a superfluid, an additional slow or 'hydrodynamic' variable is the superfluid velocity which can be introduced via the continuously broken gauge symmetry of a (superfluid) Bose system (Hohenberg and Martin, 1965). But because the superfluid velocity is curl-free (the Landau condition) and thus is given as a gradient of a scalar field, only one additional balance equation is needed.

Lastly, consider a simple crystalline solid. Unlike a simple fluid, it is not translationally invariant (although the underlying hamiltonian still is) but has the three continuous translational symmetries broken into the discrete symmetries of the lattice structure. Hence there can be three extra hydrodynamic variables which we might choose to be the three components of the vector of the local lattice distortion.

Let us then assume, in general, that in addition to mass, momentum, and energy densities a set $\{f^i\}$ of extra independent hydrodynamic variables is needed for a full macroscopic description of slow processes in a given system. The f^i's will, of course, be subject to the local balance equation (2.5). In addition, they must, as macroscopic thermodynamic variables of the system, modify the Gibbs' relation (2.65). We prefer here to rewrite that relation in the Euler (laboratory) frame as

$$T\,d(\rho s) = d(\rho u + \tfrac{1}{2}\rho v^2) - \mu\,d\rho - \mathbf{v}\cdot d(\rho\mathbf{v}) - \sum_i \Phi_i\,d(\rho f_i) \quad (2.74)$$

where
$$\rho\mu = p + \rho u - T\rho s - \tfrac{1}{2}\rho v^2 \tag{2.75}$$
introduces the hydrostatic pressure p and the ϕ_i are external forces acting on the variable f_i. Because eqn (2.74) is given in the Euler frame, we find the entropy balance according to

$$T\frac{\partial}{\partial t}(\rho s) = \frac{\partial}{\partial t}(\rho u + \tfrac{1}{2}\rho v^2) - \mu\frac{\partial \rho}{\partial t} - \mathbf{v}\cdot\frac{\partial(\rho\mathbf{v})}{\partial t} - \sum_i \Phi_i\frac{\partial \rho f_i}{\partial t} \tag{2.76}$$

by eliminating the partial time derivatives on the right-hand side using the respective balance equations. We find

$$\frac{\partial(\rho s)}{\partial t} + \nabla \mathbf{j}_s = \sigma_s \tag{2.77}$$

with

$$\begin{aligned}\mathbf{j}_s &= \rho s\mathbf{v} + \frac{\mathbf{j}_q}{T} - \sum_{\gamma=1}^c \mu_\gamma \frac{\mathbf{j}_\gamma^{\text{diff}}}{T} - \sum_i \frac{\Phi_i}{T}\mathbf{j}_{f_i} \\ \sigma_s &= \mathbf{j}_q \cdot \nabla\left(\frac{1}{T}\right) + \frac{1}{T}\sum_{\gamma=1}^c \mathbf{j}_\gamma^{\text{diff}}\cdot \mathbf{F}_\gamma - \sum_{\gamma=1}^c \mathbf{j}_\gamma^{\text{diff}}\cdot \nabla\left(\frac{\mu_\gamma}{T}\right) \\ &\quad - \frac{1}{T}\mathbf{\Pi}:(\nabla\mathbf{v}) - \frac{1}{T}\sum_{\delta=1}^\gamma j_\delta A_\delta - \sum_i \mathbf{j}_{f_i}\cdot\nabla\frac{\Phi_i}{T}\end{aligned} \tag{2.78}$$

In a solid it is convenient (see Fleming and Cohen, 1976, and also Pokrovsky and Sergeev, 1973) to introduce in lieu of the vector of lattice distortion the local strain tensor $\boldsymbol{\varepsilon}(\mathbf{r},t)$, eqn (2.45) and define its thermodynamically conjugate, extensive variable as

$$\boldsymbol{\Phi}(\mathbf{r},t) = -\frac{\partial(\rho e)}{\partial\boldsymbol{\varepsilon}}\bigg|_{s,\rho} \tag{2.79}$$

The last term in eqn (2.76) then reads

$$\sum_i \Phi_i \frac{\partial f_i}{\partial t} = Tr\boldsymbol{\Phi}\cdot\frac{\partial\boldsymbol{\varepsilon}}{\partial t} = \sum_{\alpha\beta}\Phi_{\alpha\beta}\frac{\partial \varepsilon_{\alpha\beta}}{\partial t} \tag{2.80}$$

and the balance equation for the strain tensor is

$$\frac{\partial}{\partial t}[\rho\varepsilon_{\alpha\beta}(\mathbf{r},t)] + \frac{\partial}{\partial r_\gamma}j^{(\varepsilon)}_{\alpha\beta\gamma}(\mathbf{r},t) = 0 \tag{2.81}$$

where we defined a strain flux tensor via

$$j^{(\varepsilon)}_{\alpha\beta\gamma}(\mathbf{r},t) = \tfrac{1}{2}(\delta_{\alpha\gamma}v^{(e)}_\beta + \delta_{\beta\gamma}v^{(e)}_\alpha) \tag{2.82}$$

where the local elastic deformation velocities $\mathbf{v}^{(e)} = \partial\mathbf{u}/\partial t$ have been introduced in eqn (2.47) in terms of the vector of the local lattice distortion $\mathbf{u}(\mathbf{r}, t)$.

3
Linear Phenomenological Laws

3.1. Choosing Forces and Fluxes

FROM the balance equation for entropy, eqn (2.61), we know that the entropy production σ_s, eqn (2.70), can be written as a sum of products of generalized thermodynamic fluxes Y_i and forces X_i. For the situations considered in Chapter 2,

$$\sigma_s = -\mathbf{j}_q \cdot \left(\frac{1}{T^2}\boldsymbol{\nabla} T\right) + \frac{1}{T}\sum_{\gamma=1}^{c} \mathbf{j}_\gamma^{\text{diff}} \cdot \mathbf{F}_\gamma - \sum_{\gamma=1}^{c} \mathbf{j}_\gamma^{\text{diff}} \cdot \boldsymbol{\nabla}\left(\frac{\mu_\gamma}{T}\right)$$

$$-\frac{1}{T}\boldsymbol{\Pi}:\boldsymbol{\nabla}\mathbf{v} - \frac{1}{T}\sum_{\delta=1}^{r} J_\delta A_\delta = \sum_i X_i Y_i \quad (3.1)$$

Intuitively, for each term in eqn (3.1), it is possible to separate the force and the flux. For example, for the thermal conduction term, we have $\mathbf{X}_1 \mathbf{Y}_1 = -\mathbf{j}_q \cdot [(1/T^2)\boldsymbol{\nabla} T]$ and an expression proportional to $\mathbf{j}_q$ is interpreted as the flux $\mathbf{Y}_1$, while a term proportional to $\boldsymbol{\nabla} T$ is regarded as the force $\mathbf{X}_1$. The choice of proportionality constants must be made in such a way that $\mathbf{X}_1 \mathbf{Y}_1 = (1/T^2)\mathbf{j}_q \cdot \boldsymbol{\nabla} T$. One choice would be $\mathbf{X}_1 = -\boldsymbol{\nabla} T$ and $\mathbf{Y}_1 = (1/T^2)\mathbf{j}_q$ and an alternative valid choice would be $\mathbf{X}_1' = \boldsymbol{\nabla}(1/T)$ and $\mathbf{Y}_1' = \mathbf{j}_q$. Such choices are essentially ones of convenience and have no effect on the physical interpretation.

The intuitive separation of forces and fluxes is based upon the concepts of cause and effect. In a given experimental situation, it seems possible to identify one of the factors of each term in σ_s as a constraint on the system. Temperature gradients are established by putting different boundaries of the system in contact with heat reservoirs at different temperatures; external fields are applied on the surfaces; gradients in chemical potentials are maintained by concentration differences; shear flow (gradients in **v**) is maintained by suitable boundary conditions; and nonzero affinities are created in chemical systems by adding reactants. These externally applied forces are then seen as 'causing' the corresponding fluxes. However, the clear-cut distinction between controlled external forces, such as ∇T, and the 'resulting' fluxes, such as $\mathbf{j}_q$, disappears when it is realized that ∇T itself is not controlled at all in the interior of the system, but only on its surfaces. In order to find $T(\mathbf{r}, t)$, and hence to find ∇T, it is necessary to solve a boundary value problem involving the thermal properties of the material and the heat current $\mathbf{j}_q$, as well as the boundary conditions. Hence, the applied force' ∇T implicitly involves the 'resultant flux' $\mathbf{j}_q$. (A highly unsatisfactory state of affairs!) It is obviously desirable to have a more specific criterion for separating thermodynamic forces from fluxes if these concepts are to have any real meaning. Moreover, a clear definition of force and flux is necessary to establish certain symmetry relations, known as Onsager reciprocal relations (Onsager, 1931), in the phenomenological laws which connect the forces to the fluxes.

In order to identify further thermodynamic forces and fluxes, in addition to the ones already listed in the balance equations of Chapter 2, we adopt the following definition: Let the system be locally described by a set of extensive thermodynamic variables $\alpha_1, \ldots, \alpha_n$. We then identify the fluxes in the system as the time derivatives of these variables:‡

$$Y_i = \frac{d\alpha_i}{dt} = \dot{\alpha}_i \qquad i = 1, \ldots, n \qquad (3.2)$$

They can be constructed to be either odd or even under time reversal.§ The conjugate forces X_i, of course, have to be known

‡ Special care must be taken in situations involving heat conduction, because the heat current is in general not the time derivative of an extensive thermodynamic variable. For a discussion, see Casimir (1945) and de Groot and Mazur (1969).

§ Recall that any extensive macroscopic variable $\alpha_i = \alpha_i(q_1, \ldots, q_n; p_1, \ldots, p_n)$ is, of course, a function of $2N$ microscopic canonical variables $q_1, \ldots, q_n; p_1, \ldots, p_n$ in a system of N degrees of freedom (Wigner, 1954). α_i is even (odd) under time reversal, if it does not change (changes) sign after reversal of the microscopic moments, i.e. $\alpha_i(q_1, \ldots, q_n; p_1, \ldots, p_n) = \pm \alpha_i(q_1, \ldots, q_n; -p_1, \ldots, -p_n)$. The common practice in thermodynamics is to choose extensive variables of a given parity; one chooses, for example, energy (even) and angular momentum (odd), rather than linear combinations of them which would not be of definite parity.

before the balance equations can be set up. If the system is locally described by the fundamental relation $S = S(\alpha_1, \ldots, \alpha_n)$, the forces X_i are the intensive variables conjugate to α_i:

$$X_i = \left.\frac{\partial S}{\partial \alpha_i}\right|_{\alpha_j} \tag{3.3}$$

An alternative way for their identification would be to rewrite the entropy production, eqn (3.1), in terms of the fluxes, eqn (3.2), and to identify the forces as their coefficients. Because the entropy production has to be odd under time reversal, this explicit construction of the forces ensures that they are always of the opposite time parity to their conjugate fluxes.

In recapitulating our attempts to formulate a thermodynamic theory of systems out of equilibrium, we must realize that so far we have introduced more unknown field variables than set up general balance equations. To complete the theory we now have to resort to experiment and write the constitutive equations relating fluxes and forces in a particular system. To set up such phenomenological laws, we first note that in principle each flux Y_i can depend upon all of the applied forces X_i:

$$Y_i = Y_i(X_1, \ldots, X_n) \tag{3.4}$$

if there are n separate forces.

In equilibrium, no external forces are acting and the fluxes are zero:

$$Y_i(0, 0, \ldots, 0) = 0 \tag{3.5}$$

For sufficiently small deviations from equilibrium, we linearize the dependence of the fluxes upon the forces:

$$Y_i = \sum_{k=1}^{n} L_{ik} X_k \tag{3.6}$$

where the L_{ik} are constant phenomenological coefficients.‡

This general form includes the usual linear phenomenological equations of macroscopic physics as special cases; for example, Ohm's law ($\mathbf{j} = \sigma \mathbf{E}$, where σ is the electrical conductivity), Fourier's law ($\mathbf{j}_q = -\lambda \nabla T$, where λ is the thermal conductivity), Fick's law

‡ Equation (3.6) describes a particular experimental arrangement. To obtain the time-reversed flux, we must (considering even forces only) change the sign of the force, in agreement with the requirements of definite parity discussed above.

of diffusion ($\mathbf{j}_\gamma^{\text{diff}} = -D\nabla n_\gamma$, where n_γ is the concentration of the gas component γ), and Newton's law of friction ($\sigma_{\alpha\beta} = \eta[(\partial v_\alpha/\partial r_\beta)+(\partial v_\beta/\partial r_\alpha)]$, where $\boldsymbol{\sigma}$ is the stress tensor). Examples of processes involving several forces at once are thermoelectric, electrokinetic, and galvanomagnetic phenomena.

The introduction of the phenomenological coefficients L_{ik}, however, has no predictive power if we cannot impose theoretical restrictions on them. In this section we want to list some general and fairly obvious constraints and leave the establishment of the fundamental Onsager reciprocity relations to the next section. We start from the entropy production

$$\sigma_s = \sum_i X_i Y_i = \sum_{i,k} L_{ik} X_i X_k \geq 0 \qquad (3.7)$$

In order for this bilinear form to be positive semidefinite, all diagonal elements L_{ii} must be positive or zero and the off-diagonal elements must satisfy the inequality $L_{ii}L_{kk} \geq \frac{1}{4}(L_{ik}+L_{ki})^2$.

Note that it is only the symmetric part of the matrix **L** which contributes to the entropy production. That is, if we write $\mathbf{L} = \mathbf{L}^{(a)} + \mathbf{L}^{(s)}$, where $\mathbf{L}^{(s)}$ is the symmetric part of **L** and $\mathbf{L}^{(a)}$ is its antisymmetric part, then

$$\sigma_s = \sum_{i,k} L_{ik} X_i X_k = \sum_{i,k} L_{ik}^{(s)} X_i X_k \qquad (3.8)$$

and there is freedom to add an antisymmetric part to **L** without changing σ_s. This is important when considering the dependence of the Onsager relations upon the particular choice of forces and fluxes.

In most systems, the components of the various fluxes do not depend on all of the forces. This is often a consequence of what is called *Curie's principle*. Loosely stated, Curie's principle asserts that, in an isotropic medium, fluxes and forces of different tensorial character do not couple. Some care must be taken in interpreting this principle since it is not immediately obvious what is meant by tensorial character; that is, a second-rank tensor may be regarded as being composed of a scalar (the trace of the tensor), a vector (the antisymmetric part of the tensor, an axial vector), and a symmetric tensor. For example, in an isotropic fluid, heat conduction and diffusion are vector phenomena, and viscous phenomena or chemical reactions are tensor and scalar phenomena, respectively. With proper care, this symmetry of an isotropic system can be used to reduce the number of independent relationships between the forces and fluxes. In systems that are not isotropic, it is also possible to use the symmetries of the system to reduce the number of phenomenological coefficients. For example, in crystals, it can be shown that λ, the heat conduction tensor, is symmetric in nineteen out of the thirty-two possible point symmetries.

As a consequence of the fact that different tensorial phenomena do not couple, it is found that the entropy production separates into several terms (for isotropic media), each of which is separately positive semidefinite. That is, if we write

$$\sigma_s = \sum_i X_i Y_i + \sum_j \mathbf{X}_j \cdot \mathbf{Y}_j + \sum_k \mathbf{X}_k : \mathbf{Y}_k \geq 0 \qquad (3.9)$$

where X_i, Y_i are scalars, $\mathbf{X}_j, \mathbf{Y}_j$ are vectors, and $\mathbf{X}_k, \mathbf{Y}_k$ are tensors, then it may be concluded that

$$\sum_i X_i Y_i \geq 0$$

$$\sum_j \mathbf{X}_j \cdot \mathbf{Y}_j \geq 0$$

$$\sum_k \mathbf{X}_k : \mathbf{Y}_k \geq 0 \qquad (3.10)$$

3.2. The Onsager Reciprocity Relations

We have seen in the previous section that the positive definiteness of the entropy production puts restrictions on the phenomenological coefficients L_{ik}, namely

$$L_{ii} \geq 0$$

and

$$L_{ii} L_{kk} \geq \tfrac{1}{4}(L_{ik} + L_{ki})^2 \qquad (3.11)$$

As early as 1854, Kelvin had already postulated that, for the thermoelectric effect, an additional symmetry might hold, namely

$$L_{ik} = \pm L_{ki}, \qquad (3.12)$$

LINEAR PHENOMENOLOGICAL LAWS

Similar symmetry relations were later proposed by Helmholtz for electrodiffusion processes. In 1931, Onsager advanced the idea that such relations should hold in all thermodynamic systems controlled by linear phenomenological laws. In their most general formulation, due to subsequent developments (Casimir, 1945), these Onsager reciprocity relations read

$$L_{ik}(\mathbf{B}, \boldsymbol{\omega}, \sigma_1, \ldots, \sigma_l) = \varepsilon_i \varepsilon_k L_{ki}(-\mathbf{B}, -\boldsymbol{\omega}, -\sigma_1, \ldots, -\sigma_l) \quad (3.13)$$

if the irreversible processes occur in an external magnetic field $\mathbf{B}$, in a system rotating with angular velocity $\boldsymbol{\omega}$ and depending on parameters σ_j that are odd under time reversal. In addition, $\varepsilon_i = +1$ if X_i is even under time reversal and $\varepsilon_i = -1$ if X_i is odd under time reversal.

In order to motivate these relations, Onsager (1931) first considered a transparent case, that of a simple chemical reaction triangle consisting of three chemical reactants 1, 2, and 3 with molar concentrations n_1, n_2, and n_3, respectively. If $\bar{n}_1$, $\bar{n}_2$, and $\bar{n}_3$ are their equilibrium numbers, and if the rate at which species i spontaneously transforms into species j is κ_{ji} (a constant), then the rate equations of the various species are

$$\frac{dn_1}{dt} = -(\kappa_{21} + \kappa_{31})n_1 + \kappa_{12}n_2 + \kappa_{13}n_3$$

$$\frac{dn_2}{dt} = \kappa_{21}n_1 - (\kappa_{12} + \kappa_{32})n_2 + \kappa_{23}n_3$$

$$\frac{dn_3}{dt} = \kappa_{31}n_1 + \kappa_{32}n_2 - (\kappa_{13} + \kappa_{23})n_3 \quad (3.14)$$

subject to the conditions

$$\sum_{i=1}^{3} n_i = \sum_{i=1}^{3} \bar{n}_i = n \quad (3.15)$$

and

$$\left.\frac{dn_i}{dt}\right|_{\bar{n}_i} = 0 \quad \text{for } i = 1, 2, 3 \quad (3.16)$$

If the equilibrium concentrations are known, then there are two independent constraints on the six coefficients κ_{ij}. Thus, even if the system is assumed to be in equilibrium, there are still four degrees of freedom in choosing the rate coefficients.

However, the usual procedure in chemistry is to assume detailed balance. That is, it is assumed that, in equilibrium, each simple reaction exactly balances itself:

$$\kappa_{ij}\bar{n}_j = \kappa_{ji}\bar{n}_i \tag{3.17}$$

These are three more constraints on the system, leaving one degree of freedom in the determination of the rate coefficients. This implies that without further information only relative rates can be measured. The assumption of detailed balance is based on the fact that in equilibrium all allowable states have equal a priori probabilities, and thus each time-reversed motion occurs, on the average, as often as the original motion. As a consequence, each direct reaction $1 \to 2$ must occur as often as the reverse reaction $2 \to 1$. Thus, detailed balance holds as a consequence of the time-reversal invariance of the microscopic motions.

Let us now rewrite the linear rate equations above in terms of forces and fluxes, assuming also that the three reactants can be treated as ideal gases. Then at constant pressure and temperature, the Gibbs free energy G can be written as

$$G = G_{\text{equil}} + RT \sum_{i=1}^{3} n_i \ln\left(\frac{n_i}{\bar{n}_i}\right) \tag{3.18}$$

where G_{equil} is the equilibrium (minimum) value of G. To first-order in δn_i (small changes in n_i), we have

$$\delta G \big|_{p,T} = RT \sum_{i=1}^{3} \delta n_i \ln\left(\frac{n_i}{\bar{n}_i}\right) \tag{3.19}$$

Defining $\alpha_i = n_i - \bar{n}_i$ gives the fluxes as $Y_i = \dot{\alpha}_i = \dot{n}_i$ (the dot denotes time derivative), and we obtain

$$\delta G \big|_{p,T} = -\sum_{i=1}^{3} X_i \delta \alpha_i \tag{3.20}$$

The 'forces' X_i are now seen to be

$$X_i = -RT \ln\left(\frac{n_i}{\bar{n}_i}\right) = -\frac{RT}{\bar{n}_i} \alpha_i \tag{3.21}$$

for small α_i.

The rate equations can then be written in terms of the forces X_i

and the fluxes $\dot{\alpha}_i$ as

$$Y_1 = \dot{\alpha}_1 = (\kappa_{21} + \kappa_{31})\frac{\bar{n}_1}{RT}X_1 - \frac{\kappa_{12}\bar{n}_2}{RT}X_2 - \frac{\kappa_{13}\bar{n}_3}{RT}X_3$$

$$Y_2 = \dot{\alpha}_2 = -\frac{\kappa_{21}\bar{n}_1}{RT}X_1 + (\kappa_{12} + \kappa_{32})\frac{\bar{n}_2}{RT}X_2 - \frac{\kappa_{23}\bar{n}_3}{RT}X_3 \quad (3.22)$$

$$Y_3 = \dot{\alpha}_3 = -\frac{\kappa_{31}\bar{n}_1}{RT}X_1 - \frac{\kappa_{32}\bar{n}_2}{RT}X_2 + (\kappa_{13} + \kappa_{23})\frac{\bar{n}_3}{RT}X_3$$

Comparing this with eqn (3.6), we can identify the phenomenological coefficients L_{ik} as

$$L_{ii} = \sum_{\substack{k=1 \\ k \neq i}}^{3} \kappa_{ki}\frac{\bar{n}_i}{RT}$$

$$L_{ik} = -\kappa_{ik}\frac{\bar{n}_k}{RT} \quad i \neq k$$
(3.23)

Detailed balance eqn (3.17) then implies the Onsager reciprocity relations

$$L_{ik} = L_{ki} \quad (3.24)$$

The next step is to generalize the above considerations in order to derive Onsager relations for any set of thermodynamic forces and fluxes in the near equilibrium (linear) range. In the example of a chemical reaction triangle, the individual reactions $1 \to 2$, $2 \to 3$, etc., were considered. These reactions might be thought of as progressing with nonzero rates κ_{ij}, even in equilibrium. Although in equilibrium the rates balance in such a way that the net rates are zero, at a microscopic level the separate forward and reverse reaction rates retain a clear intuitive meaning. It was considerations of microreversibility which made it possible to derive Onsager relations for the observed rates near equilibrium.

In order to generalize this derivation to any thermodynamic system, a quantity analogous to the rates is needed. This quantity must be time-reversal invariant and must be common to all thermodynamic systems. The obvious choice is the fluctuations of the thermodynamic variables near equilibrium. This idea has been most elegantly exploited by Casimir (1945), who used the assumption of local equilibrium to expand the entropy S of the system as a Taylor series about its equilibrium value. Writing

$\boldsymbol{\alpha} = (\alpha_1, \alpha_2, \ldots, \alpha_n)$ to denote the deviations of a set of linearly independent extensive thermodynamic variables‡ A_i from their equilibrium values A_i^{eq} (i.e., $\alpha_i = A_i - A_i^{eq}$), gives $\boldsymbol{\alpha} = 0$ in equilibrium and

$$S(\boldsymbol{\alpha}) = S(0) + \sum_{i=1}^{n} \frac{\partial S}{\partial \alpha_i}\bigg|_{\boldsymbol{\alpha}=0} \alpha_i + \frac{1}{2} \sum_{i,j=1}^{n} \frac{\partial^2 S}{\partial \alpha_i \partial \alpha_j}\bigg|_{\boldsymbol{\alpha}=0} \alpha_i \alpha_j + \cdots \tag{3.25}$$

Since S is a maximum in equilibrium,

$$\frac{\partial S}{\partial \alpha_i}\bigg|_{\boldsymbol{\alpha}=0} = 0 \tag{3.26}$$

and we can write

$$\Delta S = S(\boldsymbol{\alpha}) - S(0) = -\frac{1}{2} \sum_{i,j=1}^{n} S_{ij} \alpha_i \alpha_j \leq 0 \tag{3.27}$$

where

$$S_{ij} = S_{ji} = -\frac{\partial^2 S}{\partial \alpha_i \partial \alpha_j}\bigg|_{\boldsymbol{\alpha}=0} \tag{3.28}$$

The driving forces X_i conjugate to the extensive variables α_i may be defined, in the spirit of equilibrium thermodynamics, as

$$X_i = \left(\frac{\partial S}{\partial \alpha_i}\right) = -\sum_{j=1}^{n} S_{ij} \alpha_j \tag{3.29}$$

As local equilibrium has been assumed, we can take expectation values of the near equilibrium variables as averages over an ensemble. According to Einstein (1910), the probability density for the occurrence of fluctuations $\boldsymbol{\alpha}$ is given by

$$W(\boldsymbol{\alpha}) \, d\boldsymbol{\alpha} = \frac{e^{\Delta S/k_B} \, d\boldsymbol{\alpha}}{\int e^{\Delta S/k_B} \, d\boldsymbol{\alpha}'} \tag{3.30}$$

which can also be written as

$$\Delta S = k_B \ln W + \text{const} \tag{3.31}$$

‡ For simplicity, only variables which are even under time reversal will be considered initially (e.g. kinetic energies). Such variables as angular velocities and magnetic fields which change sign under time reversal will be excluded for now.

With this, we can calculate the correlation

$$\langle \alpha_i X_j \rangle_{av} = \left\langle \alpha_i \frac{\partial(\Delta S)}{\partial \alpha_j} \right\rangle_{av}$$

$$= k_B \int \alpha_i \frac{\partial \ln W}{\partial \alpha_j} W\, d\alpha_i, \ldots, d\alpha_n$$

$$= -k_B \delta_{ij} \qquad (3.32)$$

Using the fact that

$$\alpha_i = -\sum_{j=1}^{n} (S^{-1})_{ij} X_j \qquad (3.33)$$

we also find

$$\langle \alpha_i \alpha_j \rangle_{av} = k_B S_{ij}^{-1} \qquad (3.34)$$

So far we have only dealt with fluctuations around the equilibrium state of the system using Gibbs ensembles. Recalling the ergodic hypothesis according to which ensemble averages yield the same results as long-time averages over the history of a single system, we now wish to introduce the time variable explicitly, a crucial parameter in a theory of irreversible processes. We first note that for a classical N-body system with conservative forces, microscopic reversibility ('detailed balance') is a consequence of the invariance of Hamilton's equations of motion under time reversal and simply means that for every microscopic motion reversing all particle velocities also yields a solution.‡ Moreover, for a macroscopic system we know that for every microscopic motion there is an equally *probable* motion with all velocities reversed. Thus time-reversal invariance holds in that the average of a variable α_i must be the same for equal positive and negative times from some initial instant. That is,

$$\langle \alpha_i(t+\tau) \rangle_{av}^{\boldsymbol{\alpha}(t)} = \langle \alpha_i(t-\tau) \rangle_{av}^{\boldsymbol{\alpha}(t)} \qquad (3.35)$$

where the $\boldsymbol{\alpha}(t)$ superscript denotes the fact that the average is now over all states which have the initial condition $\boldsymbol{\alpha} = \boldsymbol{\alpha}(t)$ at time t. If we multiply eqn (3.35) by $\alpha_j(t)$ and average over all

‡ For a quantum-mechanical N-body system with a self-adjoint hamiltonian, time-reversal invariance simply implies that the complex conjugate ψ^* of a wavefunction ψ is also a solution of Schrödinger's equation.

possible initial conditions $\boldsymbol{\alpha}(t)$, the result being denoted by $\langle \alpha_j(t)\alpha_i(t+\tau)\rangle$, then

$$\langle \alpha_j(t)\alpha_i(t+\tau)\rangle = \langle \alpha_j(t)\alpha_i(t-\tau)\rangle = \langle \alpha_j(t+\tau)\alpha_i(t)\rangle \qquad (3.36)$$

The last line follows from the invariance of the equilibrium state under a time translation $t \to t+\tau$. Eqn (3.36) is a statement of microscopic reversibility. If the system is exactly in equilibrium, the result is moreover independent of t.

Let us now extend these results appropriately to systems in which irreversible processes occur. We recall from Chapter 1 that the assumption of local equilibrium implies that thermodynamic states of the system close to each other in space and time in some sense can be reached through local equilibrium fluctuations. This obviously allows us to retain the symmetry relation

$$\langle \alpha_j(t)\alpha_i(t-\tau)\rangle = \langle \alpha_j(t+\tau)\alpha_i(t)\rangle \qquad (3.37)$$

provided that $\tau \ll \tau_{ev}$, where τ_{ev} is a macroscopic evolution time for the system as a whole. To include this macroscopic evolution we next consider $\dot{\alpha}_i$ as thermodynamic fluxes. Their long-time averages obey linear laws of the form $\langle \dot{\alpha}_i \rangle = \sum_k L_{ik} X_k$. Fluctuations are included by adding a random driving force to arrive at a Langevin equation

$$\dot{\alpha}_i = \sum_j L_{ij} X_j + \kappa_i(t) = -\sum_j P_{ij}\alpha_j + \kappa_i(t) \qquad (3.38)$$

where $P_{ij} = \sum_k L_{ik} S_{kj}$ and $\langle \kappa_i(t) \rangle = 0$ for averages over times much greater than τ_1, the microscopic relaxation time for the system. Physically, this means that a system in local equilibrium will respond to a change in the external constraints, i.e. to a force X_i, through a local fluctuation. The overall response, however, when averaged over these fluctuations will be linear in the force. Casimir (1945) points out that it is by no means evident that a set of equations that describe the macroscopic evolution of a system can, like eqn (3.38), always be extended to include the dynamics of fluctuations. Of course, the fact that the macroscopic equations are assumed to be linear partly justifies an extrapolation to very small deviations, but in principle one may imagine a pseudo-linearity holding only at the macroscopic level. We therefore want to accept eqn (3.38) as a new hypothesis.

With the solution

$$\alpha(t) = e^{-t\mathbf{P}}\alpha(0) + e^{-t\mathbf{P}} \int_0^t e^{t'\mathbf{P}}\kappa(t')\,dt \quad (3.39)$$

to eqn (3.38), we can next calculate

$$\langle \alpha(t+\tau) - \alpha(t) \rangle_{\mathrm{av}}^{\alpha(t)} = [e^{-\tau\mathbf{P}} - 1]\langle \alpha(t) \rangle_{\mathrm{av}}^{\alpha(t)} \quad (3.40)$$

The macroscopic linear equations would only be expected to hold for times $\tau \approx \tau_{\mathrm{ev}} \gg \tau_1$. But if $\tau P_{ij} \ll 1$ (i.e., $\tau_1 \ll \tau \ll \tau_{\mathrm{ev}}$), we can write

$$\langle \alpha(t+\tau) - \alpha(t) \rangle_{\mathrm{av}}^{\alpha(t)} \simeq -\tau \mathbf{P}\langle \alpha(t) \rangle_{\mathrm{av}}^{\alpha(t)} = \tau \sum_j L_{ij}\langle X_j \rangle_{\mathrm{av}} \quad (3.41)$$

and we get

$$\langle \alpha_l(t)[\alpha_k(t+\tau) - \alpha_k(t)] \rangle = \tau \sum_j L_{kj}\langle \alpha_l X_j \rangle = k_B \tau L_{kl} \quad (3.42)$$

and similarly,

$$\langle \alpha_k(t)[\alpha_l(t+\tau) - \alpha_l(t)] \rangle = k_B \tau L_{lk} \quad (3.43)$$

However, microscopic reversibility implies that

$$\begin{aligned}
\langle \alpha_l(t)[\alpha_k(t+\tau) - \alpha_k(t)] \rangle \\
= \langle \alpha_l(t)[\alpha_k(t-\tau) - \alpha_k(t)] \rangle \\
= \langle \alpha_l(t+\tau)[\alpha_k(t) - \alpha_k(t+\tau)] \rangle \\
= \langle \alpha_k(t)[\alpha_l(t+\tau) - \alpha_l(t)] \rangle
\end{aligned} \quad (3.44)$$

This completes the derivation of the Onsager reciprocity relations (1931):

$$L_{lk} = L_{kl} \quad (3.45)$$

Let us list some of the assumptions that had to be made in this derivation. First, near equilibrium, ΔS must be a quadratic function of the α_i's and Einstein's formula holds for fluctuations. These conditions are essentially the assumption of local equilibrium. In addition, in the fluctuation range, the independent forces and fluxes have to obey a linear law or a Langevin equation and, more importantly, the coefficients in this linear law must be identical to the coefficients in the macroscopic phenomenological law. The system must also have a relaxation time τ_1 for the regression of fluctuations, in which it reaches a steady state, such that $\tau_1 \ll \tau_{ev}$, where τ_{ev} is the macroscopic evolution time of the system. Finally, the forces and fluxes considered in a real system must actually be of the form assumed in the derivation, or must be reducible to such a form.

As the Onsager reciprocity relations play such an important role in linear processes, it will be worthwhile to present an alternative geometrical formulation of Casimir's derivation (Wigner, 1954). Wigner begins with the observation that the macroscopic variables A_i ($i = 1, \ldots, n$) are functions‡ of the microscopic coordinates and moments $q_1, \ldots, q_N, p_1, \ldots, p_N$ of a system with N degrees of freedom, where $N \gg n$. The latter are the coordinates in $2N$-dimensional phase space Γ; the n-dimensional space of macroscopic variables A_i will be denoted by γ.

The probability of finding the system in a unit volume around the point $\boldsymbol{\alpha} = (\alpha_1, \ldots, \alpha_n)$ is given by Einstein's formula $W(\boldsymbol{\alpha})$, eqn (3.30). We denote by Γ_α the domain in Γ space corresponding to a unit volume around $\boldsymbol{\alpha}$ in γ space. In proper units, the volume of Γ_α is equal to $W(\boldsymbol{\alpha})$. During the time evolution of the system, points in a part of the domain Γ_α will have moved after a time interval t into the domain $\Gamma_{\alpha'}$ of volume $W(\boldsymbol{\alpha}')$, corresponding in γ space to a unit volume around $\boldsymbol{\alpha}'$. We denote this subset of Γ_α by $\Gamma_{\alpha \to \alpha'}$. The transition probability from $\boldsymbol{\alpha}$ to $\boldsymbol{\alpha}'$ is obviously given by

$$T_t(\boldsymbol{\alpha} \to \boldsymbol{\alpha}') = \frac{\text{vol } \Gamma_{\alpha \to \alpha'}}{\text{vol } \Gamma_\alpha} = \frac{\text{vol } \Gamma_{\alpha \to \alpha'}}{W(\boldsymbol{\alpha})} \qquad (3.46)$$

which is normalized

$$\int T_t(\boldsymbol{\alpha} \to \boldsymbol{\alpha}') \, d\boldsymbol{\alpha}' = 1 \qquad (3.47)$$

because every point in Γ_α will have arrived in some $\Gamma_{\alpha'}$ after time t.

Let us next apply time inversion to $\Gamma_{\alpha \to \alpha'}$, i.e., change points $q_1, \ldots, q_n, p_1, \ldots, p_n$ into points $q_1, \ldots, q_N, -p_1, \ldots, -p_N$. This new domain in Γ space we denote by $\Gamma^*_{\alpha \to \alpha'}$. It has the same

‡ We again restrict the discussion to variables that are even under the operation of time inversion.

LINEAR PHENOMENOLOGICAL LAWS

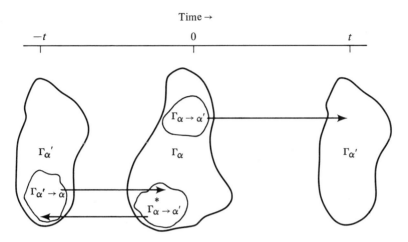

Fig. 3.1. Mappings in Γ space.

volume as $\Gamma_{\alpha\to\alpha'}$:

$$\text{vol } \Gamma^*_{\alpha\to\alpha'} = \text{vol } \Gamma_{\alpha\to\alpha'} \tag{3.48}$$

It is also contained in Γ_α because the α's are assumed to be invariant under time inversion and thus are even functions of the microscopic moments. Moreover, the points that are at time $t = 0$ in $\Gamma^*_{\alpha\to\alpha'}$ were at time $-t$ in $\Gamma_{\alpha'}$ and formed the subset $\Gamma_{\alpha'\to\alpha} \subset \Gamma_{\alpha'}$. (See Fig. 3.1.)

It follows then from Liouville's theorem that

$$\text{vol } \Gamma_{\alpha\to\alpha'} = \text{vol } \Gamma^*_{\alpha\to\alpha'} = \text{vol } \Gamma_{\alpha'\to\alpha} \tag{3.49}$$

Combined with eqn (3.46) we conclude that

$$W(\boldsymbol{\alpha})T_t(\boldsymbol{\alpha}\to\boldsymbol{\alpha}') = W(\boldsymbol{\alpha}')T_t(\boldsymbol{\alpha}'\to\boldsymbol{\alpha}) \tag{3.50}$$

i.e., that the principle of detailed balance holds in γ space. This result, similar to the statement in eqn (3.36), is based solely on the postulate of the invariance of $\boldsymbol{\alpha}$ and the microscopic hamiltonian with respect to time inversion and, of course, on Liouville's theorem.

To make use of the above relations in a derivation of Onsager's reciprocity relations, we rewrite the phenomenological linear laws, eqn (3.38), in the same language by demanding that, at least for macroscopic values of the $\boldsymbol{\alpha}$'s, their average values at time t are linear functions of their initial values, i.e.

$$\int T_t(\boldsymbol{\alpha} \to \boldsymbol{\alpha}')\alpha_i' \, d\boldsymbol{\alpha}' = \sum p_{ik}(t)\alpha_k \qquad (3.51)$$

From a purely macroscopic point of view, one would expect eqn (3.51) to hold even for infinitesimally short times. In other words, one would expect differential equations of the form

$$\dot{\bar{\alpha}}_i = -\sum P_{ik}\bar{\alpha}_k \qquad (3.52)$$

to be valid analogous to the macroscopic laws in eqn (3.38). We should, however, be aware that at the microscopic level of fluctuations such a law cannot be expected to hold because the time-inversion invariance of $\boldsymbol{\alpha}$ forces the average $\dot{\bar{\alpha}}_i$ over Γ_α to vanish identically. On the other hand, the integrated form, eqn (3.51), can be expected to hold after an induction period τ_1 (the relaxation time for the regression of fluctuations), which is so long that $\dot{\boldsymbol{\alpha}}$ has already undergone substantial changes.

Assuming now that eqn (3.51) holds even for microscopically small (fluctuating) values of $\boldsymbol{\alpha}$,‡ we can establish the Onsager reciprocity relations at once. We multiply eqn (3.51) by $W(\boldsymbol{\alpha})\alpha_l$ and integrate over $d\boldsymbol{\alpha}$ to get

$$\int d\boldsymbol{\alpha} \int d\boldsymbol{\alpha}' W(\boldsymbol{\alpha})T_t(\boldsymbol{\alpha} \to \boldsymbol{\alpha}')\alpha_l\alpha_i' = \sum p_{ik}(t)S_{kl}^{-1} \qquad (3.53)$$

where we used eqn (3.34). The principle of detailed balance, eqn (3.50), now guarantees that the left-hand side is symmetric in l

---

‡ This crucial assumption is very difficult to assess. For macroscopic values of $\boldsymbol{\alpha}$, $T_t(\boldsymbol{\alpha} \to \boldsymbol{\alpha}')$ as a function of $\boldsymbol{\alpha}'$ has a rather sharp maximum, the width of which is much smaller than its distance from $\boldsymbol{\alpha}' = 0$. If the $\boldsymbol{\alpha}$ are microscopic to begin with, most of the systems in Γ_α will have fluctuated many times through $\boldsymbol{\alpha} = 0$. After a time $t \gg \tau_1$, the system will have only a faint memory of its initial value $\boldsymbol{\alpha}$ which should, nevertheless, according to our assumption, be subject to the macroscopic laws. As an illustration of this point, let us consider the stopping of a macroscopic particle by the viscosity of the medium through which it is moving. If the velocity of the particle is many times in excess of the equilibrium average value, a Stokes's flow will establish itself around it, the properties of which will determine the energy loss of the particle. If, on the other hand, the energy of the particle is about $k_B T$, no ordered fluid motion can establish itself around the macroscopic object and no Stokes's flow will occur. Nevertheless, we assume that the energy loss of the average motion is controlled by a macroscopic law, eqn (3.51).

and i,

$$\int d\boldsymbol{\alpha} \int d\boldsymbol{\alpha}' W(\boldsymbol{\alpha}) T_t(\boldsymbol{\alpha} \to \boldsymbol{\alpha}') \alpha_l \alpha_i' = \int d\boldsymbol{\alpha} \int d\boldsymbol{\alpha}' W(\boldsymbol{\alpha}') T_t(\boldsymbol{\alpha}' \to \boldsymbol{\alpha}) \alpha_i \alpha_i'$$

$$= \int d\boldsymbol{\alpha}' \int d\boldsymbol{\alpha} W(\boldsymbol{\alpha}) T_t(\boldsymbol{\alpha} \to \boldsymbol{\alpha}') \alpha_l' \alpha_i$$

$$= \sum p_{lk}(t) S_{ki}^{-1} \qquad (3.54)$$

Going from the second to the third line we simply exchange integration variables $\boldsymbol{\alpha} \leftrightarrow \boldsymbol{\alpha}'$. Comparison of eqns (3.53) and (3.54) gives the symmetry

$$\sum P_{ik}(t) S_{kl}^{-1} = \sum P_{lk}(t) S_{ki}^{-1} \qquad (3.55)$$

which also holds, of course for the coefficients $P_{ik} = \sum_m L_{im} S_{mj}$ of eqns (3.32) and (3.52). Again we have arrived at Onsager's reciprocity relations $L_{il} = L_{li}$.‡

The assumption that the variables α_i are even under time reversal is not an essential one and the result can be generalized (Casimir, 1945). If β_i are variables which change sign under time reversal (e.g. angular momentum) then

$$\langle \beta_\lambda(t) \beta_\mu(t+\tau) \rangle = \langle \beta_\lambda(t) \beta_\mu(t-\tau) \rangle$$

and

$$\langle \alpha_i(t) \beta_\mu(t+\tau) \rangle = -\langle \alpha_i(t) \beta_\mu(t-\tau) \rangle \qquad (3.56)$$

Entropy, being an even function under time reversal, can be written as

$$\Delta S = -\frac{1}{2} \left(\sum_{i,j=1}^{n} S_{ij} \alpha_i \alpha_j + \sum_{\lambda,\mu=n+1}^{m} S_{\lambda\mu} \beta_\lambda \beta_\mu \right) \qquad (3.57)$$

and the phenomenological equations can be written as

$$\dot{\alpha}_i = \sum_{j=1}^{n} L_{ij} X_j + \sum_{\lambda=n+1}^{m} L_{i\lambda} X_\lambda \qquad \text{for } i = 1, 2, \ldots, n$$

$$\dot{\beta}_\lambda = \sum_{j=1}^{n} L_{\lambda j} X_j + \sum_{\mu=n+1}^{m} L_{\lambda\mu} X_\mu \qquad \text{for } \lambda = n+1, \ldots, m \qquad (3.58)$$

‡ The Onsager reciprocity relations can also be derived in a statistical theory assuming that the point representing the system in phase space performs a Brownian motion. For this type of theory, see Kirkwood (1946), Green (1952), Hashitzume (1952), and Onsager and Machlup (1953). See also the critical assessment of both derivations by Wigner (1954).

where

$$X_i = \sum_{j=1}^{n} S_{ij}\alpha_j$$

$$X_\lambda = \sum_{\mu=n+1}^{n} S_{\lambda\mu}\beta_\mu \qquad (3.59)$$

The corresponding generalized Onsager relations are

$$L_{ij} = L_{ji}$$
$$L_{\lambda\mu} = L_{\mu\lambda}$$
$$L_{i\lambda} = -L_{\lambda i} \qquad (3.60)$$

Also, if external magnetic fields are present, the equations of motion are invariant under time reversal if the magnetic fields are also reversed. Hence,

$$L_{ij}(\mathbf{B}) = L_{ji}(-\mathbf{B})$$
$$L_{\kappa\mu}(\mathbf{B}) = L_{\mu\lambda}(-\mathbf{B})$$
$$L_{i\lambda}(\mathbf{B}) = -L_{\lambda i}(-\mathbf{B}) \qquad (3.61)$$

is the general form of the Onsager relations, where **B** represents magnetic fields or any other forces which are odd under time reversal.‡

3.3. Transformation Properties of Onsager Relations

In the derivation of the Onsager relations, it was assumed that both the forces X_i and the fluxes Y_i separately form linearly independent sets. In choosing appropriate forces and fluxes, linear dependence might occur in two ways. First, intuition might be so faulty that what was believed to be an independent force or flux was merely a linear combination of several other forces or fluxes already considered. More likely, however, is that some constraint upon the system would be neglected. For example, if, in multicomponent diffusion, the various mass currents $\mathbf{j}_\gamma$ are defined with respect to the center of mass, there would be a constraint that $\sum_{\gamma=1}^{c} \mathbf{j}_\gamma = 0$, which would result in dependent

‡ The experimental verification of the Onsager reciprocity relations has been reviewed in great detail by Miller (1960, 1973).

fluxes. Similarly, the forces could be made dependent by a constraint that the system be in mechanical equilibrium.

If a constraint $\sum_{i=1}^{I} a_i Y_i = 0$ relates the fluxes, while the driving forces X_i are independent, then the entropy production may be written as

$$\sigma_s = \sum_{i=1}^{I} Y_i X_i = \sum_{i=1}^{I-1} Y_i \left\{ X_i - \frac{a_i}{a_n} X_n \right\} \quad (3.62)$$

With the new set of $I-1$ independent forces and fluxes and with phenomenological equations of the form

$$Y_j = \sum_{i=1}^{I-1} l_{ji} \left(X_i - \frac{s_i}{a_n} X_n \right) \quad (3.63)$$

where the coefficients l_{ji} are subject to the Onsager relations

$$l_{ij} = l_{ji} \quad (3.64)$$

we see that the matrix L_{ij} defined by

$$Y_j = \sum_{i=1}^{I} L_{ji} X_i \quad (3.65)$$

obeys Onsager's theorem

$$L_{ij} = L_{ji} \quad (3.66)$$

That is, even for the dependent set Y_i $(i = 1, 2, \ldots, n)$, the Onsager relations hold. Of course, there are additional constraints on the coefficients L_{ij} of the form

$$\sum_{i=1}^{I} a_i L_{ij} = 0 \quad (3.67)$$

Similarly, if the forces are linearly related by a constraint while the fluxes are independent, the Onsager relations remain valid.

However, in the general case where both the forces and the fluxes are dependent, no Onsager relations hold. Under such conditions, there is a certain arbitrariness in the choice of phenomenological coefficients in the dependent sets. The coefficients L_{ij} can be chosen such that $L_{ij} = L_{ji}$, but need not be.

Before considering the transformation properties of the Onsager reciprocity relations, let us recall that such relations can only be established if we choose fluxes $Y_i = \dot{\alpha}_i$ that are time derivatives of extensive thermodynamic variables α_i. They are therefore odd under time reversal, i.e. $Y_i(t) = -Y_i(-t)$ if $\alpha_i(-t) = \alpha_i(t)$ is even, and $Y_i(t) = \dot\beta_i = Y_i(-t)$ are even fluxes if $\beta_i(-t) = -\beta_i(t)$ are odd variables. The conjugate forces are then of opposite but definite time parity to ensure that the entropy production is odd under time reversal. It should be obvious that not every arbitrary transformation of these forces and fluxes is acceptable, but that a number of requirements have to be met by the admitted transformations.‡

First, the transformation must be linear in order to preserve the linear form of the phenomenological laws (3.6). Secondly, the transformed forces and fluxes must form linearly independent sets which preserve the value of the bilinear form

$$\sigma_s = \sum_i X_i Y_i \tag{3.68}$$

Since the various X_i and Y_i in eqn (3.68) are assumed to have a definite physical meaning, independently of the existence of any linear phenomenological laws relating the fluxes to the forces, we require the invariance of the bilinear form above rather than the less restrictive condition that the associated quadratic form

$$\sigma_s = \sum_{i,k} L_{ik} X_i X_k \tag{3.69}$$

be invariant. Finally, we can only expect Onsager reciprocity relations to hold for the transformed coefficients if the new forces and the new fluxes are again of a definite time parity.

We now present the detailed analysis for the simple case where all forces are even and all fluxes odd under time reversal. We must consider nonsingular transformations of the form

$$X'_i = \sum_j A_{ij} X_j$$

$$Y'_i = \sum_k B_{ik} Y_k \tag{3.70}$$

‡ Such transformations were first studied by Meixner (1943). Subsequently, a number of papers (Verschaffelt, 1951; Davies, 1952; Hooyman, de Groot, and Mazur, 1955; Coleman and Truesdell, 1960) confused this rather simple subject thoroughly. A clear analysis has since been given by Meixner (1973). In this context we should also mention the eloquent criticism of the Onsager reciprocity relations by Truesdell (1969), which in light of Meixner's rebuttal, however, has to be taken *cum grano salis*.

Transformations of the form

$$X'_i = \sum_j A_{ij} X_j + \sum_j A'_{ij} Y_j \tag{3.71}$$

are excluded since the X'_i would no longer be even under time reversal. The invariance of σ_s may be expressed as

$$\sigma_s = \sum_i X_i Y_i$$
$$= \sum_k X'_k Y'_k$$
$$= \sum_{i,j,k} A_{kj} B_{ki} X_j Y_i \tag{3.72}$$

Since the Y_i are independent, this implies that

$$X_i = \sum_{j,k} A_{kj} B_{ki} X_j \tag{3.73}$$

which, in turn, implies that

$$B_{ki} = (A^{-1})_{ik} \tag{3.74}$$

Using the phenomenological law $Y_i = \sum_j L_{ij} X_j$, where $L_{ij} = L_{ji}$, we can rewrite (3.70) as

$$Y'_i = \sum_{k,m,j} B_{ik} L_{km} (A^{-1})_{mj} X'_j, \tag{3.75}$$

or

$$Y'_i = \sum_j L'_{ij} X'_j \tag{3.76}$$

where

$$L'_{ij} = \sum_{k,m} B_{ik} (A^{-1})_{mj} L_{km} \tag{3.77}$$

Using eqn (3.74), we have

$$L'_{ij} = \sum_{k,m} B_{jm} B_{ik} L_{km} \tag{3.78}$$

from which follows, using the symmetry of L_{km}, that

$$L'_{ij} = L'_{ji} \tag{3.79}$$

That is, if we have derived the Onsager relations for one independent set of forces and fluxes, any acceptable transformation also yields forces and fluxes obeying Onsager relations. A transformation is acceptable if it produces new forces and fluxes that are even or odd under time reversal and if it keeps the bilinear form of the entropy production (3.68) invariant.

3.4. A First Example: Thermodiffusion

Consider a container separated into two equal compartments by a wall with a small hole (Casimir, 1945). Let n_1 and n_2 be the number of moles of an ideal gas in each compartment, and T_1 and T_2 be their respective temperatures.

In equilibrium, we have

$$n_1 = n_2 = n \quad \text{and} \quad T_1 = T_2 = T \tag{3.80}$$

For an ideal gas, the energy u is proportional to $n_i T_i$, so conservation of particle number and of energy gives as conditions on the fluctuations δn_1, δn_2, δT_1, and δT_2.

$$\delta n_1 + \delta n_2 = 0$$
$$\delta(n_1 T_1 + n_2 T_2) = 0 \tag{3.81}$$

Equations (3.81) imply

$$(\delta T_1 + \delta T_2) = \frac{\delta n_2}{n}(\delta T_1 - \delta T_2) \tag{3.82}$$

and, to this order of accuracy, we have

$$(\delta T_1)^2 = (\delta T_2)^2 \tag{3.83}$$

The entropy of an ideal gas is

$$S = \frac{n}{n_0} S_0 + n C_V \ln \frac{T}{T_0} + nR \ln \frac{V}{V_0} \frac{n_0}{n} \tag{3.84}$$

where the subscript 0 refers to some reference state so that we get

$$\Delta S = -\frac{nC_V}{T^2}(\delta T_2)^2 - \frac{R}{n}(\delta n_2)^2 \tag{3.85}$$

to first nonvanishing order in the fluctuations.

Transforming to new variables α_1 and α_2 defined by

$$\alpha' = \delta n_2$$

and
$$\alpha_2 = \delta u_2 = (T\delta n_2 + n\delta T_2)C_V \tag{3.86}$$
gives us
$$\Delta S = -\frac{(C_V + R)}{n}\alpha_1^2 - \frac{1}{nT^2 C_V}\alpha_2^2 + \frac{2C_V}{nT}\alpha_1\alpha_2 \tag{3.87}$$
and we can identify the conjugate forces as
$$X_1 = \frac{\partial(\Delta S)}{\partial \alpha_1} = 2\left(\frac{R}{n}\delta n_2 - \frac{C_V}{T}\delta T_2\right)$$
and
$$X_2 = \frac{\partial(\Delta S)}{\partial \alpha_2} = \frac{2\delta T_2}{T^2} \tag{3.88}$$

If diffusion and heat conduction were uncoupled processes, we would have to regard δn_2 and $\delta T_2/T^2$ as the forces causing the mass and energy flows $\delta\dot{n}_2$ and $\delta\dot{u}_2$. Intuitively, we then couple these two effects in the set of linear phenomenological equations

$$\delta\dot{n}_2 = A\,\delta n_2 + B\frac{\delta T_2}{T^2}$$

$$\delta\dot{u}_2 = Q\,\delta\dot{n}_2 + \omega\frac{\delta T_2}{T^2} \tag{3.89}$$

Here, Q is an energy convection coefficient and ω is a heat conduction coefficient. In order to apply Onsager relations, eqns (3.89) have to be cast in terms of $\dot{\alpha}_1$, $\dot{\alpha}_2$, X_1, and X_2:

$$\dot{\alpha}_1 = \frac{1}{2}\left(\frac{An}{R}\right)X_1 + \frac{1}{2}\left(\frac{AnTC_V}{R} + B\right)X_2$$

$$\dot{\alpha}_2 = \frac{1}{2}\left(\frac{AnQ}{R}\right)X_1 + \frac{1}{2}\left(\frac{AnTc_vQ}{R} + BQ + \omega\right)X_2 \tag{3.90}$$

and the Onsager relation is
$$\frac{B}{A} = \frac{(Q - TC_V)}{R}n \tag{3.91}$$

If a steady state is established, $\dot{\alpha}_1 = \delta\dot{n}_2 = 0$ while a temperature difference is maintained, then

$$\delta n_2 = -\frac{B}{A}\frac{\delta T_2}{T^2}$$

or

$$\frac{\delta n_2}{n} = -\left(\frac{Q}{T} - C_V\right)\frac{\delta T_2}{RT} \tag{3.92}$$

relating the concentration difference, caused by unequal temperatures, to the energy carried by the molecules. The value of Q may be calculated from kinetic theory. If ϕ is the diameter of the hole and l is the mean-free path of the molecules, then there are two limiting cases.

First, if $l \gg \phi$, then kinetic theory predicts $Q = (C_V + \frac{1}{2}R)T$ and so

$$\frac{\delta n_2}{n} = -\frac{1}{2}\frac{\delta T_2}{T}$$

or

$$n_2\sqrt{T_2} = \text{const} \tag{3.93}$$

which yields Knudsen's formula‡

$$P_1 T_1^{-\frac{1}{2}} = P_2 T_2^{-\frac{1}{2}} \tag{3.94}$$

Secondly, if $l \ll \phi$, then $Q = (C_V + R)T$ so that

$$\frac{\delta n_2}{n} = -\frac{\delta T_2}{T}$$

or

$$n_2 T_2 = \text{const} \tag{3.95}$$

which implies that

$$P_1 = P_2 \tag{3.96}$$

---

‡ The effusion of a gas through a small orifice was measured for the first time by Knudsen in 1908 (Knudsen, 1950). If the chambers A and B are in addition connected by an external pipe, then pressures P_A and P_B will be equalized. Keeping temperatures T_A and T_B fixed, a steady-state gas flow will establish itself which can be thought of as the analogue of the thermoelectric current. The transpiration of a gas through a porous plug (i.e. many small orifices), also obeying eqn. (3.94) has been investigated much earlier by Reynolds (1879). See also Jeans (1954).

3.5. A Second Example: Thermoelectricity

We want to consider a simple thermocouple consisting of two connected metals A and B, with an ideal capacitance C interrupting wire A according to Fig. 3.2. It is known experimentally that keeping the connections 1 and 2 at different temperatures T and $T + \Delta T$ will not only produce a heat current $\mathbf{j}_q$ but also an electrical current $\mathbf{j}_e$ in the wires A and B which, in turn, will establish a potential difference $\Delta \psi$ across the capacitance and v.v. We want to analyze these coupled thermal and electrical effects in a linear theory by writing down coupled equations

$$Y_e = L_{ee}X_e + L_{eq}X_q$$
$$Y_q = L_{qe}X_e + L_{qq}X_q \tag{3.97}$$

As discussed at the beginning of this chapter, there is some freedom in choosing thermodynamic fluxes and forces, the major requirement being that their product should be a term in the entropy production. For the heat conduction part, we choose as a thermodynamic force

$$\mathbf{X}_q = \boldsymbol{\nabla} \frac{1}{T} \tag{3.98}$$

which implies through

$$\sigma_s \bigg|_{\text{heat}} = -\frac{1}{T^2} \mathbf{j}_q \cdot (\boldsymbol{\nabla} T) = \mathbf{j}_q \cdot \mathbf{X}_q \tag{3.99}$$

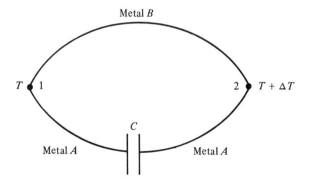

Fig. 3.2. Thermocouple.

that our heat flux is just the ordinary heat current

$$\mathbf{Y}_q = \mathbf{j}_q \tag{3.100}$$

which we now want to approximate by Fourier's law as

$$\mathbf{j}_q = -\lambda \boldsymbol{\nabla} T = \lambda T^2 \boldsymbol{\nabla} \frac{1}{T} = \lambda T^2 \mathbf{X}_q \tag{3.101}$$

The thermodynamic flux associated with electrical conduction we want to identify as the electric current

$$\mathbf{Y}_e(t) = \mathbf{j}_e(t) = \left\langle e \sum_{i=1}^{N} \dot{\mathbf{r}}_i(t) \right\rangle \tag{3.102}$$

where $\mathbf{r}_i(t)$ is the position of the ith electron at time t and N is the number of electrons. This flux is obviously the time derivative of an extensive thermodynamic variable of our system, namely its electric dipole moment

$$\mathbf{M}_e(t) = \left\langle e \sum_{i=1}^{N} \mathbf{r}_i(t) \right\rangle \tag{3.103}$$

We can rewrite eqn (3.102) as

$$\mathbf{j}_e = -\left\langle e \sum_{i=1}^{N} \dot{\mathbf{r}}_i(t) \right\rangle = -\frac{e}{m} \rho_e \mathbf{v}_e \tag{3.104}$$

where ρ_e is the mass density of electrons and $\mathbf{v}_e$ is their drift velocity relative to the ionic background, which we can assume to be at rest. Such a diffusive current enters the entropy production (2.70) as a term

$$\sigma_s \bigg|_{\text{electrical}} = -\left(\frac{e}{T} \mathbf{E} + \boldsymbol{\nabla} \frac{\mu}{T} \right) \cdot \mathbf{j}_e \tag{3.105}$$

so that we can identify the conjugate thermodynamic force as

$$\mathbf{X}_e = -\frac{e}{T} \mathbf{E} - \boldsymbol{\nabla} \frac{\mu}{T} \tag{3.106}$$

With this, eqns (3.97) take on the form

$$\mathbf{j}_e = -\frac{1}{T} L_{ee} \left(\frac{e}{T} \mathbf{E} + \boldsymbol{\nabla} \frac{\mu}{T} \right) + L_{eq} \boldsymbol{\nabla} \frac{1}{T}$$

$$\mathbf{j}_q = -\frac{1}{T} L_{qe} \left(e \mathbf{E} + \boldsymbol{\nabla} \frac{\mu}{T} \right) + L_{qq} \boldsymbol{\nabla} \frac{1}{T} \tag{3.107}$$

Observe that the isothermal electrical conductivity is given by

$$\sigma = -\left(\frac{eL_{ee}}{T}\right) \qquad (3.108)$$

and the thermal conductivity can be extracted for $\mathbf{j}_e = 0$ as

$$\lambda = \frac{L_{qq}L_{ee} - L_{qe}L_{eq}}{L_{ee}} \qquad (3.109)$$

Let us next assume that the metals A and B are thin homogeneous wires and can be treated as one-dimensional structures eliminating the need to use vector fluxes and forces. In this case we can simplify the thermodynamic forces acting in the thermocouple depicted in Fig. 3.2 as

$$X_q = -\frac{\Delta T}{T^2}$$

$$X_e = -\frac{\Delta \psi}{T} \qquad (3.110)$$

where $\Delta \psi$ is the potential difference across the capacitance C. For this system the Onsager reciprocity relation simply reads

$$L_{eq} = L_{qe} \qquad (3.111)$$

Let us first study the stationary state $\Delta T =$ const and $j_e = 0$. This implies that

$$L_{ee}X_e + L_{eq}X_q = 0$$

or

$$\frac{\Delta \psi}{\Delta T} = -\frac{L_{eq}}{L_{ee}} T \qquad (3.112)$$

This is the content of the Seebeck-effect stating what potential difference can be achieved for a given temperature difference in a thermocouple, if no electric current flows.

If we next impress a fixed potential difference $\Delta \psi =$ const across the capacitance and keep $\Delta T = 0$, then we find the dependence of the generated heat current upon the supplied electrical current, i.e.

$$\frac{j_q}{j_e} = \frac{L_{qe}}{L_{ee}} = \Pi \qquad (3.113)$$

Table 3.1. Test of the second Thomson relation for some metallic thermocouples

Couple	$T[C]$	$\dfrac{\Pi}{T}$	$-\dfrac{\Delta\psi}{\Delta T}$	$\dfrac{L_{qe}}{L_{eq}}$
		Microvolt/K		
Cu-Ni	0	18.60	20.00	0.930
	14	20.20	20.70	0.976
	22	20.50	22.30	0.919
Cu-Constantan	15	35.30	35.70	0.989
	20	37.70	38.90	1.030
	30	40.50	41.80	1.030
	40	43.20	44.60	1.030
Fe-Hg	18.4	16.72	16.66	1.004
	56.5	16.17	16.14	1.002
	99.6	15.57	15.42	1.010
	131.6	14.89	14.81	1.005
	182.3	13.88	13.74	1.011

Source: Miller (1960).

which is the Peltier effect. Using now the Onsager reciprocity relation, eqn (3.111), gives us a possibility to relate the Peltier-effect to the Seebeck-effect, i.e.

$$\frac{\Delta\psi}{\Delta T} = -\frac{\Pi}{T} \qquad (3.114)$$

which is the second Thomson relation (Kelvin, 1854). Some experimental data on the thermoelectric effect can be found in Table 3.1.

For the sake of completeness, we want to outline Kelvin's (1854) original derivation as well. He argued that the right-hand side of phenomenological laws should only contain the experimentally controlled variables $\mathbf{j}_e$ and ∇T^{-1}, i.e.

$$\frac{\nabla\mu}{T} + \frac{\mathbf{E}}{T} = L_{11}\mathbf{j}_e + L_{12}\nabla\frac{1}{T}$$

$$\mathbf{j}_q = L_{21}\mathbf{j}_e + L_{22}\nabla\frac{1}{T} \qquad (3.115)$$

where $\mathbf{E}$ is the electric field. Calculating the entropy production, he found

$$\sigma_s^* = L_{11}\mathbf{j}_e \cdot \mathbf{j}_e + (L_{12} + L_{21})\mathbf{j}_e \cdot \nabla \frac{1}{T} + L_{22}\left(\nabla \frac{1}{T}\right)^2 \quad (3.116)$$

He then observed that reversal of the current $\mathbf{j}_e$ at a fixed temperature gradient will change the sign of the middle term which, in turn, might lead to a violation of the requirement that σ_s has to be positive or zero. This, he argued cautiously, can be avoided by demanding that $L_{12} = -L_{21}$, which then leads to the establishment of the second Thomson relation. Though his result was correct, his argument cannot be trusted, because the Onsager reciprocity relation cannot be applied to the eqn (3.115) since $\mathbf{E}/T$ is not an acceptable thermodynamic flux according to our definition (3.2). Today, the second Thomson relation is regarded as an example par excellence of an Onsager reciprocity relation‡

3.6. Minimum Entropy Production

Among irreversible phenomena, there is an important class of processes that plays a role analogous to that of equilibrium states in reversible thermodynamics. These are the steady-state processes which are subject to some external constraints and characterized by time-independent forces and fluxes. Just as isothermal systems in equilibrium are characterized by a maximum of entropy, Prigogine (1945) has shown that stationary nonequilibrium states are sometimes characterized by a minimum of the entropy production.§ To prove this theorem, rather restrictive assumptions have to be made, namely that the system is described by linear phenomenological laws with constant coefficients satisfying the Onsager relations and is subject to time-independent boundary conditions.

‡ A quite complete discussion of thermoelectric effects is given by Domenicali (1954). The experimental evidence for the second Thomson relation is also reviewed by Miller (1960). See also Harman and Honig (1967).

§ Onsager (1931b) had already formulated a variational principle to find the stationary states of the system. He called it the 'principle of the least dissipation of energy,' where he defined a dissipation function which equals half the rate of entropy production. Later developments are reviewed in a monograph by Gyarmati (1970).

We present Prigogine's theorem of minimum entropy production here through several examples. First, consider transfer of matter and energy between two materials which are kept at different temperatures. Assuming linear phenomenological laws, the heat flux Y_q and mass current Y_m are related to the thermal gradient X_q and density gradient X_m by

$$Y_q = L_{11}X_q + L_{12}X_m$$
$$Y_m = L_{21}X_q + L_{22}X_m \tag{3.117}$$

where $L_{12} = L_{21}$.

The entropy production is given by

$$\sigma_s = X_q Y_q + X_m Y_m \tag{3.118}$$

and has a minimum at a constant temperature gradient, if

$$0 = \frac{\partial \sigma_s}{\partial X_m}\bigg|_{X_q} = (L_{12} + L_{21})X_q + 2L_{22}X_m = 2Y_m \tag{3.119}$$

This is nothing but the steady state of the system without any mass transport $Y_m = 0$. Thus at a fixed temperature gradient X_q, the system will establish a density gradient X_m in such a way that no mass transport occurs anymore. With respect to this one degree of freedom, namely X_m, the entropy production will be a local minimum. This extremum is not a maximum since σ_s is a positive semidefinite quadratic form.

A second example is furnished by thermal conduction in an isotropic medium. In this case, we write

$$\sigma_s = \mathbf{j}_q \cdot \mathbf{\nabla}\left(\frac{1}{T}\right) \tag{3.120}$$

and the phenomenological equation is

$$\mathbf{j}_q = L_{qq}\mathbf{\nabla}\left(\frac{1}{T}\right) \tag{3.121}$$

Therefore, the total entropy production P is

$$P \equiv \int_V \sigma_s \, dV = \int_V L_{qq}\left[\mathbf{\nabla}\left(\frac{1}{T}\right) \cdot \mathbf{\nabla}\left(\frac{1}{T}\right)\right] dV \tag{3.122}$$

which is a minimum if

$$\nabla^2\left(\frac{1}{T}\right) = 0 \tag{3.123}$$

or
$$\nabla \cdot \mathbf{j}_q = 0 \tag{3.124}$$

The local energy balance is

$$\rho \frac{\partial u}{\partial t} = \rho c_v \frac{\partial T}{\partial t} = -\nabla \cdot \mathbf{j}_q \tag{3.125}$$

where c_v is the specific heat per unit mass at constant volume. Thus, for a minimum of entropy production, we find

$$\frac{\partial u}{\partial t} = 0 \tag{3.126}$$

The system is in a steady state since heat conduction is the only process considered.

It can also be shown that these steady states with minimum entropy production are stable with respect to small local perturbations. To demonstrate this for heat conduction, we differentiate P with respect to time to get

$$\begin{aligned}\frac{\partial P}{\partial t} &= 2 \int_V L_{qq} \nabla\left(\frac{1}{T}\right) \cdot \nabla\left(\frac{\partial}{\partial t}\frac{1}{T}\right) dV \\ &= 2 \int_V \mathbf{j}_q \cdot \nabla\left(\frac{\partial}{\partial t}\frac{1}{T}\right) dV \\ &= 2 \int_\Sigma \left(\frac{\partial}{\partial t}\frac{1}{T}\right)\mathbf{j}_q \cdot d\mathbf{\Sigma} - 2 \int_V \left(\frac{\partial}{\partial t}\frac{1}{T}\right) \nabla \cdot \mathbf{j}_q \, dV \end{aligned} \tag{3.127}$$

With T fixed on the boundary Σ, we find

$$\frac{\partial P}{\partial t} = -2 \int_V \frac{\rho c_v}{T^2}\left(\frac{\partial T}{\partial t}\right)^2 dV \leq 0 \tag{3.128}$$

since ρ, c_v, and T are positive. Thus, since P decreases with time, the system evolves toward the state of minimum entropy production, and so the steady state is stable.

Glansdorff and Prigogine (1954) have generalized the principle of minimum entropy production to an evolution criterion that does not invoke linear phenomenological laws. We start with the expression for the total entropy production in the system

$$P \equiv \int_V \sigma_s \, dV = \int_V \sum_i X_i Y_i \, dV \tag{3.129}$$

and take the time derivative

$$\frac{\partial P}{\partial t} = \int_V \left(\sum_i Y_i \frac{\partial X_i}{\partial t} + \sum_i \frac{\partial Y_i}{\partial t} X_i \right) dV \qquad (3.130)$$

This may be written as

$$\frac{\partial P}{\partial t} = \int_V \left(\frac{\partial_x \sigma_s}{\partial t} + \frac{\partial_y \sigma_s}{\partial t} \right) dV \qquad (3.131)$$

where $\partial_x/\partial t$ and $\partial_y/\partial t$ denote the portions of the time derivative which result from the time variation of X_i and Y_i, respectively. Then, it can be shown that

$$\frac{\partial_x P}{\partial t} = \int_V \frac{\partial_x \sigma_s}{\partial t} dV = \int_V \sum_i Y_i \frac{\partial X_i}{\partial t} dV \leq 0 \qquad (3.132)$$

That is, the partial change in entropy production due to the changes of the thermodynamic forces will always be negative, although nothing may be said about the total change in entropy production without making more restrictive assumptions—for example, linear phenomenological laws and Onsager reciprocity relations.

4
Stability and Fluctuations

4.1. Stability Theory: An Outline

A thermodynamic theory of irreversible processes starts with a set of general balance equations for the relevant mechanical and thermodynamic quantities which have to be supplemented by a number of constitutive relations to obtain a closed set of as many equations as there are unknown functions. This last step we achieved in Chapter 3 for situations near equilibrium where linear phenomenological laws between thermodynamic forces and fluxes hold. In this linear regime, stationary states, i.e. time-independent solutions of the combined set of balance equations and constitutive equations, are characterized *cum grano salis* by a minimum of the entropy production. Such stationary states near equilibrium are, of course, also stable in the intuitive sense that a small change in external parameters will cause a small change in the response of the system due to the linearity of the constitutive equations. Near equilibrium, stability of stationary states is guaranteed by the stability of the equilibrium state.

In contrast, in situations far from equilibrium with possibly nonlinear constitutive equations, the investigation of the stability of a particular stationary or time-dependent state becomes a major and nontrivial task. Stability theory here incorporates the very important study and establishment of possible instabilities as occur, for example, in the transition of laminar to turbulent flow in a moving fluid or in the onset of convection in a fluid at rest subjected to gravity and a temperature gradient (Bénard instability).

For general time-dependent states and processes, stability of a particular solution of the balance equations somehow implies its uniqueness in the sense that other solutions starting at time t_0 from initial states 'close' to that of the first solution, will evolve in 'close proximity' to the first one for all times $t > t_0$. This last point implies that stability theory has to work in the framework of topology to make precise the meaning of neighborhoods around solutions, continuous and unique maps of such, and most importantly to give a rigorous definition of stability itself. For the sake of clarity, we present in this section an outline of abstract stability theory.‡ It is more general than needed in this book, but it seems necessary to avoid the general confusion typical for more intuitive approaches.

The object of investigation of stability theory is a dynamical system which we can identify in nonequilibrium thermodynamics as the set of solutions of a system of differential equations

$$\frac{d\mathbf{x}}{dt} = \mathbf{F}(\mathbf{x}, t) \tag{4.1}$$

subject to certain initial and boundary conditions. Here time t is the independent variable and $\mathbf{x}$ an element of a (finite or infinite) set $\mathcal{X}$ of thermodynamic and mechanical variables, e.g. position $\mathbf{r}$, velocity $\mathbf{v}$, temperature T, entropy density s, etc. Solutions of the system of differential equations (4.1) are then functions ϕ that map the real numbers $\mathcal{T}$ into $\mathcal{X}$, i.e. $\phi: \mathcal{T} \to \mathcal{X}$ or $\mathbf{x} = \phi(t)$. The dynamical system $B(\mathcal{T}, \mathcal{X}) = \{\phi\}$ is then the set of all such functions. The motion $\phi_{t_0}(t)$ is defined as the translation of ϕ, i.e.

$$\phi_{t_0}(t) = \phi(t_0 + t) \qquad \text{for } t \in \mathcal{T} \tag{4.2}$$

To give the function space $B(\mathcal{T}, \mathcal{X})$ some structure, we next define a metric $\rho(\phi_1, \phi_2)$ satisfying four basic requirements

$$\begin{aligned}
&\rho(\phi_1, \phi_2) = 0 \quad \text{if and only if } \phi_1 = \phi_2 \\
&\rho(\phi_1, \phi_2) \geq 0 \\
&\rho(\phi_1, \phi_2) = \rho(\phi_2, \phi_1) \\
&\rho(\phi_1, \phi_2) \leq \rho(\phi_1, \phi_3) + \rho(\phi_3, \phi_2)
\end{aligned} \tag{4.3}$$

‡ The historical roots of stability theory have to be found in the investigations of Dirichlet (1846), Lagrange (1853), and Klein and Sommerfeld (1897) on the stability of mechanical systems. The modern development is based on and to a large extent due to the work of Liapounoff (1892). We rely in this section very heavily on the *Handbuch* article by Knops and Wilkes (1973). A straightforward introduction to stability theory is given by Lasalle and Lefschetz (1961).

This metric implies a natural topology $B(\mathcal{T}, \mathcal{X})$ if we define a neighborhood S around ϕ of radius r as

$$S(\phi, r) = \{\psi \in B(\mathcal{T}, \mathcal{X}); \rho(\phi, \psi) < r\} \quad (4.4)$$

We are now set for a definition of stability: We say a function $\phi \in B(\mathcal{T}, \mathcal{X})$ is stable in the sense of Liapounoff (Liapounoff stable, for short) if and only if for a given initial instant t_0 and for a given $\varepsilon > 0$ we can find a $\delta(\varepsilon, t_0) > 0$ such that

$$\rho_{t_0}[\phi(t_0), \psi(t_0)] < \delta \quad (4.5)$$

implies

$$\rho[\phi_{t_0}(t)\psi_{t_0}(t)] = \sup_{t \in \mathcal{T}} \rho[\phi(t+t_0), \psi(t+t_0)] < \varepsilon$$

In other words, a solution ϕ is stable if other solutions initially close to it remain close for all times. Four points have to be stressed in this definition of stability:

1. Stability of a solution ϕ involves *all* solutions in a prescribed neighborhood.
2. It is a statement about the stability of a time-dependent motion as well as stationary and equilibrium states.
3. In linear systems, a substitution $\tilde{\phi} = \phi - \psi$ reduces the stability problem to that of the null solution, i.e. the equilibrium state.
4. Stability depends on our choice of the metric, i.e., on what we accept to be close to a certain solution. In particular, the measure for the initial data ρ_{t_0} need not be the same as that for later times. If $\mathcal{X}$ is a normable linear space, it is assumed that the metric arises from the norm defined on $\mathcal{X}$. In particular, if the system has a finite number of degrees of freedom, the metric is nothing but the euclidean distance between points in $\mathcal{X}$.‡

These points are nicely illustrated in the following example (Hadamard, 1923). Consider the two-dimensional Laplace equation for $t > 0$ and $0 \le x \le 1$

$$\frac{\partial^2 u}{\partial t^2} + \frac{\partial^2 u}{\partial x^2} = 0 \quad (4.6)$$

---

‡ If $\mathcal{X}$ is a normable finite-dimensional vector and metrics ρ_1 and ρ_2 are chosen to be norms on $\mathcal{X}$, then for any two norms $\|(\cdot)\|_1$ and $\|(\cdot)\|_2$ on $\mathcal{X}$ there exist positive real numbers α, β such that for every $x \in \mathcal{X}$, we have $\|x\|_1 \le \alpha \|x\|_2$ and $\|x\|_2 \le \beta \|x\|_1$ (see Brown and Page, 1970). Thus, in a discrete system, if stability is established for one pair of metrics ρ_{t_0} and ρ, it is established automatically with respect to any other pair of metrics.

with the boundary conditions $u(t, x = 0) = u(t, x = 1) = 0$ and the initial conditions at $t = 0$

$$u(t = 0, x) = 0 \qquad \frac{\partial}{\partial t} u(t = 0, x) = \tfrac{1}{2} \sin n\pi x \qquad (4.7)$$

with $n = 1, 2, \ldots$. The set $\mathscr{X}$ is then the subset of functions $C^2[0, 1]$ with continuous second derivatives which vanish at $x = 0, 1$. The dynamical system $B(\mathscr{F}, \mathscr{X})$ consists of all functions defined on subsets of $\mathscr{F}$ taking values in $\mathscr{X}$ such that the differential equation is satisfied. A solution to eqns (4.6) and (4.7) is

$$u(t, x) = \frac{1}{n^2 \pi} \sin n\pi x \sinh n\pi t \qquad (4.8)$$

To examine the stability of the null solution, we define a metric on the set of initial data

$$\rho_{t_0}[u(t = 0, x)] = \sup_{x \in [0, 1]} |u(t = 0, x)| = 0 \qquad (4.9)$$

If we adopt the same metric for $t > 0$, we find

$$\rho(u) = \sup_{t \in T} \sup_{x \in [0,1]} \left| \frac{1}{n^2 \pi} \sin n\pi x \sinh n\pi t \right| = \infty \qquad (4.10)$$

and the null solution is obviously unstable. However, if we measure for $t > 0$ the distance of a solution eqn (4.8) from the null solution by

$$\rho[u(t)] = \frac{1}{2} \int_0^1 \left[\left(\frac{\partial u}{\partial t} \right)^2 - \left(\frac{\partial u}{\partial x} \right)^2 \right] dx = \frac{1}{n^2} \qquad (4.11)$$

we recover stability in the sense of Liapounoff. So much for the example.

There are a great number of different or supplemental definitions of stability in the literature of which we only want to mention two. A solution $\phi \in B(\mathscr{F}, \mathscr{X})$ is said to be asymptotically stable if and only if ϕ is stable in the sense of Liapounoff and

$$\lim_{t \to \infty} \rho[\phi(t), \psi(t)] = 0 \qquad (4.12)$$

that is, if any perturbed motion $\psi(t)$ approaches the motion $\phi(t)$ arbitrarily close in the distant future.

Lasalle and Lefschetz (1961) have put forward the notion of practical stability, which is useful in oscillating systems that are unstable in the sense of Liapounoff. The solution $\phi \in B(\mathcal{F}, \mathcal{X})$ is said to be practically stable if and only if, given positive real numbers A, B, and t with $A \leq B$, it follows that

$$\rho_{t_0}[\phi(t_0), \psi(t_0)] < A \quad \text{implies} \quad \rho_t[\phi(t_0+t), \psi(t_0+t)] < B \tag{4.13}$$

Thus we still call a solution stable if perturbations of oscillatory but bounded nature originating in a given state of the system give rise to oscillations which remain within acceptable limits throughout the time interval t of interest to the operator of the system.

Finally, we mention two very useful theorems on Liapounoff stability without proof. If $\phi \in B(\mathcal{F}, \mathcal{X})$ is Liapounoff stable, it is unique. Here ϕ is unique if

$$\psi(t_0) = \phi(t_0) \quad \text{implies} \quad \psi(t_0+t) = \phi(t_0+t) \quad \text{for all } t \in \mathcal{F} \tag{4.14}$$

This theorem, in particular, implies that nonuniqueness of a solution automatically implies its instability.

The most important theorem in stability theory is related to Liapounoff's second method which, instead of examining the solution of a dynamical system directly, introduces auxiliary or Liapounoff functions whose properties are established without recourse to the solutions themselves. We first consider a discrete system with a finite number of degrees of freedom. Then $\mathcal{X} = \mathbb{R}^n$ is the n-dimensional euclidean space and the metrics ρ_{t_0} and ρ can be chosen as the euclidean norm. Liapounoff's theorem then states that a solution $\phi \in B(\mathcal{F}, \mathcal{X})$ is stable if and only if there exist continuous functions $F_{t_0,t}[\phi_{t_0}(t), \psi_{t_0}(t)]$ defined on $\mathbb{R}^n \times \mathbb{R}^n$ and dependent on the initial data such that (1)

$$F_{t_0,t}[\phi_{t_0}(t), \psi_{t_0}(t)] \equiv F(\phi, \psi) \geq 0 \tag{4.15a}$$

is positive-definite and $F(\varphi, \psi) = 0$ if and only if $\phi = \psi$, and (2)

$$\frac{d}{dt} F(\phi, \psi) \leq 0 \tag{4.15b}$$

That is, $F(\phi, \psi)$ is nonincreasing in time. If $(dF/dt) < 0$, then ϕ is asymptotically stable.

For a continuous system, the theorem is slightly more involved due to the possibility of different metrics ρ_{t_0} and ρ. It states in its most general form that $\phi \in B(\mathcal{T}, \mathcal{X})$ is stable if and only if there exist positive-definite functions $F_{t_0,t}$ defined on $\mathcal{X} \times \mathcal{X}$ for which

1. Given a real positive number ε, there exists a real positive number $\delta(\varepsilon, t_0)$ such that for $\psi \in B(\mathcal{T}, \mathcal{X})$

$$\rho_{t_0}[\phi(t_0), \psi(t_0)] < \delta$$

implies

$$F_{t_0}(\phi_{t_0}, \psi_{t_0}) = \sup_{t \in T} F_{t_0,t}[\phi_{t_0}(t), \psi_{t_0}(t)] < \varepsilon$$

2. Given a real positive number η, there exists a real positive number $\xi(\eta, t_0)$ such that

$$F_{t_0}(\phi_{t_0}, \psi_{t_0}) < \xi$$

implies

$$\rho(\phi_{t_0}, \psi_{t_0}) < \eta$$

3. If oscillatory perturbations are to be excluded one may further postulate that

$$F_{t_0,t}[\phi_{t_0}(t), \psi_{t_0}(t)] \tag{4.16}$$

are nonincreasing with respect to time.

With the help of these theorems, stability analysis reduces to the construction of suitable Liapounoff functions. We will do this in the following sections and chapters for various problems. Let us remark here that in conservative mechanical systems a constant of motion can always be used to show the stability of the null solution which, of course, can never be asymptotically stable. because $dF/dt = 0$ for all t.

4.2. Stability of Equilibrium States

Before considering the question of stability for general thermodynamic states, the classical Gibbs–Duhem theory of stability of equilibrium states will be summarized.‡ For a reversible process, the first law of thermodynamics states that the heat received

---

‡ For a good discussion, see Callen (1960). We follow here Glansdorff and Prigogine (1971).

by a simple system is

$$dQ = dU + p\, dV \tag{4.17}$$

where dU and dV are the changes in the system's internal energy and volume, respectively.

For a closed system at uniform temperature and pressure, the second law may be written as

$$T\, d_i S = T\, dS - dU - p\, dV \geq 0 \tag{4.18}$$

where the equality holds only for completely reversible processes. The inequality is also true for small but otherwise arbitrary deviations ΔE, ΔS, and ΔV, giving the condition for stability as

$$\Delta U + p\, \Delta V - T\, \Delta S > 0 \tag{4.19}$$

Thus, if no state that starts at equilibrium can change in such a way as to satisfy the second law, the system must remain in equilibrium.

In the special case of constant entropy and volume, the stability criterion is

$$\Delta U > 0 \tag{4.20}$$

That is, the internal energy is a minimum at a stable equilibrium. For an isolated system, for which $dU + p\, dV = 0$ holds, the stability criterion becomes

$$\Delta S < 0 \tag{4.21}$$

Therefore, entropy is a maximum at stable equilibrium.

We want to base our further discussion of the stability of equilibrium states on the entropy density s which, for a homogeneous system in equilibrium, is simply defined by $\rho s = S/V$ where V is the volume occupied by the system. The condition $\Delta S < 0$ now reads $\Delta s < 0$. We next want to consider the entropy density s as a function of independent thermodynamic variables (e.g., u, $v = \rho^{-1}$, $c_\gamma = \rho_\gamma/\rho$) and develop a finite change Δs in s in a Taylor series

$$\Delta s = \delta s + \tfrac{1}{2}\delta^2 s + \cdots$$

where

$$\delta s = \frac{\partial s}{\partial u}\bigg|_{v,c_\gamma} \Delta u + \frac{\partial s}{\partial v}\bigg|_{u,c_\gamma} \Delta v + \sum_{\gamma=1}^{N} \frac{\partial s}{\partial c_\gamma}\bigg|_{u,v} \Delta c_\gamma \tag{4.22}$$

is a first-order differential and $\delta^2 s$ is a second-order differential defined similarly from the Taylor expansion. Recall that $\Delta u = \delta u$, $\Delta v = \delta v$, $\Delta c_\gamma = \delta c_\gamma$, and $\delta^2 u = \delta^2 v = \delta^2 c_\gamma = 0$ for the differentials of the independent variables. The condition $\Delta s < 0$ implies that the first-order contribution vanishes, i.e. $\delta s = 0$ in an isolated system.‡ The condition $\Delta s < 0$ thus implies $\delta^2 s < 0$ and

$$\delta^2 S = \int_V \rho \delta^2 s \, dV < 0 \tag{4.23}$$

To find an explicit form for $\delta^2 s$, we start from the fundamental relation, eqn (2.65), and get

$$\delta^2 s = \delta\left(\frac{1}{T}\right)\delta u + \delta\left(\frac{p}{T}\right)\delta v - \sum_{\gamma=1}^{N} \delta\left(\frac{\mu_\gamma}{T}\right)\delta c_\gamma \tag{4.24}$$

We next introduce a new set of independent variables T, v, and c_γ and find

$$\delta^2 s = -\frac{1}{T}\left[\frac{c_v}{T}(\delta T)^2 + \frac{\rho}{\kappa_T}(\delta v)^2 + \sum_{\gamma,\gamma'=1}^{N} \mu_{\gamma\gamma'} \, \delta c_\gamma \, \delta c_{\gamma'}\right] \tag{4.25}$$

where

$$c_v = \left(\frac{\partial u}{\partial T}\right)_{v,c_\gamma}$$

$$\kappa_T = -\frac{1}{v}\left(\frac{\partial v}{\partial p}\right)_{T,c_\gamma}$$

$$\mu_{\gamma\gamma'} = \left(\frac{\partial \mu_\gamma}{\partial c_{\gamma'}}\right)_{T,p,c_\gamma} \tag{4.26}$$

Note that no crossterms occur in the differentials of the independent variables T, v, and c_γ. Thus, the stability criterion for an isolated system is

$$\int \frac{p}{T}\left[\frac{c_v}{T}(\delta T)^2 + \frac{\rho}{\kappa_T}(\delta v)^2 + \sum_{\gamma,\gamma'=1}^{N} \mu_{\gamma\gamma'} \, \delta c_\gamma \, \delta c_{\gamma'}\right] dV > 0 \tag{4.27}$$

Since the perturbations δT, δv, and δc_γ are independent, this

‡ This argument can be made more transparent if one considers a supersystem consisting of the system under investigation and a very large reservoir coupled to it. In such an approach, $\delta s = 0$ implies that T and ρ are the same for the two parts of the supersystem.

implies

$$c_v > 0 \cdot \quad \text{thermal stability}$$
$$\kappa_T > 0 \quad \text{mechanical stability}$$
$$\sum_{\gamma,\gamma'=1}^{N} \mu_{\gamma\gamma'} \, \delta c_\gamma \, \delta c_{\gamma'} > 0 \quad \begin{array}{l}\text{stability of diffusion}\\ \text{and chemical reaction}\end{array} \quad (4.28)$$

These are fairly obvious conditions for a system's stability. Having $c_v > 0$ merely means that removing energy must decrease the temperature, and $\kappa_T > 0$ simply means that the compressibility must resist any volume change, not aid it.

This classical theory of stable equilibrium works well in systems for which a thermodynamic potential can be defined. This is quite restrictive since it is known that, in an experimental situation, stability is controlled by the boundary conditions. In a solid, for example, thermal equilibrium is established by a boundary condition of either a uniform temperature or of no heat flow on the surface. This is of even more importance when considering non-equilibrium situations. Hence, it is desirable to formulate the general stability theory directly in terms of the balance equations and boundary conditions.

To this end we first reformulate the equilibrium stability theory. For the total system in a volume V, bounded by a surface Σ, we can write

$$\frac{dS}{dt} = P[S] + \phi[S] \qquad (4.29)$$

where the source term is

$$P[S] = \int_V \sigma_s \, dV = \int_V \sum_i X_i Y_i \, dV \qquad (4.30)$$

and the flow term is

$$\phi[S] = -\oint_\Sigma \mathbf{j}_s \cdot d\mathbf{\Sigma} \qquad (4.31)$$

Writing

$$S = S_0 + \delta S + \tfrac{1}{2}\delta^2 S + \cdots \qquad (4.32)$$

where δS is again the first-order differential in the deviations from equilibrium, $\delta^2 S$ is a second-order differential, etc., gives us

$$\frac{\partial S}{\partial t} = \frac{\partial(\delta S)}{\partial t} + \frac{1}{2}\frac{\partial(\delta^2 S)}{\partial t} \qquad (4.33)$$

valid to second order.

Next, we split $P[S] + \phi[S]$ into terms of first and second order. Since $X_i = Y_i = 0$ in equilibrium, we note that $P[S]$ is of second order. We can write the flow term as

$$\phi[S] = \phi^{(1)}[S] + \phi^{(2)}[S] \qquad (4.34)$$

where $\phi^{(1)}[S]$ is of first order and $\phi^{(2)}[S]$ is of second order. It is possible to find explicit expressions for $\phi^{(1)}[S]$ and $\phi^{(2)}[S]$ from eqn (2.69)

$$\phi[S] = -\int_\Sigma \left(\rho s \mathbf{v} + \frac{\mathbf{j}_q}{T} - \sum_{\gamma=1}^{c} \frac{\mu_\gamma \mathbf{j}_\gamma^{\text{diff}}}{T}\right) d\Sigma \qquad (4.35)$$

Let us write T^{-1} and $\mu_\gamma T^{-1}$ in terms of their equilibrium values as

$$T^{-1} = T_0^{-1} + \Delta T^{-1}$$

and

$$\mu_\gamma T^{-1} = (\mu_\gamma T^{-1})_0 + \Delta(\mu_\gamma T^{-1}) \qquad (4.36)$$

We then find

$$\phi^{(1)}[S] = -\int_\Sigma \left\{\frac{\mathbf{j}_q}{T_0} - \sum_{\gamma=1}^{c} \left(\frac{\mu_\gamma}{T}\right)_0 \mathbf{j}^{\text{diff}}\right\} d\Sigma \qquad (4.37)$$

and

$$\phi^{(2)}[S] = -\int_\Sigma \left\{\mathbf{j}_q \Delta T^{-1} - \sum_{\gamma=1}^{c} \Delta\left(\frac{\mu_\gamma}{T}\right) \mathbf{j}_\gamma^{\text{diff}}\right\} d\Sigma \qquad (4.38)$$

since, for a closed system, $\mathbf{v} = 0$ on the surface Σ. Equating terms of the same order gives

$$\frac{\partial(\delta S)}{\partial t} = \phi^{(1)}[S] \qquad (4.39)$$

and

$$\frac{1}{2}\frac{\partial(\delta^2 S)}{\partial t} = P[S] + \phi^{(2)}[S] \qquad (4.40)$$

valid only for systems at rest (no velocity fluctuations). In the presence of velocity fluctuations, $\partial(\delta S)/\partial t$ and $\phi^{(1)}[S]$ contain additional terms.

Starting at time $t = 0$ with a system in equilibrium, we have

$$\delta S = \int_0^t \phi^{(1)}[S]\, dt \tag{4.41}$$

Note that, if the system is isolated, we have $\phi^{(1)}[S] = 0$ and the equilibrium condition $\delta S = 0$ is recovered. Otherwise, a small change in the system's entropy must be compensated by a flow through the surface.

For the second-order equation, an assumption has to be made about $\phi^{(2)}[S]$, namely that it is possible to maintain the boundary conditions, on the average, in the presence of fluctuations. Thus we have

$$\int_0^t \Delta(T^{-1})\, dt = 0 \tag{4.42}$$

for times t large on the time scale of fluctuations. But then it is permissible to set $\phi^{(2)}[S] = 0$ and we get

$$\frac{1}{2}\frac{\partial(\delta^2 S)}{\partial t} = P[S] \geq 0 \tag{4.43}$$

This then gives the condition for stable equilibrium

$$\int_i^f P[S]\, dt \geq 0 \tag{4.44}$$

where i is the initial equilibrium state and f is the final state. The criterion can be rewritten in terms of $\delta^2 S$ since

$$\tfrac{1}{2}\delta^2 S = \int_0^t P[S]\, dt = \Delta S < 0 \tag{4.45}$$

Locally, the stability criterion reads

$$\delta^2(\rho s) < 0 \tag{4.46}$$

which immediately implies that $c_v > 0$, $\kappa_T > 0$, and

$$\sum_{\gamma,\gamma'=1}^N \mu_{\gamma\gamma'}\, \delta c_\gamma\, \delta c_{\gamma'} > 0 \tag{4.47}$$

as before.

Thus, even for the more general situations in which equilibrium is due to the boundary conditions and no thermodynamic potential exists, the necessary and sufficient conditions for stability are the same as those derived previously. Of course, the above conditions, holding only at equilibrium, cannot ensure that the equilibrium is stable against finite perturbations.

4.3. Stability of Nonequilibrium States

To generalize the above theory to include nonequilibrium processes, both the assumption of local equilibrium and the concept of Liapounoff functions are used. We have seen in the previous section that the stability of the equilibrium state of an isolated system demands that its entropy is a maximum, which implies that the second-order differential, eqn (4.25), $-\delta^2 s \geq 0$ or, likewise, from eqn (4.46), $-\delta^2(\rho s) \geq 0$; that is, stability is guaranteed if no fluctuations can satisfy the second law of thermodynamics. Considering fluctuations in the linear regime close to equilibrium we could then show in eqn (4.43) that

$$\frac{1}{2}\frac{\partial}{\partial t}\delta^2(\rho s) = \sigma_s \geq 0 \qquad (4.48)$$

That is, any fluctuation that took the system away from equilibrium to a state of smaller entropy will die out, generating that missing entropy.

To extend these ideas to systems out of equilibrium, we recall that we have restricted ourselves to systems in local equilibrium. To guarantee that these local equilibria are stable, we now demand that everywhere locally we have (Glansdorff and Prigogine, 1971)

$$\delta^2 s \leq 0 \quad \text{or} \quad \delta^2(\rho s) \leq 0 \qquad (4.49)$$

together with

$$\frac{\partial}{\partial t}\delta^2 s \geq 0 \quad \text{or} \quad \frac{\partial}{\partial t}\delta^2(\rho s) \geq 0 \qquad (4.50)$$

We should note that a second-order differential involves two states of the system that are infinitesimally close to each other and are therefore defined on the neighborhood of the state whose stability we want to establish. Thus $\delta^2 s$ or $\delta^2(\rho s)$ can be chosen as Liapounoff functions for a *linear* stability analysis involving small perturbations around a stable nonequilibrium state.‡

The conditions on the time derivatives of the second-order differentials, eqn (4.50), are now nontrivial postulates to guarantee stability in a nonequilibrium system, whereas in global equilibrium they were a direct consequence of the balance equations because entropy production was quadratic in the deviations from equilibrium.

To gain some insight into the structure of this stability criterion, let us start from the expression for $\delta^2 s$, eqn (4.25), and take the time derivative

$$\frac{1}{2}\frac{\partial}{\partial t}\delta^2 s\bigg|_{t_0} = -\frac{1}{T}\bigg[\frac{c_v}{T}\delta T\frac{\partial}{\partial t}(\delta T) + \frac{\rho}{\kappa_T}\delta v\frac{\partial}{\partial t}(\delta v)$$
$$+ \sum_{\gamma,\gamma'=1}^{N}\mu_{\gamma\gamma'}\delta c_\gamma\frac{\partial}{\partial t}(\delta c_{\gamma'})\bigg]_{t_0} \quad (4.51)$$

The subscript t_0 indicates that the coefficients in $\delta^2 s$ have to be taken at the time $t = t_0$ for which stability of the system should be established. In terms of the internal energy u, eqn. (4.51) can be written as

$$\frac{1}{2}\frac{\partial}{\partial t}(\delta^2 s)\bigg|_{t_0} = \delta\bigg(\frac{1}{T}\bigg)\frac{\partial}{\partial t}(\delta u) + \delta\bigg(\frac{p}{T}\bigg)\frac{\partial}{\partial t}(\delta v)$$
$$- \sum_{\gamma=1}^{N}\delta\bigg(\frac{\mu_\gamma}{T}\bigg)\frac{\partial}{\partial t}(\delta c_\gamma) \quad (4.52)$$

The right-hand side of this equation is a sum of products each consisting of a factor related to a thermodynamic force and controlled externally [i.e., $\delta(T^{-1})$ is the variation related to ∇T, etc.] and a factor involving the time derivative of a quantity for

---

‡ It should be noted, however, that the stability of a solution in the class of nonlinear perturbations *cannot* be judged in general from the corresponding linearized problem (Knops and Wilkes, 1973, p. 177). Dirichlet (1846) points out that linearization often conceals in itself a tautology in the stability analysis. A similar conclusion was reached by Klein and Sommerfeld (1897) for the stability of a top.

which we have established balance equations. This is an a posteriori justification for choosing $\delta^2 s$ or $\delta^2(\rho s)$ as Liapounoff functions.‡

To calculate the explicit expression for the time derivative of the Liapounoff functions $\delta^2 s$ or $\delta^2(\rho s)$ we can start from the balance equations for the relevant mechanical quantities (Section 2.1) and derive first excess balance equations for the excess mass density $\delta\rho$, excess momentum $\delta(\rho\mathbf{v})$, excess internal energy δu, etc. and insert their time derivatives in eqn (4.52).§ In a more direct approach we can start from the entropy balance, eqn (2.68), and expand both sides of the equation in a Taylor series. Collecting second-order terms we obtain the balance equation for the excess entropy $\delta^2(\rho s)$, again in the absence of velocity fluctuations, namely

$$\frac{1}{2}\frac{\partial}{\partial t}\delta^2(\rho s) = -\mathrm{div}\,(\delta^2 \mathbf{j}_s) + \sigma(\delta^2 s) \qquad (4.53)$$

with the excess entropy current

$$\delta^2 \mathbf{j}_s = \delta \mathbf{j}_q\, \delta\!\left(\frac{1}{T}\right) - \sum_{\gamma=1}^{c} \delta\!\left(\frac{\mu_\gamma}{T}\right) \delta \mathbf{j}_\gamma^{\mathrm{diff}} \qquad (4.54)$$

and the excess entropy production

$$\sigma(\delta s) = \sum_i \delta Y_i\, \delta X_i \qquad (4.55)$$

Integrating eqn (4.53) over the volume, we then arrive at a global stability criterion, namely

$$\frac{1}{2}\frac{\partial}{\partial t}\delta^2 S = P[\delta S] = \int \sum_i \delta Y_i\, \delta X_i\, dV \geq 0 \qquad (4.56)$$

where we have assumed that fluctuations can be suppressed on the surfaces (fixed boundary conditions). Otherwise, we have to

---

‡ The stability criteria in terms of $\delta^2 s$ also have a simple and straightforward interpretation if we assume that, locally, Einstein's formula for fluctuations, eqn (3.30), is valid. It now implies that a stable state is most probable in the class of states slightly perturbed from it.

§ It has to be pointed out that eqn (4.52) was derived by assuming that the entropy is a function of the independent variables u, ρ, and ρ_γ. Thus convective effects for which a further independent variable, e.g. $\rho\mathbf{v}$, has to be included in the variations are excluded. Starting from eqn (4.52), we therefore restrict ourselves for now to purely dissipative effects and exclude velocity fluctuations.

postulate in addition that the surface integral involving the excess entropy current satisfies the inequality

$$\int_{\Sigma} d\mathbf{\Sigma} \cdot \left[\sum_{\gamma} \delta\left(\frac{\mu_{\gamma}}{T}\right) \delta \mathbf{j}_{\gamma}^{\text{diff}} - \delta \mathbf{j}_q \, \delta\left(\frac{1}{T}\right) \right] \geq 0 \qquad (4.57)$$

Let us stress once more that these criteria are sufficient to test stability against infinitesimal perturbations (in the absence of velocity fluctuations) but they apply to any nonequilibrium state provided the system is in local equilibrium.

In the range of linear thermodynamic laws we can write $\delta Y_i = \sum_k L_{ik} \delta X_k$ and eqn (4.56) reads

$$P[\delta S] = \int \sum_{i,j} L_{ik} \delta X_i \, \delta X_j \, dV \geq 0 \qquad (4.58)$$

This expression for the excess entropy production has the same algebraic structure as that for the entropy production itself. The inequality is thus trivially satisfied, as it should be from general considerations in Section 4.1 because stability in the linear regime is guaranteed by the stability of the equilibrium. Moreover, stability here is also asymptotic and, indeed, exponentially asymptotic, implying that all perturbations will decay exponentially towards the steady state.

In the presence of convective velocity fluctuations, we can no longer take $-\delta^2(\rho s)$ as a Liapounoff function because it may vanish for nonvanishing velocity perturbations. We therefore add a kinetic energy term and adopt (Glansdorff and Prigogine, 1971)

$$-\delta^2(\rho \zeta) = -\varepsilon^2 \, \delta^2(\rho s) + T_0^{-1} \tau^2 \delta^2(\tfrac{1}{2} \rho \mathbf{v}^2) \geq 0 \qquad (4.59)$$

as a Liapounoff function for combined thermodynamic and hydrodynamic stability. ε^2 and τ^2 are weighting functions that can be chosen suitably to simplify a particular problem. We will treat an example of such a stability analysis when we consider Bénard convection in Chapter 6.

4.4. The General Evolution Criterion (Glansdorff and Prigogine)

So far we have only considered linear stability analysis with which the stability of a particular nonequilibrium state against small, i.e.

infinitesimal, perturbations can be established. In certain non-linear situations a linearization, however, may overlook an instability against finite perturbations.‡ We want to approach this problem in the general framework set by Glansdorff and Prigogine (1971). We start from the balance equations for partial densities, eqn (2.8), and internal energy, eqn (2.27). (We again restrict the discussion to systems *without* convection, i.e., we put $\mathbf{v} = 0$.)

$$\frac{d}{dt}\rho_\gamma = \frac{\partial}{\partial t}\rho_\gamma = \sum_{\delta=1}^{r}\nu_{\gamma\delta}J_\delta - \boldsymbol{\nabla}\cdot\mathbf{j}_\gamma^{\text{diff}}$$

$$\frac{d}{dt}(\rho u) = \frac{\partial}{\partial t}(\rho u) = \sum_{\gamma=1}^{c}\mathbf{j}_\gamma^{\text{diff}}\cdot\mathbf{F}_\gamma - \boldsymbol{\nabla}\cdot\mathbf{j}_q \quad (4.60)$$

We then multiply the first equation by $(d/dt)(\mu_\gamma T^{-1})$, the second by $(d/dt)(T^{-1})$, add the two resulting equations, and integrate over the volume of the system. If we fix the boundary conditions to be

$$\left.\frac{d\mu_\gamma}{dt}\right|_\Sigma = \left.\frac{dT}{dt}\right|_\Sigma = 0 \quad (4.61)$$

on the surface and consider only time-independent forces, we find

$$\int dV\left\{\mathbf{j}_q\cdot\frac{d}{dt}(\boldsymbol{\nabla}T^{-1}) - \sum_{\gamma=1}^{c}\mathbf{j}_\gamma^{\text{diff}}\cdot\frac{d}{dt}[\mu_\gamma\boldsymbol{\nabla}T^{-1} - T^{-1}\mathbf{F}_\gamma]\right.$$
$$+ \sum_{\delta=1}^{r}j_\delta\frac{d}{dt}(A_\delta T^{-1})\bigg\} = \int dV\frac{\rho}{T}\left[\frac{c_v}{T}\left(\frac{d}{dt}\frac{1}{T}\right)^2\right.$$
$$\left.+\frac{\rho}{\kappa_T}\left(\frac{d}{dt}\frac{1}{\rho}\right)^2_{\rho_\gamma=\text{const}} + \sum_{\gamma,\gamma'}\mu_{\gamma\gamma'}\frac{dc_\gamma}{dt}\frac{dc_{\gamma'}}{dt}\right] \leq 0 \quad (4.62)$$

Starting on the other hand from the entropy production

$$P[S] = \int\sum_i Y_iX_i\,dV \quad (4.63)$$

we can calculate

$$\frac{dP}{dt} = \int dV\sum_i Y_i\frac{dX_i}{dt} + \int dV\sum_i X_i\frac{dY_i}{dt} = \frac{d_xP}{dt} + \frac{d_yP}{dt} \quad (4.64)$$

‡ Examples can be found in hydrodynamics (see Clever and Busse, 1974, for a discussion of this difficulty in the Bénard problem) and elastic stability (Knops and Wilkes, 1973).

The left-hand side of eqn (4.62), however, is nothing but $d_x P/dt$, and we can conclude that generally

$$\frac{d_x P}{dt} \leq 0 \tag{4.65}$$

in systems in local equilibrium. This evolution criterion (Glansdorff and Prigogine, 1955) states that during the evolution of the system the thermodynamic forces X_i will change and adjust themselves in such a way that the entropy production will be minimal in the stationary state for which

$$\frac{d_x P}{dt} = \int \sum_i Y_i^{st} \frac{dX_i}{dt} = 0 \tag{4.66}$$

Note that nothing is said and can be said about the change in the fluxes or the total change in the entropy production dP/dt.

In the linear regime near equilibrium, we can write $Y_i = \sum_k L_{ik} X_k$ and get

$$\sum_i Y_i \frac{d}{dt} X_i = \sum_{i,k} L_{ik} X_k \frac{d}{dt} X_i$$

$$= \sum_k X_k \frac{d}{dt} \sum_i L_{ik} X_i = \sum_k X_k \frac{d}{dt} Y_k \tag{4.67}$$

where we have used the Onsager reciprocity relations $L_{ik} = L_{ki}$. This means that

$$\frac{d_x P}{dt} = \frac{d_y P}{dt} = \frac{1}{2} \frac{dP}{dt} \leq 0 \tag{4.68}$$

and is the theorem of minimal entropy production (Prigogine, 1945).

For the steady state we can write the evolution criterion, eqn (4.65), in terms of the finite excess fluxes $\Delta Y_i = Y_i - Y_i^{st}$ and finite excess forces $\Delta X_i = X_i - X_i^{st}$, where X_i^{st} and Y_i^{st} are the steady state forces and fluxes, namely

$$\frac{d_x}{dt} \Delta P = \int dV \sum_i \Delta Y_i \frac{d}{dt} \Delta Y_i \leq 0 \tag{4.69}$$

This relation could serve as a starting point for a nonlinear stability analysis if a finite neighborhood of the stationary state is examined. In most practical situations, one will, however, be forced to consider infinitesimal perturbations only.‡ It should be stressed that the evolution criterion, eqn (4.65), strongly hinges on the structure of the balance equations and our possibility to distinguish forces from fluxes in the system. This might be difficult to do or physically inappropriate for systems far from equilibrium, as discussed by Landauer (1975) for electrical networks. If not considered as a topological problem, the evolution criterion, eqn (4.65), is merely a consequence of the stability of a steady state and is then a very general and transparent formulation of the second law of thermodynamics.§

We want to conclude this section by considering heat conduction as an example which serves to illustrate several of the points covered so far. The equations for the heat current $\mathbf{j}_q$ and the entropy production σ_s are

$$\mathbf{j}_q = -\lambda(T)\nabla T = \mathbf{L} \cdot \mathbf{X} \tag{4.70}$$

where

$$\mathbf{X} \equiv \nabla\left(\frac{1}{T}\right)$$

$$\sigma_s = \mathbf{j}_q \cdot \mathbf{X} = \frac{\lambda(T)}{T^2}(\nabla T) \cdot (\nabla T) \tag{4.71}$$

and

$$P = \int_V \sigma_s \, dV = \int_V \frac{\lambda(T)}{T^2}(\nabla T) \cdot (\nabla T) \, dV \tag{4.72}$$

Splitting dP into $d_x P$ and $d_y P$ as in eqn (4.64) gives

$$dP = \mathbf{j}_q \cdot d\mathbf{X} + d\mathbf{j}_q \cdot \mathbf{X} = d_x P + d_y P \tag{4.73}$$

Then, we have

$$\frac{d_x P}{dt} = \int_V \mathbf{j}_q \cdot \frac{d\mathbf{X}}{dt} \, dV \tag{4.74}$$

and the energy balance equation is

$$\rho \frac{\partial u}{\partial t} = -\nabla \cdot \mathbf{j}_q \tag{4.75}$$

---

‡ A similar stability criterion for finite perturbations has been constructed by Coleman (1970) for elastic continua in the framework of rational mechanics. For a discussion, see Knops and Wilkes (1973).

§ A statistical motivation of the general evolution criterion has been attempted by Schlögl (1971a, 1971b).

Assuming that only dissipative processes are present (i.e., there is no convection), we define

$$\psi \equiv \rho \frac{\partial T^{-1}}{\partial t} \frac{\partial u}{\partial t} = -\rho \frac{c_v}{T^2}\left(\frac{\partial T}{\partial t}\right)^2 \leq 0 \qquad (4.76)$$

where the inequality holds because $c_v > 0$. Then, using eqn (4.75) we obtain

$$\psi = -\frac{\partial T^{-1}}{\partial t}\boldsymbol{\nabla}\cdot\mathbf{j}_q = \boldsymbol{\nabla}\cdot\left(-\mathbf{j}_q\frac{\partial T^{-1}}{\partial t}\right) + \mathbf{j}_q\cdot\frac{\partial}{\partial t}(\boldsymbol{\nabla}T^{-1}) \leq 0 \qquad (4.77)$$

By integrating, we obtain

$$\int_V \psi\, dV = -\int_V \boldsymbol{\nabla}\cdot\left(\mathbf{j}_q\frac{\partial T^{-1}}{\partial t}\right)dV + \int_V \mathbf{j}_q\cdot\frac{\partial}{\partial t}(\boldsymbol{\nabla}T^{-1})\, dV$$

$$= -\int_\Sigma \mathbf{j}_q\frac{\partial T^{-1}}{\partial t}\cdot d\boldsymbol{\Sigma} + \int_V \mathbf{j}_q\cdot\frac{\partial}{\partial t}(\boldsymbol{\nabla}T^{-1})\, dV \leq 0 \qquad (4.78)$$

Assuming that the boundary conditions are time-independent, we find

$$\int_V \mathbf{j}_q\cdot\frac{\partial}{\partial t}(\boldsymbol{\nabla}T^{-1})\, dV = \int_V \mathbf{j}_q\cdot\frac{\partial \mathbf{X}}{\partial t}\, dV \leq 0 \qquad (4.79)$$

Thus, in heat conduction, the thermodynamic forces change in such a way as to lower the rate of entropy production.

Assuming the quasilinear phenomenological law of eqn (4.70) holds, we get

$$\int_V \left[\lambda(T)T^2\boldsymbol{\nabla}T^{-1}\cdot\frac{\partial}{\partial t}(\boldsymbol{\nabla}T^{-1})\right] dV \leq 0 \qquad (4.80)$$

Let the steady-state solution be $T_0(\mathbf{r})$; that is, let $T_0(\mathbf{r})$ be the solution of

$$\boldsymbol{\nabla}\cdot[\lambda(T_0)\boldsymbol{\nabla}T_0] = 0 \qquad (4.81)$$

and expand

$$\lambda(T)T^2 = \lambda(T_0)T_0^2 + \delta(\lambda T)^2 + \cdots \qquad (4.82)$$

Then, to first-order in the deviations from steady state, we have

$$\frac{\partial}{\partial t}\frac{1}{2}\int_V \lambda(T_0)T_0^2(\boldsymbol{\nabla}T^{-1})\cdot(\boldsymbol{\nabla}T^{-1})\, dV \leq 0 \qquad (4.83)$$

Therefore, if the local potential ϕ is defined such that (Glansdorff and Prigogine, 1971)

$$\phi(T, T_0) = \frac{1}{2} \int_V \lambda(T_0) T_0^2 (\boldsymbol{\nabla} T^{-1}) \cdot (\boldsymbol{\nabla} T^{-1}) \, dV \qquad (4.84)$$

we obtain as a Liapounoff function

$$\begin{aligned}
\Delta\phi &\equiv \phi(T, T_0) - \delta(T_0, T_0) \\
&= \frac{1}{2} \int_V \lambda(T_0) T_0^2 [(\boldsymbol{\nabla} T^{-1}) \cdot (\boldsymbol{\nabla} T^{-1}) \\
&\quad - (\boldsymbol{\nabla} T_0^{-1}) \cdot (\boldsymbol{\nabla} T_0^{-1})] \, dV \geq 0
\end{aligned} \qquad (4.85)$$

and $\Delta\phi = 0$ if and only if $T(\mathbf{r}) = T_0(\mathbf{r})$, provided that the boundary conditions are time-independent and that the system is near its steady state, in which case we also have

$$\frac{\partial}{\partial t}(\Delta\phi) \leq 0 \qquad (4.86)$$

In a steady state, $\phi(T, T_0)$ is a minimum and

$$\phi(T_0, T_0) = \frac{1}{2} P = \frac{1}{2} \frac{d_i S}{dt} \qquad (4.87)$$

recovering the theorem of minimum entropy production.

5
Chemical Reactions

5.1. Equations of Motion and Stability

Chemically reacting systems will now be studied as examples of systems that may exhibit steady states both near and far from equilibrium as well as other quasistable states—for example, sustained oscillations which may be sensitive to perturbations.

Consider a system of chemical reactants with mole numbers n_γ ($\gamma = 1, 2, \ldots, c$). If there is only one chemical reaction occurring, then the change in n_γ is given by

$$dn_\gamma = \nu_\gamma \, d\xi \tag{5.1}$$

where ν_γ is the stoichiometric coefficient for the γth component and ξ is a parameter describing the advancement of the reaction. If there are r simultaneous reactions in our system, then

$$dn_\gamma = \sum_{\delta=1}^{r} \nu_{\gamma\delta} \, d\xi_\delta \tag{5.2}$$

where $\nu_{\gamma\delta}$ is the stoichiometric coefficient of the γth reactant in the δth reaction and ξ_δ describes the advancement of the δth reaction. If the reaction rate J_δ of the δth reaction is defined as

$$J_\delta = \frac{d\xi_\delta}{dt} \tag{5.3}$$

then

$$\frac{dn_\gamma}{dt} = \sum_{\delta=1}^{r} \nu_{\gamma\delta} J_\delta \tag{5.4}$$

To clarify the notation, consider the simultaneous reactions

$$\begin{aligned} 2C + O_2 &\rightarrow 2CO & \delta = 1 \\ C + O_2 &\rightarrow CO_2 & \delta = 2 \end{aligned} \tag{5.5}$$

Then, we have

$$dn_C = -2\,d\xi_1 - d\xi_2$$
$$dn_{O_2} = -d\xi_1 - d\xi_2$$
$$dn_{CO} = 2\,d\xi_1 \quad (5.6)$$
$$dn_{CO_2} = d\xi_2$$

If the possibility of having open chemical systems is included in order to maintain steady-state reactions, eqn (5.2) becomes

$$dn_\gamma = d_e n_\gamma + \sum_{\delta=1}^{r} \nu_{\gamma\delta}\,d\xi_\delta \quad (5.7)$$

where $d_e n_\gamma$ is the external supply of reactant γ.

The fundamental relation for a system involving chemical reactions is

$$T\,dS = dQ + \sum_{\delta=1}^{r} A_\delta\,d\xi_\delta \quad (5.8)$$

where the affinity of the δth reaction, A_δ, is defined as

$$A_\delta = \sum_{\gamma=1}^{c} \nu_{\gamma\delta}\mu_\gamma \quad (5.9)$$

with μ_γ the chemical potential of reactant γ.

Looking at the internal part of the entropy change, we find that the entropy production P:

$$P \equiv \frac{d_i S}{dt} = \frac{1}{T}\sum_{\delta=1}^{r} A_\delta J_\delta \geq 0 \quad (5.10)$$

Also, recall the relation for chemical stability, eqn (4.47),

$$\sum_{\gamma,\gamma'=1}^{c} \frac{\partial \mu_\gamma}{\partial n_{\gamma'}} \frac{dn_\gamma}{dt}\frac{dn_{\gamma'}}{dt} = -\sum_{\delta=1}^{r} J_\delta \frac{dA_\delta}{dt} > 0$$

or

$$\sum_{\delta=1}^{r} J_\delta \frac{dA_\delta}{dt} < 0 \quad (5.11)$$

If the affinities are regarded as the generalized forces and the reaction rates as the currents, then

$$T\,d_x P = \sum_{\delta=1}^{r} J_\delta\,dA_\delta \leq 0 \quad (5.12)$$

is the condition for stability. In a steady state, the forces and fluxes are constant so $d_x P = 0$, which implies that

$$\sum_{\delta=1}^{r} J_\gamma \, dA_\delta = 0 \tag{5.13}$$

The following example serves to illustrate some of these results. Consider three reactions:

$$A \rightleftarrows X \quad X \rightleftarrows B \quad X \rightleftarrows M \tag{5.14}$$

In terms of the chemical potentials, the affinities are

$$A_1 = \mu_A - \mu_X \quad A_2 = \mu_X - \mu_B \quad A_3 = \mu_X - \mu_M \tag{5.15}$$

To get a steady-state situation, we keep n_A and n_B fixed by supplying them externally. This implies that μ_A and μ_B are fixed and hence that

$$A_1 + A_2 = \mu_A - \mu_B = \text{const} \tag{5.16}$$

or

$$dA_1 + dA_2 = 0 \tag{5.17}$$

Since this is a steady-state situation, we have

$$T \, d_x P = 0 = \sum_{\delta=1}^{3} J_\delta \, dA_\delta \tag{5.18}$$

or

$$(J_1 - J_2) \, dA_1 + J_3 \, dA_3 = 0 \tag{5.19}$$

for any dA_1 and dA_3. Therefore, we must have $J_1 = J_2$ and $J_3 = 0$. These results could, of course, have been deduced simply by inspection of the reaction, remembering that the system must be in a steady state and that the reactant M is not externally supplied.

In the case of steady states very near to equilibrium, it may be assumed that the equations are linear in the deviations from equilibrium, i.e.

$$\begin{aligned} J_\delta &= \sum_{\delta'=1}^{c} \alpha_{\delta\delta'} \, d\xi_{\delta'} \\ A_\delta &= \sum_{\delta'=1}^{c} \beta_{\delta\delta'} \, d\xi_{\delta'} \end{aligned} \tag{5.20}$$

These may be combined to yield linear phenomenological laws of the form

$$J_\delta = \sum_{\delta'=1}^{c} L_{\delta\delta'} A_{\delta'} \qquad (5.21)$$

Note that in equilibrium we have $J_\delta = A_\delta = 0$. These linear laws would be expected to obey Onsager relations within the limitations described in Section 3.2.

However, due to the complex form of the chemical potentials and affinities, it is desirable to express the reaction rates directly in terms of the reactant densities n_δ. For example, for an ideal gas we have for the chemical potential

$$\mu_\delta = -RT \ln\left[\frac{(2\pi m k_B T)^{\frac{3}{2}}}{h^3 N_A} \frac{V}{n_\delta}\right] \qquad (5.22)$$

where m is the mass of a gas particle, so that in the previous example

$$A_1 = \mu_A - \mu_X = RT \ln\left[\left(\frac{m_X}{m_A}\right)^{\frac{3}{2}} \frac{n_A}{n_X}\right] \qquad (5.23)$$

In view of this complex dependence of the affinities upon the observed quantities n_δ, it is usually easier to guess the dependence of the reaction rates upon the concentrations rather than upon the affinities themselves.

For example, in considering the transformations of A particles into B particles via collisions with another B particle in a dilute gas, it is reasonable to assume that the rate at which the A particles transform into B particles is proportional to the concentration of A particles and to the concentration of B particles. Also, if there is a reverse reaction with two B's colliding to yield an A and a B, its rate should be proportional to the square of the concentration of B particles. Thus, for the reaction

$$A + B \rightleftarrows 2B \qquad (5.24)$$

we will assume that

$$\frac{dn_A}{dt} = \kappa_1 n_A n_B - \kappa_1' n_B^2 \qquad (5.25)$$

Such rate equations can, of course, be made arbitrarily complex and nonlinear. They also allow us in the remainder of this chapter to study systems of chemical reactions far from equilibrium.

5.2. A Chemical Reaction Model with a Nonequilibrium Phase Transition

We want to study a model defined by the reaction scheme (Schlögl, 1972)

$$A + 2X \rightarrow 3X$$
$$B + X \rightarrow C \tag{5.26}$$

The concentrations of A, B, and C are held constant by an external supply. Only the amount of X is allowed to vary and the steady state occurs when $dn_X/dt = 0$. The rates $r_1 = -dn_A/dt$ at which A particles disappear, and $r_2 = -dn_B/dt$ at which B particles disappear, we assume to be

$$r_1 = \kappa_1 n_A n_X^2 - \kappa_1' n_X^3$$
$$r_2 = \kappa_2 n_B n_X - \kappa_2' n_C \tag{5.27}$$

With an appropriate choice of units, we can set $\kappa_1' = 1$ and $\kappa_1 n_A = 3$. The net rate of change of n_x is then

$$\frac{dn_X}{dt} = r_1 - r_2 = \psi(n_X) = -n_X^3 + 3n_X^2 - \beta n_X + \gamma \tag{5.28}$$

Here we have $\beta = \kappa_2 n_B$ and $\gamma = \kappa_2' n_C$.

The steady states of our system are the solutions of the equation

$$\gamma = n_X^3 - 3n_X^2 + \beta n_X = \psi(n_X) \tag{5.29}$$

A plot of γ as a function of n_X for three different values of β is given in Fig. 5.1.

For $\beta \geq 3$, there is only one steady-state value of n_X for any given γ. For $\beta < 3$, however, there are three separate steady-state values $n_X^{(1)}$, $n_X^{(2)}$, and $n_X^{(3)}$, if we have $\gamma_1 < \gamma < \gamma_2$. For γ outside this range, there is only one steady state. To find out which of these states are stable, we assume that for a fixed γ a fluctuation δn_X occurs in n_X around one of the steady states $n_X^{(i)}$, i.e.

$$n_X = n_X^{(i)} + \delta n_X \tag{5.30}$$

and find from eqn (5.28)

$$\frac{d}{dt}(\delta n_X) = -\frac{d\psi}{dn_X}\bigg|_{n_X^{(i)}} \delta n_X \tag{5.31}$$

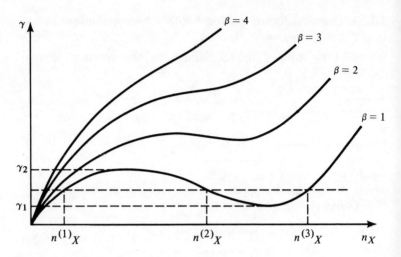

Fig. 5.1. Sketch of γ versus n_X in a chemical system. *Source:* Schlögl (1972).

Thus for $\beta > 3$, the one steady state of the system (for a given γ) is always stable because any small fluctuations will die out exponentially. However, for $\beta < 3$ the slope of $\psi(n_X)$ is negative at $n_X = n_X^{(3)}$ and we find

$$\delta n_X(t) = \delta n_X(0) \exp\left(\left|\frac{d\psi}{dn_X}\right|_{n_X^{(3)}} t\right) \qquad (5.32)$$

Thus $n_X^{(3)}$ is unstable. If the spontaneous fluctuation $\delta n_X(0)$ is positive, the system will run into the stable point $n_X^{(2)}$; if $\delta n_X(0)$ is negative, the system will wind up at the stable steady state $n_X^{(1)}$.

There also exists a critical value $\gamma = \gamma_c = 3$ for which

$$n_X^{(1)} = n_X^{(2)} = n_X^{(3)} = 1 \qquad \text{for } \gamma = \gamma_c = 1 \qquad (5.33)$$

To complete the analogy with the equilibrium states of a van der Waals gas suggested by the diagram Fig. 5.1, we replace n_X, γ, and β by the density ρ, pressure p, and RT, where T is the temperature and R is the molar gas constant. We find from eqn (5.29) the virial equation of state

$$p\rho^{-1} = RT - 3\rho + \rho^2 \qquad (5.34)$$

This analogy has been used to call the transition between the stable steady states $n_X^{(1)}$ and $n_X^{(2)}$ for $\beta < \beta_c$ a nonequilibrium phase transition of the first order.‡

So far it has been assumed that all concentrations n_A, n_B, n_C, and n_X are constant and homogeneous in space which could be achieved for example, by constantly stirring the reactants. We now want to suppose that the mobility of reactant X is much less than those of the externally supplied reactants, leading to a density gradient in $n_X = n_X(\mathbf{r})$ that will set up diffusion of component X in the system. We want to keep the concentrations n_A, n_B, and n_C homogeneous and constant in space by an appropriate feeding mechanism. In this case eqn (5.28) has to be replaced by

$$\frac{\partial n_X}{\partial t} = r_1 - r_2 + \kappa \frac{\partial^2 n_X}{\partial z^2} \tag{5.35}$$

where we have assumed that n_X depends on one coordinate z only. The three-dimensional case has also been treated by Schlögl (1972). Next we introduce a 'potential' $\Phi(n_X)$ by [see eqn (5.28)]

$$\psi(n_X) = \frac{\partial}{\partial n_X} \Phi(n_X) \tag{5.36}$$

and look for the steady states which now have to satisfy the equation

$$\kappa \frac{\partial^2 n_X}{\partial z^2} = -\frac{\partial}{\partial n_X} \Phi(n_X) \tag{5.37}$$

We have seen above that for $\beta < \beta_c$ and γ between γ_1 and γ_2 (see Fig. 5.1) the system can have two stable homogeneous steady states $n_X^{(1)}$ and $n_X^{(2)}$. We therefore want to look for a solution $n_X(z)$ of eqn (5.37) such that $n_X(+\infty) = n_X^{(1)}$ and $n_X(-\infty) = n_X^{(2)}$, in which case these two steady states are in coexistence. Obviously the 'potential' $\Phi(n_X)$ has maxima at $n_X = n_X^{(1)}$ and $n_X = n_X^{(2)}$ if $\beta < \beta_c$ and $\gamma_1 < \gamma < \gamma_2$. Coexistence can occur for a value of γ such that the two maxima are indistinguishable for the system, i.e.

$$\Phi(n_X^{(1)}) = \Phi(n_X^{(2)}) \tag{5.38}$$

‡ This point has been greatly exploited by Haken (1975).

Using eqns (5.28) and (5.36), this gives

$$0 = \Phi(n_X^{(2)}) - \Phi(n_X^{(1)}) = \int_{n_X^{(1)}}^{n_X^{(2)}} \psi(n)\, dn$$
$$= \int_{n_X^{(1)}}^{n_X^{(2)}} dn(-n^3 + 3n^2 + \beta n + \gamma) \quad (5.39)$$

This is nothing but the Maxwellian construction of the vapor pressure in a Van der Waals gas.

Thus coexistence can occur if

$$\gamma = \beta - 2 \quad (5.40)$$

We then have

$$n_X^{(1)} = n_X(+\infty) = 1 + (3-\beta)^{\frac{1}{2}}$$
$$n_X^{(2)} + n_X(-\infty) = 1 - (3-\beta)^{\frac{1}{2}} \quad (5.41)$$

The width of the layer in which $n_X(z)$ changes from $n_X^{(2)}$ to $n_X^{(1)}$ is

$$\delta z = 2\left(\frac{2\kappa}{1-\gamma}\right)^{\frac{1}{2}} \quad (5.42)$$

The diffusion-reaction model considered here thus allows the coexistence of two spatially separated phases in an open system far from equilibrium.‡ The two phases are distinguished by different concentrations of reactant X, i.e., $n_X^{(1)}$ and $n_X^{(2)}$, and thus for fixed n_A, n_B, and n_C by different rates r_1 and r_2. They have obviously nothing to do with the equilibrium properties of the system, but can only occur far from equilibrium by opening the system to an external supply to keep the concentrations of reactants A, B, and C constant. Although the system considered here is presumably too simple for experimental realization, we will see later on in this chapter that spatial and temporal structures have indeed been observed in complex chemical reactions far from equilibrium, e.g. in the Belousov-Zhabotinskii reaction.

5.3. Volterra-Lotka Model

We next study an open chemical system far from equilibrium with a steady state that is not asymptotically stable and in which sustained oscillations of the reactant concentrations are possible.

---

‡ The role of diffusion in chemical reaction kinetics was first discussed by Kramers (1940).

The extension to stochastic reaction models exhibiting nonequilibrium phase transitions has been given by Janssen (1974), Matheson, Walls, and Gardiner (1975), and Metiu, Kitahara, and Ross (1976).

We consider the following system of reactions:

$$A + X \rightleftarrows 2X \quad \text{rates } \kappa_1, \kappa_1'$$
$$X + Y \rightleftarrows 2Y \quad \text{rates } \kappa_2, \kappa_2' \quad (5.43)$$
$$Y \rightleftarrows E \quad \text{rates } \kappa_3, \kappa_3'$$

The concentrations of the initial reactant A and the final product E are maintained externally at constant values. Only the concentrations of the intermediate reactants X and Y are allowed to vary. The equations for n_X and n_Y are then

$$\frac{dn_X}{dt} = \kappa_1 n_A n_X - \kappa_1' n_X^2 - \kappa_2 n_X n_Y + \kappa_2' n_Y^2$$
$$\frac{dn_Y}{dt} = \kappa_2 n_X n_Y - \kappa_2' n_Y^2 - \kappa_3 n_Y + \kappa_3' n_E \quad (5.44)$$

where the primes denote the reverse reaction rates.

Next, we assume that the concentrations are maintained in such a way that the affinities become so large that the reactions can proceed in a forward direction only. Then, ignoring reverse reactions, eqns (5.44) become

$$\frac{dn_X}{dt} = \kappa_1 n_A n_X - \kappa_2 n_X n_Y$$
$$\frac{dn_Y}{dt} = \kappa_2 n_X n_Y - \kappa_3 n_Y \quad (5.45)$$

This system, incidentally, is isomorphic to the Lotka–Volterra model of predator–prey interactions (Lotka, 1910, 1920, 1956; Volterra, 1928, 1931, 1937).‡

---

‡ In the ecological context this set of coupled equations describes the interaction of two biological species X and Y. In the absence of species Y, the number of individuals of species X, n_X, would grow exponentially with a net birth rate $(\kappa_1 n_A)$ made possible by unlimited food resources. The population of X is, however, limited because it is the only food supply for a predator Y who, on the other hand, in the absence of X would disappear with a net death rate $(-\kappa_3)$. As n_X grows, a bigger population of Y can be supported until overhunting by too many predators leads to a decline in X, causing Y to diminish as well with a certain time lag. When n_Y is small enough, X can recover and a new cycle is set up. Such oscillations in the populations of interacting and competing biological species have indeed been observed. See d'Ancona (1954) and Elton (1942).

Eqns (5.45) have one steady-state solution apart from the trivial one $n_X = n_Y = 0$, namely

$$n_X^0 = \frac{\kappa_3}{\kappa_2} \quad \text{and} \quad n_Y^0 = \frac{\kappa_1}{\kappa_2} n_A \tag{5.46}$$

To analyze the stability of this steady state, we use a normal mode analysis for small perturbations. Assume a solution of the form

$$\begin{aligned} n_X(t) &= n_X^0 + \delta n_X e^{\omega t} \\ n_Y(t) &= n_Y^0 + \delta n_Y e^{\omega t} \end{aligned} \tag{5.47}$$

where $|\delta n_X| \ll |n_X^0|$, $|\delta n_Y| \ll |n_Y^0|$, and where ω may be complex. Then, to first-order in the perturbations, we have

$$\begin{aligned} \omega \delta n_X + \kappa_3 \delta n_Y &= 0 \\ -\kappa_1 n_A \delta n_X + \omega \delta n_Y &= 0 \end{aligned} \tag{5.48}$$

These equations together imply that

$$\omega^2 = -\kappa_1 \kappa_3 n_A \tag{5.49}$$

Therefore, ω is purely imaginary and there can be undamped periodic fluctuations about the steady state starting from infinitesimal perturbations from that steady state. Thus the perturbed system will remain in the neighborhood of the steady state, but it does not show asymptotic stability due to the absence of any restoring or dissipative forces that could damp out the fluctuations. It is interesting to calculate the excess entropy production $d_x P$ for such infinitesimal perturbations around the steady state. To rewrite the right-hand side of eqns (5.45) in term of proper thermodynamic forces, we assume that the reactants X and Y are ideal gases and calculate the change in the Gibbs free energy G due to fluctuations δn_X and δn_Y according to eqn (3.18). We obtain

$$\begin{aligned} \delta G &= RT \left[\delta n_X \ln \frac{n_X}{n_X^0} + \delta n_Y \ln \frac{n_Y}{n_Y^0} \right] \\ &= -X_X \delta n_X - X_Y \delta n_Y \end{aligned} \tag{5.50}$$

with the forces

$$\begin{aligned} X_X &= -\frac{RT}{n_X^0} \delta n_X \\ X_Y &= -\frac{RT}{n_Y^0} \delta n_Y \end{aligned} \tag{5.51}$$

in terms of which eqns (5.45) read

$$\frac{dn_X}{dt} = Y_X = \frac{n_X^0}{RT}(\kappa_2 n_Y^0 - \kappa_1 n_A) X_X + \frac{\kappa_2 n_X^0 n_Y^0}{RT} X_Y$$
$$\frac{dn_Y}{dt} = Y_Y = -\kappa_2 \frac{n_X^0 n_Y^0}{RT} X_X + \frac{n_Y^0}{RT}(\kappa_3 - \kappa_2 n_X^0) X_Y$$
(5.52)

Using the steady-state solutions, eqn (5.46), we see that only the crossterms survive, yielding

$$Y_X = \frac{\kappa_2 n_X^0 n_Y^0}{RT} X_Y$$
$$Y_Y = -\frac{\kappa_2 n_X^0 n_Y^0}{RT} X_X$$
(5.53)

for which we calculate the entropy production

$$\sigma_s = Y_X X_X + Y_Y X_Y = 0 \qquad (5.54)$$

This again proves the weak (i.e. nonasymptotic) stability of the system against small fluctuations that simply rotate around the steady state. It, moreover, shows that G is a constant of the motion in this case.

To analyze the general time-dependent solutions of eqns (5.32), we first generalize this pair of equations to the N-species set (Goel, Maitra, and Montroll, 1971)

$$\frac{dn_i}{dt} = k_i n_i + \beta_i^{-1} \sum_{j=1}^{N} a_{ij} n_i n_j \qquad (5.55)$$

with the restrictions that $a_{ij} = -a_{ji}$ and $a_{ii} = 0$.‡

The steady-state concentrations n_i^0 have to satisfy the algebraic equations

$$n_i^0 \left(k_i \beta_i + \sum_j a_{ij} n_j^0 \right) = 0 \qquad (5.56)$$

‡ The interpretation of these equations for interacting biological species is as follows: the first term describes the growth ($k_i > 0$) or decline ($k_i < 0$) of the ith species in the absence of others. The quadratic terms describe the encounters of species i and j which for $a_{ij} > 0$ will lead to a gain for species i and a loss for species j and v.v. for $a_{ij} < 0$. Hence, a_{ij} and a_{ji} must have opposite signs and can be chosen such that $a_{ij} = -a_{ji}$. The positive numbers β_i are a measure of the efficiency of the species i, i.e. β_i/β_j is the ratio of individuals i gained (or lost) per unit time to individuals j lost (or gained) in a binary encounter of i and j. For chemical reactions we must obviously have $\beta_i = 1$ for $i = 1, \ldots, N$.

We want to show first that there exists a constant of motion for eqns (5.55). We define the quantities

$$c_i = \ln \left(\frac{n_i}{n_i^0}\right) \tag{5.57}$$

which are a measure of the deviation from the steady state because as n_i approaches n_i^0, v_i tends to zero. They satisfy the rate equations

$$\beta_i \frac{dv_i}{dt} = \sum_{j=1}^{N} a_{ij} n_j^0 (e^{v_i} - 1) \tag{5.58}$$

If we multiply both sides of this equation by

$$n_i^0 (e^{v_i} - 1) \tag{5.59}$$

and sum over i we get

$$\frac{d}{dt} \sum_{i=1}^{N} \beta_i n_i^0 (e^{v_i} - v_i) = \sum_{i,j} a_{ij} n_i^0 n_j^0 (e^{v_i} - 1)(e^{v_j} - 1) = 0 \tag{5.60}$$

The double sum vanishes because $a_{ij} = -a_{ji}$. Therefore,

$$G = \sum_i \beta_i n_i^0 (e^{v_i} - v_i) = \sum_i G_i = \text{const} \tag{5.61}$$

is a constant of motion. Obviously, every individual term in G is positive and hence we have

$$G > 0 \tag{5.62}$$

and

$$\frac{dG}{dt} = 0 \tag{5.63}$$

For small deviations from the steady state, G can be used as a Liapounoff function to exhibit the weak stability of the system. The fact that $dG/dt = 0$ precludes asymptotic stability. We should add that any system with a constant of motion, i.e. a conservative system, is structurally unstable in the sense that a small change in the structure of the differential equation will destroy the constant of motion and might alter the character of the solution completely. As an example one may consider a harmonic oscillator to which a small amount of friction is added.

To return to our problem we want to use now the constant of motion G to show, in the case of two components, that away

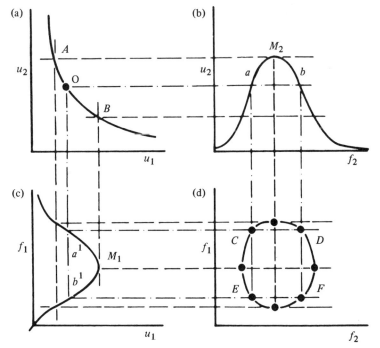

Fig. 5.2. Geometric scheme used to prove the periodicity of the solution (5.65) of the Volterra equations (5.58). *Source:* Goel, Maitra, and Montroll (1971).

from the steady state all solutions are periodic. We introduce variables

$$f_i = \frac{n_i}{n_i^0} = e^{v_i} \qquad i = 1, 2 \tag{5.64}$$

in G and exponentiate both sides of eqn (5.61) to get

$$(f_1 e^{-f_1})^{(1/\beta_1 n_1^0)} \cdot (f_2 e^{-f_2})^{(1/\beta_2 n_2^0)} = \text{const} \tag{5.65}$$

Now let

$$(f_i e^{-f_i})^{(1/\beta_i n_i^0)} = u_i \qquad i = 1, 2 \tag{5.66}$$

and eqn (5.65) becomes the equation of a hyperbola

$$u_1 u_2 = \text{const} \tag{5.67}$$

plotted in Fig. 5.2, which also shows the behavior of u_i as a function of f_i. Note that the boundedness of u_i implies that only a finite section between points A and B of the hyperbola $u_1 u_2 =$ const is relevant to the motion. As one goes from A to B, one traces out a closed curve in the $f_1 f_2$ plane. This proves that apart from the trivial initial condition $f_i(t=0) = 0$ and the steady-state initial conditions $f_i(t=0) = 1$, all solutions of the Volterra-Lotka equations are periodic. The shape of the orbits depends crucially on the initial conditions, as can be seen from Figs. 5.3 and 5.4. These oscillatory solutions cannot be stable in the sense of Liapounoff since any fluctuation will shift the system onto a new orbit with a different frequency and the perturbed and unperturbed motions will tend to separate in time.

We will close the discussion of the Volterra-Lotka model with a remark about long-time averages. We integrate the rate equations (5.55) from $t = 0$ to t_0 and divide by t_0 to get

$$\lim_{t_0 \to \infty} \frac{\beta_i}{t_0} [\ln n_i(t_0) - \ln n_i(0)] = k_i \beta_i + \sum_{j=1}^{N} a_{ij} \bar{n}_j \qquad (5.68)$$

where

$$\bar{n}_j = \lim_{t_0 \to \infty} \frac{1}{t_0} \int_0^{t_0} n_i(t)\, dt \qquad (5.69)$$

Let us next observe that the constant of motion $G = $ const prevents any v_i from increasing indefinitely. Thus $n_i(t_0)$ remains bounded and the left-hand side of eqn (5.68) vanishes as $t_0 \to \infty$ so that

$$k_i \beta_i + \sum_{j=1}^{N} a_{ij} \bar{n}_j = 0 \qquad (5.70)$$

We can multiply this equation with $\bar{n}_i$ and get

$$\bar{n}_i \left(k_i \beta_i + \sum_{j=1}^{N} a_{ij} \bar{n}_j \right) = 0 \qquad (5.71)$$

Fig. 5.3. Variation of f_1 with respect to f_2 for various values of the parameters and initial values $f_1(0)$ and $f_2(0)$. (a) $(\alpha_1/\alpha_2) = \frac{1}{2}$ with the values of $f_1(0)$ and $f_2(0)$: ——, 0.2 and 0.8; - - -, 2.0 and 0.5; · · ·, 0.5 and 2.0. (b) $(\alpha_1/\alpha_2) = 1$ with the values of $f_1(0)$ and $f_2(0)$: ——, 0.2 and 0.8; - - -, 0.5 and 2.0; · · ·, 1.0 and 0.8. (c) $(\alpha_1/\alpha_2) = 2$ with the values of $f_1(0)$ and $f_2(0)$: ——, 0.2 and 0.8; - - -, 0.2 and 2.0; · · ·, 2.0 and 0.5. $\alpha_i = \beta_i n_i^0$. (*Source:* Goel, Maitra, and Montroll, 1971).

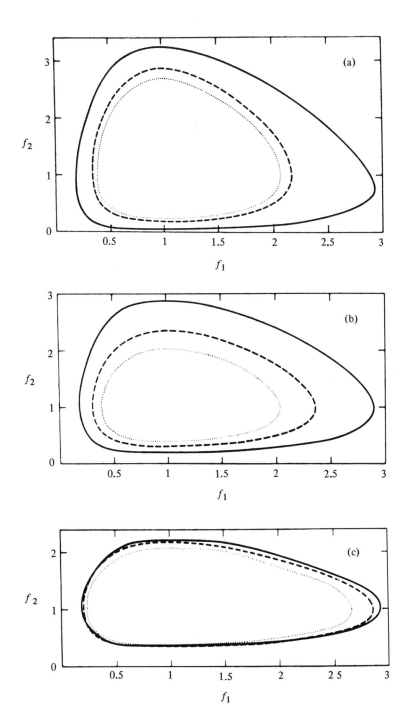

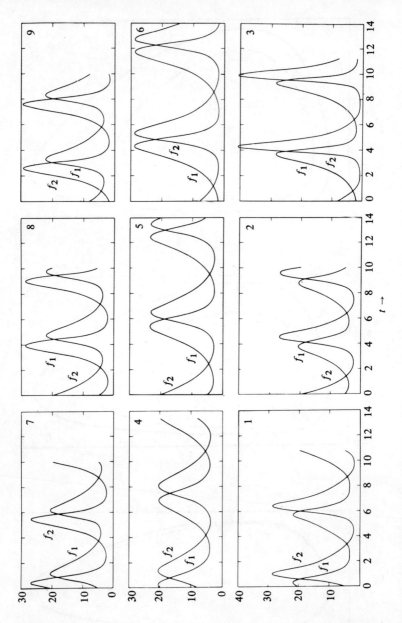

Fig. 5.4. Time variation of the two populations for several values of parameters: (1), (2), (3), $\alpha_1 = 1$, $\alpha_2 = 2$; (4), (5), (6), $\alpha_1 = \alpha_2 = 1$; (7), (8), (9), $\alpha_1 = 2$, $\alpha_2 = 1$. The initial values are the values for $t = 0$. $\alpha_i = \beta_i n_i^0$.
Source: Goel, Maitra, and Montroll (1971).

which is nothing but the equation satisfied by the steady-state concentrations n_i^0. Thus we can conclude that the long-time average of any oscillatory solution with arbitrary initial conditions is equal to the steady-state solution, i.e.

$$\bar{n}_i = n_i^0 \tag{5.72}$$

This remarkable property is again a consequence of the existence of an additive constant of motion for the system.†

5.4. Chemical Oscillations with Limit Cycle Behavior

We have seen in the previous section that the Volterra-Lotka model exhibits undamped oscillations in the reactant concentrations (or species populations) around the steady state which are strongly dependent on the initial conditions and are not stable in the sense of Liapounoff. Such a system would therefore exhibit a highly irreproducible behavior and would also be strongly affected by small external perturbations due to its structural instability. However, there are chemical systems that exhibit under well-defined conditions very complex oscillations in some of the reactant concentrations that are indeed reproducible, largely independent of the initial conditions, and stable against small external perturbations.§ One of the most widely studied chemical systems exhibiting temporal oscillations is the Belousov-Zhabotinskii reaction, which is the oxidization of analogs of malonic acid by bromate in the presence of Ce^{3+} and Ce^{4+} ions. A schematic representation of the reaction mechanism is given in Fig. 5.5.‖ Typical observed oscillations are shown in Fig. 5.6.¶

It has been argued that a number of requirements have to be met by a chemical system to exhibit oscillations in some of its

‡ Further details on the Volterra-Lotka model, including statistical aspects, have been reviewed by Goel, Maitra, and Montroll (1971). They also give a discussion of other nonlinear models of interacting populations.

§ For a review of this subject, see Nicolis and Portnov (1973), Degn (1972), or Nicolis and Prigogine (1977).

‖ A detailed analysis is given by Edelson, Field, and Noyes (1975). See also Portnov and Nicolis (1973).

¶ We should add that the Belousov-Zhabotinskii reaction also exhibits spatially periodic structures for which diffusion has to be considered as well, similar to the discussion in Section 5.2. See also Turing (1952).

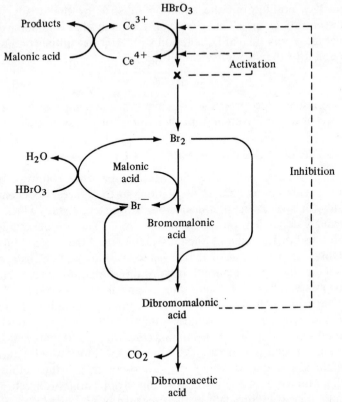

Fig. 5.5. Schematic representation of Belousov-Zhabotinskii reaction mechanism. X = bromomalonic acid, Y = dibromomalonic acid. *Source:* Degń (1972).

reactant concentrations. First, the system has to be kept far away from equilibrium. This is quite obvious because close to equilibrium the reaction rates (fluxes) can be linearized in the affinities (forces), and the theorem of minimum entropy production ensures that a time-independent steady state is approached in a monotonic way [eqn (5.12)]. We therefore have to keep the system far from equilibrium by opening it to an external supply of mass (reactants) or energy. Secondly, the reaction mechanism has to be a complex one. No simple systems are known that exhibit oscillations. Thirdly, the rate equations must contain nonlinear

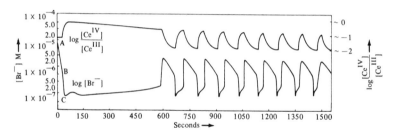

Fig. 5.6. Typical observed oscillations of $\log[Ce^{IV}]/[Ce^{III}]$ and $\log[Br^-]$ during the Belousov-Zhabotinskii reaction (in arbitrary units). After Field, Koros, and Noyes (1972).

terms. They can be due to autocatalytic reactions that, however, have to be coupled in some way to the rest of the reaction network (Field and Noyes, 1974). Degn (1972) stresses that in a sequence of reaction steps a nonlinearity should be produced through which information in the form of activating or inhibitory effects can flow.

Obviously, the Volterra-Lotka model satisfies all these requirements and, indeed, exhibits sustained oscillations. However, it has the undesired property that the periods and amplitudes of these oscillations are strongly dependent upon the initial conditions and are not stable against perturbations, whereas the known oscillating chemical systems exhibit a limit cycle behavior. In such systems the final oscillations are independent of the initial conditions and are stable against fluctuations. The limit cycle thus is a closed orbit in the appropriate phase space of the system which acts as an attractor for all close-by trajectories of motion and is thus asymptotically stable in the sense of Liapounoff. This behavior is illustrated in Fig. 5.7. A simple reaction scheme that exhibits limit cycle behavior is the so-called Brusselator:

$$\begin{aligned} A &\rightleftarrows X \\ 2X + Y &\rightleftarrows 3X \quad \text{(autocatalytic)} \\ B + Y &\rightleftarrows Y + D \\ X &\rightleftarrows E \end{aligned} \quad (5.73)$$

It has been analyzed in great detail by Glansdorff and Prigogine (1971). This is an open system, since the concentrations of A, B,

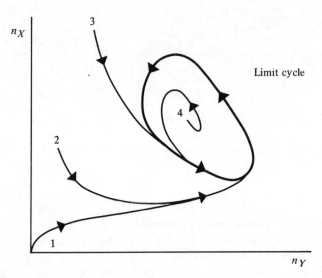

Fig. 5.7. Schematic limit cycle behavior of a system above critical affinity.

D, and E are held fixed. It is also far from equilibrium, incorporates a feedback mechanism, and has the interesting property that below a certain critical affinity for the overall reaction $A + B \rightleftarrows D + E$, the single nontrivial steady state is a point of asymptotic stability. Above the critical affinity, the system is unstable. The system exhibits instability in a manner similar to the Lotka-Volterra model of eqn (5.45). However, in this case, the orbits are not closed but rather all spiral toward a limit cycle. Thus, as $t \to \infty$, the system oscillates along the limit cycle with a frequency independent of the initial conditions.‡

---

‡ It is very suggestive that these investigations of oscillating chemical reactions will ultimately lead to a better understanding of periodic phenomena in biological systems (biological clocks), which are obviously very complex open systems far from equilibrium. This topic is unfortunately outside the scope of this book. At this stage we would like to draw attention to a deep and very stimulating article by Eigen (1971) on the self-organization of matter and the evolution of biological macromolecules.

6
Bénard Convection

6.1. Statement of the Problem

Consider a horizontal fluid layer of thickness d, uniformly heated from below (Fig. 6.1). For small temperature differences $(T_0 - T_1)$ between bottom and top surfaces, a steady state will be observed with a linear temperature gradient across the layer accompanied also by a linear density gradient due to thermal expansion. Bénard (1900) observed in addition that heating the bottom surface of a shallow layer of molten spermaceti sufficiently long with steam at 100 C, keeping the upper surface exposed to air at room temperature, the fluid started to convect in a regular pattern of hexagons tesselating the horizontal plane. The linear dimensions of the hexagons were observed to be about the thickness of the fluid layer. The fluid was rising in the center of each hexagon, flowing radially outward and descending along the sides of the hexagons.

This phenomenon can be readily understood by noting that heating the fluid from below will lead to thermal expansion in the lower fluid layer which, in the gravitational field, will experience a buoyant force and try to rise to the top. This motion is, however, prevented for small temperature gradients by viscous forces. Increasing the temperature at the bottom surface will then worsen this top heavy situation until the buoyant force can overcome viscosity in the fluid and macroscopic convection sets in.

In this chapter we will study the onset of convection in a

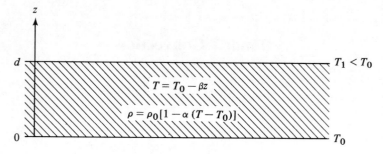

Fig. 6.1. A horizontal fluid layer heated from below.

horizontal fluid layer heated from below. The Bénard problem,‡ as it has become to be called, is an excellent example in hydrodynamics, in which a theory based on the balance equations of Chapter 2 combined with the stability analysis of Chapter 4 leads—even in the linearized version—to extraordinary agreement with superb experimental results.

Let us then recall from Chapter 2 that the dynamics of the fluid layer is controlled by a set of general balance equations. For the Bénard problem of a single component fluid, it will be sufficient to consider the balance equation for the local mass density, eqn (2.6),

$$\frac{\partial \rho}{\partial t} + \boldsymbol{\nabla} \cdot (\rho \mathbf{v}) = 0 \tag{6.1}$$

the momentum balance equation involving the velocity field, eqn (2.13),

$$\frac{\partial(\rho \mathbf{v})}{\partial t} + \boldsymbol{\nabla} \cdot (\mathbf{P} + \rho \mathbf{v}\mathbf{v}) = \rho \mathbf{g} \tag{6.2}$$

‡ The Bénard problem is dealt with extensively by S. Chandrasekhar (1961). More recent reviews have been given by Segel (1966), Koschmieder (1974) and Whitehead (1975). See also *Advances in Chemical Physics* **32** (1975), eds. I. Prigogine and S. A. Rice, and the conference proceedings on *Fluctuations, Instabilities and Phase Transitions*, ed. T. Riste, Plenum, New York, 1975. The stability of a horizontal layer of dielectric fluid under the simultaneous action of a vertical DC electric field and a vertical temperature gradient has been examined by Takashima and Aldridge (1976) following earlier work by Roberts (1969) to explain first experimental results by Gross (1967) and Gross and Porter (1966).

and the balance equation for the internal energy u

$$\frac{\partial(\rho u)}{\partial t}+\boldsymbol{\nabla}\cdot(\rho u\mathbf{v}+\mathbf{j}_q)=-\mathbf{P}:\boldsymbol{\nabla}\mathbf{v} \tag{6.3}$$

Here $\mathbf{g}$ is the (uniform) acceleration in the earth's gravitational field, $\mathbf{j}_q$ is the heat current through the fluid layer, and $\mathbf{P}$ is the fluid's pressure tensor.

It will be sufficient to consider these three equations and no more. We have argued above that the onset of convection is triggered by the thermal expansion in the lower fluid layers. Thus we must consider the continuity equation in which density changes are coupled to the velocity field. Next, we want to find out the convective motion and so must solve the balance equation for the velocity or momentum field. Third, the whole phenomenon is generated by heating the fluid from below, i.e. changing its local internal energy. Thus its balance equation has to be incorporated.

To close this set of three equations, a number of assumptions on constitutive equations have to be made. We assume that the fluid is isotropic and newtonian, i.e., that its pressure tensor is given [see eqn. (2.38)] by $\mathbf{P}=p\mathbf{I}+\mathbf{\Pi}$, where p is the hydrostatic pressure and

$$\Pi_{\alpha\beta}=-\zeta\boldsymbol{\nabla}\cdot\mathbf{v}\delta_{\alpha\beta}-\eta\left(\frac{\partial v_\alpha}{\partial r_\beta}+\frac{\partial v_\beta}{\partial r_\alpha}-\frac{2}{3}\boldsymbol{\nabla}\cdot\mathbf{v}\delta_{\alpha\beta}\right) \tag{6.4}$$

where ζ is the bulk viscosity and η is the shear viscosity. We also restrict ourselves to fluids in which Fourier's law of heat conduction holds:

$$\mathbf{j}_q=-\lambda\boldsymbol{\nabla}T \tag{6.5}$$

Assume that the thermal expansion is linear

$$\rho(T)=\rho_0[1-\alpha(T-T_0)] \tag{6.6}$$

and approximate the internal energy by

$$u=c_p T \tag{6.7}$$

where c_p is the specific heat per unit mass at constant pressure, appropriate for the experimental conditions. The coefficients α, λ, and c_p are taken to be constant.

These assumptions close our three balance equations and simplify them considerably. However, we can go one step further by observing that in most liquids thermal expansion is quite small. We then want to treat ρ as a constant everywhere except in the momentum balance where the buoyant force, resulting from thermal expansion, of course, gives rise to the onset of convection. Taking ρ constant in the continuity equation makes the velocity field divergence-free,‡ that is, $\boldsymbol{\nabla}\cdot\mathbf{v}=0$ which, in turn, simplifies the pressure tensor considerably. With these approximations (Boussinesq, 1911), the balance equations (6.1)–(6.3) read

$$\boldsymbol{\nabla}\cdot\mathbf{v}=0 \tag{6.8}$$

$$\rho_0\left[\frac{\partial \mathbf{v}}{\partial t}+\mathbf{v}\cdot(\boldsymbol{\nabla}\mathbf{v})\right]+\boldsymbol{\nabla}p=(\rho_0+\delta\rho)\mathbf{g}+\eta\nabla^2\mathbf{v} \tag{6.9}$$

$$\frac{\partial T}{\partial t}+\mathbf{v}\cdot\boldsymbol{\nabla}T=\frac{\lambda}{\rho_0 c_p}\nabla^2 T \tag{6.10}$$

where $\rho=\rho_0+\delta\rho$. We also omitted the term $\boldsymbol{\Pi}:\boldsymbol{\nabla}\mathbf{v}$ from the internal energy balance which acts as a source due to the power of compression converting kinetic energy in the fluid into internal energy. In most fluids this term is small compared to the heat conduction term in the internal energy balance, as we will show now by an order of magnitude estimate. Typical velocity gradients to be expected in the fluid layer of thickness d are of the order v/d and temperature gradients are given by $|\boldsymbol{\nabla}T|\sim\beta$ such that $\nabla^2 T\sim\beta/d$. Thus we can estimate

$$\left|\frac{-\boldsymbol{\Pi}:\boldsymbol{\nabla}\mathbf{v}}{\boldsymbol{\nabla}\cdot(\lambda\boldsymbol{\nabla}T)}\right|\approx\frac{2\eta v^2/d^2}{\lambda\beta/d}=\frac{2\eta}{\lambda\beta}\frac{v^2}{d} \tag{6.11}$$

An estimate for a typical velocity we can obtain from the momentum balance by equating

$$\rho_0\mathbf{v}\cdot(\boldsymbol{\nabla}\mathbf{v})\approx\delta\rho\mathbf{g} \tag{6.12}$$

from which we obtain, with the help of eqn (6.6),

$$\frac{v^2}{d}\approx\alpha g\Delta T\approx\alpha g\beta d$$

‡ With this approximation we also suppress sound mode excitations in the fluid, whose amplitude will be small compared to the convective motion to be studied here.

This leads to the estimate

$$\left|\frac{-\mathbf{\Pi}:\mathbf{\nabla v}}{\mathbf{\nabla}\cdot(\lambda\mathbf{\nabla}T)}\right|\approx\frac{2\eta\alpha g}{\lambda}d \qquad (6.13)$$

which for a typical liquid (see Table 6.1) of thickness $d \approx 1$ cm is of the order of 10^{-9} to 10^{-7}, except for liquid He.

We have argued above that the onset of convection in the fluid layer occurs when the buoyant forces arising from thermal expansion excel the viscous forces in the momentum balance. Let us therefore estimate by dimensional analysis the ratio

$$R=\frac{|\mathbf{F}_{\text{buoyant}}|}{|F_{\text{viscous}}|}=\frac{|\delta\rho g/\rho_0|}{|\eta\nabla^2\mathbf{v}/\rho_0|} \qquad (6.14)$$

From the internal energy balance, eqn (6.10), we find that $v \approx \lambda/(\rho_0 c_p d)$. The density change is given by $\delta\rho/\rho_0 \approx \alpha\Delta T$, so that

$$R=\frac{\alpha\beta\rho_0 c_p g}{\nu\lambda}d^4 \qquad (6.15)$$

where $\nu = \eta/\rho_0$ is the kinematic viscosity. This dimensionless combination of fluid parameters is known as Rayleigh's number (Rayleigh, 1916). Convection in the fluid layer will start if the Rayleigh number R, most often controlled experimentally through the inverse temperature gradient β, exceeds some critical value R_c. It will be our aim in the next section to calculate R_c from first principles, i.e. from the Boussinesq eqns (6.8)–(6.10), and compare it with experiments.

Table 6.1. Material constants of some fluids at 25 C

Fluid	Kinematic viscosity ν (centistokes)	Thermal expansion α (K^{-1})	Thermal diffusivity $\kappa = \lambda/(c_p\rho_0)$ (cm^2/sec)	Prandtl number Pr = ν/κ
Air	18.5	$3.37 \cdot 10^{-3}$	0.264	0.71
Water	1.0	$2.13 \cdot 10^{-4}$	$1.43 \cdot 10^{-3}$	6.7
Silicone oil 1	5	$1.05 \cdot 10^{-3}$	$0.88 \cdot 10^{-3}$	57
Silicone oil 2	1000	$0.96 \cdot 10^{-3}$	$1.18 \cdot 10^{-3}$	8500

Note: 1 centistoke = 10^{-2} cm^2/sec.

6.2. Linearized Excess Balance Equations and Boundary Conditions‡

The Boussinesq equations (6.8)–(6.10) obviously have a steady-state solution with zero velocity for any temperature difference ΔT between the top and the bottom of the fluid layer given by

$$T = T_0 - \beta z \tag{6.16}$$

$$\rho(t) = \rho_0(1 + \alpha\beta z) \tag{6.17}$$

$$p = p_0 - g\rho_0(z + \tfrac{1}{2}\alpha\beta^2) \tag{6.18}$$

We know, however, that this solution is experimentally realizable only for 'small' temperature gradients or 'small' Rayleigh numbers. At and above a critical Rayleigh number this solution becomes unstable against small perturbations, the stable solution in this regime having a nonzero velocity field.

Let us then analyze the stability of the above steady state, eqns (6.16)–(6.18), with respect to small perturbations by linearizing the Boussinesq equations in the velocity $\mathbf{v}(\mathbf{r}, t)$, fluctuations δp in the pressure, and a perturbed temperature distribution

$$T' = T_0 - \beta z + \theta(\mathbf{r}, t) \tag{6.19}$$

In keeping with our approximations, we do not consider independent fluctuations in the density $\rho(\mathbf{r}, t)$, but only those that are induced by $\theta(\mathbf{r}, t)$; that is, we write

$$\rho(\mathbf{r}, t) = \rho_0[1 + \alpha(\beta z - \theta(\mathbf{r}, t))] \tag{6.20}$$

The excess balance equations then become

$$\nabla \cdot \mathbf{v} = 0 \tag{6.21}$$

$$\frac{\partial \mathbf{v}}{\partial t} = -\nabla\left(\frac{\delta p}{\rho_0}\right) + \frac{\eta}{\rho_0}\nabla^2 \mathbf{v} + g\alpha\theta\hat{z} \tag{6.22}$$

$$\frac{\partial \theta}{\partial t} = \beta v_z + \frac{\lambda}{\rho_0 c_p}\nabla^2 \theta \tag{6.23}$$

where $\hat{z}$ is a unit vector in the z direction. We can eliminate the pressure fluctuations by taking the curl of eqn (6.22). Calling the vorticity of the velocity field $\boldsymbol{\omega} = \nabla \times \mathbf{v}$, we obtain for its z

‡ In this section we follow closely the treatment given by Chandrasekhar (1961).

component

$$\frac{\partial \omega_z}{\partial t} = \nu \nabla^2 \omega_z \qquad (6.24)$$

and taking the curl twice on the momentum balance eqn (6.22) we find

$$\frac{\partial}{\partial t}(\nabla^2 v_z) = g\alpha\left(\frac{\partial^2 \theta}{\partial x^2} + \frac{\partial^2 \theta}{\partial y^2}\right) + \nu \nabla^4 v_z \qquad (6.25)$$

where $\nu = \eta/\rho_0$ is the kinematic viscosity. Equations. (6.21), (6.23), (6.24), and (6.25) are the required excess balance equations for the fluctuations of the velocity and temperature fields around the steady-state solution (6.16)–(6.18). We are interested in solutions of these equations satisfying a number of boundary conditions. Because we keep the temperature constant at the top and the bottom surface of the fluid layer and do not allow the fluid to move out of this layer, we must demand that

$$\theta = v_z = 0 \qquad \text{at } z = 0 \quad \text{and} \quad z = d \qquad (6.26)$$

The further boundary conditions depend on whether the surface bounding the fluid layer is rigid or free. For a rigid surface we must demand that $v_x = v_y = 0$ and since this is true everywhere in the plane, this implies that $\partial v_x/\partial x = \partial v_x/\partial y = \partial v_y/\partial x = \partial v_y/\partial y = 0$. Thus we conclude that

$$\omega_z = 0 \qquad (6.27)$$

and from $\nabla \cdot \mathbf{v} = 0$ that

$$\frac{\partial v_z}{\partial z} = 0 \qquad (6.28)$$

on a rigid surface. For a free surface, the shear stresses must vanish, i.e.

$$P_{xz} = \eta\left(\frac{\partial v_x}{\partial z} + \frac{\partial v_z}{\partial x}\right) = 0 = P_{yz} = \eta\left(\frac{\partial v_y}{\partial z} + \frac{\partial v_z}{\partial y}\right) \qquad (6.29)$$

But since $v_z = 0$ on the whole surface, its derivatives $\partial v_z/\partial x = \partial v_z/\partial y = 0$ must also be zero here. We conclude from eqn (6.29) that

$$\frac{\partial v_x}{\partial z} = 0 = \frac{\partial v_y}{\partial z} \qquad (6.30)$$

Taking the derivative of $\mathbf{\nabla} \cdot \mathbf{v} = 0$ with respect to z implies that

$$\frac{\partial^2 v_z}{\partial z^2} = 0 \tag{6.31}$$

From the definition of the vorticity $\boldsymbol{\omega}$ we find similarly

$$\frac{\partial \omega_z}{\partial z} = 0 \tag{6.32}$$

on a free surface.

6.3. Normal Mode Analysis in the Linear Theory

The solutions of the excess balance equations (6.21) and (6.23)–(6.25) with appropriate boundary conditions are best analyzed in terms of normal modes. Let us therefore assume that the perturbations are of the form

$$v_z = V(z) \exp[i(k_x x + k_y y) + ft]$$
$$\omega_z = W(z) \exp[i(k_x x + k_y y) + ft] \tag{6.33}$$
$$\theta = \theta(z) \exp[i(k_x x + k_y y) + ft]$$

where the frequency f may be complex but the wavenumbers k_x and k_y are real if we make the plausible assumptions that the x and y dependence is periodic, as suggested by experiment. With this ansatz eqns (6.23)–(6.25) become

$$f\theta(z) = \beta V(z) + \frac{\lambda}{\rho_0 c_p}\left(\frac{d^2}{dz^2} - k^2\right)\theta(z) \tag{6.34}$$

$$fW(z) = \nu\left(\frac{d^2}{dz^2} - k^2\right)W(z) \tag{6.35}$$

$$f\left(\frac{d^2}{dz^2} - k^2\right)V(z) = -g\alpha k^2 \theta(z) + \nu\left(\frac{d^2}{dz^2} - k^2\right)^2 V(z) \tag{6.36}$$

where $k^2 = k_x^2 + k_y^2$. Acceptable solutions of this set of differential equations have to satisfy boundary conditions inferred from eqns (6.26)–(6.32), namely $\theta(0) = \theta(d) = V(0) = V(d) = 0$ and $W = dV/dz = 0$ on a rigid surface or $dW/dz = d^2V/dz^2 = 0$ on a free surface, respectively.

To simplify the notation let us measure all lengths in units of the layer thickness d, times in units of $d^2\rho_0 c_p/\lambda$, and temperature in units of $\beta d/R$, where

$$R = \frac{g\alpha\beta\rho_0 c_p d^4}{\nu\lambda} \tag{6.37}$$

is the Rayleigh number introduced in eqn (6.15). We also define a dimensionless wavenumber $a = kd$ and a dimensionless frequency $\sigma = fd^2\rho_0 c_p/\lambda$. With the notation $D = d/dz$, eqns (6.34)–(6.36) become

$$(D^2 - a^2 - \sigma)\theta(z) = -RV(z) \tag{6.38}$$

$$\sigma W(z) = \Pr(D^2 - a^2)W(z) \tag{6.39}$$

$$[\sigma(D^2 - a^2) - \Pr(D^2 - a^2)^2]V(z) = -\Pr a^2 \theta(z) \tag{6.40}$$

where we also introduced the Prandtl number

$$\Pr = \frac{\nu\rho_0 c_p}{\lambda} \tag{6.41}$$

Eliminating $\theta(z)$ gives us

$$(D^2 - a^2)(D^2 - a^2 - \sigma)\left(D^2 - a^2 - \frac{\sigma}{\Pr}\right)V(z) = -Ra^2 V(z) \tag{6.42}$$

The same equation also holds for $\theta(z)$.

Before we proceed to find solutions of the boundary value problem, eqn (6.42), with appropriate boundary conditions, let us quickly prove that the parameter σ controlling the time evolution of the perturbations, eqns (6.33), is real (Rayleigh, 1916; Pellew and Southwell, 1940).

To this end, we introduce auxiliary functions

$$G(z) = (D^2 - a^2)V(z) \tag{6.43}$$

$$F(z) = \left(D^2 - a^2 - \frac{\sigma}{\Pr}\right)G(z) \tag{6.44}$$

Equation (6.40) now reads

$$F(z) = a^2 \theta(z) \tag{6.45}$$

implying the boundary condition $F(z=0) = F(z=1) = 0$. Next we write eqn (6.42) as

$$(D^2 - a^2 - \sigma)F(z) = -Ra^2 V(z) \tag{6.46}$$

multiply both sides by $F^*(z)$, the complex conjugate of $F(z)$, and integrate over the fluid layer

$$\int_0^1 F^*(z)(D^2 - a^2 - \sigma)F(z)\,dz = -Ra^2 \int_0^1 F^*(z)V(z)\,dz \quad (6.47)$$

We integrate the first term on the left by parts

$$\int_0^1 F^*(z)D^2F(z)\,dz = F^*(z)DF(z)\Big|_0^1 - \int_0^1 |DF|^2\,dz \quad (6.48)$$

and observe that the integrated part is zero due to the boundary conditions on $F^*(z)$. Next, we write the right-hand side of eqn (6.47) as

$$\int_0^1 V(z)F^*(z)\,dz = \int_0^1 V(z)\left(D^2 - a^2 - \frac{\sigma^*}{\text{Pr}}\right)G^*(z)\,dz$$

$$= \int_0^1 G^*(z)\left(D^2 - a^2 - \frac{\sigma^*}{\text{Pr}}\right)V(z)\,dz \quad (6.49)$$

where the last line follows after two partial integrations. Using the definition of $G(z)$ this can also be written as

$$\int_0^1 G^*(z)G(z)\,dz - \frac{\sigma^*}{\text{Pr}} \int_0^1 V(z)G^*(z)\,dz$$

$$= \int_0^1 G^*(z)G(z)\,dz - \frac{\sigma^*}{\text{Pr}} \int_0^1 V(z)(D^2 - a^2)V^*(z)\,dz$$

$$= \int_0^1 G^*(z)G(z)\,dz + \frac{\sigma^*}{\text{Pr}} \int_0^1 (|DV(z)|^2 + a^2|V(z)|^2)\,dz \quad (6.50)$$

Combining eqns (6.48) and (6.50) with (6.47), we obtain

$$-\int_0^1 dz[|DF(z)|^2 + (a^2 + \sigma)|F(z)|^2]$$

$$= -Ra^2 \int_0^1 dz\left\{|G(z)|^2 + \frac{\sigma^*}{\text{Pr}}[|DV(z)|^2 + a^2|V(z)|^2]\right\} \quad (6.51)$$

Taking the imaginary part we find

$$\text{Im}\,(\sigma)\int_0^1 dz\left\{|F(z)|^2 + \frac{R}{\text{Pr}}a^2[|DV(z)|^2 + a^2|V(z)|^2]\right\} = 0 \quad (6.52)$$

Because the integrand is obviously positive definite we can conclude that $\text{Im}(\sigma) = 0$ and σ is real.‡

The fact that σ is real has been referred to by Eddington as the principle of exchange of stabilities, because as the critical Rayleigh number is crossed the system moves from one stable configuration (the fluid at rest) to another stationary state (the convective mode) in an aperiodic (exponential) fashion. This is in contrast to systems with overstability which typically show an oscillatory approach to and damped oscillations around the new stationary state. In Planck's descriptive terminology (Planck, 1896), we would say that the irreversible processes involved in an exchange of stabilities are 'consumptive' whereas those in systems exhibiting overstability are 'conservative' in nature.

With this information let us return to eqns (6.33) and recall that for small Rayleigh numbers we expect the stationary solution, eqns (6.16)–(6.18), to be stable; that is, any fluctuations described by eqns (6.33) should die out as a function of time. Hence $\sigma < 0$ implies stability of the fluid at rest. On the other hand, for large Rayleigh numbers $R > R_c$ the fluid will, at the slightest disturbance (e.g. the very increase to the temperature gradient at that particular R), start to convect. That is, perturbations are amplified from the level of fluctuations to the size of the convective motion. Hence we expect $\sigma > 0$ signalling the instability of the stationary solution against infinitesimal perturbations. The borderline $\sigma = 0$ can then be called the state of marginal stability at which the onset of convection occurs in a careful experiment.

To determine the critical Rayleigh number let us then put $\sigma = 0$ in eqn (6.42). This reduces then to

$$(D^2 - a^2)^3 V(z) = -Ra^2 V(z) \tag{6.53}$$

Observe that the state of marginal stability is independent of the Prandtl number. Together with appropriate boundary conditions [in addition to the ones specified below eqn (6.36)], we also must demand that $(D^2 - a^2)^2 V(z) = 0$ on either surface. This boundary

---

‡ Recall that the Boussinesq approximation suppresses sound waves. If we had kept independent density fluctuations in the continuity equation, eqn (6.1), σ would now be complex, the imaginary part being needed to describe the propagation of sound. However, the amplitude of v in a sound wave would be negligible compared to that of the macroscopic motion of the convecting fluid.

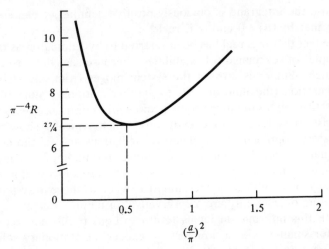

Fig. 6.2. Plot of the initial Rayleigh number, eqn (6.55), versus the wavenumber a, for $n = 1$, under free-free boundary conditions.

value problem can be solved exactly. If the top and bottom surface of the fluid layer are both free,‡ the solution is simply

$$V(z) = V_0 \sin n\pi z \tag{6.54}$$

and the Rayleigh number can only take eigenvalues

$$R = \frac{(n^2\pi^2 + a^2)^3}{a^2} \tag{6.55}$$

where $n = 1, 2, 3, \ldots$. The lowest value of R must have $n = 1$ and is found by minimizing R with respect to a^2 (see Fig. 6.2), yielding the critical Rayleigh number

$$R_c = \frac{27}{4}\pi^4 \approx 657.511 \tag{6.56}$$

at a critical value of the dimensionless wavenumber

$$a_c^2 = \frac{\pi^2}{2} \tag{6.57}$$

---

‡ This case, considered by Rayleigh (1916), is a theorist's dream and cannot easily be subjected to experimental tests because a free fluid layer, like a soap film between a rigid frame, is controlled by its surface tension, which has not been considered in our theory at all. The most convincing experiments have actually been carried out in systems where both surfaces of the fluid layer were rigid. The normal mode analysis for this case is given by Chandrasekhar (1961).

Table 6.2. Critical Rayleigh numbers and wavenumbers for Bénard convection under various boundary conditions

Boundary conditions	R_c	a_c
Free-free	657.511	2.2214
Rigid-free	1100.650	2.6820
Rigid-rigid	1707.762	3.1170

corresponding to a critical wavelength

$$L_c = \frac{2\pi d}{a_c} = 2^{\frac{3}{2}} d \tag{6.58}$$

The critical Rayleigh numbers and wavenumbers for the various boundary conditions are summarized in Table 6.2.

The picture emerging from the normal mode analysis is then as follows. For 'small' Rayleigh numbers, i.e. for $R < R_c$, the fluid layer will establish itself in the stationary state described by eqns (6.16)–(6.18). As R approaches and exceeds R_c as a consequence of raising the temperature at the bottom surface, this state becomes unstable and the system will start to convect. The convection structures will not grow gradually but will immediately assume a finite size a measure of which is $L_c = 2\pi d/a_c$, where d is the thickness of the fluid layer.‡ Having proceeded this far, we now will address four questions. What kind of convection cells do emerge for $R > R_c$? Up to which Rayleigh numbers are these structures stable? What kind of instabilities do we have to expect for larger R still? And what is observed experimentally?

6.4. Convection Cell Patterns from the Linear Theory

The linearized theory presented in the previous section allows us to calculate a variety of convection patterns as we will show presently but, because it is a theory valid at the state of marginal

‡ This abrupt appearance of a finite-size convective structure has been termed a nonequilibrium phase transition. This nomenclature might be useful as a unifying concept but has so far not led to any new insight.

stability only, it cannot be used to predict which patterns will actually occur in experiments. To decide that and to answer the question up to which Rayleigh numbers these patterns are stable we must invoke the full nonlinear theory (still based on the Boussinesq approximation however) of Section 6.6. Armed with a number of stability diagrams, we will then be able to convey in detail what is observed experimentally.

Let us then proceed with the calculation of some convection patterns in the linearized theory. First, we note that after solving the boundary value problem, eqn (6.53), or, in full, eqns (6.38)–(6.40), we know the functions $V(z)$, $\theta(z)$, and $W(z)$ as well as, according to eqns (6.33), the z components of the velocity v_z and the vorticity ω_z, and the deviation $\theta(x, y, z, t)$ from the steady-state temperature field. We still must calculate the x and y components of the velocity field v_x and v_y. Recall that the velocity field $\mathbf{v}$ is assumed to be solenoidal, i.e. $\boldsymbol{\nabla} \cdot \mathbf{v} = 0$ according to eqn (6.21). It therefore can be expressed in terms of two scalar fields $\phi(x, y, z, t)$ and $\psi(x, y, z, t)$ as

$$\mathbf{v} = \boldsymbol{\delta}\phi + \boldsymbol{\varepsilon}\psi \tag{6.59}$$

where the operators $\boldsymbol{\delta}$ and $\boldsymbol{\varepsilon}$ are defined as

$$\boldsymbol{\delta}\phi = \boldsymbol{\nabla} \times (\boldsymbol{\nabla} \times \hat{z}\phi)$$
$$\boldsymbol{\varepsilon}\psi = \boldsymbol{\nabla} \times \hat{z}\psi \tag{6.60}$$

and $\hat{z}$ is a unit vector in the z direction. Equation (6.59) implies, with $\tilde{\phi} = \partial\phi/\partial z$, that

$$v_x = \frac{\partial \tilde{\phi}}{\partial x} + \frac{\partial \psi}{\partial y}$$
$$v_y = \frac{\partial \tilde{\phi}}{\partial x} - \frac{\partial \psi}{\partial y} \tag{6.61}$$

Equation (6.21) can then be written as

$$-\frac{\partial v_z}{\partial z} = \frac{\partial v_x}{\partial x} + \frac{\partial v_y}{\partial y} = \left(\frac{\partial^2}{\partial x^2} + \frac{\partial^2}{\partial y^2}\right)\tilde{\phi} = -a^2 \tilde{\phi} \tag{6.62}$$

The last line follows because according to our ansatz eqns (6.33) all fields are assumed to depend on x and y, such as

$$e^{i(a_x x + a_y y)} \tag{6.63}$$

(Remember that x and y are dimensionless and measured in units of d.) Similarly, we find from the definition of the vorticity $\boldsymbol{\omega} = \boldsymbol{\nabla} \times \mathbf{v}$ that

$$d\omega_z = \frac{\partial v_y}{\partial x} - \frac{\partial v_x}{\partial y} = -a^2 \psi \tag{6.64}$$

Because we know v_z and ω_z explicitly after having solved the boundary value problem, we can now calculate

$$\tilde{\phi} = \frac{1}{a^2} \frac{\partial v_z}{\partial z} \tag{6.65}$$

and

$$\psi = -\frac{d}{a^2} \omega_z \tag{6.66}$$

and have thus determined v_x and v_y according to eqns (6.61). The functions ϕ and ψ, of course, satisfy the membrane equations

$$\left(\frac{\partial^2}{\partial x^2} + \frac{\partial^2}{\partial y^2}\right)\phi + a^2 \phi = 0$$

$$\left(\frac{\partial^2}{\partial x^2} + \frac{\partial^2}{\partial y^2}\right)\psi + a^2 \psi = 0 \tag{6.67}$$

which have to be solved under appropriate boundary conditions to find v_x and v_y.

The simplest convection pattern that can emerge in a fluid layer that extends indefinitely in the x and y directions is a sequence of parallel rolls (parallel to the y axis, say) in which all quantities depend on only one of the two horizontal components, namely x in this case. Intuitive appeal to symmetry would then suggest that identical rolls should repeat themselves in a spatially periodic pattern. We therefore assume that

$$v_z = V(z) \cos \frac{2\pi}{L} x \tag{6.68}$$

Let us stress again that in a linearized theory there is no compelling reason to assume this particular x dependence. Once postulated it fixes via eqns (6.61) and (6.65) the x and y component of the velocity

$$v_x = -\frac{1}{a^2} \frac{2\pi}{L} DV(z) \sin \frac{2\pi}{L} x$$

$$v_y = 0 \tag{6.69}$$

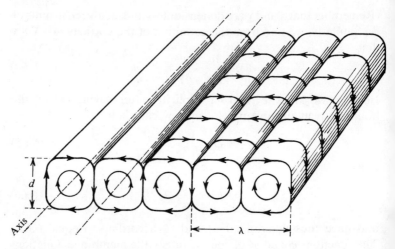

Fig. 6.3. Schematic view of two-dimensional convection rolls for free-free boundary conditions with streamlines in the cross sections calculated according to eqn (6.71).

Thus there is no fluid motion along the rolls. Also $v_x = 0$ for $x = \pm(n+\tfrac{1}{2})L$ and $\pm nL$ for all integer n. The width of the rolls is therefore $L = 2\pi/a$. In addition, neighbouring rolls are separated by a vertical plane of symmetry in which the normal gradient of the vertical velocity vanishes, i.e. $\partial v_z/\partial x = 0$ for $x = \pm(n+\tfrac{1}{2})L$ and $x = \pm nL$. The streamlines (i.e., in an incompressible fluid the trajectories along which some marked fluid elements will move) can be obtained by integrating the equations

$$\frac{dx}{v_x} = \frac{dz}{v_y} \tag{6.70}$$

to yield for free-free boundary conditions

$$\sin ax \sin \pi z = \text{const} \tag{6.71}$$

A schematic view of convection rolls is given in Fig. 6.3.

Apart from these two-dimensional rolls the system could also adjust itself in three-dimensional convection cells. Appeal to symmetry would again demand that there should be a basic unit cell that repeats itself periodically, tesselating the xy plane completely. The horizontal cross section of a unit cell therefore has to be a polygon. If m polygons of n sides meet at any vertex, each polygon sharing an angle $\pi[1-(2/n)]$, complete tesselation of the plane demands that $m\pi[1-(2/n)]=2\pi$. This can be satisfied only for $n = 3, 4, 6$ and $m = 6, 4, 3$, respectively. Thus three-dimensional convection cells must have triangles, rectangles or hexagons as horizontal cross sections. Symmetry would also

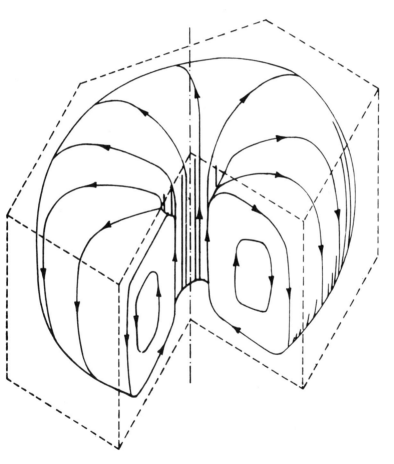

Fig. 6.4. Streamlines in a hexagonal convection cell. After Avsec (1939).

demand that the walls of a unit cell should be vertical planes of symmetry in which the normal gradient of the vertical velocity is again zero.‡ A cross section through a calculated hexagonal cell is shown in Fig. 6.4. Observe that the fluid will rise in the center of the hexagon, flow radially outward, and descend along the walls. Figure 6.5 gives a top view of a hexagonal cell showing lines of constant vertical velocity.

6.5. Experimental Results

Experiments on the Bénard problem have a twofold aim: (1) the precise measurement of the critical Rayleigh number R_c for the onset of convection and for any higher instabilities, and (2) the visual observation of the convection cells as a function of R and as a function of the Prandtl number Pr for $R > R_c$. A typical experimental arrangement is shown in Fig. 6.6 for rigid boundary conditions both at the bottom and at the top. A fluid layer of horizontal dimensions 50×50 cm and between 0.5 and 5 cm thick, floats on a perfectly level metal block (aluminum or copper, level to 0.0001 in 1 in). The high thermal conductivity of the metal (about 10^3 times that of the fluids to be examined) ensures a uniform temperature at the bottom surface which is controlled by an electric heater—a fine mesh of resistance wire attached to the metal block. The fluid layer is topped by a similar metal block that is cooled by passing a coolant through a number of channels. The top and bottom temperatures are measured by thermocouples embedded in the metal blocks. Visual observation of the convection cells is made by either replacing the top metal block by a transparent material or by side-on viewing with a tricky camera design (Krishnamurti, 1968b). In both cases the fluid motion is made visible by suspending small particles in the fluid, e.g. Al flakes in oil or smoke in air.

Let us first review the measurement of the critical Rayleigh number R_c. The electric heater in the bottom metal block will generate heat proportional to the square of the heating current and raise the temperature at the bottom surface. Part of this heat

‡ To find v_x, x_y and v_z explicitly, we must solve the membrane eqns (6.67) under the appropriate boundary conditions at the cell walls. This was done for a hexagonal cell by Christopherson (1940). (See also Chandrasekhar's book.)

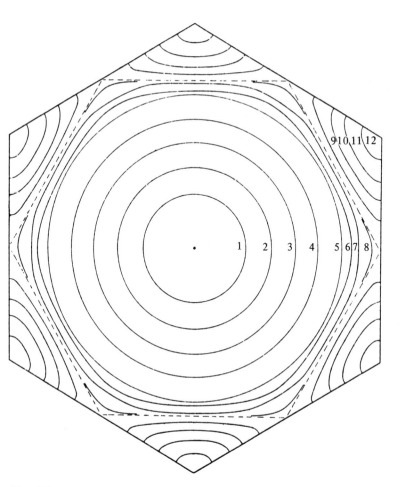

Fig. 6.5. Horizontal cross section through a hexagonal convection cell with lines of constant vertical velocity. The numbers label the upward velocity v_z. On the inscribed hexagon (dashed line) $v_z = -\frac{1}{3}$; After Chandrasekhar (1961).

is transferred through the fluid layer to the top block and carried away by the coolant.

For $R < R_c$ the only heat transfer mechanism through the fluid layer is by conduction according to Fourier's law $\mathbf{j}_q = -\lambda \nabla T$. For $R \geq R_c$ additional heat will be transferred through the fluid by its convective motion. Thus for the same $\mathbf{j}_q$ a smaller temperature difference can be established between the top and bottom of the fluid. If we therefore plot the Nusselt number

$$\text{Nu} = \frac{|\mathbf{j}_q| d}{\lambda \Delta T} \qquad (6.72)$$

which is a measure of the deviation from pure conduction as a function of R, we should find Nu = 1 for $R < R_c$ and a dramatic change to values Nu > 1 for $R \geq R_c$. This was indeed observed by Schmidt and Milverton (1935) and later in high precision measurements by Silveston (1958). The results are summarized for a variety of fluids in Fig. 6.7 and lead to a value $R_c = 1700 \pm 51$ at the onset of convection under rigid-rigid boundary conditions.

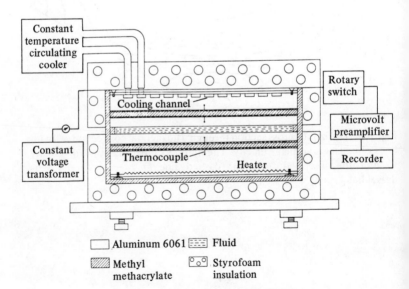

Fig. 6.6. A schematic diagram of the apparatus for studying horizontal convection. After Krishnamurti (1968a).

BÉNARD CONVECTION

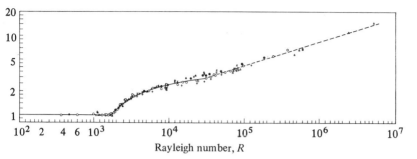

Fig. 6.7. Onset of convection for a number of different fluids under rigid-rigid boundary conditions. After Silveston (1958).

This is in remarkable agreement with the theoretical value $R_c = 1707.162$ of the linearized theory.‡

What kind of convective structures are observed for Rayleigh numbers R slightly above R_c? In almost all experiments with rigid-rigid boundary conditions, one has seen two-dimensional rolls. If the vessel containing the fluid is circular, the rolls will be concentric rings. The number of these rings, of course, has to be commensurate with the radius of the vessel. As R is increased, the size of the rolls increases, leading to a smaller number of rings. (See Fig. 6.8 from Koschmieder, 1966.) Experiments in rectangular vessels show straight rolls with the sides parallel to the shorter side of the rectangle. Straight rolls emerge if the lateral dimensions of the fluid layer are more than about six times its thickness. At this aspect ratio, the critical Rayleigh number in a finite box approaches that of a fluid layer of infinite horizontal extension! For aspect ratios less than one-sixth, the critical Rayleigh number rises sharply (Davis, 1967; Charlson and Sani, 1970). This was confirmed experimentally by Stork and Müller (1972).

Where are Bénard's hexagons then? To make a long story short, Block (1956) showed almost conclusively that the hexagonal cells in Bénard's original experiment and in many that

‡ High precision measurements have also been conducted in liquid He around 3 K by Ahlers (1974, 1975).

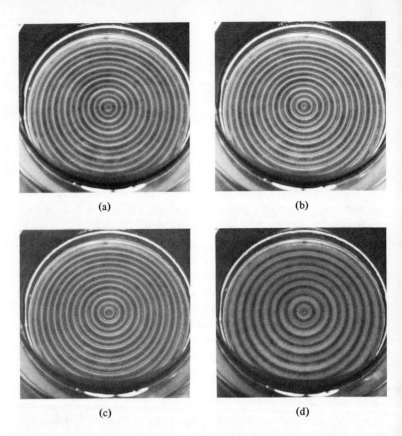

Fig. 6.8. The increase of the wavelength of convective motion with increasing Rayleigh number in a quasi-steady state. Uniformly heated from below, uniformly cooled glass lid. (a) Thirteen concentric rings, just critical. (b) Center ring disappearing, $R = 2.0 R_c$. The fine bright circles are caused by aluminum powder settled at the bottom under the location of ascending motion. (c) Twelfth ring shrinking, $R = 2.88 R_c$. (d) Nine rings left, $R = 7.23 R_c$. After Koschmieder (1974).

followed in the intervening fifty years were *not* caused by the buoyancy instability but are almost certainly due to surface tension at a free upper surface. He removed Bénard cells in shallow layers of hydrocarbons by covering the fluid with silicone monolayers, thus almost removing surface tension. He also observed Bénard cells in layers only 50 μ thick for which R is only a fraction of R_c.‡ Moreover, he observed Bénard cells under rigid-free boundary conditions where the fluid was cooled from below, thus stabilizing the system in a gravitational field. That hexagonal Bénard convection cells can indeed be driven by surface tension alone has been shown by Grodzka and Bannister (1972) in an experiment aboard the Apollo XIV spaceship in essentially zero gravitational field (10^{-6} g).

To understand surface-tension-driven convection, imagine that a spot in the upper (free) surface is locally heated, e.g., through the arrival of a fluid element from below via a velocity fluctuation. In most fluids, surface tension at this hotspot will decrease considerably and neighboring surface elements will pull the fluid away from the hotspot. A depression will appear which will cause fluid from below to rise. Convection has started! Surface-tension-driven convection should therefore show a depression in the free surface above rising fluid as, indeed, was already observed by Bénard. The buoyancy-driven convection with a free upper surface, on the other hand, should show bulging above the ascending fluid (Jeffreys, 1951).§ A beautiful example of hexagonal (and some triangular) convection cells under rigid-free boundary conditions is given in Fig. 6.9.

The question still remains whether in the buoyancy-driven Bénard convection hexagonal cells can be seen under rigid-free or rigid-rigid boundary conditions. This has been finally answered affirmatively in experiments by Sommerscales and Dougherty (1970), in which related work by Silveston (1958), Koschmieder (1966), and Krishnamurti (1968) is also analyzed. The outcome is that in a fluid in which the material properties like viscosity,

---

‡ Bénard himself was not aware that a critical temperature gradient is needed before convection sets in. This insight is due to Rayleigh who set up a theory totally inappropriate for the experiments he wanted to explain and started a whole new field in fluid dynamics.

§ For more details, see the review by Koschmieder (1974).

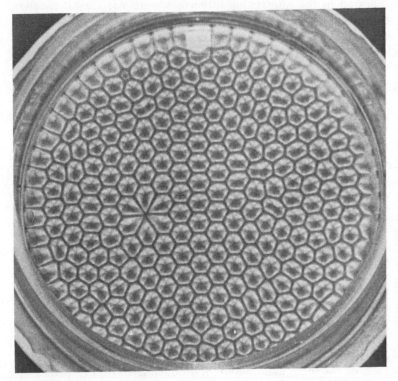

Fig. 6.9. Convection cells in silicone oil under an air surface. Visualization is caused by aluminum powder. The dark lines indicate vertical motion. Bright areas indicate predominantly horizontal motion. After Koschmieder (1974).

density, specific heat, and thermal conductivity vary considerably over the depth of the layer, the additional nonlinearities, i.e. the deviations from the Boussinesq approximation,‡ have a stabilizing effect on the formation of hexagonal convection cells over a small range of Rayleigh numbers. These, however, shrink to zero in the limit of the Boussinesq approximation, i.e. constant fluid properties. Theories to this effect were carried through by Busse (1962), Segel (1966), and Palm, Ellingsen, and Gjerik (1967).

‡ Deviations from the Boussinesq approximation can also be achieved by time-dependent heating (Krishnamurti, 1968b).

The theoretically predicted and experimentally confirmed stability diagram is given in Fig. 6.10. The parameter

$$R_{11} = 15.28 \frac{\nu_H - \nu_C}{\nu_M} + 14.85 \frac{\rho_C - \rho_H}{\rho_M} - 34.55 \frac{c_H - c_C}{c_M} + 16.18 \frac{\lambda_H - \lambda_C}{\lambda_M}$$

is a measure for the overall temperature dependence of the fluid properties. The subscripts H, C, and M refer to the temperature at the hot and cold surfaces and their mean, respectively. We see that a sector in the R vs. $-R_{11}$ plane bounded by R_c and $R^{(1)} \approx R_c + 0.8\, R_{11}^2$ allows stable hexagons. Above this region up to a line $R^{(2)} = R_c + 2.75\, R_{11}^2$, a mixture of hexagons and rolls is possible. From $R^{(2)}$ to the line $R_{11} = 0$, including fluids with temperature-independent properties (apart from thermal expansion of course) only rolls are stable as observed in most experiments. We should also mention here that Graham (1933) observed that gases also exhibit polygonal convection patterns in which the fluid, however, descends in the center of the polygon close to the critical Rayleigh number, in contrast to convection in liquids (with temperature-dependent material properties) where we find ascending motion in the center of the polygons. Graham rightly attributed this to the fact that viscosity in gases generally increases with temperature, whereas in liquids it decreases. This idea was verified by Tippelskirch (1956) using liquid sulfur which has a minimum in viscosity around 153 C.

6.6. Nonlinear Theory

In the discussion of the experiments on Bénard convection, we have at several occasions referred to nonlinear effects. In particular, we have always argued that the stability of the various convection patterns that are feasible in the linearized theory must be analyzed in a nonlinear theory. Let us therefore return to the balance equations (6.1)–(6.3) and rewrite them in dimensionless form, measuring, as before, lengths in units of d, times in units of $d^2 \rho_0 c_p / \lambda$, and temperature in units of $\beta d / R$. We get for the dimensionless velocity field $\mathbf{v}$ and the dimensionless deviation θ

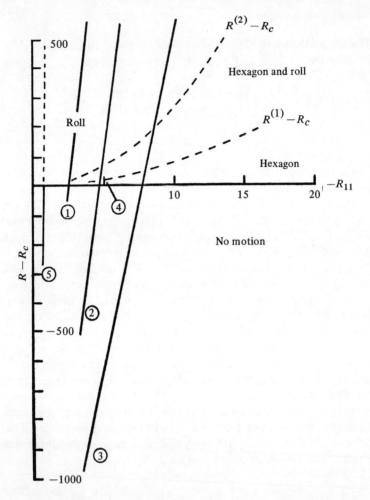

Fig. 6.10. Diagnostic diagram (after Sommerscales and Dougherty, 1970) for Bénard convection in a fluid layer with temperature-dependent properties. Experimental curves 1, 2, and 3 are from Silveston (1958) and curves 4 and 5 are from Koschmieder (1966).

from the static temperature distribution eqn (6.19)

$$\nabla \cdot \mathbf{v} = 0 \tag{6.73}$$

$$\frac{\partial \mathbf{v}}{\partial t} + \mathbf{v} \cdot (\nabla \mathbf{v}) = \mathrm{Pr}\,(\nabla^2 \mathbf{v} + \hat{z}\theta - \nabla \tilde{p}) \tag{6.74}$$

$$\frac{\partial \theta}{\partial t} + \mathbf{v} \cdot \nabla \theta = \nabla^2 \theta + R\hat{\mathbf{z}} \cdot \mathbf{v} \tag{6.75}$$

with the Prandtl number Pr and the Rayleigh number R characterizing the physical properties of the fluid layer. $\hat{z}$ is a unit vector in the vertical z-direction, showing up. Using the representation eqn (6.59) for the solenoidal velocity field, we obtain from eqns (6.74) and (6.75) the following equations for the scalar fields ϕ, ψ and θ:

$$\nabla^4 \Delta_2 \phi - \Delta_2 \theta = \frac{1}{\mathrm{Pr}}\left\{\boldsymbol{\delta}\cdot[(\boldsymbol{\delta}\phi + \boldsymbol{\varepsilon}\psi)\cdot\nabla(\boldsymbol{\delta}\phi + \boldsymbol{\varepsilon}\psi)] + \frac{\partial}{\partial t}\nabla^2\Delta_2\phi\right\} \tag{6.76}$$

$$\nabla^2 \Delta_2 \psi = \frac{1}{\mathrm{Pr}}\left\{\boldsymbol{\varepsilon}\cdot[(\boldsymbol{\delta}\phi + \boldsymbol{\varepsilon}\psi)\cdot\nabla(\boldsymbol{\delta}\phi + \boldsymbol{\varepsilon}\psi)] + \frac{\partial}{\partial t}\Delta_2\psi\right\} \tag{6.77}$$

$$\nabla^2 \theta - R\Delta_2 \phi = [(\boldsymbol{\delta}\phi + \boldsymbol{\varepsilon}\psi)\cdot\nabla]\theta + \frac{\partial \theta}{\partial t} \tag{6.78}$$

where $\Delta_2 = \partial^2/\partial x^2 + \partial^2/\partial y^2$ is the laplacian in two dimensions. The boundary conditions at rigid surfaces are given by $\phi = \partial\phi/\partial z = \psi = \theta = 0$ at $z = \pm\frac{1}{2}$, putting the origin of the coordinate system into the middle of the fluid layer.

In the linearized theory in Sections 6.2 and 6.3, we have found solutions for rigid-rigid boundary conditions with vanishing vertical vorticity, i.e. $\psi = 0$, for $R > R_c = 1707.67$. They also have to be solutions of the nonlinear eqns (6.76)–(6.78), which for $\psi = 0$ reduce to

$$\partial_y(\nabla^4\phi - \theta) = \mathrm{Pr}^{-1}(\partial^2_{yz}\partial^4_{yyzz}\phi - \partial^2_{yy}\phi\partial^4_{yzzz}\phi + \partial^2_{yz}\partial^4_{yyyy}\phi - \partial^2_{yy}\phi\partial^4_{yyyz}\phi) \tag{6.79}$$

$$\nabla^2\theta - R\partial^2_{yy}\phi = \partial^2_{yz}\phi\partial_y\theta - \partial^2_{yy}\phi\partial_z\theta \tag{6.80}$$

where $\partial^i_{x_1,\ldots,x_i}$ indicates the ith partial derivative with respect to the variables in the subscripts. To solve these equations for the case of steady two-dimensional convection rolls (a task that can ultimately only be done numerically), expand θ and ϕ in terms of orthogonal functions,

$$\phi = \sum_{m,n} a_{mn} e^{imay} g_n(z) = \sum_{m,n} a_{mn} \phi_{mn} \tag{6.81}$$

$$\theta = \sum_{m,n} b_{mn} e^{imay} f_n(z) = \sum_{m,n} b_{mn} \theta_{mn} \tag{6.82}$$

where a is the wavenumber characterizing a two-dimensional convection roll. The functions

$$g_n(z) = \frac{\sinh(\beta_{\frac{1}{2}n} z)}{\sinh(\frac{1}{2}\beta_{\frac{1}{2}n})} - \frac{\sin(\beta_{\frac{1}{2}n} z)}{\sin(\frac{1}{2}\beta_{\frac{1}{2}n})} \qquad \text{for } n \text{ even} \tag{6.83}$$

$$= \frac{\cosh(\lambda_{\frac{1}{2}(n+1)} z)}{\cosh(\frac{1}{2}\lambda_{\frac{1}{2}(n+1)})} - \frac{\cos(\lambda_{\frac{1}{2}(n+1)} z)}{\cos(\frac{1}{2}\lambda_{\frac{1}{2}(n+1)})} \qquad \text{for } n \text{ odd} \tag{6.84}$$

$$f_n(z) = \sin[n\pi(z + \tfrac{1}{2})] \tag{6.85}$$

satisfy the boundary condition for ϕ and θ, respectively. The eigenvalues β_n and λ_n are positive roots of

$$\coth \tfrac{1}{2}\beta_n - \cot \tfrac{1}{2}\beta_n = 0$$
$$\tanh \tfrac{1}{2}\lambda_n - \tan \tfrac{1}{2}\lambda_n = 0 \tag{6.86}$$

and can be found in Chandrasekhar's book (1961).

To calculate the expansion coefficients a_{mn} and b_{mn}, insert eqns (6.81) and (6.82) into eqns (6.79) and (6.80), multiply them by ϕ_{mn} and θ_{mn}, respectively, and integrate over the fluid layer to obtain a set of coupled algebraic equations

$$I^{(1)}_{klmn} a_{mn} + I^{(2)}_{klmn} b_{mn} + \text{Pr}^{-1} I^{(3)}_{klmnij} a_{mn} a_{ij} = 0$$
$$I^{(4)}_{klmn} b_{mn} + R I^{(5)}_{klmn} a_{mn} + I^{(6)}_{klmnij} a_{mn} b_{ij} = 0 \tag{6.87}$$

where, for example,

$$I^{(1)}_{klmn} = \left\langle \phi_{kl} \frac{\partial}{\partial y} \nabla^4 \phi_{mn} \right\rangle = \int_0^1 dz \int_{-\infty}^{+\infty} dy\, \phi_{kl} \frac{\partial}{\partial y} \nabla^4 \phi_{mn}$$

$$I^{(2)}_{klmn} = -\langle \phi_{kl} \theta_{mn} \rangle \tag{6.88}$$

The nonlinear terms in eqns (6.87) are quadratic in the expansion coefficients a_{mn} and b_{mn} and contain an odd number of z derivatives. Therefore an independent subset of these equations can be singled out which couples only variables with $(m+n)$ even, containing the two-dimensional convection rolls close to R_c.

To determine the detailed structure of these rolls, eqns (6.87) have to be truncated at sufficiently high indices and solved numerically (see Clever and Busse, 1974). Let us assume this difficult job has been done and let us next formulate the stability problem by superimposing on this solution small disturbances, i.e. by replacing ϕ, ψ, and θ by $\phi + \tilde{\phi}$, $\psi + \tilde{\psi}$, and $\theta + \tilde{\theta}$. Their linearized excess balance equations can be obtained from eqns (6.76)–(6.78) to be

$$\nabla^4 \Delta_2 \tilde{\phi} - \Delta_2 \tilde{\theta} = \mathrm{Pr}^{-1}\Big\{ \boldsymbol{\delta} \cdot [(\boldsymbol{\delta}\tilde{\phi} + \boldsymbol{\varepsilon}\tilde{\psi}) \cdot \boldsymbol{\nabla}(\boldsymbol{\delta}\phi)$$
$$+ (\boldsymbol{\delta}\phi) \cdot \boldsymbol{\nabla}(\boldsymbol{\delta}\tilde{\phi} + \boldsymbol{\varepsilon}\tilde{\psi})] \frac{\partial}{\partial t} \nabla^2 \Delta_2 \tilde{\phi} \Big\}$$

$$\nabla^2 \Delta_2 \tilde{\psi} = \mathrm{Pr}^{-1}\Big\{ \boldsymbol{\varepsilon} \cdot [(\boldsymbol{\delta}\tilde{\phi} + \boldsymbol{\varepsilon}\tilde{\psi}) \cdot \boldsymbol{\nabla}(\boldsymbol{\delta}\phi) + (\boldsymbol{\delta}\phi) \cdot \boldsymbol{\nabla}(\boldsymbol{\delta}\tilde{\phi} + \boldsymbol{\varepsilon}\tilde{\psi})]$$
$$+ \frac{\partial}{\partial t} \Delta_2 \tilde{\psi} \Big\}$$

$$\nabla^2 \tilde{\theta} - R\Delta_2 \tilde{\phi} = (\partial^2_{yz}\tilde{\phi} - \partial_x \tilde{\psi})\partial_y \theta - (\Delta_2 \tilde{\phi})\partial_z \theta$$
$$+ \partial^2_{yz}\phi \partial_y \tilde{\theta} - \partial^2_{yy}\phi \partial_z \tilde{\theta} + \frac{\partial \tilde{\theta}}{\partial t} \quad (6.89)$$

To find solutions of this set of equations, expand again in terms of the orthogonal functions $g_n(z)$ and $f_n(z)$ but this time include explicitly a possible dependence on x, y, and t as we did in our linear normal mode analysis eqns (6.33). Thus we write

$$\tilde{\phi} = \sum_{m,n} \tilde{a}_{mn} e^{imay} g_n(z) \exp[i(dy+bx)+\sigma t]$$
$$\tilde{\psi} = \sum_{m,n} \tilde{c}_{mn} e^{imay} f_n(z) \exp[i(dy+bx)+\sigma t] \quad (6.90)$$
$$\tilde{\theta} = \sum_{m,n} \tilde{b}_{mn} e^{imay} f_n(z) \exp[i(dy+bx)+\sigma t]$$

The excess field $\tilde{\psi}$ has been expanded in terms of $f_n(z)$ because it satisfies the same boundary conditions as $\tilde{\theta}$. Substituting eqns (6.90) into (6.89) and proceeding similarly to what was done to go from eqns (6.79)–(6.80) to (6.87) in the steady-state problem,

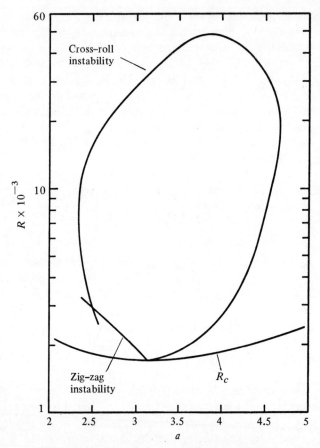

Fig. 6.11. The stability of two-dimensional convection in a water layer (Pr = 7.0). Convection rolls are stable in the closed region. After Clever and Busse (1974).

we arrive at a coupled set of algebraic equations for the expansion coefficients $\tilde{a}_{mn}$, $\tilde{b}_{mn}$ and $\tilde{c}_{mn}$ which, however, now is an eigenvalue problem for σ. To examine the stability of the solution ϕ and θ, we must calculate the Rayleigh number R, as a function of wavenumber a‡ and Prandtl number Pr, for the state of marginal stability at which the real part of the eigenvalue σ with largest real part vanishes. [Recall similar arguments in the linear normal mode analysis eqns (6.52).]

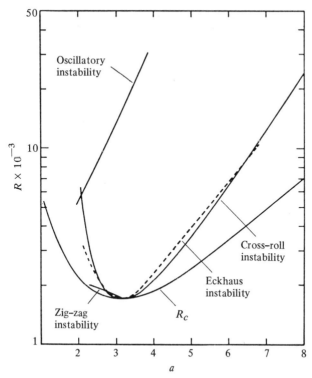

Fig. 6.12. The stability of two-dimensional convection in an air layer (Pr = 0.7). Convection rolls are stable for R below the oscillatory stability boundary and above the Eckhaus or cross-roll boundary. After Clever and Busse (1974).

The results of the numerical analysis of this problem by Clever and Busse (1974) is summarized in Fig. 6.11 for a water layer with Prandtl number 7.0, where the Rayleigh number R is plotted against the wavenumber a. The line labeled R_c is, of

---

‡ The wavenumber a, characteristic of a particular convection pattern, cannot be controlled externally in an experiment. It depends on numerous secondary effects, e.g. on the rate at which the Rayleigh number is increased. (This is the reason why experiments designed to show whether hexagonal convection cells can be stable have to be done over many hours, e.g. 110 hours in a particular experiment reported by Sommerscales and Dougherty, 1970.) Busse and Whitehead (1971) have shown that nonlinear effects from the sidewalls can influence the wavenumber, and Krishnamurti (1970) has even observed hysteresis effects in the R vs. a plane.

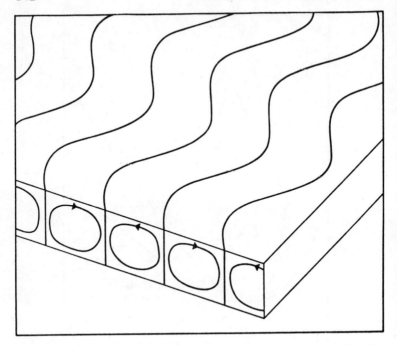

Fig. 6.13. A qualitative sketch of oscillating convection rolls. The bending of these rolls propagates along the roll axis in time. After Clever and Busse (1974).

course, identical to the one in Fig. 6.2 and has a minimum $R_c = 1707.67$ at $a = 3.11$. Two-dimensional convection rolls of the type given in Fig. 6.3 are stable in the closed region.

If we cross the zig-zag instability line from higher to lower a, the system readjusts itself by deforming the boundaries of the, so far, straight convection rolls in zig-zag fashion. This shortens the effective wavelength of the rolls and achieves the same stability as for larger wavenumber and straight rolls.

Traversing the cross-roll instability line for increasing R, cross rolls at right angles to the original straight rolls appear, stabilizing the system. This second transition from two-dimensional convection rolls to a three-dimensional cellular flow pattern has been observed as a break of the heat flux as a function of R by Krishnamurti (1970) at about $R \sim 10 R_c$ to $13 R_c$ for various fluids (water and silicone oil) with Prandtl number between 6.7 and 8500.

For a fluid with a low Prandtl number, e.g. air with $\mathrm{Pr} = 0.71$,

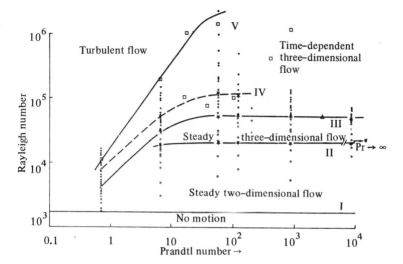

Fig. 6.14. The regime diagram of free convection under rigid-rigid boundary conditions. The data points represent various experiments from which the diagram is reconstructed. After Krishnamurti (1968).

an additional instability of the two-dimensional convection rolls is predicted (labeled oscillatory instability in Fig. 6.12) in which a periodic zig-zag motion of the convection rolls propagates along the roll axis as illustrated in Fig. 6.13. This has been observed experimentally by Busse and Whitehead (1974).

In Fig. 6.14, we finally reproduce the stability diagram for Bénard convection in the R vs. Pr plane. In addition to the convection patterns discussions above, it also shows, for high enough Rayleigh numbers a transition to turbulent flow, i.e. a chaotic space and time-dependent fluid motion with a continuous wavenumber and frequency spectrum rather than a discrete one for the regular and periodic convection flows.‡

‡ There are two conflicting views in the literature about the onset of turbulence. Landau has proposed that the turbulent state of a fluid results from a large number of discrete transitions, each of which causes the velocity field to oscillate with a different frequency f_i until, for sufficiently large i, the motion appears chaotic (see Landau and Lifshitz, 1959). This view has been challenged recently by Ruelle and Takens (1971) who argue that the motion becomes aperiodic with exponentially damped time correlation functions after a few discrete transitions. This second view seems to be confirmed by Bénard convection and also by recent experiments on the onset of turbulence in Couette flow in a rotating fluid. See Gollub and Swinney (1975) and Siggia (1977).

6.7. Variational Principles and Stability Criterion of Glansdorff and Prigogine‡

Since the normal mode approach of Section 6.1 provides an exact determination of the instability point, the Bénard problem is a suitable example to which we can apply the stability theory of Chapter 4. In this section we will summarize the results of applying Glansdorff and Prigogine's stability criterion and will discuss the results and their relationship to the exact solution.

In the spirit of Section 4.3, we require thermodynamic stability in the sense that entropy fluctuations must regress. This can be expressed as a requirement that locally [see eqns (4.49) and (4.50)]

$$\frac{\partial(\delta^2 s)}{\partial t} > 0 \qquad (6.91)$$

However, this criterion does not immediately apply to systems in which there are convective motions, as we mentioned at the end of Section 4.3. In this case, velocity fluctuations must also regress and so we must have

$$\frac{\partial(\delta v)^2}{\partial t} < 0 \qquad (6.92)$$

where δv is the fluctuating fluid velocity. Using the balance equations in the Boussinesq approximation, the criterion (6.30) can be written as ($\langle \cdots \rangle$ implies averaging by integrating over the horizontal plane)

$$\int_0^d \left[\frac{\lambda}{\rho_0 c_v}\langle(\boldsymbol{\nabla}\theta)\cdot(\boldsymbol{\nabla}\theta)\rangle - \beta\langle\theta v_z\rangle\right]dz > 0 \qquad (6.93)$$

This thermodynamic stability criterion implies that instability occurs in the fluid layer at the minimum temperature gradient at which a balance can be steadily maintained between the entropy generated through heat conduction by the temperature fluctuations and the corresponding entropy flow carried away by the

‡ We follow closely the treatment in their book *Thermodynamic Theory of Structure, Stability and Fluctuations*, chapter 11.

velocity fluctuations. Likewise, eqn (6.92) yields in the Boussinesq approximation

$$\int_0^d \left[\nu \left\langle \left(\frac{\partial v_\alpha}{\partial x_\beta}\right)^2 \right\rangle - g\alpha \langle \theta v_z \rangle \right] dz > 0 \qquad (6.94)$$

This hydrodynamic stability criterion was already proposed by Chandrasekhar (1954). It implies that instability occurs at the minimum temperature gradient at which a balance can be steadily maintained between the kinetic energy dissipated by viscosity and the internal energy released by the buoyance force.

The two criteria (6.91) and (6.92) can be combined by defining as a Liapounoff function

$$\zeta^2 \equiv \tau^2 (\delta v)^2 - \varepsilon^2 \delta^2 s \qquad (6.95)$$

where ε^2 and τ^2 are arbitrary positive weighting functions with $\delta \tau^2 = \delta \varepsilon^2 = 0$. The resulting sufficient condition for stability is

$$\frac{\partial \zeta^2}{\partial t} < 0 \qquad (6.96)$$

Combining eqns (6.93) and (6.94) gives the explicit form of criterion (6.96) as

$$\int_0^d \varepsilon^2 \left[\frac{\lambda}{\rho_0 c_v} \langle (\boldsymbol{\nabla} \theta) \cdot (\boldsymbol{\nabla} \theta) \rangle - \beta \langle \theta v_z \rangle \right]$$
$$+ \tau^2 \left[\nu \left\langle \left(\frac{\partial v_\alpha}{\partial x_\beta}\right)^2 \right\rangle - g\alpha \langle \theta v_z \rangle \right] dz > 0 \quad (6.97)$$

For convenience, Glansdorff and Prigogine (1971) choose for the weighting functions

$$\varepsilon^2 = (T^\dagger)^{-1} T_0^2$$
$$\tau^2 = T_0 \qquad (6.98)$$

where

$$T^\dagger = \frac{T_0 \beta}{\alpha g} \qquad (6.99)$$

With this choice, by replacing the inequality in eqn (6.97) by an equality, we obtain the marginal stability condition

$$\int_0^d \left[R^{-1} \frac{(g\alpha)^2 d^4}{\nu} \langle \boldsymbol{\nabla}\theta \cdot \boldsymbol{\nabla}\theta \rangle + \nu \left\langle \left(\frac{\partial v_\alpha}{\partial x_\beta}\right)^2 \right\rangle - 2g\alpha \langle \theta v_z \rangle \right] dz = 0 \tag{6.100}$$

This can be solved for R to give

$$R = \frac{(g\alpha)^2 d^4 \int_0^d \langle \boldsymbol{\nabla}\theta \cdot \boldsymbol{\nabla}\theta \rangle \, dz}{2g\alpha\nu \int_0^d \langle \theta v_z \rangle \, dz - \nu^2 \int_0^d \left\langle \left(\frac{\partial v_\alpha}{\partial x_\beta}\right)^2 \right\rangle dz} \tag{6.101}$$

The lowest value of this ratio for nonzero solutions of the perturbation equations (6.21)–(6.23) represents the critical Rayleigh number characterizing the onset of convection. If we write

$$R = \frac{I_1}{I_2} \tag{6.102}$$

where I_1 and I_2 are taken from eqn (6.101), then, if we vary the functions θ and v_α, the variation in R is

$$\delta R = \frac{1}{I_2} (\delta I_1 - R \delta I_2) \tag{6.103}$$

and so the condition for a variational minimum is

$$\delta I_1 - R \delta I_2 = 0 \tag{6.104}$$

We next choose trial functions in dimensionless form, eqns (6.33),

$$\theta = \alpha_\theta \exp\left[i(a_x x + a_y y)\right] \phi(z)$$

$$\delta v_x = \alpha_1 \exp\left[i(a_x x + a_y y)\right] \frac{d\phi}{dz} \tag{6.105}$$

$$\delta v_y = \alpha_2 \exp\left[i(a_x x + a_y y)\right] \frac{d\phi}{dz}$$

$$\delta v_z = \alpha_3 \exp\left[i(a_x x + a_y y)\right] \phi(z)$$

and let ϕ be a polynomial of the form‡

$$\phi(z) = A_1 z + A_3 z^3 + A_5 z^5 \tag{6.106}$$

‡ Chandrasekhar (1961) chooses for rigid-rigid boundary conditions the more obvious trial functions $\phi = \sum A_m \cos\left[(2m+1)\pi(z - \tfrac{1}{2})\right]$ or $\phi = \sum A_m \sin\left[2m\pi(z - \tfrac{1}{2})\right]$.

Then the boundary conditions, assuming free surfaces for simplicity, are

$$\theta = \delta v_z = 0 \qquad P_{xz} = P_{zx} = 0 \qquad \text{at } z = 0, d \qquad (6.107)$$

which imply that

$$\phi(0) = \phi(d) = \phi''(0) = \phi''(d) = 0 \qquad (6.108)$$

These conditions are just sufficient to determine $\phi(z)$ as

$$\phi(z) = 7\left(\frac{z}{d}\right) - 10\left(\frac{z}{d}\right)^3 + 3\left(\frac{z}{d}\right)^5 \qquad (6.109)$$

With this expression we can perform the integrals over z in eqn (6.101) and then minimize the resulting expression with respect to each of the parameters α_θ, α_1, α_2, and α_3. The consistency requirement for the existence of a solution to the four resulting homogeneous equations, i.e. the vanishing of their determinant, is

$$R = \frac{(A + a^2 C)(4A + B/a^2 + a^2 C + 2D)}{C^2} \qquad (6.110)$$

where $a^2 = a_x^2 + a_y^2$ is the square of the wave vector and where

$$A = \int_0^1 \left[\frac{d\phi\left(\frac{z}{d}\right)}{d\left(\frac{z}{d}\right)}\right] d\left(\frac{z}{d}\right) \approx 27.43$$

$$B = \int_0^1 \frac{d^2\phi\left(\frac{z}{d}\right)}{d\left(\frac{z}{d}\right)^2} d\left(\frac{z}{d}\right) \approx 274.3 \qquad (6.111)$$

$$C = \int_0^1 \phi^2\left(\frac{z}{d}\right) d\left(\frac{z}{d}\right) \approx 2.752$$

$$D = \int_0^1 \phi\left(\frac{z}{d}\right) \frac{d^2\phi\left(\frac{z}{d}\right)}{d\left(\frac{z}{d}\right)^2} d\left(\frac{z}{d}\right) \approx 27.4$$

If we now minimize expression (6.110) with respect to a^2, we find approximate values for a_c^2 and R_c^i, namely

$$a_c^2 \approx 4.99$$

and

$$R_c^i \approx 672.15 \tag{6.112}$$

which compare well to the exact values of $\pi^2/2 \approx 4.935$ and 657 found in the normal mode analysis. We would like to point out that for the Bénard problem the extended variational principle eqn (6.95) of Glansdorff and Prigogine (1971) does not lead to an improvement over the simpler variational principle, eqn (6.94), due to Chandrasekhar (1961). In addition, the good results of the former are tied to the choice, eqns (6.98)–(6.99), of the weighting functions ε^2 and τ^2, for which no a priori reason can be given. It goes without saying that in systems more complicated than the Bénard problem, the availability of several variational principles in stability theory is always welcome.

7
Classical Statistical Mechanics and Kinetic Theory

7.1. Introduction

IN THE first part of this book we have set up and studied in examples the general theory of nonequilibrium thermodynamics. It was conceived as a phenomenological macroscopic field theory concerned with states and processes in systems out of equilibrium. In the remaining chapters we will study and expose the microscopic foundations of nonequilibrium thermodynamics which are obviously to be found in nonequilibrium statistical mechanics. Before we set out to do this, let us mention a few simple ideas connected with the subject.

It is customary to divide physics into two branches: macroscopic physics—phenomenological in nature, such as nonequilibrium thermodynamics—and microscopic or atomic (and subatomic) physics in which we see the basis and explanation of macroscopic physics. Considering, for example, a container filled with a neutral gas, macroscopic physics (nonequilibrium thermodynamics) would describe this system in terms of a few macrovariables like p, T, ρ, and S that are amenable directly to measurements, whereas microscopic physics would treat the same gas as a collection of N interacting molecules. The microscopic dynamics of this system manifests itself in the fact that these molecules can scatter off each other, at the rate of some 10^9 collisions per second for any one of the molecules in a gas at room temperature and atmospheric pressure. Knowing the details of the molecular dynamics, i.e. the two-body scattering cross sections, it would then be our task to derive, say in a dilute gas with frequent collisions but no correlations between particles, the ideal gas law. It was Maxwell and Boltzmann who, in the second half of the 19th century, stressed that such laws of macroscopic physics can be deduced from the microscopic dynamics of the system, if statistical considerations are invoked.

Statistical mechanics is then the theory that calculates macrophysical properties of a system like ρ, p, T, and S as statistical averages over the detailed microscopic motions of its atomic constituents. In doing so, statistical mechanics eliminates most of the microscopic degrees of freedom (some 10^{23} in a mole of gas, for example) introducing, instead, a few relevant macrovariables. The price to be paid in such a procedure is the fact that the statements of thermodynamics are no longer precisely deterministic but only true to the effect that a predicted behavior of a macroscopic object will occur with overwhelming probability. That such probabilities are, however, extremely close to one we know from the fact that (relative) fluctuations in macrovariables—a manifestation of the statistical basis of our theory—turn out to be typically of the order of $N^{-\frac{1}{2}}$, indeed a small number for macroscopic objects.

So far our comments are mainly aimed at systems in equilibrium. Let us now try to delineate the role and importance of statistical mechanics for systems away from equilibrium. To do this let us follow schematically the history of such a system. Let us suppose that a thermodynamic system (a gas, say) has been prepared initially in an equilibrium state characterized by certain macrovariables ρ, p, T, etc. At some time $t=0$, the system is disturbed externally by applying a temperature gradient or any other external field across it. The system will immediately try to respond and adjust to this external stimulus by evolving toward a new macroscopic state that is compatible with this external constraint.

We typically expect the following time evolution, again illustrated in a gas (see Bogolyubov, 1946, 1970). Initially (i.e. in times of the order of the duration of a binary collision in the gas),‡ the system will change very rapidly and a detailed dynamical description of the system is needed. We call this initial time

---

‡ Classically, the time scale is typically given by $\tau_0 = r_0/\bar{v}$, where r_0 is the range of the two-body interaction and $\bar{v}$ is a typical velocity in the gas. Taking helium gas at room temperature as an example, we have $r_0 \approx 3$ Å and $\bar{v} \approx 1.2 \times 10^5$ cm/sec, and τ_0 turns out to be 2.5×10^{-13} sec. If it is necessary to treat the two-body dynamics quantum mechanically, some other time scales may become important

evolution the transient or statistical regime because here we need a full knowledge of the time-dependent particle distribution functions or the time-dependent density matrix.

By the time molecules in the gas have undergone on the average one or two collisions, i.e. in times of the order of the collision time $\tau_{\text{coll}} \approx 10^{-10}$ sec, the system has evolved to such complexity that a kinetic description becomes feasible. In most cases, it is done in terms of a single-particle distribution function. This is the kinetic stage in the time evolution of a system.

Finally, for times $t \gg \tau_{\text{coll}}$, particles in the gas have undergone so many collisions that locally equilibrium may have been established. In this hydrodynamic regime, local macroscopic variables like mass density $\rho(\mathbf{r}, t)$, local temperature $T(\mathbf{r}, t)$, etc. may be defined and nonequilibrium thermodynamics can be invoked to describe the system's further time evolution and, in most cases, its approach to the new stationary, i.e. time-independent, state.

The task of nonequilibrium statistical mechanics is thus threefold. In the transient regime, if it should be of interest, the theory must provide us with a complete description of the system's time evolution. This, of course, amounts to an exact solution of an initial value problem and can only rarely be done, e.g., in exactly soluble models (see Chapter 12). Fortunately, this initial statistical stage is in most cases not amenable to experiments and thus needs no theoretical investigation.‡

For the kinetic regime nonequilibrium statistical mechanics must provide us with rigorous derivations of kinetic equations—rigorous in the sense that it must be possible to assess the validity of the approximations necessary in such derivations and consequently to delimit the range of validity of any kinetic equation proposed. Boltzmann (1872), of course, started this development

‡ An exception would be computer simulation studies of the molecular dynamics of comparatively few (10^2 to 10^3) particles. For a review, see Alder (1973).

---

arising basically from the uncertainty principle. Connected with the thermal energy is a thermal time $\hbar/k_B T$ which at room temperature is 2.5×10^{-14} sec. With the interaction energy V we can associate a dynamic time scale $\hbar/V$ which in He is 6.6×10^{-13} sec if we identify $V = \varepsilon$ from the Lenard-Jones potential (Fig. 7.1). In addition, we have $\lambda_0/\bar{v}$, where $\lambda_0 = \hbar/(4mV)^{\frac{1}{2}}$ is the de Broglie wavelength of V; in He $\lambda_0 \sim 0.5$ Å $\ll r_0$. All these times are less than or of the order of $r_0/\bar{v}$, which can therefore be taken as the relevant microscopic time scale in similar situations.

with his famous equation. Efforts at its generalization continue to the present. Such kinetic equations are also needed to establish criteria for the existence of local equilibrium in a system, to derive, based on it, the balance equations of nonequilibrium thermodynamics and to calculate the phenomenological transport coefficients in the linear constitutive laws.

This complete program of nonequilibrium statistical mechanics has so far only been carried out successfully for dilute gases, in which case the theory is based on the Boltzmann equation.‡ In the general case, the balance equations of nonequilibrium thermodynamics can always be derived assuming (but not assessing) local equilibrium, which as we have seen in Chapter 1 lies at the heart of nonequilibrium thermodynamics. We should also mention that general expressions for transport coefficients can be derived in linear response theory, but in most cases are not evaluated without recourse to kinetic equations (see Chapter 9). Attempts at generalized kinetic equations will be discussed in Section 7.6 and Chapter 10.

Let us now start to set out the general principles of nonequilibrium statistical mechanics. The starting point is the set of equations of motion for the N objects (atoms, molecules, spins, etc.) that make up the macroscopic system under study. Although all mechanical systems obey the laws of quantum mechanics, it is still worthwhile to study the general statistical mechanical problem by purely classical methods. One reason is that in many physical systems quantum effects are negligible. The study of dilute systems, high-temperature systems, and much of macroscopic hydrodynamics and continuum mechanics are essentially classical problems. Also, many formal methods of solution and some exact results are easily carried over to quantum mechanical problems. Historically, some of these methods were actually derived well after the establishment of quantum physics. In addition, many general arguments concerning irreversibility itself apply equally well to the classical and quantum cases, although they were formulated for the classical case. In later chapters we will also discuss nonequilibrium statistical mechanics based on quantum mechanics, a necessary procedure when details of the interaction and the internal structure of the microscopic objects become

‡ Many ideas and arguments in this chapter can be traced to Green's (1952) classic book, *The Molecular Theory of Fluids*.

important, as is the case in most solids and in certain liquids at lower temperature.

Let us then look first at a classical N-body system and define an ensemble probability function $\rho(\xi_1, \xi_2, \ldots, \xi_N, t)$ where $\xi_i = (\mathbf{r}_i, \mathbf{p}_i)$ are the six-dimensional vectors consisting of the space coordinates $\mathbf{r}_i$ and the momentum $\mathbf{p}_i$ of the ith particle such that $\rho(\xi_1, \ldots, \xi_N, t) d^6\xi_1, \ldots, d^6\xi_N$ is the fraction of the ensemble of replicates of the system to be found at time t in a volume element about the point $(\xi_1, \xi_2, \ldots, \xi_N)$ in the $6N$-dimensional Γ space of the system. The equation of motion for ρ can be derived from Hamilton's dynamic equations. For a closed holonomic, classical mechanical system, the latter are

$$\dot{\mathbf{r}}_i = \nabla_{p_i} H_N$$

and

$$\dot{\mathbf{p}}_i = -\nabla_{r_i} H_N \quad \text{for } i = 1, 2, \ldots, N \tag{7.1}$$

where $\nabla_{r_i} = \partial/\partial \mathbf{r}_i$ and $\nabla_{p_i} = \partial/\partial \mathbf{p}_i$ are the gradients with respect to the position $\mathbf{r}_i$ and momentum $\mathbf{p}_i$ of the ith particle, H_N is the hamiltonian for the system, and the dot denotes the time derivative. Since members of the ensemble described by the probability density ρ are neither created nor destroyed, the density ρ obeys a continuity equation

$$\frac{\partial \rho}{\partial t} + \nabla_X \cdot (\rho \dot{\mathbf{X}}) = 0 \tag{7.2}$$

where $(\nabla_X \cdot)$ represents the $6N$-dimensional divergence and $\mathbf{X}$ represents the $6N$ coordinates and momenta $(\mathbf{r}_1, \mathbf{p}_1, \mathbf{r}_2, \mathbf{p}_2, \ldots, \mathbf{r}_N, \mathbf{p}_N)$. Writing eqn (7.2) as

$$\frac{\partial \rho}{\partial t} - \dot{\mathbf{X}} \cdot (\nabla_X \rho) + \rho (\nabla_X \cdot \dot{\mathbf{X}}) = 0 \tag{7.3}$$

and using Hamilton's equations (7.1) gives us

$$\nabla_X \cdot \dot{\mathbf{X}} = \sum_{i=1}^{N} \left(\frac{\partial \dot{\mathbf{r}}_i}{\partial \mathbf{r}_i} + \frac{\partial \dot{\mathbf{p}}_i}{\partial \mathbf{p}_i} \right)$$

$$= \sum_{i=1}^{N} \left(\frac{\partial^2 H_N}{\partial \mathbf{r}_i \, \partial \mathbf{p}_i} - \frac{\partial^2 H_N}{\partial \mathbf{p}_i \, \partial \mathbf{r}_i} \right) = 0 \tag{7.4}$$

Therefore, eqn (7.3) becomes

$$\frac{\partial \rho}{\partial t} + \sum_{i=1}^{N} [\dot{\mathbf{r}}_i \cdot (\nabla_{r_i} \rho) + \dot{\mathbf{p}}_i \cdot (\nabla_{p_i} \rho)] = 0 \tag{7.5}$$

Using Hamilton's equations once again leads to Liouville's equation which we write in standard form as

$$\frac{\partial \rho}{\partial t} = \{H_N, \rho\} \tag{7.6}$$

where $\{A, B\}$ is the Poisson bracket of A and B, defined by

$$\{A, B\} = \sum_{i=1}^{N} [(\boldsymbol{\nabla}_{r_i} A) \cdot (\boldsymbol{\nabla}_{p_i} B) - (\boldsymbol{\nabla}_{p_i} A) \cdot (\boldsymbol{\nabla}_{r_i} B)] \tag{7.7}$$

The Liouville equation (7.6) still contains all the microscopic information about the system. To use it as a basis for a statistical theory of the behavior of large systems, we must find ways to extract from this differential equation in $(6N+1)$ variables a description involving only a few macrovariables. Such a procedure, by necessity, will involve a number of additional assumptions and approximations which imply a loss of some dynamical information about the system in order to make an approximate determination of the quantities of interest. A systematic scheme for doing this is embodied in the BBGKY hierarchy of equations for reduced distribution functions introduced independently by Bogolyubov (1946), Born and Green (1946), Kirkwood (1946), and Yvon (1937).‡ As this chapter is fairly long it might be useful to briefly outline its contents. In the next section we derive from the Liouville equation the BBGKY hierarchy for reduced particle distribution functions. In the following section this is used for an exact microscopic derivation of the balance equations of nonequilibrium thermodynamics. This set of (now) macroscopic equations is then closed in Section 7.4 by deriving the necessary number of linear constitutive laws. It will also be pointed out that the so derived formal expressions for linear transport coefficients can only be evaluated if suitable kinetic equations can be set up for the particular system under study. This will be done in Section 7.5 in which we derive the Vlasov

---

‡ A second possibility to extract information about a few macrovariables from Liouville's equation is available if the initial conditions imposed on the system are 'simple.' e.g. that the system be in equilibrium until an external field is switched on. This may simplify the dynamics of the system to the extent that we can solve Liouville's equation either exactly or approximately. Linear response theory does the latter for 'weak' external fields. Exactly soluble models, e.g. those presented in Chapter 12, are in the former category.

equation for a rarefied plasma and the Boltzmann equation for a dilute gas. The difficulties with kinetic equations for dense gases and liquids are pointed out. The last two sections are concerned with a derivation of balance equations and constitutive laws from the Boltzmann equation. The Chapman-Enskog solutions will be constructed to calculate transport coefficients and to discuss the validity of the local equilibrium assumption in a dilute gas.

7.2. The BBGKY Hierarchy

In this section we first want to define reduced distribution functions and then derive the BBGKY hierarchy of equations for them. To do this we introduce the full N-particle distribution function $f_N(\xi_1, \ldots, \xi_N, t)$ such that

$$V^{-N} f_N(\xi_1, \ldots, \xi_N, t) d^6\xi_1, \ldots, d^6\xi_N$$

gives the probability of finding the system in the volume element $d\xi_1, \ldots, d\xi_N$ in the 6N-dimensional phase space or Γ space spanned by $(\xi_1, \ldots, \xi_N)$. Obviously $f_N(\xi_1, \ldots, \xi_N, t)$ is proportional to the ensemble density ρ but normalized so that

$$\int_\Gamma f_N(\xi_1, \ldots, \xi_N, t) \, d^6\xi_1, \ldots, d^6\xi_N = V^N \qquad (7.8)$$

where V is the volume occupied by our system and where $\xi_i = (\mathbf{r}_i, \mathbf{p}_i)$ with the integral extending over all Γ space. Next we define the reduced l-particle distribution function by

$$f_l(\xi_1, \ldots, \xi_l, t) = V^{(l-N)} \int f_N(\xi_1, \ldots, \xi_N, t) \, d^6\xi_{l+1}, \ldots, d^6\xi_N$$

$$(7.9)$$

where again the integrations are over the entire range of ξ_i. Since the particles are identical, f_N, and hence the reduced distribution function f_l are symmetric functions of the ξ_i's, for any choice of l particles. Thus, the expression

$$V^{-l} f_l(\xi_1, \ldots, \xi_l, t) \, d^6\xi_1, \ldots, d^6\xi_l \qquad (7.10)$$

gives the probability that any l particles are located in the volumes $d^6\xi_1, d^6\xi_2, \ldots, d^6\xi_l$ about the points $\xi_1, \xi_2, \ldots, \xi_l$ at time t.‡

To derive the BBGKY hierarchy, consider a system of N identical particles in a volume V, influenced by an external potential $\Phi(\mathbf{r}_i)$ and interacting via a two-body potential $V_{ij} = V(\mathbf{r}_i, \mathbf{r}_j)$. The hamiltonian of this system is

$$H_N = \sum_{i=1}^{N} \left[\frac{p_i^2}{2m} + \Phi(\mathbf{r}_i)\right] + \sum_{1 \leq i < j \leq N} V_{ij} \qquad (7.11)$$

Note that since f_N is normalized, it must vanish rapidly at infinity, i.e.

$$\int \left\{\left[\frac{p_i^2}{2m} + \Phi(\mathbf{r}_i)\right], f_N\right\} d^6\xi_i$$

$$= \int \left(\boldsymbol{\nabla}_{r_i}\Phi \cdot \boldsymbol{\nabla}_{p_i} f_N - \frac{\mathbf{p}_i}{m} \boldsymbol{\nabla}_{r_i} f_N\right) d^3r_i \, d^3p_i$$

$$= \int (\boldsymbol{\nabla}_{r_i}\Phi) f_N \bigg|_{\mathbf{p}_i=-\infty}^{\mathbf{p}_i=\infty} d^3r_i - \int \frac{\mathbf{p}_i}{m} f_N \bigg|_{\mathbf{r}_i=-\infty}^{\mathbf{r}_i=\infty} d^3p_i = 0 \qquad (7.12)$$

Similarly, we could show that

$$\int \{V_{ij}, f_N\} d^6\xi_i \, d^6\xi_j = 0 \qquad (7.13)$$

Thus, integrating Liouville's equation (7.6) over $(N-l)$ coordi-

---

‡ The one-particle distribution function $f_1(\mathbf{r}, \mathbf{p}, t)$, and higher distribution functions, can also be introduced without reference to an ensemble (Grad, 1958; Lanford, 1975). We partition the single-particle phase space γ into cells i of size $\Delta_i = \Delta\mathbf{r}\Delta\mathbf{p}$ around a point $(\mathbf{r}, \mathbf{p})$ and define a distribution function $\tilde{f}_1(\mathbf{r}, \mathbf{p}, t)$ such that the occupation number in the ith cell is given by

$$n_i = \int_{\Delta_i} \tilde{f}(\mathbf{r}, \mathbf{p}, t) \, d\mathbf{r} \, d\mathbf{p}$$

If the cells Δ_i are small enough, $\tilde{f}_1(\mathbf{r}, \mathbf{p}, t)$ will be a smooth function over several cells. In this approach $\tilde{f}_1$ is a measure of the real density of particles at point $(\mathbf{r}, \mathbf{p})$ in γ space, whereas in the ensemble approach in the text, f_1 refers to the probability of finding a certain number of systems at point $(\mathbf{r}, \mathbf{p})$ in repeated experiments. Although f_1 and $\tilde{f}_1$ have a different physical meaning, they (and their multiparticle equivalents) obey the same set of equations, namely the BBGKY hierarchy.

nates gives us, for $l = 1, \ldots, N$,

$$\begin{aligned}
\frac{\partial f_l}{\partial t} &= V^{l-N} \frac{\partial}{\partial t} \int f_N(\xi_1, \ldots, \xi_N, t) \, d^6\xi_{l+1}, \ldots, d^6\xi_N \\
&= V^{l-N} \int \left\{ \sum_{i=1}^{N} \left[\frac{p_i^2}{2m} + \Phi(\mathbf{r}_i) \right], f_N \right\} d^6\xi_{l+1}, \ldots, d^6\xi_N \\
&\quad + V^{l-N} \int \left\{ \sum_{1 \leq i < j \leq N} V_{ij}, f_N \right\} d^6\xi_{l+1}, \ldots, d^6\xi_N \\
&= V^{l-N} \int \left\{ \sum_{i=1}^{l} \left[\frac{p_i^2}{2m} + \Phi(\mathbf{r}_i) \right], f_N \right\} d^6\xi_{l+1}, \ldots, d^6\xi_N \\
&\quad + V^{l-N} \int \left\{ \sum_{1 \leq i < j \leq l} V_{ij}, f_N \right\} d^6\xi_{l+1}, \ldots, d^6\xi_N \\
&\quad + V^{l-N} \int \sum_{i=1}^{l} \sum_{j=l+1}^{N} (\boldsymbol{\nabla}_{r_i} V_{ij})(\boldsymbol{\nabla}_{p_i} f_N) \, d^6\xi_{l+1}, \ldots, d^6\xi_N \\
&= \left\{ \left[\sum_{i=1}^{l} \left(\frac{p_i^2}{2m} + \Phi(\mathbf{r}_i) \right) + \sum_{1 \leq i < j \leq l} V_{ij} \right], f_l \right\} \\
&\quad + (N-l) V^{l-N} \int \sum_{i=1}^{l} (\boldsymbol{\nabla}_{r_i} V_{i,l+1})(\boldsymbol{\nabla}_{p_i} f_N) \, d^6\xi_{l+1}, \ldots, d^6\xi_N
\end{aligned}$$

(7.14)

where the sum over j in the last term was carried out using the fact that f_N is symmetric in the $\boldsymbol{\xi}_i$. This can be rewritten as

$$\frac{\partial f_l}{\partial t} = \{H_l, f_l\} + \frac{(N-1)}{V} \int \sum_{i=1}^{l} (\boldsymbol{\nabla}_{r_i} V_{i,l+1}) \cdot (\boldsymbol{\nabla}_{p_i} f_{l+1}) \, d^6\xi_{l+1} \quad (7.15)$$

where

$$H_l = \sum_{i=1}^{l} \left[\frac{p_i^2}{2m} + \Phi(\mathbf{r}_i) \right] + \sum_{1 \leq i < j \leq l} V_{ij} \quad (7.16)$$

is the hamiltonian for an l-particle system with the same dynamics as eqn (7.11).

This set of coupled equations is known as the BBGKY hierarchy, and is exactly equivalent to Liouville's equation. These equations, however, are still too general in this form, since in order to calculate the single-particle distribution function f_1, we must know f_2 which, in turn, is coupled implicitly to all higher correlation functions. However, in many physical situations it is possible to ignore or approximate the second or some higher correlation function. This, in effect, truncates the BBGKY hierarchy and results in a closed and finite set of coupled equations. Examples of such kinetic equations are Vlasov's and Boltzmann's equations. Before we proceed to their derivation in Section 7.5, we want to present in the next section the derivation of the mechanical balance equation of nonequilibrium thermodynamics from the first two members of the BBGKY hierarchy.

7.3. Microscopic Derivation of Balance Equations

We now want to use the BBGKY hierarchy (7.15) to derive the macroscopic balance equations of nonequilibrium thermodynamics introduced phenomenologically in Chapter 2. To do so, we must first express the relevant macroscopic densities in terms of the microscopic distribution functions. We define the macroscopic mass density

$$\rho(\mathbf{r}, t) = m\left(\frac{N}{V}\right) \int f_1(\mathbf{r}, \mathbf{p}, t) \, d^3p \tag{7.17}$$

the fluid velocity

$$\mathbf{v}(\mathbf{r}, t) = \frac{1}{\rho}\left(\frac{N}{V}\right) \int \mathbf{p} f_1(\mathbf{r}, \mathbf{p}, t) \, d^3p \tag{7.18}$$

the thermal or internal kinetic energy density

$$\varepsilon^{(K)}(\mathbf{r}, t) = \frac{1}{\rho}\left(\frac{N}{V}\right) \int \left(\frac{p^2}{2m} - \frac{1}{2} mv^2\right) f_1(\mathbf{r}, \mathbf{p}, t) \, d^3p$$

$$= \frac{1}{\rho}\left(\frac{N}{V}\right) \int \frac{1}{2m} (\mathbf{p} - m\mathbf{v})^2 f_1(\mathbf{r}, \mathbf{p}, t) \, d^3p \tag{7.19}$$

and the mean interparticle potential energy density

$$\varepsilon^{(V)}(\mathbf{r}, t) = \frac{1}{2}\frac{1}{\rho}\left(\frac{N}{V}\right) \int V(\mathbf{r}, \mathbf{r}_1) f_2(\mathbf{r}, \mathbf{p}, \mathbf{r}_1, \mathbf{p}_1, t) \, d^3p \, d^3r_1 \, d^3p_1 \tag{7.20}$$

in agreement with the intuitive microscopic interpretation of these quantities. As in Chapter 2, the energy densities above are specific densities (per unit mass). A similar definition can be given for the angular momentum density which we, however, do not want to consider here. The above are all the mechanical quantities for which we postulated phenomenological balance equations in Chapter 2. Note also that mass, momentum, energy, and angular momentum are those mechanical quantities for which we can, in an isolated system, derive microscopic conservation laws as a consequence of isotropy and homogeneity of space and time.

The task of deriving balance equation for the macroscopic quantities (7.17)–(7.20) is considerably simplified by the fact that their definitions only involve the single- and two-particle distribution functions. It therefore suffices to consider the first two members of the BBGKY hierarchy, the first one being [setting $(N-1)/V \approx N/V$]‡

$$\left[\frac{\partial}{\partial t}+\frac{\mathbf{p}}{m}\cdot\mathbf{\nabla}_r + m\mathbf{F}(\mathbf{r})\cdot\mathbf{\nabla}_p\right]f_1(\mathbf{r},\mathbf{p},t)$$
$$=\left(\frac{N}{V}\right)\int[\mathbf{\nabla}_r V(\mathbf{r},\mathbf{r}_1)]\cdot\mathbf{\nabla}_p f_2(\mathbf{r},\mathbf{p},\mathbf{r}_1,\mathbf{p}_1,t)\,d^3r_1\,d^3p_1 \quad (7.21)$$

where the negative gradient of the external potential $-\mathbf{\nabla}_r\Phi(\mathbf{r})$ has been replaced by the force density $\mathbf{F}(\mathbf{r})$ per unit mass to conform to the notation of Chapter 2.

7.3.1. Mass Balance

To get a balance equation for the mass density $\rho(\mathbf{r},t)$, eqn (7.17), we simply integrate eqn (7.21) with respect to $\mathbf{p}$ giving

$$\int\frac{\partial}{\partial t}f_1\,d^3p + \int\frac{\mathbf{p}}{m}\cdot\mathbf{\nabla}_r f_1\,d^3p + m\int\mathbf{F}(\mathbf{r})\cdot(\mathbf{\nabla}_p f_1)\,d^3p$$
$$-\left(\frac{N}{V}\right)\int(\mathbf{\nabla}_r V(\mathbf{r},\mathbf{r}_1))\cdot(\mathbf{\nabla}_p f_2)\,d^3r_1\,d^3p_1\,d^3p = 0 \quad (7.22)$$

which may also be written, after multiplication with $m(N/V)$ as

$$\frac{\partial}{\partial t}\left[m\left(\frac{N}{V}\right)\int f_1\,d^3p\right]$$
$$+\mathbf{\nabla}_r\left[m\left(\frac{N}{V}\right)\int\frac{\mathbf{p}}{m}f_1\,d^3p\right] + m^2\left(\frac{N}{V}\right)\int\mathbf{\nabla}_p\cdot(\mathbf{F}f_1)\,d^3p$$
$$-m\left(\frac{N}{V}\right)^2\int\mathbf{\nabla}_p[(\mathbf{\nabla}_r V)f_2]\,d^3p\,d^3r_1\,d^3p_1 = 0 \quad (7.23)$$

‡ Note that the difference between $(N-1)/V$ and N/V is, however, significant in some calculations, e.g., for isothermal compressibility and density fluctuations. See Baxter (1971).

Using the divergence theorem and the asymptotic behavior of f_1 and f_2 as $\mathbf{p} \to \infty$ [see eqn (7.12)], we can see that the last two integrals are zero. We then obtain, using definitions (7.17) and (7.18),

$$\frac{\partial \rho}{\partial t} + \nabla \cdot (\rho \mathbf{v}) = 0 \tag{7.24}$$

which is simply the continuity equation for a macroscopic medium, eqn (2.6).

7.3.2. Momentum Balance

To obtain the momentum balance equation, we multiply eqn (7.21) by $(N/V)\mathbf{p}$ and then integrate with respect to $\mathbf{p}$. This yields

$$\frac{\partial}{\partial t}\left[\left(\frac{N}{V}\right)\int f_1 \mathbf{p}\, d^3p\right] + \nabla_r \cdot \left[\left(\frac{N}{V}\right)\int \frac{\mathbf{pp}}{m} f_1\, d^3p\right]$$

$$+ m\left(\frac{N}{V}\right)\int \nabla_p \cdot (\mathbf{p}\mathbf{F}f_1)\, d^3p - m\left(\frac{N}{V}\right)\int f_1 \mathbf{F} \cdot (\nabla_p \mathbf{p})\, d^3p$$

$$- \left(\frac{N}{V}\right)^2 \int \mathbf{p}[\nabla_r V(\mathbf{r}, \mathbf{r}_1)] \cdot (\nabla_p f_2)\, d^3r_1\, d^3p_1\, d^3p = 0 \tag{7.25}$$

which can be written as

$$\frac{\partial(\rho\mathbf{v})}{\partial t} + \nabla_r \cdot \left[\left(\frac{N}{V}\right)\int \frac{\mathbf{pp}}{m} f_1\, d^3p\right]$$

$$- \left(\frac{N}{V}\right)^2 \int \mathbf{p}(\nabla_r V) \cdot (\nabla_p f_2)\, d^3r_1\, d^3p_1\, d^3p = \rho\mathbf{F} \tag{7.26}$$

where the divergence theorem was used again to show that

$$\int \nabla_p \cdot (\mathbf{p}\mathbf{F}f_1)\, d^3p = 0 \tag{7.27}$$

Next, we multiply eqn (7.24) by $\mathbf{v}$ to obtain

$$\mathbf{v}\frac{\partial \rho}{\partial t} + \nabla_r \cdot (\rho\mathbf{v}\mathbf{v}) - \rho(\mathbf{v} \cdot \nabla_r)\mathbf{v} = 0 \tag{7.28}$$

Subtracting this from eqn (7.26) gives us

$$\rho\left[\frac{\partial \mathbf{v}}{\partial t} + (\mathbf{v} \cdot \nabla_r)\mathbf{v}\right] + \nabla_r \cdot \left[\left(\frac{N}{V}\right)\int \left(\frac{\mathbf{pp}}{m} - m\mathbf{v}\mathbf{v}\right) f_1(\mathbf{r}, \mathbf{p}, t)\, d^3p\right]$$

$$+ \left(\frac{N}{V}\right)^2 \int \nabla_r V(\mathbf{r}, \mathbf{r}_1) f_2(\mathbf{r}, \mathbf{p}, \mathbf{r}_1, \mathbf{p}_1, t)\, d^3r_1\, d^3p_1\, d^3p = \rho\mathbf{F} \tag{7.29}$$

We can rewrite the second term above as

$$\nabla_r \cdot \left[\left(\frac{N}{V}\right) \int \left(\frac{\mathbf{pp}}{m} - m\mathbf{vv}\right) f_1(\mathbf{r}, \mathbf{p}, t) \, d^3p\right]$$

$$= \nabla_r \cdot \left(\frac{N}{V}\right) \int f_1(\mathbf{r}, \mathbf{p}, t) \frac{(\mathbf{p} - m\mathbf{v})(\mathbf{p} - m\mathbf{v})}{m} \, d^3p \quad (7.30)$$

Thus, defining a kinetic pressure tensor $\mathbf{P}^{(K)}$ by

$$\mathbf{P}^{(K)}(\mathbf{r}, t) = \left(\frac{N}{V}\right) \int f_1(\mathbf{r}, \mathbf{p}, t) \frac{(\mathbf{p} - m\mathbf{v})(\mathbf{p} - m\mathbf{v})}{m} \, d^3p \quad (7.31)$$

and also introducing the hydrodynamic derivative $D/Dt = \partial/\partial t + \mathbf{v} \cdot \nabla$, eqn (2.8), we obtain from eqn (7.29)

$$\rho \frac{D\mathbf{v}}{Dt} = \rho \mathbf{F} - \nabla_r \cdot \mathbf{P}^{(K)} - \left(\frac{N}{V}\right)^2 \int (\nabla_r V) f_2(\mathbf{r}, \mathbf{p}, \mathbf{r}_1, \mathbf{p}_1, t) \, d^3r_1 \, d^3p_1 \, d^3p \quad (7.32)$$

To cast this equation into the form of the phenomenological momentum balance eqn (2.12) for a one-component system, we rewrite the last term as part of the tensor divergence of the full pressure tensor of the system. We now perform the necessary sequence of transformations, using a simplified notation (Clarke and Rice, 1970). Write

$$-\left(\frac{N}{V}\right)^2 \int [\nabla_r V(\mathbf{r}, \mathbf{r}_1)] f_2(\mathbf{r}, \mathbf{p}, \mathbf{r}_1, \mathbf{p}_1, t) \, d^3r_1 \, d^3p_1 \, d^3p$$

$$= \int [\nabla_r V(\mathbf{r}, \mathbf{r}_1)] f(\mathbf{r}, \mathbf{r}_1) \, d^3r_1$$

$$= \int [\nabla_r V(\mathbf{r}', \mathbf{r}_1)] f(\mathbf{r}', \mathbf{r}_1) \delta(\mathbf{r} - \mathbf{r}') \, d^3r_1 \, d^3r' \quad (7.33)$$

Next let us assume that the two-body interaction is spherically symmetric, i.e. dependent on only the relative separation $|\mathbf{r} - \mathbf{r}_1|$ of two particles $V(\mathbf{r}', \mathbf{r}_1) = V(|\mathbf{r}' - \mathbf{r}_1|)$, as it is actually the case for the dominant part of most molecular interactions. In this case we have

$$\frac{d}{d\mathbf{r}'} V(|\mathbf{r}' - \mathbf{r}_1|) = -\frac{d}{d\mathbf{r}_1} V(|\mathbf{r}' - \mathbf{r}_1|) \quad (7.34)$$

and we can rewrite the last line of eqn (7.33) as

$$\frac{1}{2} \int d^3r_1 \, d^3r' f(\mathbf{r}', \mathbf{r}_1) \delta(\mathbf{r} - \mathbf{r}')(\nabla_{r'} - \nabla_{r_1}) V(|\mathbf{r}' - \mathbf{r}_1|) \quad (7.35)$$

Changing $\mathbf{r}_1$ into $\mathbf{r}'$ in the last term gives identically

$$\frac{1}{2} \int d^3r_1 \, d^3r' f(\mathbf{r}', \mathbf{r}_1)[\delta(\mathbf{r}-\mathbf{r}') - \delta(\mathbf{r}-\mathbf{r}_1)]\nabla_{\mathbf{r}'} V(|\mathbf{r}'-\mathbf{r}_1|) \quad (7.36)$$

because $f(\mathbf{r}', \mathbf{r}_1) = f(\mathbf{r}_1, \mathbf{r}')$ is symmetric as well. In eqn (7.36) we now use the identity

$$\delta(\mathbf{r}-\mathbf{r}') - \delta(\mathbf{r}-\mathbf{r}_1) = -\nabla_r \cdot \int_0^1 d\lambda (\mathbf{r}'-\mathbf{r}_1)\delta[\mathbf{r}-\lambda\mathbf{r}'-(1-\lambda)\mathbf{r}_1] \quad (7.37)$$

and obtain

$$\nabla_r \cdot \left\{ -\frac{1}{2} \int_0^1 d\lambda \int d^3r_1 \, d^3r' f(\mathbf{r}', \mathbf{r}_1)(\mathbf{r}'-\mathbf{r}_1) \right.$$
$$\left. \times \delta[\mathbf{r}-\lambda\mathbf{r}'-(1-\lambda)\mathbf{r}_1]\nabla_r V(|\mathbf{r}'-\mathbf{r}_1|) \right\}$$
$$= \nabla_r \cdot \left[-\frac{1}{2} \int_0^1 d\lambda \int d^3r_1 \, d^3r'_{12} f(\mathbf{r}_1+\mathbf{r}'_{12}, \mathbf{r}_1) \right.$$
$$\left. \times \mathbf{r}'_{12}\delta(\mathbf{r}-\mathbf{r}_1-\lambda\mathbf{r}'_{12})\nabla_{r'_{12}} V(r'_{12}) \right\rangle \quad (7.38)$$

where we introduced the variables $\mathbf{r}'_{12} = \mathbf{r}-\mathbf{r}_1$ and $r'_{12} = |\mathbf{r}'_{12}|$ in the second line. Thus we can, indeed, write the last term in eqn (7.32) as

$$-\left(\frac{N}{V}\right)^2 \int [\nabla_r V(\mathbf{r}, \mathbf{r}_1)] f_2(\mathbf{r}, \mathbf{p}, \mathbf{r}_1, \mathbf{p}_1, t) \, d^3r_1 \, d^3p_1 \, d^3p = -\nabla \cdot \mathbf{P}^{(V)} \quad (7.39)$$

with the potential energy contribution to the pressure given by

$$\mathbf{P}^{(V)} = -\frac{1}{2}\left(\frac{N}{V}\right)^2 \int_0^1 d\lambda \int d^3r'_{12} \, d^3p \, d^3p_1 \frac{\mathbf{r}'_{12}\mathbf{r}'_{12}}{r'_{12}} \frac{dV(r'_{12})}{dr'_{12}}$$
$$\times f_2[\mathbf{r}+(1-\lambda)\mathbf{r}'_{12}, \mathbf{p}, \mathbf{r}-\lambda\mathbf{r}'_{12}, \mathbf{p}, t] \quad (7.40)$$

The total pressure tensor is then the sum of eqns (7.31) and (7.40), namely

$$\mathbf{P} = \mathbf{P}^{(K)} + \mathbf{P}^{(V)} \quad (7.41)$$

and eqn (7.32) reduces to the familiar momentum balance [see eqns (2.12) and (2.13)]

$$\rho \frac{D\mathbf{v}}{Dt} + \nabla \cdot \mathbf{P} = \rho \mathbf{F}$$

or

$$\frac{\partial(\rho\mathbf{v})}{\partial t}+\mathbf{\nabla}(\rho\mathbf{v}\mathbf{v}+\mathbf{P}) = \rho\mathbf{F} \quad (7.42)$$

We want to stress that these equations are an exact result. So are the microscopic expressions (7.31) and (7.40) for the pressure tensor, which we will now discuss briefly.

As implied by its name, $\mathbf{P}^{(K)}$ is a purely kinetic pressure and would, for example, be present in an ideal gas. The term $\mathbf{P}^{(V)}$ is a potential energy pressure representing the force exerted on a fluid element due to the molecular two-body interactions. We can see the structure of $\mathbf{P}$ more clearly by looking at the hydrostatic pressure p, given by

$$p(\mathbf{r}, t) = \frac{1}{3} Tr(\mathbf{P})$$

$$= \frac{1}{3}\left(\frac{N}{V}\right) \int \frac{(\mathbf{p}-m\mathbf{v})^2}{m} f_1(\mathbf{r}, \mathbf{p}, t) \, d^3p$$

$$- \frac{1}{6}\left(\frac{N}{V}\right)^2 \int_0^1 d\lambda \, d^3r'_{12} \, d^3p \, d^3p_1$$

$$\times r'_{12} \frac{dV(r'_{12})}{dr'_{12}} f_2[\mathbf{r}+(1-\lambda)\mathbf{r}'_{12}, \mathbf{p}, \mathbf{r}-\lambda\mathbf{r}'_{12}, \mathbf{p}_1, t] \quad (7.43)$$

which can be written, using the definition (7.19), as

$$p(\mathbf{r}, t) = \frac{2}{3}\rho\varepsilon^{(K)}(\mathbf{r}, t) - \frac{1}{6}\left(\frac{N}{V}\right)^2 \int_0^1 d\lambda \, d^3r'_{12} \, d^3p \, d^3p_1$$

$$\times r'_{12} \frac{dV(r'_{12})}{dr'_{12}} f_2[\mathbf{r}+(1-\lambda)\mathbf{r}'_{12}, \mathbf{p}, \mathbf{r}-\lambda\mathbf{r}'_{12}, \mathbf{p}_1, t] \quad (7.44)$$

In an ideal gas in local equilibrium we can assume that the single-particle distribution function f_1 is a local Maxwellian, i.e.

$$f_1^{(0)}(\mathbf{r}, \mathbf{p}, t) = \frac{1}{[2\pi m k_B T(\mathbf{r}, t)]^{3/2}} \exp\left[-\frac{(\mathbf{p}-m\mathbf{v})^2}{2m k_B T(r, t)}\right] \quad (7.45)$$

for which the thermal kinetic energy (per unit mass) is given by

$$\varepsilon^{(K)}(\mathbf{r}, t) = \frac{3}{2m} k_B T(\mathbf{r}, t) \quad (7.46)$$

where $T(\mathbf{r}, t)$ is the local temperature in the gas. With the second term in eqn (7.44) absent in an ideal gas, we find

$$p(\mathbf{r}, t) = \frac{\rho}{m} k_B T(\mathbf{r}, t) \qquad (7.47)$$

which is the local equation of state.

We should mention that the potential energy contribution to the pressure becomes important in the liquid state and is generally slowly decreasing with increasing temperature. At low densities it is proportional to the square of the density in marked contrast to the kinetic pressure, eqn (7.47). Whereas in gases momentum is transferred through collisions, it happens in liquids through the correlations established by the intermolecular forces. The former become more frequent at higher temperatures and so their contribution increases with rising temperatures, whereas the latter weaken at higher temperatures (and above the liquid-gas transition become unimportant at low densities), leading to the observed decrease in the potential energy contribution to the pressure with rising temperature.

7.3.3. Energy Balance

Let us next proceed to the derivation of the balance equation for the internal energy of the system. A balance equation for the kinetic energy follows, of course, straightforwardly from the momentum balance. Multiplying eqn (7.42) by $\mathbf{v}$ and rearranging terms, we obtain

$$\frac{\partial}{\partial t}(\tfrac{1}{2}\rho v^2) + \boldsymbol{\nabla} \cdot (\tfrac{1}{2}\rho v^2 \mathbf{v} + \mathbf{P} \cdot \mathbf{v}) = \mathbf{P} : \boldsymbol{\nabla}\mathbf{v} + \rho \mathbf{F} \cdot \mathbf{v} \qquad (7.48)$$

which is nothing but eqn (2.18) for a one-component system.

We can then concentrate on the internal energy density

$$u = \varepsilon^{(K)} + \varepsilon^{(V)} \qquad (7.49)$$

noting that $\varepsilon^{(K)}$, eqn (7.19), is simply the contribution to the internal energy of the thermal motion within the system. To calculate the effect of $\varepsilon^{(K)}$ on the internal energy balance we multiply eqn (7.21) by $(N/V)(\mathbf{p} - m\mathbf{v})^2/2m$ and integrate over $\mathbf{p}$, obtaining

$$\frac{\partial(\rho\varepsilon^{(K)})}{\partial t} - \int f_1(\mathbf{r}, \mathbf{p}, t) \frac{\partial}{\partial t}\left[\frac{N}{V}\frac{(\mathbf{p}-m\mathbf{v})^2}{2m}\right] d^3p$$

$$+ \left(\frac{N}{V}\right) \int \frac{(\mathbf{p}-m\mathbf{v})^2}{2m} \frac{\mathbf{p}}{m} \cdot [\nabla_r f_1(\mathbf{r}, \mathbf{p}, t)] d^3p$$

$$+ \left(\frac{N}{V}\right) m \int \frac{(\mathbf{p}-m\mathbf{v})^2}{2m} \mathbf{F}(\mathbf{r}) \cdot [\nabla_p f_1(\mathbf{r}, \mathbf{p}, t)] d^3p$$

$$= \left(\frac{N}{V}\right)^2 \int \frac{(\mathbf{p}-m\mathbf{v})^2}{2m} [\nabla_r V(|\mathbf{r}-\mathbf{r}_1|)]$$

$$\cdot [\nabla_p f_2(\mathbf{r}, \mathbf{p}, \mathbf{r}_1, \mathbf{p}_1, t)] d^3r_1 d^3p_1 d^3p \quad (7.50)$$

where we immediately performed a partial differentiation on the first term. The second term above is obviously zero and the third one can be transformed by partial differentiation into

$$\nabla_r \cdot [\rho\varepsilon^{(K)}\mathbf{v} + \mathbf{j}_q^{(K)}(\mathbf{r}, t)] - \left(\frac{N}{V}\right) \int f_1(\mathbf{r}, \mathbf{p}, t) \nabla_r \cdot \left[\frac{\mathbf{p}}{m}\frac{(\mathbf{p}-m\mathbf{v})^2}{2m}\right] d^3p$$

where (7.51)

$$\mathbf{j}_q^{(K)}(\mathbf{r}, t) = \left(\frac{N}{V}\right) \int \frac{\mathbf{p}-m\mathbf{v}}{m} \frac{(\mathbf{p}-m\mathbf{v})^2}{2m} f_1(\mathbf{r}, \mathbf{p}, t) d^3p \quad (7.52)$$

The last term in eqn (7.51) can be written as

$$\left(\frac{N}{V}\right) \int f_1(\mathbf{r}, \mathbf{p}, t) \sum_{i,k=1}^{3} \frac{p_i - mv_i + mv_i}{m}(p_k - mv_k)\frac{\partial v_k}{\partial x_i} d^3p$$

$$= \sum_{i,k=1}^{3} \mathbf{P}_{ik}^{(K)} \frac{\partial v_k}{\partial x_i} = \mathbf{P}^{(K)}:(\nabla\mathbf{v}) \quad (7.53)$$

where we introduced the kinetic pressure tensor, eqn (7.31). By partial integration we can show that the fourth term in eqn (7.50) is zero which, indeed, must be the case on physical grounds because the external force cannot effect changes in the internal energy except through diffusion processes that, however, are absent in a one-component system.

Finally we transform the term on the right-hand side of eqn (7.50) by partial integration with respect to $\mathbf{p}$ into

$$-\left(\frac{N}{V}\right)^2 \int f_2(\mathbf{r}, \mathbf{p}, \mathbf{r}_1, \mathbf{p}_1, t) \frac{\mathbf{p}-m\mathbf{v}}{m} \cdot \nabla_r V(|\mathbf{r}-\mathbf{r}_1|) d^3r_1 d^3p_1 d^3p$$

(7.54)

Eqn (7.50) then reads

$$\frac{\partial(\rho\varepsilon^{(K)})}{\partial t}+\boldsymbol{\nabla}[\rho\varepsilon^{(K)}\mathbf{v}+\mathbf{j}_q^{(K)}(\mathbf{r},t)]=-\mathbf{P}^{(K)}:(\boldsymbol{\nabla}\mathbf{v})$$

$$-\left(\frac{N}{V}\right)^2\int[\boldsymbol{\nabla}_r V(|\mathbf{r}-\mathbf{r}_1|)]\cdot\frac{\mathbf{p}-m\mathbf{v}}{m}f_2(\mathbf{r},\mathbf{p},\mathbf{r}_1,\mathbf{p}_1,t)\,d^3r_1\,d^3p_1\,d^3p \quad (7.55)$$

To obtain the balance equation for the internal energy (ρu) we must add to eqn (7.55) the balance equation for $\rho\varepsilon^{(V)}$, which we obtain from the second member of the BBGKY hierarchy [setting $(N-2)/V \approx N/V$],

$$\left\{\frac{\partial}{\partial t}-\boldsymbol{\nabla}_r[\Phi(\mathbf{r})+V(|\mathbf{r}-\mathbf{r}_1|)]\cdot\boldsymbol{\nabla}_p-\boldsymbol{\nabla}_{r_1}[\Phi(\mathbf{r}_1)+V(|\mathbf{r}-\mathbf{r}_1|)]\cdot\boldsymbol{\nabla}_{p_1}\right.$$

$$\left.+\frac{\mathbf{p}}{m}\cdot\boldsymbol{\nabla}_r+\frac{\mathbf{p}_1}{m}\cdot\boldsymbol{\nabla}_{r_1}\right\}f_2(\mathbf{r},\mathbf{p},\mathbf{r}_1,\mathbf{p}_1,t)$$

$$=\left(\frac{N}{V}\right)\int[\boldsymbol{\nabla}_r V(|\mathbf{r}-\mathbf{r}_2|)\cdot\boldsymbol{\nabla}_p+\boldsymbol{\nabla}_{r_1}V(|\mathbf{r}_1-\mathbf{r}_2|)\cdot\boldsymbol{\nabla}_{p_1}]$$

$$\times f_3(\mathbf{r},\mathbf{p},\mathbf{r}_1,\mathbf{p}_1,\mathbf{r}_2,\mathbf{p}_2,t)\,d^3r_2\,d^3p_2 \quad (7.56)$$

We multiply this equation by $(N/V)^2 V(|\mathbf{r}-\mathbf{r}_1|)$, integrate over $d^3p\,d^3r_1\,d^3p_1$ and observe immediately that the right-hand side is zero. (Perform partial integrations with respect to $\mathbf{p}$ for the first term and with respect to $\mathbf{p}_1$ for the second term on the right-hand side.) This ensures that only single-particle and two-particle distribution functions enter the macroscopic balance equations explicitly. The left-hand side of eqn (7.56) can be handled similarly to the way leading from eqn (7.50) to (7.55), yielding eventually

$$\frac{\partial(\rho\varepsilon^{(V)})}{\partial t}+\boldsymbol{\nabla}(\rho\varepsilon^{(V)}\mathbf{v}+\mathbf{j}_q^{(V_1)})=\frac{1}{2}\left(\frac{N}{V}\right)^2\int\boldsymbol{\nabla}_r V(|\mathbf{r}-\mathbf{r}_1|)$$

$$\cdot\frac{\mathbf{p}-\mathbf{p}_1}{m}f_2(\mathbf{r},\mathbf{p},\mathbf{r}_1,\mathbf{p}_1,t)\,d^3r_1\,d^3p_1\,d^3p \quad (7.57)$$

where

$$\mathbf{j}_q^{(V_1)}(\mathbf{r},t)=\frac{1}{2}\left(\frac{N}{V}\right)^2\int\frac{\mathbf{p}-m\mathbf{v}}{m}V(|\mathbf{r}-\mathbf{r}_1|)$$

$$\times f_2(\mathbf{r},\mathbf{p},\mathbf{r}_1,\mathbf{p}_1,t)\,d^3r_1\,d^3p_1\,d^3p \quad (7.58)$$

Adding this to eqn (7.55) we obtain

$$\frac{\partial(\rho u)}{\partial t} - \boldsymbol{\nabla} \cdot (\rho u \mathbf{v} + \mathbf{j}_q^{(K)} + \mathbf{j}_q^{(V_1)})$$

$$= -\mathbf{P}^{(K)} : (\boldsymbol{\nabla}\mathbf{v}) - \frac{1}{2}\left(\frac{N}{V}\right)^2 \int \boldsymbol{\nabla}_r V(|\mathbf{r}-\mathbf{r}_1|)$$

$$\times \frac{\mathbf{p}+\mathbf{p}_1-2m\mathbf{v}}{m} f_2(\mathbf{r},\mathbf{p},\mathbf{r}_1,\mathbf{p}_1,t)\, d^3r_1\, d^3p_1\, d^3p \quad (7.59)$$

It remains to convert the last term into the divergence of an additional vector flux and the missing potential pressure contribution in the first term on the right-hand side. This can be done using the trick leading from eqn (7.33) to (7.38). First write

$$-\frac{1}{2}\left(\frac{N}{V}\right)^2 \int \boldsymbol{\nabla}_r V(|\mathbf{r}-\mathbf{r}_1|) \cdot \frac{\mathbf{p}+\mathbf{p}_1}{m} f_2(\mathbf{r},\mathbf{p},\mathbf{r}_1,\mathbf{p}_1,t)\, d^3r_1\, d^3p_1\, d^3p$$

$$= \boldsymbol{\nabla}_r \cdot \left\{\frac{1}{4}\left(\frac{N}{V}\right)^2 \int_0^1 d\lambda \int d^3r'_{12} \frac{1}{r'_{12}} \frac{dV(r'_{12})}{dr'_{12}} \int d^3p\, d^3p_1 \frac{\mathbf{p}+\mathbf{p}_1}{m} \right.$$

$$\left. \cdot (\mathbf{r}'_{12}\mathbf{r}'_{12}) f_2[\mathbf{r}+(1-\lambda)\mathbf{r}'_{12},\mathbf{p},\mathbf{r}-\lambda\mathbf{r}'_{12},\mathbf{p}_1,t]\right\} \quad (7.60)$$

and next

$$\left(\frac{N}{V}\right)^2 \mathbf{v} \cdot \int \boldsymbol{\nabla}_r V(|\mathbf{r}-\mathbf{r}_1|) f_2(\mathbf{r},\mathbf{p},\mathbf{r}_1,\mathbf{p}_1,t)\, d^3r_1\, d^3p_1\, d^3p$$

$$= \mathbf{v} \cdot \boldsymbol{\nabla}_r \left\{-\frac{1}{2}\left(\frac{N}{V}\right)^2 \int_0^1 d\lambda \int d^3r'_{12} \frac{\mathbf{r}'_{12}\mathbf{r}'_{12}}{r'_{12}} \frac{dV(r'_{12})}{dr'_{12}} \right.$$

$$\left. \times f_2[\mathbf{r}+(1-\lambda)\mathbf{r}'_{12},\mathbf{p},\mathbf{r}-\lambda\mathbf{r}'_{12},\mathbf{p}_1,t]\, d^3p\, d^3p_1\right\}$$

$$= \mathbf{v} \cdot (\boldsymbol{\nabla}_r \mathbf{P}^{(V)})$$

$$= \boldsymbol{\nabla}_r \cdot (\mathbf{v}\mathbf{P}^{(V)}) - \mathbf{P}^{(V)} : (\boldsymbol{\nabla}\mathbf{v}) \quad (7.61)$$

The first term in the last line above can next be combined with the last line of eqn (7.60) as $-\boldsymbol{\nabla}_r \mathbf{j}_q^{(V_2)}$ with

$$\mathbf{j}_q^{(V_2)} = -\frac{1}{4}\left(\frac{N}{V}\right)^2 \int_0^1 d\lambda \int d^3r'_{12} \frac{\mathbf{r}'_{12}\mathbf{r}'_{12}}{r'_{12}} \cdot \frac{\mathbf{p}+\mathbf{p}_1-2m\mathbf{v}}{m}$$

$$\times f_2[\mathbf{r}+(1-\lambda)\mathbf{r}'_{12},\mathbf{p},\mathbf{r}-\lambda\mathbf{r}'_{12},\mathbf{p}_1,t]\, d^3p\, d^3p_1 \quad (7.62)$$

and eqn (7.59) evolves into its final form

$$\frac{\partial(\rho u)}{\partial t} + \nabla \cdot (\rho u \mathbf{v} + \mathbf{j}_q) = -\mathbf{P} : (\nabla \mathbf{v}) \tag{7.63}$$

where $\mathbf{P} = \mathbf{P}^{(K)} + \mathbf{P}^{(V)}$ is the full pressure tensor and

$$\mathbf{j}_q = \mathbf{j}_q^{(K)} + \mathbf{j}_q^{(V_1)} + \mathbf{j}_q^{(V_2)} \tag{7.64}$$

is the internal energy or heat current.

Thus we have managed not only to derive the phenomenological balance equation, eqn (2.27), for the internal energy from first principles (the Liouville equation) but have also found a microscopic expression for the heat current. Two different mechanisms contribute to heat conduction. In $\mathbf{j}_q^{(K)}$ heat is conducted on account of the thermal motion of the molecules. Bear in mind, however, that this mode is not associated with any net mass transfer, the latter being only the case in the internal energy convection term $\rho u \mathbf{v}$. The two terms $\mathbf{j}_q^{(V_1)}$ and $\mathbf{j}_q^{(V_2)}$ represent heat conduction via the two-body interaction of the molecules. It is therefore dominant in liquids and solids and negligible in dilute gases where molecules are only rarely within their spheres of mutual interaction.

Simple mechanical pictures might help to clarify these two mechanisms of heat conduction. For the kinetic heat conduction $\mathbf{j}_q^{(K)}$, we model a system of identical elastic spheres aligned in a straight line. Shooting an additional sphere onto this chain along its direction of alignment will transfer its kinetic energy along the chain in successive elastic collisions without actually transferring mass, i.e. without convection. For the potential heat conduction $(\mathbf{j}_q^{(V_1)} + \mathbf{j}_q^{(V_2)})$, we model a similar chain of spheres but this time they are coupled together with elastic springs. Energy imparted into the chain at one end will now be transferred along the chain as elastic deformation energy in the springs.

7.3.4. Entropy Balance

From the Liouville equation we have derived the balance equations for the mechanical quantities mass density ρ, momentum density $\rho \mathbf{v}$, and internal energy density ρu, and have given microscopic expressions for the relevant currents and sources. It remains to consider the entropy balance. To proceed as before, we would first start with a microscopic expression for a time- and space-dependent nonequilibrium entropy. Such a quantity can, however, not be written in general, except *cum grano salis* in systems described by the Boltzmann equation. This special case will be discussed at length in Section 7.6. For the general situation, our only option is to resort to some phenomenological arguments. This can be done readily for systems in local equilibrium, because then we know from Section 2.3 that the fundamental relation, eqn (2.65),

$$Tds = du + pd\rho^{-1} \tag{7.65}$$

is valid locally, i.e. in the continuum limit even pointwise, and can be used to determine the entropy balance from the balance equations of mechanical quantities. Therefore, for systems in local equilibrium, we take eqns (2.69)–(2.71)

$$\frac{\partial(\rho s)}{\partial t} + \nabla \cdot \mathbf{j}_s = \sigma_s \tag{7.66}$$

as the local entropy balance with

$$\mathbf{j}_s = \rho s \mathbf{v} + \frac{1}{T} \mathbf{j}_q \tag{7.67}$$

and

$$\sigma_s = \mathbf{j}_q \cdot \nabla\left(\frac{1}{T}\right) - \frac{1}{T}(\mathbf{P} - p\mathbf{I}) \tag{7.68}$$

with $\mathbf{j}_q$, $\mathbf{P}$, and p given now microscopically by eqns (7.64), (7.41) and (7.43), respectively.

If the system is described statistically by a local Maxwell single-particle distribution, eqn (7.45), as is the case in a dilute gas, then the local temperature $T(\mathbf{r}, t)$ is connected with the thermal kinetic energy via eqn (7.46) and with the hydrostatic pressure $p(\mathbf{r}, t)$ via eqn (7.47). In the general case, it will be necessary to establish constitutive laws, connecting among others, the heat current with the temperature gradient (Fourier's law); in this case, the local temperature can be calculated from the internal energy equation as a boundary condition problem.

7.4. Derivation of Constitutive Laws

As we have mentioned in Chapters 2 and 3, the balance equations themselves do not form a closed set of equations for all quantities occurring in them but must be supplemented by an appropriate number of constitutive laws to obtain closure. For a wide range of physical laws, linear constitutive relations between thermodynamic forces and fluxes are sufficient, as argued in Chapter 3. There we had postulated linear laws and introduced in each of them a set of transport coefficients that, in a phenomenological theory, have to be determined experimentally. Having now based our theory on microscopic foundations, we are in a position to deduce these linear laws and to calculate, in principle, the transport coefficients from microscopic expressions. We will now deal with these matters for viscosity and thermal conduction.

7.4.1. Viscosity

If a nonuniform velocity field is impressed on a fluid, there will be reactive forces set up within the fluid that will tend to even out such inhomogeneities. Two separate physical effects take place. The first one, dominant in gases, is collisional. Suppose that two adjacent volume elements are moving at different speeds. The molecules arriving from the faster moving cell in the slower one will have, on the average, a higher speed than the molecules in the slow moving cell. They will then deposit their excess energy in a sequence of collisions into the slow cell, thus speeding it up. Similarly, molecules from the slower cell will reduce the average speed of a faster cell upon arrival there.

The second effect that leads to a reduction of velocity inhomogeneities in a moving fluid arises from the intermolecular forces and is therefore dominant in liquids. In a liquid, neighboring molecules are always within the range of their mutual interaction. Their mean-free path is practically cut to zero and the collisional source of viscosity is absent. However, in a liquid, neighboring volume elements, moving at different velocities, are linked together by the interactions of their surface molecules and produce a certain molecular stickiness of the liquid which is the second source of viscosity.

From this elementary discussion it is apparent that the distribution functions in a fluid must reflect, in a basic way, its local state of motion. In particular, in a fluid with constant density and constant temperature moving, in addition, at a constant velocity,

CLASSICAL STATISTICAL MECHANICS AND KINETIC THEORY 171

the single-particle distribution function $f_1(\mathbf{r}, \mathbf{p}, t)$ will be independent of the position $\mathbf{r}$ in the fluid and only a function of $|\mathbf{p}|$. Indeed, it will be nothing but the equilibrium distribution function, e.g. eqn (7.45) in a Maxwell gas. Allowing now for 'small' inhomogeneities in the velocity field, we expect that f_1 can be expanded in a Taylor series in the velocity gradients

$$f_1(\mathbf{r}, \mathbf{p}, t) = f_1^{(0)}(\mathbf{r}, |\mathbf{p}|, t)$$
$$+ A_1\left[\frac{(\mathbf{p}-m\mathbf{v})\cdot\mathbf{\Lambda}\cdot(\mathbf{p}-m\mathbf{v})}{(\mathbf{p}-m\mathbf{v})^2} - \frac{1}{3}\mathbf{\nabla}_r\cdot\mathbf{v}\right] + \cdots \quad (7.69)$$

where $A_1 = A_1(|\mathbf{p}-m\mathbf{v}|)$ can still be a function of $|\mathbf{p}-m\mathbf{v}|$ and

$$\Lambda_{\alpha\beta} = \frac{1}{2}\left(\frac{\partial v_\alpha}{\partial r_\beta} + \frac{\partial v_\beta}{\partial r_\alpha}\right) \quad (7.70)$$

is the symmetric rate of strain tensor, eqn (2.47).

The peculiar form of the second term in eqn (7.69) arises from the requirement that it has to be a scalar, linear in the velocity gradient, and that its integral over $\mathbf{p}$ has to vanish to ensure that f_1 is still normalized to one if $f_1^{(0)}$ is. If temperature and density inhomogeneities are present in the system, their respective gradients will also show up in the expansion (7.69) and A_1 can also be a function of position.

With eqn (7.69) we can enter eqn (7.31) and evaluate the kinetic pressure tensor up to terms linear in the velocity gradients. We obtain

$$\mathbf{P}^{(K)}_{\alpha\beta} = \left(\frac{N}{V}\right)\frac{1}{m}\int f_1(\mathbf{r}, \mathbf{p}, t)(p_\alpha - mv_\alpha)(p_\beta - mv_\beta)\,d^3p$$
$$= \frac{1}{3}\left(\frac{N}{V}\right)\int f_1^{(0)}(\mathbf{r}, \mathbf{p}, t)\frac{(\mathbf{p}-m\mathbf{v})^2}{m}d^3p\,\delta_{\alpha\beta} + \left(\frac{N}{V}\right)\frac{1}{m}\int d^3p$$
$$\times A_1(|\mathbf{p}-m\mathbf{v}|)(p_\alpha - mv_\alpha)(p_\beta - mv_\beta)$$
$$\times \left[\sum_{\gamma,\delta=1}^{3}(p_\gamma - mv_\gamma)\Lambda_{\alpha\beta}(p_\delta - mv_\delta) - \frac{1}{3}\mathbf{\nabla}\cdot\mathbf{v}\right] \quad (7.71)$$

But we can verify that for a traceless tensor, the identity

$$\int \mathbf{p}\cdot\left\{\mathbf{T}(|\mathbf{p}|) - \frac{1}{3}Tr[\mathbf{T}(|\mathbf{p}|)]\mathbf{I}\right\}\cdot\mathbf{p}\,\frac{\mathbf{pp}}{|\mathbf{p}|^2}\,d^3p$$
$$= \frac{2}{15}\int\left\{\mathbf{T}(|\mathbf{p}|) - \frac{1}{3}Tr(\mathbf{T}(|\mathbf{p}|))\mathbf{I}\right\}|\mathbf{p}|^2\,d^3p \quad (7.72)$$

exists. It follows from eqn (7.71) that
$$\mathbf{P}^{(K)}(\mathbf{r}, t) = p^{(K)}(\mathbf{r}, t)\mathbf{I} - 2\eta^{(K)}[\mathbf{\Lambda} - \tfrac{1}{3}(\mathbf{\nabla}_r \cdot \mathbf{v})\mathbf{I}] \quad (7.73)$$
where
$$p^{(K)}(\mathbf{r}, t) = \left(\frac{N}{V}\right) \int f_1^{(0)}(\mathbf{r}, \mathbf{p}, t) \frac{(\mathbf{p} - m\mathbf{v})^2}{m} d^3p \quad (7.74)$$

is the kinetic energy part of the (local) equilibrium hydrostatic pressure and
$$\eta^{(K)} = -\frac{2}{15}\left(\frac{N}{V}\right) \int A_1(|\mathbf{q}|) \frac{q^2}{m} d^3q \quad (7.75)$$

is a part of the coefficient of shear viscosity.

To evaluate the potential energy contribution $\mathbf{P}^{(V)}$, eqn (7.40), to the pressure tensor, we need some information on the two-particle distribution f_2. It is, however, more convenient to consider instead the reduced distribution function
$$n_2(\mathbf{r}, \mathbf{r}'_{12}, t) = \left(\frac{N}{V}\right)^2 \int_0^1 d\lambda \int d^3p\, d^3p_1 f_2[\mathbf{r} + (1-\lambda)\mathbf{r}'_{12}, \mathbf{p}, \mathbf{r} - \lambda \mathbf{r}'_{12}, \mathbf{p}_1, t] \quad (7.76)$$

In an equilibrium system, n_2 can only depend on the relative separation $\mathbf{r}'_{12} = \mathbf{r}' - \mathbf{r}_1$ of the two particles involved and is proportional to the radial correlation function. In the presence of 'small' velocity gradients in the fluid, we then expand n_2 as
$$n_2(\mathbf{r}, \mathbf{r}'_{12}, t) = n_2^{(0)}(r'_{12}, t)\mathbf{r}'_{12} \cdot \mathbf{\Lambda} \cdot \mathbf{r}'_{12}/(r'_{12})^2 + [B_2(r'_{12}, t) - \tfrac{1}{3}B_1(r'_{12}, t)]\mathbf{\nabla}_r \cdot \mathbf{v} + \cdots \quad (7.77)$$

where in a nonuniform fluid, $n_2^{(0)}$, B_1, and B_2 could still depend on the position $\mathbf{r}$ in the fluid. With this expansion, we enter eqn (7.40) for $\mathbf{P}^{(V)}$ and obtain
$$\mathbf{P}^{(V)}(\mathbf{r}, t) = -\frac{1}{2}\int d^3r'_{12}\, n_2(\mathbf{r}, \mathbf{r}'_{12}, t) \frac{\mathbf{r}'_{12}\mathbf{r}'_{12}}{(r'_{12})^2} \frac{dV(r'_{12})}{dr'_{12}}$$
$$= (p^{(V)} - \zeta \mathbf{\nabla}_r \cdot \mathbf{v})\mathbf{I} - 2\eta^{(V)}[\mathbf{\Lambda} - \tfrac{1}{3}(\mathbf{\nabla}_r \cdot \mathbf{v})\mathbf{I}] \quad (7.78)$$

where
$$p^{(V)} = -\frac{1}{6}\int d^3r'_{12}\, n_2^{(0)}(r'_{12})r'_{12}\frac{dV(r'_{12})}{dr'_{12}}$$
$$\eta^{(V)} = \frac{1}{15}\int B_1(r'_{12}, t)r'_{12}\frac{dV(r'_{12})}{dr'_{12}} d^3r'_{12}$$
$$\zeta = \frac{1}{2}\int B_2(r'_{12}, t)r'_{12}\frac{dV(r'_{12})}{dr'_{12}} d^3r'_{12} \quad (7.79)$$

We thus obtain for the complete pressure tensor
$$\mathbf{P} = (p - \zeta \mathbf{\nabla} \cdot \mathbf{v})\mathbf{I} - 2\eta[\mathbf{\Lambda} - \tfrac{1}{3}(\mathbf{\nabla} \cdot \mathbf{v})\mathbf{I}] \tag{7.80}$$
where
$$p = p^{(K)} + p^{(V)} \tag{7.81}$$
is the hydrostatic pressure, $\eta = \eta^{(K)} + \eta^{(V)}$ is the shear viscosity, and ζ is the bulk viscosity, as we can infer by comparing eqn (7.80) with the earlier phenomenological ansatz eqns (2.37) and (2.38).

It should be stressed at this stage that the derivation of eqn (7.80) was purely formal in nature. We managed to derive the structure of the pressure tensor; the resultant expressions for p, η and ζ are, however, not very useful unless means are found to calculate the expansions, eqns (7.69) and (7.77), explicitly. This is the task of kinetic theory, as we will see in Section 7.6, where we will calculate the coefficients of viscosity explicitly for a dilute gas obeying Boltzmann's equation.

Let us at this stage develop a handwaving argument, following Green (1952, 1969) to show that the second of eqns (7.79) will yield a positive shear viscosity in a liquid. In this case $\eta^{(K)}$, eqn (7.75), should be negligible because viscosity in a liquid arises from particle correlations rather than collisions. Now, we must remember that the expansion coefficients A_1, B_1, and B_2 in eqns (7.69) and (7.77) measure the deformation of the distribution functions from their equilibrium form. But then one would expect that the B_i's are, for small deviations, proportional to $n_2^{(0)}(r'_{12})$. The latter quantity, being a measure for the probability of finding two particles a distance r'_{12} apart, is nearly zero for r'_{12} less than a molecular diameter d due to the mutual hard core repulsion between molecules, and, of course, by definition positive for larger r'_{12}. On the other hand, we infer from the general shape of the two-body interaction between two molecules (Fig. 7.1) that $dV(r'_{12})/dr'_{12} > 0$ for $r'_{12} > r_0$ and rapidly decreasing. Thus $\eta^{(V)}$ and ζ will be roughly proportional to $n_2^{(0)}$, which we might approximate by $(N/V)^2 \exp[-\beta V(r_0)]$. But then we find that $\eta \sim \exp[-V(r_0)/k_B T]$, a relation that is the basis of Andrade's theory of viscosity (1934). This temperature dependence is, indeed, exhibited by many liquids over a wide range of temperatures. It can also be understood in the framework of Eyring's (1936)

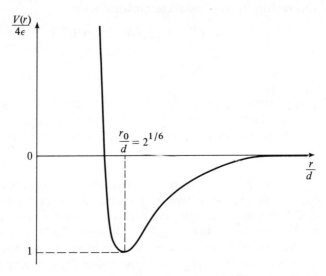

Fig. 7.1. Van der Waals potential between two neutral molecules according to Lennard-Jones; $V(r)/4\varepsilon = (d/r)^{12} - (d/r)^6$.

	He	A	H_2	O_2
ε/k_B [K]	10.22	124.0	38.0	113.0
d [Å]	2.576	3.441	2.948	3.433

After Hirschfelder, Curtiss, and Bird (1954).

theory of transport processes (see Eyring and Jhon 1969), in which case $\eta \sim \exp[\Delta G^{\mathrm{act}}/k_B T]$ where ΔG^{act} is the Gibbs free energy of activation per molecule.

7.4.2. Thermal Conduction

We next derive a constitutive law for the heat current, eqn (7.64), to close the balance equation (7.63) for the internal energy. Again we restrict ourselves to the linear regime where the heat current depends linearly on the temperature gradient according to Fourier's law. The procedure will obviously be quite analogous to what we just performed for viscosity.

To simplify the treatment we deal with a fluid in which only a temperature gradient is imposed externally. Secondary effects like thermal expansion resulting in a density gradient can, indeed, be neglected under isobaric conditions, in which case temperature and density gradients are related thermodynamically by

$$\nabla \rho(\mathbf{r}, t) = -\frac{(\partial p/\partial T)_\rho}{(\partial p/\partial \rho)_T} \nabla T \qquad (7.82)$$

Let us then add to the expansion (7.69) for the single particle distribution function f_1 a term linear in ∇T, namely

$$A_2 \frac{\mathbf{p} - m\mathbf{v}}{|\mathbf{p} - m\mathbf{v}|} \cdot \nabla T \qquad (7.83)$$

where the expansion coefficient $A_2 = A_2(|\mathbf{p} - m\mathbf{v}|)$ can still be a function of $|\mathbf{p} - m\mathbf{v}|$. Inserted in eqn (7.52), this yields

$$\mathbf{j}_q^{(K)}(\mathbf{r}, t) = \left(\frac{N}{V}\right) \int \frac{\mathbf{p} - m\mathbf{v}}{m} \frac{(\mathbf{p} - m\mathbf{v})^2}{2m} A_2 \frac{(\mathbf{p} - m\mathbf{v})}{|\mathbf{p} - m\mathbf{v}|} \cdot \nabla T \, d^3p$$

$$= \frac{1}{6}\left(\frac{N}{V}\right) \int \frac{|\mathbf{q}|^3}{m^2} A_2(|\mathbf{q}|, r) \, d^3q \nabla T(\mathbf{r}, t) \qquad (7.84)$$

Next we consider the first potential energy contribution to the heat current, eqn (7.58), after symmetrizing it with respect to $\mathbf{p}$ and $\mathbf{p}_1$ as follows:

$$\mathbf{j}_q^{(V_1)} = \frac{1}{4}\left(\frac{N}{V}\right)^2 \frac{1}{m} \int V(r_{12})[(\mathbf{p} - \mathbf{p}_1) + (\mathbf{p} + \mathbf{p}_1 - 2m\mathbf{v})]$$

$$\times f_2(\mathbf{r}, \mathbf{p}, \mathbf{r} + \mathbf{r}_{12}, \mathbf{p}_1, t) \, d^3p \, d^3p_1 \, d^3r_{12} \qquad (7.85)$$

where we also introduced a new integration variable $\mathbf{r}_{12} = \mathbf{r}_1 - \mathbf{r}$. Observe now that

$$\int \frac{\mathbf{p} - \mathbf{p}_1}{m} f_2(\mathbf{r}, \mathbf{p}, \mathbf{r} + \mathbf{r}_{12}, \mathbf{p}_1, t) \, d^3p \, d^3p_1 \qquad (7.86)$$

is proportional to the average relative velocity of two molecules separated by $\mathbf{r}_{12}$. This quantity is zero in equilibrium and, away from equilibrium, will have an expansion

$$C_1 \frac{\mathbf{r}_{12}}{r_{12}} \cdot (\mathbf{\Lambda} - \tfrac{1}{3}\nabla \cdot \mathbf{v} \, \mathbf{I}) + \frac{C_2}{r_{12}^2} \mathbf{r}_{12} \cdot (\mathbf{\Lambda} - \tfrac{1}{3}\nabla \cdot \mathbf{v} \, \mathbf{I}) \cdot \mathbf{r}_{12}\mathbf{r}_{12} \qquad (7.87)$$

linear in the velocity gradients, but with no terms linear in the temperature gradients. Equation (7.87), however, does not contribute to $\mathbf{j}_q^{(V_1)}$ due to the angular integration left in eqn (7.85). For the remaining terms in (7.85) we first notice that

$$\int \frac{\mathbf{p}+\mathbf{p}_1}{m} f_2(\mathbf{r}, \mathbf{p}, \mathbf{r}+\mathbf{r}_{12}, \mathbf{p}_1, t) \, d^3p \, d^3p_1 \qquad (7.88)$$

is the average resultant velocity of two molecules with center of mass at $\mathbf{r}$ and separated by $\mathbf{r}_{12}$. If f_2 is taken to be the equilibrium two-particle distribution function, this term is equal to

$$2\mathbf{v} \int f_2^{(0)}(\mathbf{r}, \mathbf{p}, \mathbf{r}+\mathbf{r}_{12}, \mathbf{p}_1, t) \, d^3p \, d^3p_1 \qquad (7.89)$$

With a temperature gradient present, eqn (7.88) will make an additional contribution accounting for the fact that two molecules within each other's sphere of attraction and alligned parallel to the temperature gradient will drift together along the latter as a result of the net force exerted by the surrounding molecules. To see that this is so, recall that $\mathbf{j}_q^{(V_1)}$ and $\mathbf{j}_q^{(V_2)}$ are the most important contributions (compared with $\mathbf{j}_q^{(K)}$) in liquids, where the pressure decreases with increasing temperature. Thus our pair of molecules will experience a smaller pressure at its high temperature side, and thus a net force in the direction of the temperature gradient.

We can then write the following expansion linear in the temperature gradient

$$\left(\frac{N}{V}\right)^2 \frac{1}{m} \int (\mathbf{p}+\mathbf{p}_1 - 2m\mathbf{v}) f_2(\mathbf{r}, \mathbf{p}, \mathbf{r}+\mathbf{r}_{12}, \mathbf{p}_1, t) \, d^3p \, d^3p_1$$

$$= D_1 \left[(\mathbf{r}_{12} \cdot \boldsymbol{\nabla} T) \frac{\mathbf{r}_{12}}{r_{12}^2} - \frac{1}{3} \boldsymbol{\nabla} T \right] \quad (7.90)$$

where $D_1 = D_1(|\mathbf{r}_{12}|)$. This expansion is dictated by the requirement that integrated over $\mathbf{r}_{12}$, it must yield a zero mean resultant velocity averaged over all pairs of molecules with center of mass at position $\mathbf{r}$. Obviously, eqn (7.90) inserted in eqn (7.85) yields $\mathbf{j}_q^{(V_1)} = 0$.

However, if we use the same expansion (7.90) in eqn (7.62) for $\mathbf{j}_q^{(V_2)}$, we find

$$\mathbf{j}_q^{(V_2)} = -\frac{1}{18} \int D_1(r_{12}) r_{12} \frac{dV(r_{12})}{dr_{12}} \, d^3r_{12} \boldsymbol{\nabla} T \qquad (7.91)$$

and have formally derived Fourier's law of heat conduction
$$\mathbf{j}_q = -\lambda \boldsymbol{\nabla} T \tag{7.92}$$
with
$$\lambda = \frac{1}{18} \int D_1(r_{12}) r_{12} \frac{dV(r_{12})}{dr_{12}} d^3 r_{12} - \frac{1}{6}\left(\frac{N}{V}\right) \int \frac{|\mathbf{q}|^3}{m^2} A_2(|\mathbf{q}|) d^3 q \tag{7.93}$$

We have thus achieved the goal of this section and closed the set of balance equations by deriving linear constitutive equations for the pressure tensor and the heat current. It is obvious that the methods developed here can be used to derive nonlinear constitutive laws as well. It has to be stressed, however, that these derivations are purely formal in nature and, indeed, do not allow us to determine the transport coefficients, introduced in them, explicitly. To do this, it is necessary to set up kinetic equations for the particular system under study so that the expansion (7.69) and (7.77) for the distribution functions and their generalizations can be calculated explicitly. This program has so far only been completed for a dilute gas in which Boltzmann's equation can be used. The remainder of this chapter is devoted to it.

7.5. Simple Kinetic Equations: Vlasov and Boltzmann

7.5.1. Preliminaries and a Derivation of Vlasov's Equation

As we have stressed repeatedly in foregoing sections, a satisfactory a priori calculation of transport coefficients presupposes a knowledge of nonequilibrium single-particle and two-particle distribution functions. This problem can, of course, not be solved in general but only in certain simplified models, one of which is the Boltzmann gas. This is a dilute gas of rarely interacting neutral molecules, described by Boltzmann's (1872) kinetic equation for the single-particle distribution function. Rather than following Boltzmann's intuitive approach to arrive at his equation, we want to present here a derivation from the BBGKY hierarchy as given first by Bogolyubov (1946), Born and Green (1946), and Kirkwood (1946).

Our starting point is again the first member of the BBGKY hierarchy, eqn (7.15), namely

$$\left[\frac{\partial}{\partial t} + \frac{\mathbf{p}}{m} \cdot \boldsymbol{\nabla}_r + m \mathbf{F}(r) \cdot \boldsymbol{\nabla}_p \right] f_1(\mathbf{r}, \mathbf{p}, t) = \left(\frac{\partial f_1}{\partial t}\right)_{\text{coll}} \tag{7.94}$$

where

$$\left(\frac{\partial f}{\partial t}\right)_{\text{coll}} = \left(\frac{N}{V}\right)\int [\boldsymbol{\nabla}_r V(\mathbf{r}, \mathbf{r}')] \cdot [\boldsymbol{\nabla}_p f_2(\mathbf{r}, \mathbf{p}, \mathbf{r}_1, \mathbf{p}_1, t)] \, d^3 r_1 \, d^3 p_1 \tag{7.95}$$

is commonly referred to as the collision integral. It will now be our task to find suitable approximations to the collision integral so as to render an equation involving the single-particle distribution function f_1 alone, decoupled from the higher-order members of the BBGKY hierarchy.

Before we proceed with the derivation of the Boltzmann equation, we will consider the simplest such approximation in which we assume that $f_2(\mathbf{r}, \mathbf{p}, \mathbf{r}_1, \mathbf{p}_1, t)$, the probability of finding particles at $(\mathbf{r}, \mathbf{p})$ and $(\mathbf{r}_1, \mathbf{p}_1)$ at the same time, is simply the product of the single-particle probability of finding a particle at $(\mathbf{r}, \mathbf{p})$ times the single-particle probability of finding a particle at $(\mathbf{r}_1, \mathbf{p}_1)$. In this case, we write

$$f_2(\mathbf{r}, \mathbf{p}, \mathbf{r}_1, \mathbf{p}_1, t) = f_1(\mathbf{r}, \mathbf{p}, t) f_1(\mathbf{r}_1, \mathbf{p}_1, t) \tag{7.96}$$

and obtain for the collision integral

$$\left(\frac{\partial f}{\partial t}\right)_{\text{coll}} = [\boldsymbol{\nabla}_r \bar{\Phi}(\mathbf{r}, t)] \cdot [\boldsymbol{\nabla}_p f_1(\mathbf{r}, \mathbf{p}, t)] \tag{7.97}$$

where

$$\bar{\Phi}(\mathbf{r}, t) = \left(\frac{N}{V}\right)\int V(\mathbf{r}, \mathbf{r}_1) f_1(\mathbf{r}_1, \mathbf{p}_1, t) \, d^3 r_1 \, d^3 p_1 \tag{7.98}$$

is called the mean field potential and acts as an external potential in the resulting mean field or Vlasov (1938) kinetic equation

$$\left\{\frac{\partial}{\partial t} + \frac{\mathbf{p}}{m} \cdot \boldsymbol{\nabla}_r + [m\mathbf{F}(\mathbf{r}) - \boldsymbol{\nabla}_r \bar{\Phi}(\mathbf{r}, t)] \cdot \boldsymbol{\nabla}_p \right\} f_1(\mathbf{r}, \mathbf{p}, t) = 0 \tag{7.99}$$

We can see now what the approximation (7.96) entails. It presumes that the particles move independently of each other or uncorrelated, the effect of their mutual interactions being such that any one particle experiences an average potential field produced by all the others.‡ The latter point is reflected in the dependence of $\bar{\Phi}$ on f_1 itself. This obviously makes eqn (7.99)

---

‡ The Vlasov approximation has its analogue in the Hartree (1928) approximation in quantum-mechanical many-body theory and in the Debye–Hückel (1923) theory of dilute electrolytes.

nonlinear and demands that a solution be consistent with the mean field $\bar{\Phi}$ calculated with it. It might also be useful to perform the integration over $\mathbf{p}_1$ in eqn (7.98) formally by introducing the mass density $\rho(\mathbf{r}, t)$, eqn (7.17). This then gives for the mean field

$$\bar{\Phi}(\mathbf{r}, t) = \frac{1}{m} \int V(\mathbf{r}, \mathbf{r}_1)[\rho(\mathbf{r}_1, t) - \rho_{eq}] \, d^3 r_1 \qquad (7.100)$$

where we have also subtracted a constant term, the equilibrium density ρ_{eq}, to make clear that it is only the local nonequilibrium excess density that contributes to the mean force $\nabla \cdot \bar{\Phi}$.

A physical system for which the Vlasov equation can yield a useful description is a dilute plasma. The long-range Coulomb forces act between any two charged particles no matter how remote they are from each other. Of course, it is weaker the further apart they are. Taking then for $V(|\mathbf{r} - \mathbf{r}_1|)$ the Coulomb potential, we find that $-\nabla \cdot \bar{\Phi}$ from eqn (7.100) is simply the local electrical field of a nonzero charge distribution produced by the local deviations from charge neutrality in a nonequilibrium plasma. It should be stressed that electromagnetic retardation is not included in eqn (7.99).

The Vlasov or mean field equation is still reversible, i.e. changing t into $-t$ and $\mathbf{p}$ into $-\mathbf{p}$ results in an identical equation for $f_1(\mathbf{r}, -\mathbf{p}, -t)$. It can therefore not account for any irreversible behavior of macroscopic systems, and indeed is only valid and useful in the initial phase of the time evolution of a dilute plasma for times short compared to the macroscopic evolution time for the approach to equilibrium. This last point comes out very clearly in Balescu's (1963) derivation of the Vlasov equation, using diagrammatic techniques to resum a subset of the time-dependent perturbation series in the two-body interaction. Let us also note that in the limit where the particle charge $q \to 0$ and the particle $m \to 0$ tend to zero with $q/m = \text{const}$ and where $N/V \to \infty$ such that $(N/V)q = \text{const}$, Debye-Hückel screening becomes complete and renders collisions ineffective so that the Vlasov equation becomes exact (Braun and Hepp, 1977).

7.5.2. Derivation of Boltzmann's Equation
Let us now start with the derivation of the Boltzmann equation. We restrict ourselves to a dilute gas of neutral molecules interacting via short-range van der Waals forces (Fig. 7.1). Such a system

suggests a number of possible simplifications. First and foremost, we can neglect triple collisions because it will be very unlikely that, once two particles are within their (short-range) sphere of mutual interaction, a third particle will be there as well during the short-time interval of the two-body encounter. Apart from those rare events when two particles circle around their common center of mass in closed orbits, the time of interaction will be of the order of $\tau_0 = r_0/\bar{v}$, where r_0 is the range of the interaction and $\bar{v}$ is the average speed of a molecule. In He gas, we take $r_0 = 3$ Å and for $\bar{v}$, we choose the thermal speed at room temperature, i.e. $\bar{v} \sim 10^5$ cm/sec. We then find that $\tau_0 \sim 2.2 \times 10^{-13}$ sec as compared to the collision time $\tau_{\text{coll}} \sim 2.2 \times 10^{-10}$ sec (i.e. the inverse collision frequency) at atmospheric pressure. This implies that at most a fraction $\tau_0/\tau_{\text{coll}} \sim 10^{-3}$ of all collisions are triple encounters. They can therefore safely be ignored in a chemically non-reacting gas.‡

But this implies that we can neglect the right-hand side of eqn (7.56) for the two-particle distribution function f_2 if $|\mathbf{r} - \mathbf{r}_1| \leq r_0$ which, with this approximation, now satisfies the simple equation

$$\left\{ \frac{\partial}{\partial t} - \boldsymbol{\nabla}_r [\Phi(\mathbf{r}) + V(|\mathbf{r} - \mathbf{r}_1|)] \cdot \boldsymbol{\nabla}_p - \boldsymbol{\nabla}_{r_1} [\Phi(\mathbf{r}_1) + V(|\mathbf{r} - \mathbf{r}_1|)] \cdot \boldsymbol{\nabla}_{p_1} \right.$$
$$\left. + \frac{\mathbf{p}}{m} \cdot \boldsymbol{\nabla}_r + \frac{\mathbf{p}_1}{m} \cdot \boldsymbol{\nabla}_{r_1} \right\} f_2(\mathbf{r}, \mathbf{p}, \mathbf{r}_1, \mathbf{p}_1, t) = 0 \quad (7.101)$$

effectively truncating and thus closing the BBGKY hierarchy at this level.

We can simplify eqn (7.101) immediately even further by realizing that $f_2(\mathbf{r}, \mathbf{p}, \mathbf{r}_1, \mathbf{p}_1, t)$ will be needed in the collision integral (7.95) for only such values of $\mathbf{r}_1$ which are within the range of the molecular interaction from $\mathbf{r}$. Over such distances, we can assume that the external potential $\Phi(\mathbf{r})$ is constant as compared to the two-body potential in eqn (7.10), saying in effect

---

‡ This, of course, cannot be done in a chemically reacting gas. There, for example, two hydrogen atoms can only combine into an H_2 molecule in a collision if a third collision partner is present simultaneously to carry away the excess energy freed in binding two atoms into a molecule. In hydrogen gas at atmospheric pressure and room temperature, one out of 10^3 collisions is a triple collision and every tenth of the latter leads to the formation of an H_2 molecule (Eyring, 1962). A self-consistent kinetic theory for a dilute gas in which two-body bound states can be formed has been attempted by Green (1971).

CLASSICAL STATISTICAL MECHANICS AND KINETIC THEORY 181

that an external force field has no appreciable effect on the dynamics of a two-body collision (mediated by short-range forces) during the time of interaction. Of course, once outside each other's range of interaction, the scattering partners will be affected in their free trajectories by the external field. But this is taken care of by the presence of $\Phi(\mathbf{r})$ in eqn (7.94) for f_1.

Equation (7.101) can also be obtained as a result of Liouville's theorem. We observe that f_2 depends implicitly on time through its dependence on the positions $\mathbf{r} = \mathbf{r}(t)$ and $\mathbf{r}_1 = \mathbf{r}_1(t)$ and momenta $\mathbf{p} = \mathbf{p}(t)$ and $\mathbf{p}_1 = \mathbf{p}_1(t)$ which themselves are, of course, changing in time along the particle trajectories as a consequence of Hamilton's equations of motion, eqns (7.1). Neglecting triple collisions implies that during a binary encounter two molecules move unaffected by the rest of the gas. But then the probability

$$f_2(\mathbf{r}(t), \mathbf{p}(t), \mathbf{r}_1(t), \mathbf{p}_1(t), t) \, d\mathbf{r}(t) \, d\mathbf{p}(t) \, d\mathbf{r}_1(t) \, d\mathbf{p}_1(t) \quad (7.102)$$

of finding the molecules in the elementary volume elements along their (causal) trajectories will remain unchanged in time. On the other hand, the volume element

$$d\mathbf{r}(t) \, d\mathbf{p}(t) \, d\mathbf{r}_1(t) \, d\mathbf{p}_1(t) \quad (7.103)$$

occupied by the two molecules in phase space remains unchanged due to Liouville's theory. Thus we have

$$\frac{d}{dt} f_2(\mathbf{r}, \mathbf{p}, \mathbf{r}_1, \mathbf{p}_1, t) = 0 \quad (7.104)$$

during a binary collision event on account of the neglect of ternary interactions in a dilute gas. Here d/dt is the total time derivative given in the curly brackets in (7.101). We can furthermore argue that the partial time derivative $\partial/\partial t$ in (7.101) can be dropped. It accounts for the explicit time dependence in f_2 introduced by the overall evolution of the gas over times t_{coll}. In the collision integral we follow f_2, however, only over times τ_0 of the duration of a two-body encounter. We can thus set $\partial f_2/\partial t \approx 0$ and eqn (7.101) reduces to

$$\left\{ [\boldsymbol{\nabla}_r V(|\mathbf{r}-\mathbf{r}_1|)] \cdot \boldsymbol{\nabla}_p + [\boldsymbol{\nabla}_{r_1} V(|\mathbf{r}-\mathbf{r}_1|)] \cdot \boldsymbol{\nabla}_{p_1} \right.$$

$$\left. - \frac{\mathbf{p}}{m} \cdot \boldsymbol{\nabla}_r - \frac{\mathbf{p}_1}{m} \cdot \boldsymbol{\nabla}_{r_1} \right\} f_2(\mathbf{r}, \mathbf{p}, \mathbf{r}_1, \mathbf{p}_1, t) = 0 \quad (7.105)$$

This can now be used to rewrite the collision integral (7.95) as

$$\left(\frac{\partial f_1}{\partial t}\right)_{\text{coll}} = \left(\frac{N}{V}\right)\int d\mathbf{r}_1\, d\mathbf{p}_1 \left(\frac{\mathbf{p}}{m}\cdot\nabla_r + \frac{\mathbf{p}_1}{m}\cdot\nabla_{r_1}\right)f_2(\mathbf{r},\mathbf{p},\mathbf{r}_1,\mathbf{p}_1,t)$$

(7.106)

The second term in eqn (7.105) does not contribute, as one can show by partial integration with respect to $\mathbf{p}_1$.

Yet another consequence of the neglect of triple collisions is the fact that f_2 can depend on $\mathbf{r}_1$ only through the difference $\mathbf{r}_{12} = \mathbf{r}_1 - \mathbf{r}$. We can therefore write eqn (7.106) as

$$\left(\frac{\partial f_1}{\partial t}\right)_{\text{coll}} = \left(\frac{N}{V}\right)\int d^3r_{12}\, d^3p_1 \left(\frac{\mathbf{p}_1}{m} - \frac{\mathbf{p}}{m}\right)\cdot\nabla_{r_{12}} f_2(\mathbf{r},\mathbf{p},\mathbf{r}_{12},\mathbf{p}_1,t)$$

(7.107)

Here we left an explicit $\mathbf{r}$-dependence in f_2 which may enter through an external force field on the right-hand side of eqn (7.95) or through position-dependent initial conditions.

For a last manipulation of the collision integral we recall that the two particles entering f_2 in eqn (7.107) interact via short-range potential. We can therefore find a separation distance $R_0 > r_0$, larger than the range r_0 of the potential, such that their interaction has essentially ceased for $|\mathbf{r}_{12}| \geq R_0$. Outside R_0 the two particles will therefore move independently of each other and we can write for $|\mathbf{r}_{12}| \geq R_0$

$$f_2(\mathbf{r},\mathbf{p},\mathbf{r}_1,\mathbf{p}_1,t) = f_1(\mathbf{r},\mathbf{p},t)f_1(\mathbf{r}_1,\mathbf{p}_1,t) + \Lambda(\mathbf{r},\mathbf{p},\mathbf{r}_1,\mathbf{p}_1,t)$$

(7.108)

The factorized part is Boltzmann's celebrated *Stosszahlansatz*, or the assumption of molecular chaos. It must be emphasized that it can only be correct in precollision configurations, because the interaction between the particles induces dynamical correlations in the postcollisional state. This difference in precollisional and postcollisional behavior is the origin of irreversibility. In eqn (7.108) Λ is a correction term (Green, 1952 and 1969) that takes care of the fact that molecular interactions, though short-ranged, still extend to infinity. In physical terms, Λ includes all the grazing scattering events at large impact parameters. It is, of course, zero for a potential of truly finite range, e.g. for hard spheres.

The factorization in eqn (7.108) is not quite as straightforward as argued above. Indeed, we only know that two particles, interacting with each other at time t at position $\mathbf{r}(t)$ and $\mathbf{r}_1(t)$ with momenta $\mathbf{p}(t)$ and $\mathbf{p}_1(t)$, respectively, were at some time $t_0 < t$ separated by more than R_0, i.e. not interacting, with the consequence that—remember eqn (7.104)—

$$f_2[\mathbf{r}(t), \mathbf{p}(t), \mathbf{r}_1(t), \mathbf{p}_1(t), t]$$

$$= f_2[\mathbf{r}(t_0)\mathbf{p}(t_0), \mathbf{r}_1(t_0), \mathbf{p}_1(t_0), t_0]$$

$$= f_1[\mathbf{r}(t_0)\mathbf{p}(t_0), t_0]f_1[\mathbf{r}_1(t_0), \mathbf{p}_1(t_0), t_0] \quad (7.109)$$

with the arguments in the functions f_1 different from those in eqn (7.108). This exact statement is called the superposition principle. Equations (7.108) and (7.109) can be easily related to each other by integrating Hamilton's equations of motion (7.1) for just two particles backward in time from t to t_0. But observe that the time difference $(t - t_0)$ is of the order of the interaction time τ_0 and thus microscopic. In deriving a kinetic equation we have declared our willingness to sacrifice a complete knowledge of the time evolution of the system for the simplicity of an equation for f_1 alone, valid now, as we can see, on a time scale large compared to τ_0.

This consideration gives us also an upper limit on R_0 because $(t - t_0)$ must be small compared to the collision time τ_{coll} to avoid that an independent interaction with a third particle becomes likely. This implies that $r_0 < R_o \ll l$, where l is the mean-free path between collisions which takes over as the length scale in the system, if described by a kinetic equation. On the scale of l we therefore have $\mathbf{r}(t) \approx \mathbf{r}(t_0)$ and $\mathbf{r}_1(t) \approx \mathbf{r}_1(t_0)$. With this approximation and the further one that $t \approx t_0$, the superposition principle goes over into Boltzmann's *Stosszahlansatz*. Nothing as simple can be said about the particle momenta at times t and t_0 as one can quickly realize by visualizing large angle scattering. We postpone this problem for the present.

Returning to the collision integral (7.107), we use Gauss' theorem to convert the $\mathbf{r}_{12}$ volume integration into a surface integral over a sphere $\Sigma(R_0)$ of radius R_0, where we know that the essential part of f_2 factorizes according to eqn (7.108). We

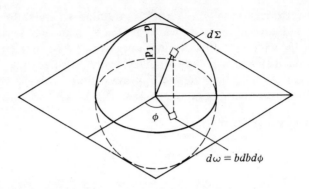

Fig. 7.2. The projection of the sphere $\Sigma(R_0)$ onto the diametrical plane ω perpendicular to $(\mathbf{p}_1 - \mathbf{p})$.

obtain

$$\left(\frac{\partial f}{\partial t}\right)_{\text{coll}} = \left(\frac{N}{V}\right) \int d^3p_1 \int_{\Sigma(R_0)} d\mathbf{\Sigma} \cdot \frac{\mathbf{p}_1 - \mathbf{p}}{m} f_1(\mathbf{r}, \mathbf{p}', t) f_1(\mathbf{r}, \mathbf{p}_1', t)$$
$$+ \frac{N}{V} \int d^3p_1 \, d^3r_1 \frac{\mathbf{p}_1 - \mathbf{p}}{m} \cdot \nabla_{r_1} \Lambda(\mathbf{r}, \mathbf{p}, \mathbf{r}_1, \mathbf{p}_1, t) \quad (7.110)$$

Concentrating on the first term we still must specify the momenta $\mathbf{p}'$ and $\mathbf{p}_1'$ in terms of $\mathbf{p}$ and the integration variables. To do this, we introduce as a parameter domain for the integration over the sphere $\Sigma(R_0)$ that diametrical plane ω of the sphere which is perpendicular to the direction $(\mathbf{p}_1 - \mathbf{p})$ (See Fig. 7.2). We introduce polar coordinates (b, ϕ) in this plane with $0 < b < R_0$ and $0 < \phi < 2\pi$ and write

$$d\omega = b \, db \, d\phi \quad (7.111)$$

for its area element; ϕ is measured from an arbitrary axis in ω. This projection of $\Sigma(R_0)$ onto ω obviously covers ω twice. For the upper hemisphere, referring to particles leaving the interaction volume enclosed by $\Sigma(R_0)$ with final momenta $\mathbf{p}^f$ and $\mathbf{p}_1^f$, we have

$$(\mathbf{p}_1 - \mathbf{p}) \cdot d\mathbf{\Sigma} = |\mathbf{p}_1 - \mathbf{p}| \, d\omega > 0 \quad (7.112)$$

For the lower hemisphere, referring to particles entering the interaction volume for the first time with initial momenta $\mathbf{p}^i$ and

$\mathbf{p}_1^i$, we obtain

$$(\mathbf{p}_1-\mathbf{p})\cdot d\mathbf{\Sigma} = -|\mathbf{p}_1-\mathbf{p}|\,d\omega < 0 \tag{7.113}$$

The first term in eqn (7.110) therefore takes on the form

$$\left(\frac{N}{V}\right)\int d^3p_1 \int_0^{R_0} b\,db \int_0^{2\pi} d\phi\,|\mathbf{p}_1-\mathbf{p}|\,[f_1(\mathbf{r},\mathbf{p}^f,t)f_1(\mathbf{r},\mathbf{p}_1^f,t)$$
$$- f_1(\mathbf{r},\mathbf{p}^i,r)f_1(\mathbf{r},\mathbf{p}_1^i,t)] \tag{7.114}$$

According to the superposition principle (7.109), the momentum arguments in the f_1's are those that the particles had at time t_0 before the scattering process commenced. This leads trivially to the identification $\mathbf{p}^i = \mathbf{p}$ and $\mathbf{p}_1^i = \mathbf{p}_1$. It remains then to express $\mathbf{p}^f$ and $\mathbf{p}_1^f$ in terms of a set of initial momenta (at time t_0). To do this we invoke—for elastic scattering of equal mass particles—momentum conservation

$$\mathbf{p} + \mathbf{p}_1 = \mathbf{p}^f + \mathbf{p}_1^f \tag{7.115}$$

and energy conservation

$$(\mathbf{p})^2 + (\mathbf{p}_1)^2 = (\mathbf{p}^f)^2 + (\mathbf{p}_1^f)^2$$

These four equations (one vector and one scalar) allow a two-parameter family of solutions for the six unknown components of $\mathbf{p}^f$ and $\mathbf{p}_1^f$. A useful parametrization (Waldmann, 1958) is in terms of a unit vector

$$\mathbf{e}' = \frac{\mathbf{p}^f - \mathbf{p}_1^f}{|\mathbf{p}^f - \mathbf{p}_1^f|} \tag{7.116}$$

which reads

$$\begin{aligned}\mathbf{p}^f &= \tfrac{1}{2}(\mathbf{p}+\mathbf{p}_1+|\mathbf{p}-\mathbf{p}_1|\,\mathbf{e}')\\ \mathbf{p}_1^f &= \tfrac{1}{2}(\mathbf{p}+\mathbf{p}_1-|\mathbf{p}-\mathbf{p}_1|\,\mathbf{e}')\end{aligned} \tag{7.117}$$

It still remains to connect the parameter $\mathbf{e}'$ with the integration variables b and ϕ in eqn (7.114). This can be done via the relation‡

$$\int_0^{2\pi}\int_0^{R_0}\ldots,d\phi\,b\,db = \int\ldots,\sigma(\chi,|\mathbf{p}-\mathbf{p}_1|)\,d\mathbf{e}' \tag{7.118}$$

‡ For a derivation, see Waldmann (1958) or Grad (1958), in which other equivalent representations of the collision integral can be found.

where $\sigma(\chi, |\mathbf{p}-\mathbf{p}_1|)$ is the scattering cross section which, for a spherical interaction, is a function of the magnitude of the relative initial momentum $|\mathbf{p}-\mathbf{p}_1|$ and the scattering angle χ defined by

$$\cos \chi = \frac{\mathbf{p}-\mathbf{p}_1}{|\mathbf{p}-\mathbf{p}_1|} \cdot \frac{\mathbf{p}^f-\mathbf{p}_1^f}{|\mathbf{p}^f-\mathbf{p}_1^f|} \tag{7.119}$$

The final form of the Boltzmann equation then reads

$$\left[\frac{\partial}{\partial t}+\frac{\mathbf{p}}{m}\cdot\nabla_r + m\mathbf{F}(\mathbf{r})\cdot\nabla_p\right]f_1(\mathbf{r},\mathbf{p},t)$$

$$= \left(\frac{N}{V}\right)\int |\mathbf{p}_1-\mathbf{p}|\,\sigma(\chi,|\mathbf{p}_1-\mathbf{p}|)[f_1(\mathbf{r},\mathbf{p}^f,t)f_1(\mathbf{r},\mathbf{p}_1^f,t)$$

$$- f_1(\mathbf{r},\mathbf{p},t)f_1(\mathbf{r},\mathbf{p}_1,t)]\,d^3p_1\,d\mathbf{e}' \tag{7.120}$$

7.5.3. Discussion of the Boltzmann Equation

The Boltzmann equation (7.120) has an intuitive physical interpretation which, indeed, originally guided Boltzmann to postulate it. The probability $f_1(\mathbf{r},\mathbf{p},t)\,d^3r\,d^3p$ of finding gas particles in a volume element $d^3r\,d^3p$ around the point $(\mathbf{r},\mathbf{p})$ in the single-particle phase space γ can change on account of four effects. There can be an explicit time dependence in the problem, originating from initial conditions and resulting in the term $\partial f_1/\partial t$. Associated with spatial inhomogeneities there will be a drift of particles into $d^3r\,d^3p$ described by $\mathbf{p}/m \cdot \nabla_r f_1$. Writing $\mathbf{F}(\mathbf{r}) = \dot{\mathbf{p}}/m$, we see that $\dot{\mathbf{p}} \cdot \nabla_p f_1$ describes a 'drift' in momentum space due to inhomogeneities there. These drift terms can be absorbed into a generalized barycentric derivative

$$\frac{\tilde{D}}{\tilde{D}t} = \frac{\partial}{\partial t}+\frac{\mathbf{p}}{m}\cdot\nabla_r + \dot{\mathbf{p}}\cdot\nabla_p \tag{7.121}$$

in terms of which Boltzmann's equation reads

$$\frac{\tilde{D}f_1}{\tilde{D}t} = \left(\frac{\partial f_1}{\partial t}\right)_{\text{coll}} \tag{7.122}$$

stating that, in addition, f_1 will change on account of the net flux of particles into and out of $d^3r\,d^3p$ due to two-particle collisions.

It might be useful to recapitulate the basic assumptions necessary to establish the Boltzmann equation. First and foremost is the neglect of triple collisions with which we achieve the truncation and closure of the BBGKY hierarchy at the level of the two-particle distribution function. We are thus restricted to deal with dilute gases of neutral molecules interacting via short-range forces in which the range of interaction r_0 is much smaller than the average separation between gas particles and their mean-free path between collisions. The state of the gas and the processes therein are therefore completely controlled by binary collisions.

The second major assumption is contained in the *Stosszahlansatz* relating the two-particle distribution function f_2 to the one-particle distribution function f_1, symbolically written as $f_2 = f_1 f_1$. To make this relation of any use in the evaluation of the collision integral, two approximations had to be invoked in the superposition principle: a coarse-graining of space over distance of the order of the range r_0 of the interaction and a coarse-graining of time over intervals of the order of the duration of a collision τ_0. This coarse-graining implies that we regard the details of the molecular motion over distances r_0 and times τ_0 as irrelevant for the statistical evolution of the gas. The superposition-ansatz with this coarse-graining of space and time is equivalent to Boltzmann's *Stosszahlansatz*, i.e. his assumption of molecular chaos. Physically, it means that all two-body encounters in the gas are statistically independent. Thus no particle entering a collision carries any information about a previous encounter and its memory about dynamical correlations from previous collisions is wiped out before a new collision starts. This statistical *Stosszahlansatz* introduces a distinction between the event 'before a collision' and the event 'after a collision' and is the source of irreversibility in the Boltzmann equation.

In a dilute gas, a statistical description in terms of a simple kinetic equation, for the single-particle distribution function, namely Boltzmann's equation, becomes possible at the expense of a loss of dynamical information. This implies, in particular, that the Boltzmann equation is only valid over a coarse-grained space-time manifold; i.e. it describes the 'smoothed' behavior of the system over times large compared to the interaction time τ_0 and over distances large compared to the range r_0 of the interaction.‡

---

‡ Kirkwood (1946) performs a coarse-graining explicitly by averaging the distribution functions over times τ_0. However, as can be seen in the above derivation of Boltzmann's equation, this time averaging is not necessary and should therefore be irrelevant for the physics of a dilute gas, as H. S. Green and M. Born have pointed out repeatedly.

It is worthwhile at this stage to stress the fundamental difference between Vlasov's and Boltzmann's equations. The former, we have seen, is valid in the transient phase of the time evolution of the system and is still reversible. The latter describes the statistical evolution over a coarse-grained manifold only and is irreversible. In this connection it is also satisfying to see that the information about the two-particle dynamics enters the Boltzmann equation in a statistical statement, namely as the scattering cross section in the collision integral. The Boltzmann equation is a nonlinear one and is valid for arbitrarily large deviation from equilibrium even over a mean-free path l and for times short compared to the collision time τ_{coll}.

Grad (1958) has argued that the Boltzmann equation should be an exact statement in a gas of N particles of mass m interacting via a potential that scales according to

$$V(r) = d^2 \hat{V}\left(\frac{r}{d}\right) \tag{7.123}$$

in the limit $N \to \infty$, $m \to 0$, $d \to 0$ with

$$Nm = \text{const} \tag{7.124}$$

and

$$Nd^2 = \text{const} \tag{7.125}$$

In this Grad-limit,‡ all macroscopic quantities derived from f_1 (like ρ, p, $\mathbf{v}$, etc.) remain constant as will the mean-free path l and the collision time τ_{coll}. In the case of hard-sphere molecules, the volume occupied by them will tend to zero, i.e.

$$Nd^3 \to 0 \tag{7.126}$$

which implies that the imperfect gas correction in the equation of state tends to zero. Thus the potential energy part $\varepsilon^{(V)}$ of the internal energy drops out, i.e. the gas will in equilibrium obey the ideal gas law $p = (N/V)k_B T$, but the collisional effect of the intermolecular forces is retained to ensure irreversible behavior out of equilibrium. Lanford (1975) has recently proved that for a gas of hard-sphere molecules the Boltzmann equation is exact in

‡ The Grad, or hydrodynamic, limit is obviously not identical with the thermodynamic limit because the volume V occupied by our system is held fixed.

the Grad limit at least for an initial time interval of the order of a fraction of the collision time. That is to say, given at time $t = 0$ initial values of the n-particle distribution function f_n that factorize

$$f_n(\mathbf{r}_1, \mathbf{p}_1, \mathbf{r}_2, \mathbf{p}_2, \ldots, \mathbf{r}_n, \mathbf{p}_n, t) = \prod_{i=1}^{n} f_1(\mathbf{r}_i, \mathbf{p}_i, t) \qquad n = 1, \ldots, N$$

(7.127)

there will be solutions of the BBGKY hierarchy (in the Grad limit) with these initial conditions that remain factorized for the above time period, and $f_1(\mathbf{r}, \mathbf{p}, t)$ will also be a solution of the Boltzmann equation. Although an extension of these results for all times seems plausible, no proof has been given yet.[‡]

Our final comments are concerned with an extension of kinetic theory to dense gases and liquids. First corrections to the Boltzmann equation were obtained by Green (1952 and 1969) and Bogolyubov (1946) in the spirit of the Λ term in eqn (7.110).[§] The latter also generalizes Boltzmann's *Stosszahlansatz* by postulating in the kinetic regime a general time-dependent functional relationship expressing all higher n-particle distribution functions f_n in terms of f_1 alone. His argument is as follows: In the initial statistical stage the system will evolve rapidly on a time scale τ_0, and the complete BBGKY hierarchy is needed for an adequate description. In the kinetic stage, i.e. on the time scale of τ_{coll}, it seems reasonable to assume that all f_n for $n \geq 2$ have relaxed and are synchronized with f_1. Thus Bogolyubov suggests for $\tau_0 < t < \tau_{\text{coll}}$

$$f_n(\mathbf{r}_1, \mathbf{p}_1, \ldots, \mathbf{r}_n, \mathbf{p}_n, t) = \mathcal{F}_n(f_1) \qquad n = 2, \ldots, N \quad (7.128)$$

where any time dependence only enters implicitly via f_1. This implies that we can evaluate the time derivative of $\mathcal{F}_n$ as a functional derivative as follows:

$$\frac{\partial \mathcal{F}_n}{\partial t} = \frac{\delta \mathcal{F}_n}{\delta f_1} \frac{\partial f_1}{\partial t} \qquad (7.129)$$

where the first member of the BBGKY hierarchy should be used

[‡] Lanford's result shows that in the Grad limit there is no contradiction between the reversibility of molecular dynamics and the irreversibility of the Boltzmann equation. We will come back to this point in Chapter 11.

[§] See also Green (1971) and Blatt and Opie (1974).

to eliminate $\partial f_1/\partial t$. Next it is suggested that a density expansion is attempted, i.e., to write symbolically

$$\mathscr{F}_n = \mathscr{F}_n^{(0)} + \left(\frac{N}{V}\right)\mathscr{F}_n^{(1)} + \left(\frac{N}{V}\right)^2 \mathscr{F}_n^{(2)} + \cdots \qquad (7.130)$$

which is inserted into the BBGKY hierarchy after eliminating $\partial \mathscr{F}_n/\partial t$ by means of eqn (7.129). Equating powers of (N/V), one arrives at a new hierarchy of equations that can be truncated and closed trivially at whatever maximum power of (N/V) one chooses. Keeping only the term $\mathscr{F}_n^{(0)}$, one can derive Bogolyubov's generalization of the Boltzmann equation. We will not pursue this approach because the density expansion is nonconvergent and, indeed, all terms beyond $(N/V)^2$ are infinite or divergent. To understand this divergence, recall that the Boltzmann equation only accounts for binary collision, as does the term $\mathscr{F}_n^{(0)}$ from which the Boltzmann equation can be derived. $\mathscr{F}_n^{(1)}$ will then take triple collisions into account, as shown in great detail by Choh and Uhlenbeck (1958), Dorfman and Cohen (1965), and Cohen (1966). Similarly the volume in phase space of those particles 3 and 4 which contribute to $\mathscr{F}_2^{(2)}$ increases in three dimensions proportional to $\ln(t/\tau_{\text{coll}})$ and in two dimensions proportional to (t/τ_{coll}). In general, the phase volumes of the particles 3, 4, ..., l which contribute to $\mathscr{F}_2^{(l)}$ for $t \gg \tau_{\text{coll}}$ increase with time: $(t/\tau_{\text{coll}})^{l-4}$ in three dimensions and $(t/\tau_{\text{coll}})^{l-3}$ in two dimensions for $l \geq 5$. This implies that the expansion (7.130) contains secular terms and that the coefficients $\mathscr{F}_2^{(l)}$ do not exist beyond $l = 1$ in three dimensions. Simple-minded density expansions in the BBGKY hierarchy intended to arrive at kinetic equations in a dense gas have failed, and the whole field is currently under active review. (See Ernst, Haines, and Dorfman, 1969; Rice, Boon, and Davis, 1968; Alder, 1973.) These difficulties are highlighted in an exactly soluble model studied by Anstis, Green, and Hoffman (1973).

7.6. Balance Equations from the Boltzmann Equation

We again turn our attention to a dilute gas as described by the Boltzmann equation (7.120). Our first aim is a derivation of the balance equations for density, momentum, energy, and entropy, the first three being defined by eqns (7.17)–(7.19).

As a first step we show that

$$\int \psi_i(\mathbf{p}) \left(\frac{\partial f}{\partial t}\right)_{\text{coll}} d^3p = 0 \qquad i = 0, 1, \ldots, 4 \qquad (7.131)$$

for the five summational invariants $\psi_0 = 1$, $(\psi_1, \psi_2, \psi_3) = \mathbf{p}$, and $\psi_4 = \mathbf{p}^2$ of a two-body scattering process. For this we have, referring to eqn (7.115),

$$\psi_i(\mathbf{p}^i) + \psi_i(\mathbf{p}_1^i) - \psi_i(\mathbf{p}^f) + \psi_i(\mathbf{p}_1^f) \qquad (7.132)$$

where the superscripts i and f refer to initial and final scattering momenta. To show this, we multiply the right-hand side of eqn (7.120) by $\psi_i(\mathbf{p})$ and integrate over $\mathbf{p}$ to obtain

$$\int \psi_i(\mathbf{p}) \sigma(\chi, |\mathbf{p} - \mathbf{p}_1|) |\mathbf{p}_1 - \mathbf{p}| [f_1(\mathbf{r}, \mathbf{p}^f, t) f_1(\mathbf{r}, \mathbf{p}_1^f, t)$$

$$- f_1(\mathbf{r}, \mathbf{p}, t) f_1(\mathbf{r}, \mathbf{p}_1, t)] d\mathbf{e}' \, d^3p_1 \, d^3p \qquad (7.133)$$

In the first term we change $(\mathbf{p}, \mathbf{p}_1)$ to $(\mathbf{p}^f, \mathbf{p}_1^f)$, use Liouville's theorem with $\mathbf{e} = (\mathbf{p}_1 - \mathbf{p})/|\mathbf{p}_1 - \mathbf{p}|$,

$$d\mathbf{e} \, d^3p_1^f \, d^3p_1^f = d\mathbf{e}' \, d^3p_1 \, d^3p \qquad (7.134)$$

and find that eqn (7.133) is equal to

$$\int [\psi_i(\mathbf{p}^f) - \psi_i(\mathbf{p})] \sigma(\chi, |\mathbf{p}_1 - \mathbf{p}|) |\mathbf{p}_1 - \mathbf{p}| \, f_1(\mathbf{r}, \mathbf{p}, t) f_1(\mathbf{r}, \mathbf{p}_1, t) \, d\mathbf{e}' \, d^3p \, d^3p_1 \qquad (7.135)$$

A further change of $(\mathbf{p}, \mathbf{p}_1, \mathbf{p}^f, \mathbf{p}_1^f)$ into $(\mathbf{p}_1, \mathbf{p}, \mathbf{p}_1^f, \mathbf{p}^f)$ leads to

$$\frac{1}{2} \int [\psi_i(\mathbf{p}^f) + \psi_i(\mathbf{p}_1^f) - \psi_i(\mathbf{p}) - \psi_i(\mathbf{p}_1)] \sigma(\chi, |\mathbf{p}_1 - \mathbf{p}|) |\mathbf{p}_1 - \mathbf{p}|$$

$$\times f_1(\mathbf{r}, \mathbf{p}, t) f_1(\mathbf{r}, \mathbf{p}_1, t) \, d\mathbf{e}' \, d^3p \, d^3p_1 = 0 \qquad (7.136)$$

on account of eqn (7.132) for the five summational invariants of the two-body problem; i.e. eqn (7.131) is proved.

Armed with relation (7.131) it is very easy to derive the balance equations for mass, momentum, and energy. Indeed, nothing remains that has not been done already in the general situation in Section 7.3 starting from the first member of the BBGKY hierarchy, because the latter only differs from the Boltzmann equation in its right-hand side which, now, however, does not contribute anything. We simply collect the relevant formulae from Section 7.3. The continuity equation (7.24) reads

$$\frac{\partial \rho}{\partial t} + \boldsymbol{\nabla} \cdot (\rho \mathbf{v}) = 0 \tag{7.137}$$

the momentum balance equation (7.42) is

$$\frac{\partial (\rho \mathbf{v})}{\partial t} + \boldsymbol{\nabla} \cdot (\rho \mathbf{v} \mathbf{v} + \mathbf{P}) = \rho \mathbf{F} \tag{7.138}$$

and the internal energy balance states that

$$\frac{\partial (\rho u)}{\partial t} + \boldsymbol{\nabla} \cdot (\rho u \mathbf{v} + \mathbf{j}_q) = -\mathbf{P} : (\boldsymbol{\nabla} \mathbf{v}) \tag{7.139}$$

where the pressure tensor eqn (7.31)

$$\mathbf{P} = \mathbf{P}^{(K)} = \left(\frac{N}{V}\right) \int \frac{(\mathbf{p} - m\mathbf{v})(\mathbf{p} - m\mathbf{v})}{m} f_1(\mathbf{r}, \mathbf{p}, t) \, d^3p \tag{7.140}$$

and the heat current eqn (7.52)

$$\mathbf{j}_q = \mathbf{j}_q^{(K)} = \left(\frac{N}{V}\right) \int \frac{\mathbf{p} - m\mathbf{v}}{m} \frac{(\mathbf{p} - m\mathbf{v})^2}{2m} f_1(\mathbf{r}, \mathbf{p}, t) \, d^3p \tag{7.141}$$

consist of the kinetic energy contribution only due to the vanishing of the collision integral terms.

As a preliminary step for the derivation of the entropy balance, we want to derive a balance equation for a local $\mathcal{H}$ quantity, defined by

$$\rho h(\mathbf{r}, t) = \left(\frac{N}{V}\right) \int f_1(\mathbf{r}, \mathbf{p}, t) \ln \tilde{f}_1(\mathbf{r}, \mathbf{p}, t) \, d^3p \tag{7.142}$$

where $\tilde{f}_1 = (N/V)f_1$. We multiply the Boltzmann equation (7.120) by $(N/V)[\ln \tilde{f}_1(\mathbf{r}, \mathbf{p}, t) + 1]$ and integrate over $\mathbf{p}$. For the left-hand side we note that

$$\begin{aligned}\frac{\partial}{\partial t}(\tilde{f}_1 \ln \tilde{f}_1) &= (\ln \tilde{f}_1 + 1) \frac{\partial \tilde{f}_1}{\partial t} \\ \boldsymbol{\nabla}_r(\tilde{f}_1 \ln \tilde{f}_1) &= (\ln \tilde{f}_1 + 1) \boldsymbol{\nabla}_r \tilde{f}_1\end{aligned} \tag{7.143}$$

and obtain

$$\frac{\partial}{\partial t}[\rho h(\mathbf{r}, t)] + \boldsymbol{\nabla}_r \cdot \mathbf{j}_h(\mathbf{r}, t) = \sigma_h \tag{7.144}$$

where the h current is given by

$$\mathbf{j}_h(\mathbf{r}, t) = \left(\frac{N}{V}\right) \int \frac{\mathbf{p}}{m} f_1(\mathbf{r}, \mathbf{p}, t) \ln \tilde{f}_1(\mathbf{r}, \mathbf{p}, t) \, d^3p \qquad (7.145)$$

and the source density of h can be written as

$$\begin{aligned}
\sigma_h &= \left(\frac{N}{V}\right)^2 \int |\mathbf{p}_1 - \mathbf{p}| \, \sigma(\chi, |\mathbf{p}_1 - \mathbf{p}|)[\ln \tilde{f}_1(\mathbf{r}, \mathbf{p}, t) + 1] \\
&\quad \times [f_1(\mathbf{r}, \mathbf{p}^f, t) f_1(\mathbf{r}, \mathbf{p}_1^f, t) - f_1(\mathbf{r}, \mathbf{p}, t) f_1(\mathbf{r}, \mathbf{p}_1, t)] \, d\mathbf{e}' \, d^3p_1 \, d^3p \\
&= -\frac{1}{4}\left(\frac{N}{V}\right)^2 \int |\mathbf{p}_1 - \mathbf{p}| \, \sigma(\chi, |\mathbf{p}_1 - \mathbf{p}|)[f_1(\mathbf{r}, \mathbf{p}^f, t) f_1(\mathbf{r}, \mathbf{p}_1^f, t) \\
&\quad - f_1(\mathbf{r}, \mathbf{p}, t) f_1(\mathbf{r}, \mathbf{p}_1, t)] \ln \frac{\tilde{f}_1(\mathbf{r}, \mathbf{p}^f, t) \tilde{f}_1(\mathbf{r}, \mathbf{p}_1^f, t)}{\tilde{f}_1(\mathbf{r}, \mathbf{p}, t) \tilde{f}_1(\mathbf{r}, \mathbf{p}_1, t)} \, d\mathbf{e}' \, d^3p_1 \, d^3p
\end{aligned}$$
$$(7.146)$$

To go from the first to the second expression for σ_h, we dropped the 1 because it is a summational invariant and then used the symmetries of the scattering process with respect to the momenta $\mathbf{p}$, $\mathbf{p}_1$, $\mathbf{p}^f$, and $\mathbf{p}_1^f$ in a similar way to what was done to go from eqns (7.133) to (7.136). Obviously, σ_h is nonpositive because

$$(ab - cd) \ln \frac{ab}{cd} \geq 0 \qquad (7.147)$$

for any positive numbers a, b, c, and d.

Let us next integrate eqn (7.144) over the volume V occupied by the gas and define a total $\mathcal{H}$ quantity (Boltzmann, 1872)

$$\mathcal{H}(t) = \int_V \rho(\mathbf{r}, t) h(\mathbf{r}, t) \, d^3r \qquad (7.148)$$

for which we obtain

$$\frac{\partial \mathcal{H}(t)}{\partial t} = -\int_\Sigma \mathbf{j}_h(\mathbf{r}, t) \cdot d\mathbf{\Sigma} + \int_V \sigma_h(\mathbf{r}, t) \, d^3r \qquad (7.149)$$

where we used Gauss' theorem to convert the volume integral over the divergence of $\mathbf{j}_h$ into an integral over the surface Σ enclosing V. Let us, for the moment, suppose that the gas is thermally isolated. Then there will be no flux of h through the surface, and the surface integral in eqn (7.149) vanishes. Under this condition $\mathcal{H}(t)$ will be a monotonically decreasing function of time which vanishes if

$$f_1^{(0)}(\mathbf{r}, \mathbf{p}^f, t) f_1^{(0)}(\mathbf{r}, \mathbf{p}_1^f, t) = f_1^{(0)}(\mathbf{r}, \mathbf{p}, t) f_1^{(0)}(\mathbf{r}, \mathbf{p}_1, t) \qquad (7.150)$$

This is the contents of Boltzmann's famous $\mathcal{H}$ theorem. It implies that $\ln f_1^{(0)}(\mathbf{r}, \mathbf{p}, t)$ is a collisional invariant and must therefore be expresssible as

$$\ln f_1^{(0)}(\mathbf{r}, \mathbf{p}, t) = \Phi_0(\mathbf{r}, t) + \Phi_1(\mathbf{r}, t) \cdot \mathbf{p} + \Phi_2(\mathbf{r}, t)(\mathbf{p})^2 \qquad (7.151)$$

Inserting this into the definitions (7.17)–(7.19) for the local macroscopic mass density, velocity, and internal energy density, we can specify the functions Φ_0, Φ_1, and Φ_2 and obtain for $f_1^{(0)}$ a locally maxwellian distribution function, eqn (7.45),

$$f_1^{(0)}(\mathbf{r}, \mathbf{p}, t) = [2\pi m k_B T(\mathbf{r}, t)]^{-\frac{3}{2}} \exp\left[-\frac{(\mathbf{p} - m\mathbf{v})^2}{2m k_B T(\mathbf{r}, t)}\right] \qquad (7.152)$$

where we used eqn (7.46) to introduce a temperature field $T(\mathbf{r}, t)$. For this distribution function we can also identify up to a constant the local entropy density

$$\begin{aligned}\rho s^{(0)}(\mathbf{r}, t) &= -k_B \rho h^{(0)}(\mathbf{r}, t) \\ &= \frac{N}{V} k_B \left[\frac{5}{2} + \ln\left(\frac{(2\pi m k_B T)^{\frac{3}{2}} V}{h^3 N}\right)\right] + \frac{N}{V} k_B \ln\left(\frac{h^3}{e}\right)\end{aligned} \qquad (7.153)$$

the first part being the well-known expression from equilibrium statistical mechanics. We can then calculate the local $\mathcal{H}$ flux

$$\mathbf{j}_h^{(0)}(\mathbf{r}, t) = \rho \mathbf{v} h(\mathbf{r}, t) \qquad (7.154)$$

or the local entropy current

$$\mathbf{j}_s^{(0)}(\mathbf{r}, t) = \rho s \mathbf{v} \qquad (7.155)$$

and see that both are purely convective in a system in strict local equilibrium.‡ Dissipative effects are therefore produced by deviations from strict local equilibrium, a fact that we have already used extensively in our general derivation of constitutive laws in Section 7.4.

‡ We call a system described by a distribution function $f_1^{(0)}$, eqn (7.152), in strict local equilibrium, in contrast to systems in local equilibrium, in which infinitesimally small deviations from strict local equilibrium do occur that are controlled by the balance equations of Chapter 2.

One would have hoped at this stage that a relation $s(\mathbf{r}, t) = -k_B h(\mathbf{r}, t)$ with $h(\mathbf{r}, t)$ given by eqn (7.142) could be used in general to introduce a local nonequilibrium entropy density at least in the dilute gas described by the Boltzmann equation. Although such a procedure can always be adopted as a matter of definition, it seems neither useful nor convincing for a number of reasons. The first is that

$$-k_B \mathbf{j}_h(\mathbf{r}, t) \neq -k_B \rho h(\mathbf{r}, t)\mathbf{v} + \frac{1}{T}\mathbf{j}_q \tag{7.156}$$

That is, adopting definitions (7.141) for the heat current and (7.142) for $h(\mathbf{r}, t)$, we find that the sum of the right side cannot be reduced to the expression on the left with $\mathbf{j}_h(\mathbf{r}, t)$ defined by (7.145). Such an equality, however, has to hold according to the statement (2.69) for reasons of consistency of the set of all balance equations. According to eqns (2.69) and (2.70), we must have

$$\mathbf{j}_s(\mathbf{r}, t) = \rho s \mathbf{v} + \frac{1}{T}\mathbf{j}_q \tag{7.157}$$

and

$$\sigma_s = \mathbf{j}_q \cdot \nabla\left(\frac{1}{T}\right) - \frac{1}{T}(\mathbf{P} - \tfrac{1}{3}Tr(\mathbf{P})\mathbf{I}) : (\nabla \mathbf{v}) \tag{7.158}$$

in order that the balance equations for ρ, $\mathbf{v}$, u and s are consistent with the fundamental relation

$$Tds = du + pd\rho^{-1} \tag{7.159}$$

The appearance of temperature in these expressions obviously causes all the problems because, in general, the quantities $(-k_B \mathbf{j}_h)$ and $(-k_B \sigma_h)$ that we wish to identify with the left-hand sides of eqns (7.157) and (7.158), respectively, do not even contain a temperature. The concept of temperature is one that makes sense rigorously only in systems in equilibrium, and can be extended in a meaningful way in systems that are at least locally close to equilibrium. This is the basic assumption on which nonequilibrium thermodynamics rests. But then we can hope that $-k_B h(\mathbf{r}, t)$ can be identified as the local entropy density for systems in which the solutions of the Boltzmann equation can be, in some sense, expanded, around the state of local equilibrium eqn (7.152) as

$$f_1 = f_1^{(0)} + f_1^{(1)} + f_1^{(2)} + \cdots \qquad (7.160)$$

Such solutions will be developed shortly in a rigorous way. At this stage, we can already anticipate that this will give us a means to determine rigorously, in the case of a dilute gas, the range of validity of the local equilibrium assumption by looking at the first contributions from eqn (7.160) that result in deviations in eqns (7.145) and (7.146) from the canonical forms (7.157) and (7.158).

Two points must be stressed again. First, because a kinetic description (Boltzmann equation) is more detailed than a hydrodynamic one, we should not expect that all the kinetic concepts can be reduced to thermodynamic ones. The balance equation for the local $\mathcal{H}$ quantity is always valid, necessary, and useful, even in situations which bear absolutely no resemblance to local equilibrium, as long as the gas is describable by the Boltzmann equation. Secondly, in systems in local equilibrium, the fundamental relation of thermodynamics is absolutely necessary for the introduction of entropy because it is the operational definition for its measurement.

7.7. Constitutive Laws and Transport Coefficients from the Boltzmann Equation

In Sections 7.3 and 7.6 we introduced the macroscopic density ρ, the velocity $\mathbf{v}$, and the internal energy u and derived the general balance equations they are subject to. We have seen that this can always be done without any restriction on the systems considered. This, of course, implies that a description of the system's evolution in terms of ρ, $\mathbf{v}$, and u alone is not complete, in particular not in the transient and kinetic stage of the time evolution in which drastic changes can occur over distances of the mean-free path l and times of the order of the collision time τ_{coll}. Obviously, additional information is contained in the distribution function $f_1(\mathbf{r}, \mathbf{p}, t)$, as is made apparent in our discussion of the local $\mathcal{H}$ quantity $h(\mathbf{r}, t)$ in the last section.

In the course of the time evolution of our system, a gas say, the stage of rapid evolution over times τ_{coll} and distance l will in most systems eventually go over into a smooth behavior over characteristic distances large compared to l and characteristic times large compared to τ_{coll}. In this hydrodynamic stage the system's state will be more and more completely characterized by the local macroscopic fields ρ, $\mathbf{v}$, and u. In a gas that, in the kinetic stage, is well described by the Boltzmann equation, this implies that the single-particle distribution function $f_1(\mathbf{r}, \mathbf{p}, t)$ must be a function of ρ, $\mathbf{v}$, and u. If we are seeking a description of the system's time evolution in the hydrodynamic regime, we can turn this argument around and insist that we should in this case only look for so-called normal solutions of the Boltzmann equation that show such a functional dependence on ρ, $\mathbf{v}$, and u. This can be achieved by invoking an additional limiting process (Grad, 1958) in which we let l and τ_{coll} approach zero in the Boltzmann equation. To this end we can introduce a small parameter ε in the Boltzmann equation

$$\left[\frac{\partial}{\partial t}+\frac{\mathbf{p}}{m}\cdot\nabla_r + m\mathbf{F}(\mathbf{r})\cdot\nabla_p\right]f_1(\mathbf{r}, \mathbf{p}, t) = \frac{1}{\varepsilon}\left(\frac{\partial f_1}{\partial t}\right)_{\text{coll}} \quad (7.161)$$

and look for solutions that can be expanded in ε (Hilbert, 1924), i.e.

$$f_1 = f_1^{(0)} + \varepsilon f_1^{(1)} + \varepsilon^2 f_1^{(2)} + \cdots \quad (7.162)$$

Inserting this in eqn (7.161) and equating powers, we obtain

$$\left(\frac{\partial f_1^{(0)}}{\partial t}\right)_{\text{coll}} = 0 \quad (7.163)$$

$$2J(f_1^{(0)}, f_1^{(n)}) = \dot{f}_1^{(n-1)} - \theta(n-1)\sum_{m=1}^{n-1} J(f_1^{(m)}, f_1^{(n-m)}) \quad (7.164)$$

where we defined

$$J(f, g) = \frac{1}{2}\frac{N}{V}\int |\mathbf{p}_1 - \mathbf{p}|\,\sigma(\chi, |\mathbf{p}_1 - \mathbf{p}|)$$
$$\times [f(\mathbf{r}, \mathbf{p}^f, t)g(\mathbf{r}, \mathbf{p}_1^f, t) + f(\mathbf{r}, \mathbf{p}_1^f, t)g(\mathbf{r}, \mathbf{p}^f, t)$$
$$- f(\mathbf{r}, \mathbf{p}, t)g(\mathbf{r}, \mathbf{p}_1, t) - f(\mathbf{r}, \mathbf{p}_1, t)g(\mathbf{r}, \mathbf{p}, t)]\,d^3p\,d\mathbf{e}' \quad (7.165)$$

Equation (7.163) implies that $f_1^{(0)}$ is a local maxwellian distribution, eqn (7.152). Thus the Hilbert expansion is one around the state of local equilibrium. The higher corrections $f_1^{(n)}$ all satisfy the same type of integral equations (7.164) provided the lower-order terms $f_1^{(1)}, \ldots, f_1^{(n-1)}$ have been determined. Rather than pursuing Hilbert's approach we will present below the Chapman-Enskog theory (Chapman, 1916; Enskog, 1917).‡ Important to remember are the facts that the Hilbert expansion constructs those normal solutions of the Boltzmann equation that vary smoothly on the scales τ_{coll} and l. Information about the initial evolution over times of the order to τ_{coll} is suppressed as are boundary effects, e.g. at physical walls or in shock waves, where fluid properties change drastically over a mean-free path l.

To calculate the first correction to the local equilibrium distribution, the so-called hydrodynamic approximation, we write

$$f_1(\mathbf{r}, \mathbf{p}, t) = f_1^{(0)}(\mathbf{r}, \mathbf{p}, t)[1 + \Phi^{(1)}(\mathbf{r}, \mathbf{p}, t)] \qquad (7.166)$$

and insist that the first term alone gives the correct local values of density, velocity, and internal energy. That is, we demand that

$$\rho(\mathbf{r}, t) = m\left(\frac{N}{V}\right) \int f_1^{(0)}(\mathbf{r}, \mathbf{p}, t) \, d^3p$$

$$\rho\mathbf{v}(\mathbf{r}, t) = \left(\frac{N}{V}\right) \int \mathbf{p} f_1^{(0)}(\mathbf{r}, \mathbf{p}, t) \, d^3p$$

$$\rho\varepsilon^{(K)}(\mathbf{r}, t) = \rho u(\mathbf{r}, t) = \frac{3}{2}\frac{\rho}{m}k_B T(\mathbf{r}, t) \qquad (7.167)$$

$$= \left(\frac{N}{V}\right) \int \frac{(\mathbf{p} - m\mathbf{v})^2}{2m} f_1^{(0)}(\mathbf{r}, \mathbf{p}, t) \, d^3p$$

This implies that $f_1^{(0)}\Phi^{(1)}$ must locally satisfy the five conditions

$$\int \psi_i(\mathbf{p}) f_1^{(0)}(\mathbf{r}, \mathbf{p}, t) \Phi^{(1)}(\mathbf{r}, \mathbf{p}, t) \, d^3p = 0 \qquad (7.168)$$

for the five summational invariants $\psi_0 = 1$, $\psi_i = p_i$ for $i = 1, 2, 3$, and $\psi_4 = (\mathbf{p})^2$.

Looking next at the pressure tensor, eqn (7.140), and the heat current, eqn (7.141), one can easily show that $f_1^{(0)}$ only contributes to the local hydrostatic pressure

$$p(\mathbf{r}, t) = \tfrac{1}{3} Tr\left[\left(\frac{N}{V}\right) \int \frac{(\mathbf{p} - m\mathbf{v})(\mathbf{p} - m\mathbf{v})}{m} f_1^{(0)}(\mathbf{r}, \mathbf{p}, t) \, d^3p\right] = \frac{\rho}{m} k_B T(\mathbf{r}, t) \qquad (7.169)$$

‡ A critical assessment of the Hilbert and Chapman-Enskog expansions has been given by Grad (1963), who also advances generalizations thereof in terms of extended sets of macroscopic state variables in a generalized fluid dynamical theory.

so that we have

$$\mathbf{P}(\mathbf{r}, t) = p\mathbf{I} + \left(\frac{N}{V}\right) \int \frac{(\mathbf{p} - m\mathbf{v})(\mathbf{p} - m\mathbf{v})}{m} f_1^{(0)}(\mathbf{r}, \mathbf{p}, t) \Phi^{(1)}(\mathbf{r}, \mathbf{p}, t) \, d^3p \tag{7.170}$$

and

$$\mathbf{j}_q(\mathbf{r}, t) = \left(\frac{N}{V}\right) \int \frac{\mathbf{p} - m\mathbf{v}}{m} \frac{(\mathbf{p} - m\mathbf{v})^2}{2m} f_1^{(0)}(\mathbf{r}, \mathbf{p}, t) \Phi^{(1)}(\mathbf{r}, \mathbf{p}, t) \, d^3p \tag{7.171}$$

The ansatz (7.166) for f_1 is therefore such that the first term $f_1^{(0)}$ determines the local thermodynamic properties ρ, $\mathbf{v}$, and u, whereas the second term $f_1^{(0)}\Phi^{(1)}$ controls the transport processes.

Let us then insert (7.166) into the Boltzmann equation (7.122) and, assuming that $\Phi^{(1)} \ll 1$, linearize the collision integral in $\Phi^{(1)}$. We obtain

$$\frac{\tilde{D}}{\tilde{D}t} f_1^{(0)}(\mathbf{r}, \mathbf{p}, t) = -\left(\frac{N}{V}\right) I[\Phi^{(1)}] \tag{7.172}$$

with

$$I[\Phi^{(1)}] = \int \sigma(\chi, |\mathbf{p}_1 - \mathbf{p}|) |\mathbf{p}_1 - \mathbf{p}| f_1^{(0)}(\mathbf{r}, \mathbf{p}, t) f_1^{(0)}(\mathbf{r}, \mathbf{p}_1, t)$$

$$\times [\Phi^{(1)}(\mathbf{r}, \mathbf{p}, t) + \Phi^{(1)}(\mathbf{r}, \mathbf{p}_1, t) - \Phi^{(1)}(\mathbf{r}, \mathbf{p}^{(f)}, t)$$

$$- \Phi^{(1)}(\mathbf{r}, \mathbf{p}_1^{(f)}, t] \, de' \, d^3p_1 \tag{7.173}$$

Boltzmann's integro-differential equation is thus transformed into a linear inhomogeneous integral equation with the left-hand side of eqn (7.172) a known function. Any solution $\Phi^{(1)}$ is still subject to the five conditions (7.168). They can be implemented by multiplying eqn (7.172) by one of the five summational invariants $\psi_i(\mathbf{p})$ listed below eqn (7.168) and integrating over $\mathbf{p}$. Because the resulting right-hand side is zero, we obtain

$$m\left(\frac{N}{V}\right) \int \psi_i(\mathbf{p}) \frac{\tilde{D}}{\tilde{D}t} f_1^{(0)}(\mathbf{r}, \mathbf{p}, t) \, d^3p = 0 \tag{7.174}$$

as five necessary and sufficient conditions to ensure a solution of eqn (7.172). With $f_1^{(0)}$ given by eqn (7.152) as a local maxwellian, these relations can be worked out explicitly:

$$\frac{\partial \rho}{\partial t} = -\boldsymbol{\nabla} \cdot (\rho \mathbf{v})$$

$$\frac{\partial (\rho \mathbf{v})}{\partial t} = -\boldsymbol{\nabla} p - \rho \mathbf{v}(\boldsymbol{\nabla} \cdot \mathbf{v}) + \rho \mathbf{F} \qquad (7.175)$$

$$\frac{\partial}{\partial t}\left(\frac{3}{2}\frac{\rho}{m}k_B T\right) = p\boldsymbol{\nabla}\cdot\mathbf{v} - \boldsymbol{\nabla}\cdot\left(\frac{3}{2}\frac{\rho}{m}k_B T\mathbf{v}\right)$$

These are just the balance equations for mass, momentum, and internal energy in lowest order, i.e. in strict local equilibrium with dissipative effects absent. They can be used to eliminate the time derivatives in the left-hand side of eqn (7.172), if we note that

$$\frac{\partial f_1^{(0)}(\mathbf{r},\mathbf{p},t)}{\partial t} = \frac{\partial f_1^{(0)}}{\partial T}\frac{\partial T}{\partial t} + \frac{\partial f_1^{(0)}}{\partial \mathbf{v}}\frac{\partial \mathbf{v}}{\partial t} + \frac{\partial f_1^{(0)}}{\partial \rho}\frac{\partial \rho}{\partial t} \qquad (7.176)$$

We then proceed with straightforward algebra

$$\frac{\tilde{D}f_1^{(0)}}{\tilde{D}t} = f_1^{(0)}(\mathbf{r},\mathbf{p},t)\left\{\left(\frac{(\mathbf{p}-m\mathbf{v})^2}{2mk_B T}-\frac{5}{2}\right)\frac{\mathbf{p}-m\mathbf{v}}{m}\cdot\left(\frac{1}{T}\boldsymbol{\nabla}T\right)\right.$$
$$\left. + \frac{1}{k_B T}\left[\frac{(\mathbf{p}-m\mathbf{v})(\mathbf{p}-m\mathbf{v})}{m}\cdot(\boldsymbol{\nabla}\mathbf{v}) - \frac{1}{3}\frac{(\mathbf{p}-m\mathbf{v})^2}{2m}(\boldsymbol{\nabla}\cdot\mathbf{v})\right]\right\} \qquad (7.177)$$

and eqn (7.172) takes on the form

$$f_1^{(0)}(\mathbf{r},\mathbf{p},t)\left[\left(\frac{mv^2}{2k_B T}-\frac{5}{2}\right)\frac{(\mathbf{p}-m\mathbf{v})}{m}\cdot\left(\frac{1}{T}\boldsymbol{\nabla}T\right)\right.$$
$$\left. + \frac{1}{mk_B T}\left[(\mathbf{p}-m\mathbf{v})(\mathbf{p}-m\mathbf{v}) - \frac{1}{3}(\mathbf{p}-m\mathbf{v})^2\mathbf{I}\right]:\mathbf{\Lambda} = -\left(\frac{N}{V}\right)I[\Phi^{(1)}] \qquad (7.178)$$

where we introduced the rate of strain tensor $\mathbf{\Lambda}$, eqn (2.47), and eliminated $\boldsymbol{\nabla}\rho$ and $\boldsymbol{\nabla}p$ terms by the local equation of state $p = \rho k_B T_m$. A solution $\Phi^{(1)}$ of this equation must obviously be of the form

$$\Phi^{(1)}(\mathbf{r},\mathbf{p},t) = -\frac{1}{\rho}\left\{A(|\mathbf{p}-m\mathbf{v}|)(\mathbf{p}-m\mathbf{v})\cdot\left(\frac{1}{T}\boldsymbol{\nabla}T\right)\right.$$
$$+ B(|\mathbf{p}-m\mathbf{v}|)\frac{1}{mk_B T}[(\mathbf{p}-m\mathbf{v})(\mathbf{p}-m\mathbf{v})$$
$$\left. -\frac{1}{3}(\mathbf{p}-m\mathbf{v})^2\mathbf{I}]:\mathbf{\Lambda}\right\} \qquad (7.179)$$

where the functions A and B still have to be determined by inserting this solution into eqn (7.178) and equating the factors multiplying the independent 'parameter' functions ∇T and Λ, i.e.

$$f_1^{(0)}(\mathbf{r}, \mathbf{p}, t)\left[\frac{1}{k_B T}\frac{(\mathbf{p}-m\mathbf{v})^2}{2m}-\frac{5}{2}\right](\mathbf{p}-m\mathbf{v}) = \rho I[(\mathbf{p}-m\mathbf{v})A]$$

$$f_1^{(0)}(\mathbf{r}, \mathbf{p}, t)[(\mathbf{p}-m\mathbf{v})(\mathbf{p}-m\mathbf{v})-\tfrac{1}{3}(\mathbf{p}-m\mathbf{v})\mathbf{I}]$$
$$= \rho I[((\mathbf{p}-m\mathbf{v})(\mathbf{p}-m\mathbf{v})-\tfrac{1}{3}(\mathbf{p}-m\mathbf{v})^2\mathbf{I})B] \quad (7.180)$$

All constraints (7.174) on the solutions $\Phi^{(1)}$ are automatically satisfied, except the three for the function A arising from the collisional invariants $\psi_i = p_i - mv_i$ which read

$$\int f_1^{(0)}(\mathbf{r}, \mathbf{q}+m\mathbf{v}, t)A(|\mathbf{q}|)\mathbf{q}\mathbf{q}\, d^3q$$
$$= \frac{1}{3}\int f_1^{(0)}(\mathbf{r}, \mathbf{q}+m\mathbf{v}, t)q^2\, d^3q\, \mathbf{I}$$
$$= 0 \quad (7.181)$$

Before we attempt an explicit calculation of the functions A and B, let us return to the problem of transport coefficients. Inserting the solution (7.166) with $\Phi^{(1)}$ from eqn (7.179) into the definition of the heat current, eqn (7.141), we obtain

$$\mathbf{j}_q(\mathbf{r}, t) = -\lambda \nabla T(\mathbf{r}, t) \quad (7.182)$$

where the coefficient of heat conduction is now given by

$$\lambda = \frac{2}{3}\frac{m}{T}\int \left(\frac{\mathbf{q}}{m}\right)^4 f_1^{(0)}(\mathbf{r}, \mathbf{q}+m\mathbf{v}, t)A(|\mathbf{q}|)\, d^3q \quad (7.183)$$

or, using eqn (7.181),

$$\lambda = \frac{1}{3}k_B \int \left(\frac{q^2}{2mk_B T}-\frac{5}{2}\right)f_1^{(0)}A(|\mathbf{q}|)\left(\frac{\mathbf{q}}{m}\right)^2 d^3q \quad (7.184)$$

or, using the first of eqns (7.180),

$$\lambda = \frac{1}{3}k_B \sum_{i=1}^{3}\int I[A(|\mathbf{q}|)q_i]A(|\mathbf{q}|)q_i\, d^3q \quad (7.185)$$

From the last expression, we see in particular that λ is positive in accordance with the second law of thermodynamics.

Similarly, we can determine the pressure tensor by inserting the solution eqn (7.166) into the definition eqn (7.140). We obtain

$$\mathbf{P} = p\mathbf{I} - 2\eta\mathbf{\Lambda} \qquad (7.186)$$

where the shear viscosity η is given by

$$\eta = \frac{2}{15}\frac{1}{mk_B T}\int\left(\frac{\mathbf{q}}{m}\right)^4 f_1^{(0)}(\mathbf{r}, \mathbf{q}+m\mathbf{v}, t) B(|\mathbf{q}|)\, d^3q \qquad (7.187)$$

or

$$\eta = \frac{1}{10}\frac{1}{k_B T}\int \sum_{i,j=1}^{3} I[B(|\mathbf{q}|)(q_i q_j - \tfrac{1}{3}q^2 \delta_{ij})]$$
$$\times B(|\mathbf{q}|)(q_i q_j - \tfrac{1}{3}q^2 \delta_{ij}) f_1^{(0)}(\mathbf{r}, \mathbf{q}+m\mathbf{v}, t)\, d^3q \qquad (7.188)$$

Again, we see that $\eta > 0$.

The set of linear integral eqns (7.180)–(7.181) can be solved exactly for a Maxwell gas.‡ in which the two-body force is

$$|\mathbf{F}_2(\mathbf{r})| = \frac{\kappa}{r^5} \qquad (7.189)$$

One finds

$$A(|\mathbf{q}|) = \frac{1}{\omega_{o2}}\frac{N}{V}\left(\frac{q^2}{mk_B T} - \frac{5}{2}\right)$$
$$B(|\mathbf{q}|) = \frac{4}{3}\frac{\rho}{\omega_{o2}} \qquad (7.190)$$

where the inverse relaxation time ω_{o2} is given approximately by

$$\omega_{o2} = 0.436\frac{N}{V}\sqrt{\frac{2\kappa}{m}} \qquad (7.191)$$

For the coefficient of heat conduction, we then obtain from eqn (7.183)

$$\lambda = \frac{5}{2}\frac{k_B}{m}\frac{p}{\omega_{o2}} \qquad (7.192)$$

---

‡ For other force laws, approximate solutions can be found by expanding in terms of the eigenfunctions of the Maxwell gas, so-called Sonine polynomials. Details can be found in Chapman and Cowling (1939) or Waldmann (1958).

and for the shear viscosity we find from eqn (7.187)

$$\eta = \frac{2}{3}\frac{p}{\omega_{o2}} \qquad (7.193)$$

where $p = (N/V)k_B T$. Eliminating p we find a relation between λ and η, namely

$$\lambda = \frac{5}{2} c_V \eta \quad \text{with } c_V = \frac{3}{2}\frac{k_B}{m} \qquad (7.194)$$

or for the Prandtl number

$$\Pr = \frac{\eta c_p}{\lambda} = \frac{2}{3} \qquad (7.195)$$

In particular, this last relation is well satisfied by monatomic gases at high enough temperatures, i.e. for He, Ne, and A at atmospheric pressure for temperatures between 20 K and 600 K, $\Pr^{-1} = .69$ to $.66$ (see Waldmann, 1958‡). This suggests that the Maxwell model is quite appropriate for the calculation of transport coefficients in dilute gases, as is indeed verified by explicit calculations of corrections for more realistic potential models.

In terms of λ and η, the normal solution of the Boltzmann equation, linear in temperature and velocity gradients, then reads

$$f_1(\mathbf{r}, \mathbf{p}, t) = f_1^{(0)}(\mathbf{r}, \mathbf{p}, t)$$
$$\times \left\{ 1 - \frac{1}{(k_B T)^2} \left[\frac{2}{5} \lambda \left(\frac{(\mathbf{p} - m\mathbf{v})^2}{2mk_B T} - \frac{5}{2} \right) (\mathbf{p} - m\mathbf{v}) \cdot \nabla T \right. \right.$$
$$\left. \left. + \eta[(\mathbf{p} - m\mathbf{v})(\mathbf{p} - m\mathbf{v}) - \tfrac{1}{3}(\mathbf{p} - m\mathbf{v})^2 \mathbf{I}] : \mathbf{\Lambda} \right] \right\} \qquad (7.196)$$

Let us recall that this solution was constructed in such a way that the first term yields the correct local values of ρ, $\mathbf{v}$, and u, and the second term determines the transport properties. To justify the linearization of the collision integral, we further assumed that the second and third terms in the braces are small for the relevant range of momenta $\mathbf{p}$. The last proviso must obviously be made because these terms can become arbitrarily large by simply increasing $\mathbf{p}$. But such contributions are unimportant in the whole distribution due to the exponential pre-factor $f_1^{(0)}$. The latter actually determines the relevant **p** range to values such that

$$\frac{(\mathbf{p}-m\mathbf{v})^2}{2mk_BT}<1 \qquad (7.197)$$

Concentrating, for an estimate, on the term involving the temperature gradient only (i.e. putting $\Lambda = 0$), we find

$$|\Phi^{(1)}|=\frac{2}{5}\frac{\lambda}{2m\rho(k_BT)^2}\left|\left(\frac{(\mathbf{p}-m\mathbf{v})^2}{2mk_BT}-\frac{5}{2}\right)(\mathbf{p}-m\mathbf{v})\cdot\boldsymbol{\nabla} T\right|$$

$$\leq \frac{7}{5}\frac{\lambda}{2m\rho(k_BT)^2}\sqrt{2mk_BT}\,|\boldsymbol{\nabla} T| \qquad (7.198)$$

We then recall from elementary kinetic theory that λ is connected with the mean-free path l (up to a constant of order unity) by

$$\lambda \approx k_B\rho v l \approx k_B\rho l\sqrt{2mk_BT} \qquad (7.199)$$

and find that

$$|\Phi^{(1)}|\leq \frac{7}{5}\frac{l}{T}|\boldsymbol{\nabla} T|\ll 1 \qquad (7.200)$$

as the condition for the validity of the linearization of the Boltzmann equation. This implies that our solution, and with it the assumption of local equilibrium and linear transport laws, is valid for temperature gradients that satisfy

$$|\boldsymbol{\nabla} T|\ll \frac{T}{l} \qquad (7.201)$$

That is, local equilibrium can be assumed in dilute gases in which the temperature changes little over a mean-free path.‡ This then also suggests that the mean-free path is the dimension of a local equilibrium cell, as we already argued in the introductory chapter where the condition eqn (7.201) was derived from arguments of local fluctuations.

The picture emerging from this analysis of the Boltzmann equation is even more beautiful and self-consistent than presented so far. Indeed, a simple calculation shows that the solution

‡ As an example, the hot zone of a H_2 flame has a temperature of some 2000 K and is about 0.2 mm thick; eqn (7.201) is still satisfied because $l \sim 10^{-6}$ cm (Clusius, Kölsch, and Waldmann, 1941). On the other hand, this condition can be violated in extremely rarefied gases.

eqn (7.196) inserted into eqns (7.145) and (7.146) yields

$$\mathbf{j}_h(\mathbf{r}, t) = \rho h(\mathbf{r}, t)\mathbf{v} - \frac{1}{k_B T}\mathbf{j}_q(\mathbf{r}, t) \qquad (7.202)$$

and

$$\sigma_h(\mathbf{r}, t) = -\frac{1}{k_B}\left\{\mathbf{j}_q \cdot \boldsymbol{\nabla}\left(\frac{1}{T}\right) - \frac{1}{T}[\mathbf{P} - \tfrac{1}{3}Tr(\mathbf{P})\mathbf{I}] : (\boldsymbol{\nabla}\mathbf{v})\right\} \qquad (7.203)$$

respectively (Enskog, 1929; Meixner, 1941 and 1943; Prigogine, 1949). This now compels us to the identification of a local entropy density via

$$s(\mathbf{r}, t) = -k_B h(\mathbf{r}, t) \qquad (7.204)$$

for which we also have a balance equation from eqn (7.144), namely

$$\frac{\partial(\rho s)}{\partial t} + \boldsymbol{\nabla} \cdot \mathbf{j}_s = \sigma_s \qquad (7.205)$$

with

$$\sigma_s \geq 0 \qquad (7.206)$$

This completes the program of nonequilibrium statistical mechanics for a dilute gas. We have (1) established a kinetic equation, namely, Boltzmann's; (2) derived from it balance equations for ρ, $\mathbf{v}$, and u; (3) solved the kinetic equation in the hydrodynamic regime; (4) constructed linear transport laws to achieve closure; (5) calculated transport coefficients explicitly; and (6) given conditions on the validity of the local equilibrium assumption under which a local entropy can be introduced and the Gibbs relation made valid locally.‡

In closing this section, let us mention that higher terms in the Chapman–Enskog theory, i.e. those involving higher derivatives of the field gradients and powers of such, lead to generalized

---

‡ Deviations from local equilibrium can be seen in an expanding gas, where a thin plate will measure a different temperature whether it is parallel or perpendicular to the direction of expansion. Although the temperature difference is small, this suggests that a single kinetic temperature is not sufficient to describe the local anisotropy of an expanding gas (Meissner and Meissner, 1939). See also Cercignani (1975) for a comprehensive treatment of the theory and application of the Boltzmann equation.

hydrodynamic equations—the so-called Burnett and super-Burnett equations (Burnett, 1935). Their usefulness seems doubtful. A precise upper limit for the correctness of the Navier-Stokes hydrodynamic theory, i.e. of linear constitutive laws, with respect to the kinetic theory has been given by Truesdell (1969). We should also point out that the linear analysis in gas mixtures produces the correct diffusion currents and also shows that for cross-couplings between different thermodynamic fluxes Onsager reciprocity relations are valid (Waldmann, 1958).

7.8. Derivation of Ohm's Law from the Boltzmann Equation

As a last example, let us use Boltzmann's equation to study electrical conductivity in a metal. Since any attempt to solve the Boltzmann equation for the general problem would involve a detailed description of electron-electron scattering, lattice effects, and impurity structure, we approximate the collision integral rather drastically by writing

$$\left(\frac{\partial f_1}{\partial t}\right)_{\text{coll}} = -\frac{(f_1 - f_1^{(0)})}{\tau} \qquad (7.207)$$

where τ is a constant relaxation time and $f_1^{(0)}$ is the equilibrium value of f_1 for an ideal gas of conduction electrons in the absence of external fields. This relaxation time approximation is perhaps justified for a system in which the conduction electrons are essentially free (e.g. in a metal) and where the dominant collisions are with impurities and lattice sites which can be treated as essentially random so that a given electron will have a mean-free time $\tau = \tau_{\text{coll}}$ between collisions. If each collision randomizes the direction of motion of a conduction electron and if there are no external fields, the system will return to equilibrium monotonically in a time roughly equal to τ. This may be expressed as

$$f_1(t) - f_1^{(0)} = [f_1(t=0) - f_1^{(0)}] e^{-t/\tau} \qquad (7.208)$$

which is equivalent to eqn (7.207).

For simplicity, we assume that the metal is spatially homogeneous ($\nabla_r f_1 = 0$), in a steady state ($\partial f_1/\partial t = 0$), and in a constant external electric field F in the z-direction. Boltzmann's equation then reduces to

$$eF\frac{\partial f_1}{\partial p_z} = \frac{(f_1 - f_1^{(0)})}{\tau} \quad (7.209)$$

or

$$\frac{\partial f_1}{\partial p_z} - \frac{f_1}{eF\tau} = -\frac{f_1^{(0)}}{eF\tau} \quad (7.210)$$

This simple equation is readily solved by the method of variation of the constant of the homogeneous solution to obtain (Trofimenkoff and Kreuzer, 1973)

$$f_1 = \int_{p_z}^{\infty} dp_z' \frac{f_1^{(0)}}{eF\tau} \exp\left(-\int_{p_z}^{p_z'} \frac{dp_z''}{eF\tau}\right) \quad (7.211)$$

where the boundary conditions have been included.

For the special case $T = 0$ (electrons, of course, satisfy Fermi-Dirac statistics!), we have $f_1^{(0)} = \theta(E_F - E)$ where E_F is the Fermi energy and $\theta(x) = 1$ for $x > 0$; $\theta(x) = 0$ for $x < 0$. With the relationship

$$\begin{aligned}f_1^{(0)}(E_r, p_z; T=0) &= \theta(E_F - E)\\ &= \{\theta[2m(E_F - E_r)]^{\frac{1}{2}} + p_z\}\theta\{[2m(E_F - E_r)]^{\frac{1}{2}} - p_z\}\end{aligned}$$

(7.212)

where a parabolic energy-momentum relationship is assumed for the electrons and $E_r = (p_x^2 + p_y^2)/2m$, $f_1(E_r, p_z; T=0)$ is easily computed from eqn (9.28). The result is

$$f_1(E_r, p_z; T=0) = \theta\{[2m(E_F + E_r)]^{\frac{1}{2}} - p_z\}\theta\{[2m(E_F - E_r)]^{\frac{1}{2}} + p_z\}$$
$$\times \left[1 - \exp\left(\frac{p_z - [2m(E_F - E_r)]^{\frac{1}{2}}}{eF\tau}\right)\right] + 2\theta\{-[2m(E_F - E_r)]^{\frac{1}{2}} - p_z\}$$
$$\times \exp\left(\frac{p_z}{eF\tau}\right) \sinh\left(\frac{2m(E_F - E_r)^{\frac{1}{2}}}{eF\tau}\right) \quad (7.213)$$

Note that $f_1 = 0$ for $p_z \geq [2m(E_F - E_r)]^{\frac{1}{2}}$. The first term on the right of eqn (7.123) is nonzero only for

$$-[2m(E_F - E_r)]^{\frac{1}{2}} < p_z < [2m(E_F - E_r)]^{\frac{1}{2}} \quad (7.214)$$

and the second term is nonzero only for $p_z < -[2m(E_F - E_r)]^{\frac{1}{2}}$. At $p_z = -[2m(E_F - E_r)]^{\frac{1}{2}}$, the two terms (without the θ functions)

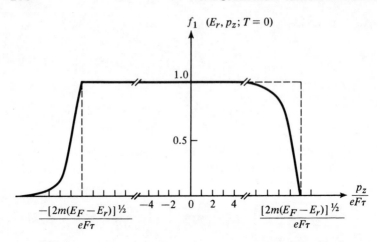

Fig. 7.3. Variation of f_1 with p_z for a constant E_r at $T = 0$.

take the same value. In Fig. 7.3, $f_1(E_r, p_z; T = 0)$ is shown for a constant value of E_r.

A straightforward calculation gives

$$\frac{4\pi m}{(2\pi\hbar)^3} \int_0^{E_f} dE_r \int_{-\infty}^{\infty} dp_z f_1(E_r, p_z; T = 0) = \frac{N}{V} \quad (7.215)$$

where (N/V) is the conduction electron density ($p_F^3 = 3\pi^2\hbar^3 N/V$) and

$$j = -\frac{4\pi em}{(2\pi\hbar)^3} \int_0^{E_f} dE_r \int_{-\infty}^{\infty} dp_z v_z f(E_r, p_z; T = 0)$$
$$= (N/V)(ne^2\tau/m)F \quad (7.216)$$

is the current density. Note that j is linear in F for all values of F; for a constant relaxation time, Ohm's law is the only steady-state solution of Boltzmann's equation in a homogeneous system. This rather surprising result is obviously a consequence of the fact that the constant relaxation time approximation amounts to a linearization of the collision integral.‡

---

‡ Guth and Meyerhöfer (1940) estimated that deviations from Ohm's law at the 1 percent level occur only at current densities exceeding 10^9 A/cm^2. The validity of the local equilibrium assumption in inhomogeneous metals has been assessed in great detail, particularly for the case of thermionic emission by Herring and Nichols (1949).

8
Microscopic Derivation of Balance Equations: Quantum Mechanical Theory

8.1. Outline of Approach and Introduction of Reduced Density Matrices

IN THE previous chapter we presented the basics of classical kinetic theory in order to be able to derive the macroscopic balance equations of nonequilibrium thermodynamics starting from an N-body system satisfying classical hamiltonian mechanics. Such a derivation is certainly satisfactory for dilute gases at high enough temperatures and low enough densities where the details of the particle statistics and the exact nature of the particle dynamics do not play a decisive role. If statistics and dynamics, however, are important, a satisfactory derivation of macroscopic balance equations has to start from the quantum mechanical N-body problem. Moreover, because quantum mechanics is a richer dynamical theory than classical mechanics, certain macroscopic phenomena can only be understood at this level, in particular all of quantum hydrodynamics of superfluids and superconductors and hydrodynamics of spin systems and solids.

In quantum mechanics the dynamics of an N-body system, consisting of one kind of spinless particles only, is controlled by a hamiltonian operator

$$\hat{H} = \int \psi^\dagger(\mathbf{x}) \hat{H}(\mathbf{x}) \psi(\mathbf{x}) \, d\mathbf{x} \tag{8.1}$$

The hamiltonian density

$$\hat{H}(x) = \hat{T}(\mathbf{x}) + \hat{V}(\mathbf{x}) + \Phi(\mathbf{x}, t) \tag{8.2}$$

is made up of three contributions: the kinetic energy operator

$\hat{T}(\mathbf{x}) = -(\hbar^2/2m)\boldsymbol{\nabla}^2$, the two-body interaction energy

$$\hat{V}(\mathbf{x}) = \frac{1}{2} \int \psi^\dagger(\mathbf{y}) V(\mathbf{x}, \mathbf{y}) \psi(\mathbf{y}) \, d^3y \tag{8.3}$$

where $V(\mathbf{x}, \mathbf{y}) = V(\mathbf{y}, \mathbf{x})$ is symmetric, and a (possibly time-dependent) external potential field $\Phi(\mathbf{x}, t)$. Throughout this chapter we will make use of the formalism of second quantization. Thus $\psi(\mathbf{x})$ and its hermitian conjugate $\psi^\dagger(\mathbf{x})$ are field operators subject to Heisenberg's equation of motion

$$i\hbar \frac{\partial}{\partial t} \psi = [\psi, \hat{H}] \tag{8.4}$$

If the ψ's describe bosons (fermions), they have to satisfy equal-time commutation (anticommutation) relations

$$[\psi(\mathbf{x}, t), \psi^\dagger(\mathbf{x}', t)]_\mp = \delta(\mathbf{x} - \mathbf{x}')$$

and (8.5)

$$[\psi(\mathbf{x}, t), \psi(\mathbf{x}', t)]_\mp = [\psi^\dagger(\mathbf{x}, t), \psi^\dagger(\mathbf{x}', t)]_\mp = 0$$

where the minus sign refers to bosons and the plus sign to fermions. The generalization to multicomponent systems is straightforward: For each additional component, we must introduce a new field operator $\psi_i(\mathbf{x}, t)$ and add appropriate terms to the hamiltonian. As an example, if the particles carry spin s, the index i will refer to the $(2s+1)$ independent spin states. If internal excitations in the particles become important, i might label particles in the various excited states etc.

Equations (8.1)–(8.5) are, apart from boundary and initial conditions, the basic ingredients of the theory. We want to stress that nonequal-time commutation or anticommutation relations, e.g.

$$[\psi(\mathbf{x}, t), \psi^\dagger(\mathbf{x}', t')]_\mp = ? \tag{8.6}$$

which, in the following, will play important roles in one form or the other, are not given a priori but must be calculated from the solution of the equation of motion, and thus contain already all the dynamical information of the N-body problem.

To derive macroscopic balance equations from the microscopic theory, we have two avenues open to us. In a first attack, we could try to keep the theory as long as possible at the operator level and only introduce the necessary approximations, e.g. the ones connected with our all-pervasive local equilibrium assumption, at a later stage when passing from the exact microscopic description to an approximate macroscopic description in terms of a few averaged fields like mass, momentum, and energy densities to recall the more important ones. Such an approach (Bogolyubov, 1970; Kreuzer, 1975) will be taken in Sections 8.3 and 8.4.

In an alternative approach (Fröhlich, 1967) to the derivation of macroscopic balance equations, we proceed right away from the operator description of the quantum-mechanical N-body description to a still exact c-number theory by taking appropriate expectation values in the Liouville–von Neumann equation

$$i\hbar\dot{\hat{\rho}} = [\hat{H}, \hat{\rho}] \qquad (8.7)$$

for the statistical operator or the density matrix (operator) $\hat{\rho}$. These expectation values should, of course, be statistical averages rather than expectation values between pure states. The most elegant way to do this is in the reduced density matrix formalism (Husimi, 1940; see also ter Haar (1961) for a review of the theory and applications of the density matrix) which has been set up analogous to the construction of reduced distribution functions in the classical theory.

The statistical time evolution of a large system is contained in the statistical operator $\hat{\rho}(t)$ which is, through eqn (8.7), subject to the microscopic hamiltonian $\hat{H}$ of the system. Let us then define time-dependent statistical averages of field operator products by

$$\rho_n(\mathbf{x}'_1, \ldots, \mathbf{x}'_n; \mathbf{x}''_1, \ldots, \mathbf{x}''_n; t)$$
$$= Tr[\psi^\dagger(\mathbf{x}''_1) \cdots \psi^\dagger(\mathbf{x}''_n) \psi(\mathbf{x}'_n) \cdots \psi(\mathbf{x}'_1) \hat{\rho}(t)] \qquad (8.8)$$

where the field operators are time-independent and defined at time $t = t_0$ at which initial values $\hat{\rho}(t_0)$ are also specified. We call ρ_n the nth order reduced density matrix. The index n runs from 1 to N. The ρ_n's are functions of time t, of $2n$ space points, and possibly of $2nI$ additional parameters, if each particle carries I internal quantum numbers or labels in a system of I components.

To understand the definition (8.8), let us briefly look at a simple nonequilibrium situation for illustration. Assume that a system is initially subject to a hamiltonian $\hat{H}_0$ and prepared in thermal equilibrium according to the canonical statistical operator $\hat{\rho}_0 = \hat{\rho}(t=0) = \exp[-\beta\hat{H}_0]/Tr\exp[-\beta\hat{H}_0]$, say. At time $t = 0$, an additional interaction $\hat{H}_1$ is switched on in the system so that its full hamiltonian now reads

$$\hat{H} = \hat{H}_0 + \theta(t)\hat{H}_1 \qquad (8.9)$$

From eqn (8.7), this leads to the formal solution

$$\hat{\rho}(t) = e^{-i\hat{H}t/\hbar}\hat{\rho}(t=0)e^{i\hat{H}t/\hbar} \qquad (8.10)$$

which we want to insert into the definition of ρ_1. We obtain

$$\rho_1(\mathbf{x}'; \mathbf{x}''; t) = Tr[\psi^\dagger(\mathbf{x}'')e^{-i\hat{H}t/\hbar}e^{i\hat{H}t/\hbar}\psi(\mathbf{x}')e^{-i\hat{H}t/\hbar}\hat{\rho}_0 e^{i\hat{H}t/\hbar}] \qquad (8.11)$$

where we have inserted a decomposition of unity between $\psi^\dagger$ and ψ and where we have used the fact that

$$Tr\hat{\rho}(t) = Tr(e^{-i\hat{H}t/\hbar}\hat{\rho}_0 e^{i\hat{H}t/\hbar}) = Tr\hat{\rho}_0 = 1 \qquad (8.12)$$

due to the invariance of the trace under cyclic permutation. Now observe that

$$e^{i\hat{H}t/\hbar}\psi(\mathbf{x}')e^{-i\hat{H}t/\hbar} = \psi(\mathbf{x}', t) \qquad (8.13)$$

is the formal solution of eqn (8.4). We have thus managed to shift the time evolution from the statistical operator onto the field operators and can equally well use as a definition of the nth order reduced density matrix

$$\rho_n(\mathbf{x}'_1, \ldots, \mathbf{x}'_n; \mathbf{x}''_1, \ldots, \mathbf{x}''_n; t)$$
$$= Tr[\psi^\dagger(\mathbf{x}''_1, t)\cdots\psi^\dagger(\mathbf{x}''_n, t)\psi(\mathbf{x}'_n, t)\cdots\psi(\mathbf{x}'_1, t)\hat{\rho}_0] \qquad (8.14)$$

Here $\hat{\rho}_0$ is *any* statistical operator describing the initial state of the system. To check that the definitions (8.8) and (8.14) are, indeed, equivalent, one can calculate the time derivative starting from eqn (8.8)

$$\begin{aligned}
i\hbar\frac{\partial}{\partial t}\rho_n &= Tr\left\{\psi^\dagger\cdots\psi i\hbar\frac{\partial}{\partial t}\hat{\rho}(t)\right\} \\
&= Tr\{\psi^\dagger\cdots\psi[\hat{H}, \hat{\rho}]\} \\
&= Tr\{[\psi^\dagger\cdots\psi, \hat{H}]\hat{\rho}\} \\
&= i\hbar\frac{\partial}{\partial t}Tr\{\psi^\dagger(\mathbf{x}, t)\cdots\psi(\mathbf{x}, t)\hat{\rho}_0\}
\end{aligned} \qquad (8.15)$$

q.e.d.

To go from line 2 to 3 we again used the invariance of the trace under cyclic permutation of its operator arguments.

Let us continue with the discussion of some of the properties of the reduced density matrices. Their normalization can be calculated easily to be

$$\int \rho_1(\mathbf{x}; \mathbf{x}; t)\, d\mathbf{x} = N$$

$$\int \rho_2(\mathbf{x}', \mathbf{y}; \mathbf{x}'', \mathbf{y}; t)\, d\mathbf{y} = (N-1)\rho_1(\mathbf{x}', \mathbf{x}''; t)$$

.
.
.

$$\int \rho_{n+1}(\mathbf{x}'_1, \ldots, \mathbf{x}_n, \mathbf{y}; \mathbf{x}''_1, \ldots, \mathbf{x}''_n, \mathbf{y}; t)\, d\mathbf{y} = (N-n)$$
$$\times \rho_n(\mathbf{x}'_1, \ldots, \mathbf{x}'_n; \mathbf{x}''_1, \ldots, \mathbf{x}''_n; t) \quad (8.16)$$

and also

$$\int \rho_2(\mathbf{x}_1, \mathbf{x}_2; \mathbf{x}_1, \mathbf{x}_2; t)\, d\mathbf{x}_1\, d\mathbf{x}_2 = N(N-1)$$

.
.
.

$$\int \rho_N(\mathbf{x}_1, \ldots, \mathbf{x}_N; \mathbf{x}_1, \ldots, \mathbf{x}_N; t)\, d\mathbf{x}_1, \ldots, d\mathbf{x}_N = N!$$
$$(8.17)$$

The ρ_n's also satisfy certain symmetry relations that are direct consequences of the commutation (anticommutation) relations between the field operators. In particular, we have for $n = 1$ and 2

$$\rho_1(\mathbf{x}'; \mathbf{x}''; t) = \rho_1^*(\mathbf{x}''; \mathbf{x}'; t)$$
$$\rho_2(\mathbf{x}'_1, \mathbf{x}'_2; \mathbf{x}''_1, \mathbf{x}''_2; t) = \rho_2^*(\mathbf{x}''_1, \mathbf{x}''_2; \mathbf{x}'_1, \mathbf{x}'_2; t) = \pm \rho_2(\mathbf{x}'_2, \mathbf{x}'_1; \mathbf{x}''_2, \mathbf{x}''_1; t)$$
$$(8.18)$$

where the plus sign holds for bosons and the minus sign for fermions.

Starting from Heisenberg's equation of motion, eqn (8.4), we can derive an exact hierarchy of coupled equations of motion for the reduced density matrices as defined in eqn (8.14)—in fact, using the third line of eqn (8.15)—

$$\left\{i\hbar\frac{\partial}{\partial t}+\frac{\hbar^2}{2m}\sum_{k=1}^{n}\left[\left(\frac{\partial}{\partial \mathbf{x}'_k}\right)^2-\left(\frac{\partial}{\partial \mathbf{x}''_k}\right)^2\right]-\frac{1}{2}\sum_{i,k=1}^{n}[V(\mathbf{x}'_i,\mathbf{x}'_k)-V(\mathbf{x}''_k,\mathbf{x}''_i)]\right\}$$
$$\times \rho_n(\mathbf{x}'_1,\ldots,\mathbf{x}'_n;\mathbf{x}''_1,\ldots,\mathbf{x}''_n;t)=W_n(\mathbf{x}'_1,\ldots,\mathbf{x}'_n;\mathbf{x}''_1,\ldots,\mathbf{x}''_n;t)$$
(8.19)

where

$$W_n(\mathbf{x}'_1,\ldots,\mathbf{x}'_n;\mathbf{x}''_1,\ldots,\mathbf{x}''_n;t)$$
$$=\int\sum_{k=1}^{n}[V(\mathbf{x}'_k,\mathbf{y})-V(\mathbf{x}''_k,\mathbf{y})]$$
$$\times \rho_{n+1}(\mathbf{x}'_1,\ldots,\mathbf{x}'_n,\mathbf{y};\mathbf{x}''_1,\ldots,\mathbf{x}''_n,\mathbf{y};t)\,d\mathbf{y} \quad (8.20)$$

couples ρ_n to ρ_{n+1} for $n=1,\ldots,N$. This hierarchy is the analogue of the BBGKY hierarchy, eqn (7.15), for the reduced distribution functions of classical statistical mechanics.

8.2. Fröhlich's Derivation of the Equations of Hydrodynamics in the Reduced Density Formalism

In this section we derive the balance equations for mass and momentum of a fluid within the reduced density matrix formalism closely following Fröhlich's work (Fröhlich, 1967 and 1973). Because mass and momentum density, like most hydrodynamic quantities, are averages over single-particle operators that are quadratic in field operators, we can restrict the discussion to the first-order reduced density matrix $\rho_1(\mathbf{x}';\mathbf{x}'';t)$ and its equation of motion:

$$\left\{i\hbar\frac{\partial}{\partial t}+\frac{\hbar^2}{2m}\left[\left(\frac{\partial}{\partial \mathbf{x}'}\right)^2-\left(\frac{\partial}{\partial \mathbf{x}''}\right)^2\right]\right\}\rho_1(\mathbf{x}';\mathbf{x}'';t)$$
$$=W_1(\mathbf{x}';\mathbf{x}'';t)=\int[V(|\mathbf{x}'-\mathbf{y}|)-V(|\mathbf{x}''-\mathbf{y}|)]$$
$$\times \rho_2(\mathbf{x}',\mathbf{y};\mathbf{x}'',\mathbf{y};t)\,d^3\mathbf{y} \quad (8.21)$$

From now on we will assume that the two-body interaction $V(\mathbf{x},\mathbf{y})$ is a central potential which only depends on the relative separation $|\mathbf{x}-\mathbf{y}|$ of two interacting particles. Let us observe that the first of the symmetry relations (8.18) allows us to write $\rho_1(\mathbf{x}';\mathbf{x}'';t)$ in terms of two real fields

$$\rho_1(\mathbf{x}';\mathbf{x}'';t)=\sigma(\mathbf{x}';\mathbf{x}'';t)e^{i\chi(\mathbf{x}';\mathbf{x}'';t)} \quad (8.22)$$

where the amplitude function

$$\sigma(\mathbf{x}'; \mathbf{x}''; t) = \sigma(\mathbf{x}''; \mathbf{x}'; t) \tag{8.23}$$

is symmetric and the phase function

$$\chi(\mathbf{x}'; \mathbf{x}''; t) = -\chi(\mathbf{x}''; \mathbf{x}'; t) \tag{8.24}$$

is antisymmetric. This, in particular, implies that $\chi(\mathbf{x}; \mathbf{x}; t) = 0$.

Let us recall that the macroscopic mass density in a system is usually defined in terms of field operators as

$$\rho(\mathbf{x}, t) = m Tr[\psi^\dagger(\mathbf{x})\psi(\mathbf{x})\hat{\rho}(t)] \tag{8.25}$$

and the macroscopic mass current density as

$$\mathbf{j}(\mathbf{x}, t) = Tr[\hbar \hat{\mathbf{j}}(\mathbf{x})\hat{\rho}(t)] \tag{8.26}$$

where the Schrödinger current operator is given by

$$\hat{\mathbf{j}}(\mathbf{x}) = \frac{1}{2i} \{\psi^\dagger(\mathbf{x})\boldsymbol{\nabla}\psi(\mathbf{x}) - [\boldsymbol{\nabla}\psi^\dagger(\mathbf{x})]\psi(\mathbf{x})\} \tag{8.27}$$

In a straightforward manner we can rewrite the definitions (8.25) and (8.26) in terms of the first-order reduced density matrix. We first introduce a center-of-mass variable $\mathbf{x}$, a nonlocal variable $\boldsymbol{\xi}$, and their respective derivative operators by

$$\mathbf{x} = \tfrac{1}{2}(\mathbf{x}' + \mathbf{x}'')$$

$$\boldsymbol{\xi} = \mathbf{x}' - \mathbf{x}''$$

$$\frac{\partial}{\partial \boldsymbol{\xi}} = \frac{1}{2}\left(\frac{\partial}{\partial \mathbf{x}'} - \frac{\partial}{\partial \mathbf{x}''}\right)$$

$$\boldsymbol{\nabla} = \frac{\partial}{\partial \mathbf{x}} = \frac{\partial}{\partial \mathbf{x}'} + \frac{\partial}{\partial \mathbf{x}''} \tag{8.28}$$

We then find for the mass density

$$\rho(\mathbf{x}, t) = m \lim_{\boldsymbol{\xi} \to 0} \rho_1(\mathbf{x}'; \mathbf{x}''; t) = m\sigma(\mathbf{x}; \mathbf{x}; t) \tag{8.29}$$

whereas the current density is given by

$$\mathbf{j}(\mathbf{x}, t) = \frac{\hbar}{i} \lim_{\boldsymbol{\xi} \to 0} \frac{\partial}{\partial \boldsymbol{\xi}} \rho_1(\mathbf{x}'; \mathbf{x}''; t) = \rho(\mathbf{x}, t)\mathbf{v}(\mathbf{x}, t) \tag{8.30}$$

Here we have used the fact that, in the local limit $\boldsymbol{\xi} \to 0$,

$$\lim_{\boldsymbol{\xi} \to 0} \frac{\partial}{\partial \boldsymbol{\xi}} \sigma(\mathbf{x}'; \mathbf{x}''; t) = 0 \qquad (8.31)$$

due to the symmetry of $\sigma(\mathbf{x}'; \mathbf{x}''; t)$. We also identified a macroscopic velocity field

$$\mathbf{v}(\mathbf{x}, t) = \frac{\hbar}{m} \lim_{\boldsymbol{\xi} \to 0} \frac{\partial}{\partial \boldsymbol{\xi}} \chi(\mathbf{x}'; \mathbf{x}''; t) \qquad (8.32)$$

At first sight it might seem strange to see $\hbar$ in the definition of the velocity $\mathbf{v}$ which, after all, is a macroscopic variable, but we must observe that $\partial/\partial\boldsymbol{\xi}$ is the derivative with respect to the relative separation $\boldsymbol{\xi}$ of space points $\mathbf{x}'$ and $\mathbf{x}''$ in the operator product $\psi^{\dagger}(\mathbf{x}'')\psi(\mathbf{x}')$, which is a microscopic quantity actually entering in the limit $\boldsymbol{\xi} \to 0$ only. But then the differential operator on the right of eqn (8.31) is, up to factors, nothing but the quantum-mechanical momentum operator $(-i\hbar\partial/\partial\boldsymbol{\xi})$. Observe also that with the definition (8.31) we luckily avoided the (impossible?) task of defining a quantum-mechanical velocity operator altogether.

To continue with the derivation of the macroscopic balance equations of hydrodynamics, we first rewrite eqn (8.21) in terms of the center-of-mass variable $\mathbf{x}$ and the nonlocal variable $\boldsymbol{\xi}$

$$\left(i\hbar \frac{\partial}{\partial t} - \frac{\hbar^2}{m} \boldsymbol{\nabla} \cdot \frac{\partial}{\partial \boldsymbol{\xi}}\right) \rho_1(\mathbf{x}+\tfrac{1}{2}\boldsymbol{\xi}; \mathbf{x}-\tfrac{1}{2}\boldsymbol{\xi}; t) = W_1(\mathbf{x}+\tfrac{1}{2}\boldsymbol{\xi}; \mathbf{x}-\tfrac{1}{2}\boldsymbol{\xi}; t) \qquad (8.33)$$

Taking the local limit $\boldsymbol{\xi} \to 0$ in this equation we immediately obtain with the definitions, eqn (8.29), for the mass density $\rho(\mathbf{x}, t)$ and eqn (8.30) for the mass current density $\mathbf{j}(\mathbf{x}, t) = \rho(\mathbf{x}, t)\mathbf{v}(\mathbf{x}, t)$,

$$\frac{\partial}{\partial t} \rho(\mathbf{x}, t) + \boldsymbol{\nabla} \cdot [\rho(\mathbf{x}, t)\mathbf{v}(\mathbf{x}, t)] = 0 \qquad (8.34)$$

because $W_1(\mathbf{x}, \mathbf{x}, t) = 0$. Equation (8.34) is nothing but the continuity equation (2.6), i.e. the law of mass conservation.

We next take the derivative of eqn (8.33) with respect to the nonlocal variable $\boldsymbol{\xi}$ and proceed to the limit $\boldsymbol{\xi} \to 0$. On the right-hand side we are led to adopt the definition

$$\mathbf{F}_V(\mathbf{x}, t) = \lim_{\boldsymbol{\xi} \to 0} \frac{\partial}{\partial \boldsymbol{\xi}} W_1(\mathbf{x} + \tfrac{1}{2}\boldsymbol{\xi}; \mathbf{x} - \tfrac{1}{2}\boldsymbol{\xi}; t)$$

$$= -\int \frac{\partial}{\partial \mathbf{x}} V(|\mathbf{x} - \mathbf{y}|)\rho_2(\mathbf{x}, \mathbf{y}; \mathbf{x}, \mathbf{y}; t)\, d\mathbf{y} \qquad (8.35)$$

of a force density due to the two-body interactions in the system.

To proceed with the left-hand side let us first calculate

$$-\frac{\hbar^2}{m} \lim_{\boldsymbol{\xi} \to 0} \frac{\partial}{\partial \xi_k} \frac{\partial}{\partial \xi_l} \rho_1(\mathbf{x} + \tfrac{1}{2}\boldsymbol{\xi}; \mathbf{x} - \tfrac{1}{2}\boldsymbol{\xi}; t)$$

$$= -\frac{\hbar^2}{m} \lim_{\boldsymbol{\xi} \to 0} e^{i\chi} \left\{ \frac{\partial^2 \sigma}{\partial \xi_k \partial \xi_l} + i \frac{\partial \chi}{\partial \xi_k} \frac{\partial \sigma}{\partial \xi_l} \right.$$

$$\left. + i \frac{\partial \sigma}{\partial \xi_k} \frac{\partial \chi}{\partial \xi_l} + i\sigma \frac{\partial^2 \chi}{\partial \xi_k \partial \xi_l} - \sigma \frac{\partial \chi}{\partial \xi_k} \frac{\partial \chi}{\partial \xi_l} \right\} \qquad (8.36)$$

and observe that all terms except the first and the last vanish in the limit $\boldsymbol{\xi} \to 0$ due to the symmetry properties of χ and σ, eqn (8.23) and (8.24). Also, using eqn (8.32), we obtain for the last term

$$\frac{\hbar^2}{m} \lim_{\boldsymbol{\xi} \to 0} \sigma \frac{\partial \chi}{\partial \xi_k} \frac{\partial \chi}{\partial \xi_l} = \rho(\mathbf{x}, t) v_k(\mathbf{x}, t) v_l(\mathbf{x}, t) \qquad (8.37)$$

The first term in eqn (8.36) is a new quantity for which we introduce a tensor $\mathbf{T}^{(0)}(\mathbf{x}, t)$ with components

$$T^{(0)}_{kl}(\mathbf{x}, t) = -\frac{\hbar^2}{2m} \lim_{\boldsymbol{\xi} \to 0} \frac{\partial^2 \sigma}{\partial \xi_k \partial \xi_l} \qquad (8.38)$$

With this we obtain from the left-hand side of eqn (8.33)

$$-\lim_{\boldsymbol{\xi} \to 0} \frac{\partial}{\partial \boldsymbol{\xi}} \left(i\hbar \frac{\partial}{\partial t} + \frac{\hbar^2}{m} \boldsymbol{\nabla} \cdot \frac{\partial}{\partial \boldsymbol{\xi}} \right) \rho_1(\mathbf{x} + \tfrac{1}{2}\boldsymbol{\xi}; \mathbf{x} - \tfrac{1}{2}\boldsymbol{\xi}; t)$$

$$= \frac{\partial}{\partial t}(\rho \mathbf{v}) + \boldsymbol{\nabla} \cdot (2\mathbf{T}^{(0)} + \rho \mathbf{v}\mathbf{v}) \qquad (8.39)$$

Combined with eqn (8.35), this yields a balance equation for momentum

$$\frac{\partial}{\partial t}[\rho(\mathbf{x}, t)\mathbf{v}(\mathbf{x}, t)] + \boldsymbol{\nabla} \cdot [2\mathbf{T}^{(0)}(\mathbf{x}, t) + \rho(\mathbf{x}, t)\mathbf{v}(\mathbf{x}, t)\mathbf{v}(\mathbf{x}, t)] = F_V(\mathbf{x}, t)$$

$$(8.40)$$

This is the complete analogue of eqn (7.26) of the classical theory. We can, in particular, identify $2\mathbf{T}^{(0)}(\mathbf{x}, t)$ as the kinetic pressure tensor.‡ Following the classical treatment, we can also introduce a potential energy pressure tensor according to eqn (7.40):

$$\mathbf{P}_{kl}^{(V)}(\mathbf{x}, t) = -\frac{1}{2}\int d^3\xi \frac{\xi_k \xi_l}{\xi^2}\frac{dV(\xi)}{d\xi}\int_0^\xi d\lambda \rho_2(\mathbf{y}', \mathbf{y}''; \mathbf{y}', \mathbf{y}''; t) \quad (8.41)$$

where $\xi = |\boldsymbol{\xi}|$, $\mathbf{y}' = \mathbf{x} + (\lambda - \xi)\boldsymbol{\xi}/\xi$ and $\mathbf{y}'' = \mathbf{x} + \lambda \boldsymbol{\xi}/\xi$, such that again

$$\mathbf{F}_V(\mathbf{x}, t) = -\boldsymbol{\nabla} \cdot \mathbf{P}^{(V)}(\mathbf{x}, t) \quad (8.42)$$

The total pressure tensor is then given by

$$\mathbf{P}(\mathbf{x}, t) = 2\mathbf{T}^{(0)}(\mathbf{x}, t) + \mathbf{P}^{(V)}(\mathbf{x}, t) \quad (8.43)$$

and the macroscopic balance equation for momentum reads, in the absence of external forces,

$$\frac{\partial}{\partial t}(\rho \mathbf{v}) + \boldsymbol{\nabla}(\rho \mathbf{v}\mathbf{v} + \mathbf{P}) = 0 \quad (8.44)$$

in agreement with eqns (7.42) and (2.13).

So far we have kept out theory completely general. Any further, detailed information about the particular system to be studied must obviously be used in the calculation of the pressure tensor **P**. As it stands, such a calculation still involves the complete coupled hierarchy of ρ_n equations (8.19). It would however, defeat our purpose if we were to try to solve this hierarchy in order to calculate **P**. Rather we should look into the structure of **P** and try to find approximations that yield constitutive equations expressing $\mathbf{P}(\mathbf{x}, t)$ as a function of $\rho(\mathbf{x}, t)$ and $\mathbf{v}(\mathbf{x}, t)$. Such a procedure would be in complete analogy to the phenomenological macroscopic theory of Chapters 2 and 3, where we also had to supplement the set of general balance equations by an appropriate number of constitutive laws to form a closed set involving as many equations as unknown fields. We should emphasize here that we will now, however, be in a position to derive such constitutive laws from first principles.

‡ We follow here the standard notation in quantum mechanics but want to stress that $2\mathbf{T}^{(0)}$ and $\mathbf{P}^{(K)}$ play, of course, the same role in the macroscopic equations, although their explicit computation starts from different microscopic theories.

To derive from eqn (8.44) the Navier–Stokes equation of hydrodynamics, we have to express $\mathbf{P}(\mathbf{x}, t)$ as a linear functional of $\rho(\mathbf{x}, t)$ and $\mathbf{v}(\mathbf{x}, t)$, and also introduce a scalar hydrostatic pressure $p(\mathbf{x}, t)$. We proceed in analogy to the classical case in Section 7.4. For this purpose let us first look at an isotropic fluid in equilibrium. It is characterized not only by a zero velocity field and a constant density $\rho(\mathbf{x}, t) = \rho_0 = \text{const}$ due to the translational and rotational symmetries of the system, but also by the fact that the second-order reduced density matrix‡ $\rho_2(\mathbf{x}, \mathbf{y}; \mathbf{x}, \mathbf{y}; t)$ is a function of the relative distance $|\mathbf{x} - \mathbf{y}|$ only, i.e. $\rho_2(\mathbf{x}, \mathbf{y}; \mathbf{x}, \mathbf{y}; t) = \rho_2^{(0)}(|\mathbf{x} - \mathbf{y}|)$. But then we find from eqn (8.35) after a change of variables $\mathbf{r} = \mathbf{x} - \mathbf{y}$

$$\mathbf{F}_V(\mathbf{x}) = -\int \frac{\partial}{\partial \mathbf{x}} V(|\mathbf{x} - \mathbf{y}|) \rho_2^{(0)}(|\mathbf{x} - \mathbf{y}|) \, d\mathbf{y}$$

$$= -\int \frac{\mathbf{r}}{r} \frac{\partial V(r)}{\partial r} \rho_2^{(0)}(r) \, d\mathbf{r} = 0 \qquad (8.45)$$

due to the angular integration; i.e. a fluid in equilibrium does not experience any net forces arising from its intermolecular interactions. The latter, nevertheless, contribute to the constant equilibrium pressure in the system, which we can calculate from eqn (8.43) as

$$p_{\text{eq}} = \tfrac{1}{3} Tr(\mathbf{P}) = \tfrac{2}{3} T^{(0)} - \tfrac{1}{6} \int d^3\xi \frac{dV(\xi)}{d\xi} \xi \rho_2^{(0)}(\xi) \qquad (8.46)$$

where

$$T^{(0)} = \sum_k T_{kk}^{(0)}. \qquad (8.47)$$

Let us consider a fluid away from its equilibrium state but still in local equilibrium. Obviously an expression like eqn (8.46) for the hydrostatic pressure now holds locally

$$p(\mathbf{x}, t) = \tfrac{2}{3} T^{(0)}(\mathbf{x}, t) - \tfrac{1}{6} \int d^3\xi \frac{dV}{d\xi} \xi \rho_2^{(0)}(\xi; \mathbf{x}; t) \qquad (8.48)$$

but with a space and time dependence dictated by the local equilibrium averages on the right-hand side. This pressure can

‡ In equilibrium statistical mechanics, this quantity is also referred to as the two-body correlation function.

therefore also be calculated from the local equation of state for a given mass density $\rho(\mathbf{x}, t)$ and energy density $u(\mathbf{x}, t)$.

To calculate the dissipative contributions to the pressure tensor **P**, we must consider deviations from local equilibrium. It will be sufficient to include the modification of the second-order reduced density matrix due to the small variation of the macroscopic velocity field $\mathbf{v}(\mathbf{x}, t)$ over the microscopic range of the two-body interaction. We therefore try an expansion of ρ_2, linear in **v** and its gradient; i.e.

$$\rho_2(\mathbf{y}', \mathbf{y}''; \mathbf{y}', \mathbf{y}''; t) = \rho_2^{(0)}(\boldsymbol{\xi}, \mathbf{x}, t)$$
$$+ [\mathbf{v}(\mathbf{x}+\boldsymbol{\xi}) - \mathbf{v}(\mathbf{x})] \cdot \boldsymbol{\xi} f_1(\boldsymbol{\xi}, \mathbf{x}, t)$$
$$+ \tfrac{1}{2}[\boldsymbol{\nabla} \cdot \mathbf{v}(\mathbf{x}+\boldsymbol{\xi}) + \boldsymbol{\nabla} \cdot \mathbf{v}(\mathbf{x})] f_2(\boldsymbol{\xi}, \mathbf{x}, t)$$
$$+ \left[\frac{\partial \mathbf{v}_m(\mathbf{x}+\boldsymbol{\xi})}{\partial x_k} + \frac{\partial \mathbf{v}_m(\mathbf{x})}{\partial x_k}\right] \xi_k \xi_m f_3(\boldsymbol{\xi}, \mathbf{x}, t)$$
$$+ \cdots$$
$$= \rho_2^{(0)}(\boldsymbol{\xi}, \mathbf{x}, t) + \boldsymbol{\nabla} \cdot \mathbf{v}(\mathbf{x}) f_2(\boldsymbol{\xi}, \mathbf{x}, t)$$
$$+ \frac{\partial \mathbf{v}_m}{\partial x_n} \xi_n \xi_m [f_1(\boldsymbol{\xi}, \mathbf{x}, t) + 2 f_3(\boldsymbol{\xi}, \mathbf{x}, t)] + \cdots \quad (8.49)$$

Here the first form is dictated by the symmetries (8.18). The last form is then obtained by developing $\mathbf{v}(\mathbf{y})$ near $\mathbf{v}(\mathbf{x})$. The **x**- and t-dependence of the functions $f_i(\boldsymbol{\xi}, \mathbf{x}, t)$ reflects the fact that the local equilibrium averages may vary throughout the system. This dependence is usually small and can be neglected in most cases. Let us next insert eqn (8.49) into (8.43) using eqn (8.48):

$$\mathbf{P}_{kl}(\mathbf{x}, t) = \delta_{kl} p(\mathbf{x}, t)$$
$$- \frac{1}{2} \int d^3\xi \frac{\xi_k \xi_l}{\xi} \frac{dV(\xi)}{d\xi} \left\{ \boldsymbol{\nabla} \cdot \mathbf{v}(\mathbf{x}) f_2(\xi, \mathbf{x}, t) \right.$$
$$\left. + \xi_m \xi_n \frac{\partial v_m(\mathbf{x}, t)}{\partial x_n} [f_1(\xi, \mathbf{x}, t) + 2 f_3(\xi, \mathbf{x}, t)] \right\} \quad (8.50)$$

Using the formulas

$$\int d^3\xi \, \xi_k \xi_l g(\xi) = \tfrac{1}{3} \delta_{kl} \int d^3\xi \, \xi^2 g(\xi)$$
$$\int d^3\xi \, \xi_k \xi_l \xi_m \xi_n g(\xi) = (\delta_{kl}\delta_{mn} + \delta_{km}\delta_{ln} + \delta_{kn}\delta_{lm}) \tfrac{1}{15} \int d^3\xi \, \xi^4 g(\xi)$$
$$(8.51)$$

we finally obtain the well-known expression for the pressure tensor [see eqns. (2.37) and (2.38)],

$$\mathbf{P}_{kl}(\mathbf{x}, t) = \delta_{kl}p(\mathbf{x}, t) - \zeta \mathbf{\nabla} \cdot \mathbf{v}\, \delta_{kl} - \eta\left(\frac{\partial v_k}{\partial x_l} + \frac{\partial v_l}{\partial x_k} - \frac{2}{3}\delta_{kl}\mathbf{\nabla} \cdot \mathbf{v}\right) \quad (8.52)$$

where the shear and bulk viscosities are given by

$$\eta = \frac{1}{30} \int d^3\xi \frac{dV(\xi)}{d\xi} \xi^3 [f_1(\xi, \mathbf{x}, t) + 2f_3(\xi, \mathbf{x}, t)]$$

$$\zeta = \frac{1}{6} \int d^3\xi \frac{dV(\xi)}{d\xi} \xi f_2(\xi, \mathbf{x}, t) + \tfrac{5}{3}\eta \quad (8.53)$$

respectively. A number of comments on these results are in order. We see that the $\mathbf{x}$ and t dependence of the expansion function $f_i(\xi, \mathbf{x}, t)$ leads to a like dependence in the viscosity coefficients which, however, in most practical cases can be neglected. In this circumstance we can rewrite eqn (8.44) with eqn (8.52) as

$$\frac{\partial(\rho \mathbf{v})}{\partial t} + \mathbf{\nabla} p = \eta \nabla^2 \mathbf{v} + (\zeta + \tfrac{1}{3}\eta)\mathbf{\nabla}(\mathbf{\nabla} \cdot \mathbf{v}) \quad (8.54)$$

which is the Navier–Stokes equation of hydrodynamics.

The extension of the above derivation of hydrodynamics to two-component systems has been presented by Terreaux (1969). Fröhlich (1967, 1968, 1969, 1973, and 1974) has used the reduced density formalism in conjunction with the concept of off-diagonal long-range order (Onsager, 1951; Penrose and Onsager, 1956; and Yang, 1962) to derive the hydrodynamics of superfluids and superconductors.

8.3. Operator Balance Equations

In this and the following section, we present an alternative derivation of the balance equations of nonequilibrium thermodynamics. Unlike the reduced density matrix formalism of the previous section, we will keep the theory at the operator level as long as possible and only introduce the necessary approximations (the ones connected with our all-pervasive local equilibrium assumption) at a later stage when passing from the exact microscopic description to an approximate macroscopic picture in terms of a few averaged fields such as mass, momentum, and energy densities. We will follow closely the approach taken by Kreuzer (1975).‡ What we gain in this approach, apart from a further elucidation of the microscopic foundations of hydrodynamics, is twofold. We will be able to derive a new hierarchy of balance equations for nth-rank tensor operators and their statistical averages different from the reduced density matrix hierarchy. Moreover, by a careful study of the lower-order members of this hierarchy, we will be led to a natural generalization of hydrodynamics to include viscoelastic effects (see Section 2.2 for the phenomenological macroscopic theory).

Our theory is again based on the hamiltonian (8.1–8.3) for a quantum mechanical N-body system.§ We will focus our attention on the nonlocal operator

$$\hat{\rho}(\mathbf{x}', \mathbf{x}'', t) = \psi^\dagger(\mathbf{x}'', t)\psi(\mathbf{x}'t) \qquad (8.55)$$

where $\psi^\dagger(\mathbf{x}'', t)$ and $\psi(\mathbf{x}', t)$ are field operators subject to Heisenberg's equation of motion (8.4) and satisfy commutation relations for bosons and anticommutation relations for fermions, eqns (8.5). The operators $\hat{\rho}(\mathbf{x}', \mathbf{x}'', t)$, however, satisfy the commutation relations

$$[\hat{\rho}(\mathbf{x}', \mathbf{x}'', t), \hat{\rho}(\mathbf{y}', \mathbf{y}'', t)]_-$$
$$= \delta^3(\mathbf{x}' - \mathbf{y}'')\hat{\rho}(\mathbf{y}', \mathbf{x}'', t) - \delta^3(\mathbf{x}'' - \mathbf{y}')\hat{\rho}(\mathbf{x}', \mathbf{y}'', t) \qquad (8.56)$$

for either statistics.‖

Let us briefly show the connection of our nonlocal operators $\hat{\rho}(\mathbf{x}', \mathbf{x}'', t)$ with various other distribution functions and operators used in kinetic theory. Obviously, the trace over any statistical ensemble (simply denoted by $\langle \cdots \rangle$)

$$\langle \hat{\rho}(\mathbf{x}', \mathbf{x}'', t) \rangle = \rho_1(\mathbf{x}', \mathbf{x}'', t) \qquad (8.57)$$

‡ The derivation of hydrodynamic balance equations in the presence of an external electromagnetic field using the same formalism has been given by Kreuzer and Zasada (1976).

§ Kreuzer (1975) and Kreuzer and Zasada (1976) have written the two-body interaction term as $\int V(\mathbf{x}', \mathbf{x}'')\psi^\dagger(\mathbf{x}')\psi(\mathbf{x}')\psi^\dagger(\mathbf{x}'')\psi(\mathbf{x}'') \, d\mathbf{x}' \, d\mathbf{x}''$, absorbing contributions arising from normal ordering into the external potential term Φ. We will use here, however, the more conventional form of the many-body hamiltonian in second quantized form.

‖ These commutation relations are obviously antisymmetric $[x, y] + [y, x] = 0$, additive $[x, \alpha y + \beta z] = \alpha[x, y] + \beta[x, z]$, and satisfy the Jacobi identity $[x, [y, z]] + [y, [z, x]] + [z, [x, y]] = 0$. The symbols x, y, z stand for $\rho(\mathbf{x}', \mathbf{x}'', t)$, $\rho(\mathbf{y}', \mathbf{y}'', t)$, and $\rho(\mathbf{z}', \mathbf{z}'', t)$, respectively. Thus the nonlocal operators $\hat{\rho}(\mathbf{x}', \mathbf{x}'', t)$ form a Lie algebra.

is the first-order reduced density matrix studied extensively in the last section. Taking the Fourier transform of $\hat{\rho}(\mathbf{x}', \mathbf{x}'', t)$ with respect to the nonlocal variable $\boldsymbol{\xi} = \mathbf{x}' - \mathbf{x}''$, introduced in eqn (8.28) together with the center-of-mass variable $\mathbf{x} = \frac{1}{2}(\mathbf{x}' + \mathbf{x}'')$, we obtain

$$\hat{f}^{(1)}(\mathbf{x}, \mathbf{p}, t) = (2\pi)^{-3} \int d\boldsymbol{\xi} e^{-i\mathbf{p}\cdot\boldsymbol{\xi}/\hbar} \psi^{\dagger}(\mathbf{x} + \tfrac{1}{2}\boldsymbol{\xi}, t) \psi(\mathbf{x} - \tfrac{1}{2}\boldsymbol{\xi}, t) \quad (8.58)$$

whose thermal expectation value is Wigner's one-particle distribution function (Wigner, 1932).

$$f_W^{(1)}(\mathbf{x}, \mathbf{p}, t) = \langle \hat{f}^{(1)}(\mathbf{x}, \mathbf{p}, t) \rangle$$

$$= (2\pi)^{-3} \int d\boldsymbol{\xi} e^{-i\mathbf{p}\cdot\boldsymbol{\xi}/\hbar} \rho_1(\mathbf{x} - \tfrac{1}{2}\boldsymbol{\xi}; \mathbf{x} + \tfrac{1}{2}\boldsymbol{\xi}; t) \quad (8.59)$$

which for $\hbar \to 0$ goes into Boltzmann's distribution function $f_1(\mathbf{x}, \mathbf{p}, t)$, eqn (9.2). The operator $\hat{f}^{(1)}(\mathbf{x}, \mathbf{p}, t)$ is also the quantum analogue of the exact microscopic one-particle density functional (or operator in phase space)

$$\hat{f}_K^{(1)}(\mathbf{x}, \mathbf{p}, t) = \sum_{i=1}^{N} \delta[\mathbf{x} - \mathbf{x}_i(t)] \delta[\mathbf{p} - \mathbf{p}_i(t)] \quad (8.60)$$

introduced in classical theory by Klimontovich (1958). Here $\mathbf{x}_i(t)$ and $\mathbf{p}_i(t)$ are the position and momentum of the ith particle at time t.‡

Returning to our nonlocal operator $\hat{\rho}(\mathbf{x}', \mathbf{x}'', t)$, we can easily show from eqn (8.4) that its equation of motion reads (repeated indices are summed from 1 to 3 from now on)

$$\left(i\hbar \frac{\partial}{\partial t} + \frac{\hbar^2}{m} \frac{\partial}{\partial x_i} \frac{\partial}{\partial \xi_i}\right) \hat{\rho}(\mathbf{x}', \mathbf{x}'', t)$$

$$= [\Phi(\mathbf{x}') - \Phi(\mathbf{x}'')] \hat{\rho}(\mathbf{x}', \mathbf{x}'', t) + \hat{W}(\mathbf{x}', \mathbf{x}'', t) \quad (8.61)$$

with

$$\hat{W}(\mathbf{x}', \mathbf{x}'', t) = \int [V(\mathbf{x}', \mathbf{y}) - V(\mathbf{x}'', \mathbf{y})]$$

$$\times \psi^{\dagger}(\mathbf{x}'', t) \psi^{\dagger}(\mathbf{y}, t) \psi(\mathbf{y}, t) \psi(\mathbf{x}', t) \, d\mathbf{y}$$

$$= \int [V(\mathbf{x}', \mathbf{y}) - V(\mathbf{x}'', \mathbf{y})] \{\hat{\rho}(\mathbf{y}, \mathbf{y}, t) \hat{\rho}(\mathbf{x}', \mathbf{x}'', t)\} \, d\mathbf{y} \quad (8.62)$$

‡ For a discussion of these distribution functions and functionals, see Brittin and Chappell (1962).

where the curly brackets indicate symmetrized products, i.e. $\{AB\} = (AB + BA)/2$. Starting from eqn (8.61), we can derive a hierarchy of operator equations by expanding the nonlocal operator $\hat{\rho}(\mathbf{x}', \mathbf{x}'', t)$ around the center-of-mass coordinate $\mathbf{x}$:

$$\hat{\rho}(\mathbf{x}', \mathbf{x}'', t) = \sum_{n=0}^{\infty} \frac{1}{n!} \xi_{i_1}, \ldots, \xi_{i_n} \times \frac{\partial^n}{\partial \xi_{i_1}, \ldots, \partial \xi_{i_n}} \hat{\rho}(\mathbf{x}', \mathbf{x}'', t) \bigg|_{\mathbf{x}' = \mathbf{x}''} \quad (8.63)$$

In the local limit $\mathbf{x}' = \mathbf{x}'' = \mathbf{x}$, we find the continuity equation for the operator $\hat{\rho}(\mathbf{x}, t) \equiv \hat{\rho}(\mathbf{x}, \mathbf{x}, t)$

$$\hbar \frac{\partial \hat{\rho}(\mathbf{x}, t)}{\partial t} + \frac{\hbar^2}{m} \frac{\partial}{\partial x_i} \hat{j}_i(\mathbf{x}, t) = 0 \quad (8.64)$$

where

$$\hat{j}_k(\mathbf{x}, t) = -i \lim_{\boldsymbol{\xi} \to 0} \frac{\partial \hat{\rho}(\mathbf{x}', \mathbf{x}'', t)}{\partial \xi_k} \quad (8.65)$$

is the kth cartesian component of the current operator, eqn (8.27).‡ Collecting terms proportional to $\boldsymbol{\xi}$, we obtain the equation of motion

$$\hbar \frac{\partial \hat{j}_k(\mathbf{x}, t)}{\partial t} - \frac{\hbar^2}{m} \frac{\partial \hat{T}_{kl}(\mathbf{x}, t)}{\partial x_l} = -\left[\frac{\partial \Phi(\mathbf{x})}{\partial x_k}\right] \hat{\rho}(\mathbf{x}, t) \\ - \int \left[\frac{\partial V(\mathbf{x}, \mathbf{y})}{\partial x_k}\right] \{\hat{\rho}(\mathbf{y}, t) \hat{\rho}(\mathbf{x}, t)\} \, d^3 \mathbf{y} \quad (8.66)$$

where the kinetic energy tensor is given by

$$\hat{T}_{kl}(\mathbf{x}, t) = \lim_{\boldsymbol{\xi} \to 0} \frac{\partial}{\partial \xi_k} \frac{\partial \hat{\rho}(\mathbf{x}', \mathbf{x}'', t)}{\partial \xi_l} \quad (8.67)$$

Equation (8.66) can also be written, following the procedure leading to eqn (7.42), as

$$\hbar \frac{\partial \hat{\mathbf{j}}(\mathbf{x}, t)}{\partial t} - \frac{\hbar^2}{m} \boldsymbol{\nabla} \hat{\mathbf{P}}(\mathbf{x}, t) = -\left[\frac{\partial \Phi(\mathbf{x})}{\partial \mathbf{x}}\right] \hat{\rho}(\mathbf{x}, t) \quad (8.68)$$

‡ This definition of $\hat{\mathbf{j}}(\mathbf{x}, t)$ arises naturally from the Taylor expansion (8.63) without a factor $\hbar$. The latter we will include later when we introduce a macroscopic current via thermal expectation values of this operator. This comment can be generalized to the nth rank tensor operators to be introduced below, e.g. eqns (8.67) and (8.69), which will have to be multiplied by $\hbar^n$ to make them macroscopic.

where we introduced

$$\hat{\mathbf{P}}(\mathbf{x}, t) = \hat{\mathbf{T}}(\mathbf{x}, t) - \frac{1}{2}\int_0^1 d\lambda \int d^3z \frac{\mathbf{z}\mathbf{z}}{z}\frac{dV(z)}{dz}$$
$$\times \{\hat{\rho}(\mathbf{x} - \lambda\mathbf{z}, t)\hat{\rho}(\mathbf{x} + (1-\lambda)\mathbf{z}, t)\} \quad (8.69)$$

where $z = |\mathbf{z}|$. Terms with ξ^2 yield

$$\hbar \frac{\partial \hat{T}_{kl}(\mathbf{x}, t)}{\partial t} + \frac{\hbar^2}{m}\frac{\partial \hat{F}_{ikl}(\mathbf{x}, t)}{\partial x_i} = \left[\frac{\partial \Phi(\mathbf{x})}{\partial x_k}\right]\hat{j}_l(\mathbf{x}, t)$$
$$+ \left[\frac{\partial \Phi(\mathbf{x})}{\partial x_l}\right]\hat{j}_k(\mathbf{x}, t) + \int \left[\frac{\partial V(\mathbf{x}, \mathbf{y})}{\partial x_k}\right]\{\hat{\rho}(\mathbf{y}, t)\hat{j}_l(\mathbf{x}, t)\} d^3y$$
$$+ \int \left[\frac{\partial V(\mathbf{x}, \mathbf{y})}{\partial x_l}\right]\{\hat{\rho}(\mathbf{y}, t)\hat{j}_k(\mathbf{x}, t)\} d^3y \quad (8.70)$$

where

$$\hat{F}_{ikl}(\mathbf{x}, t) = -i \lim_{\boldsymbol{\xi} \to 0} \frac{\partial}{\partial \xi_i}\frac{\partial}{\partial \xi_k}\frac{\partial}{\partial \xi_l} \hat{\rho}(\mathbf{x}', \mathbf{x}'', t) \quad (8.71)$$

which will, in turn, couple in its balance equation to yet higher tensor operators, namely

$$\hbar \frac{\partial \hat{F}_{ikl}(\mathbf{x}, t)}{\partial t} - \frac{\hbar^2}{m}\frac{\partial \hat{G}_{jikl}(\mathbf{x}, t)}{\partial x_j}$$
$$= -\left[\frac{\partial \Phi(\mathbf{x})}{\partial x_i}\right]\hat{T}_{kl}(\mathbf{x}, t) - \left[\frac{\partial \Phi(\mathbf{x})}{\partial x_k}\right]\hat{T}_{il}(\mathbf{x}, t)$$
$$- \left[\frac{\partial \Phi(\mathbf{x})}{\partial x_l}\hat{T}_{ik}(\mathbf{x}, t)\right] - \frac{1}{4}\left[\frac{\partial}{\partial x_i}\frac{\partial}{\partial x_j}\frac{\partial \Phi(\mathbf{x})}{\partial x_k}\right]\hat{\rho}(\mathbf{x}, t)$$
$$- \int \left[\frac{\partial V(\mathbf{x}, \mathbf{y})}{\partial x_i}\right]\{\hat{\rho}(\mathbf{y}, t)\hat{T}_{kl}(\mathbf{x}, t)\} d^3y$$
$$- \int \left[\frac{\partial V(\mathbf{x}, \mathbf{y})}{\partial x_k}\right]\{\hat{\rho}(\mathbf{y}, t)\hat{T}_{il}(\mathbf{x}, t)\} d^3y$$
$$- \int \left[\frac{\partial V(\mathbf{x}, \mathbf{y})}{\partial x_l}\right]\{\hat{\rho}(\mathbf{y}, t)\hat{T}_{ik}(\mathbf{x}, t)\} d^3y$$
$$- \frac{1}{4}\int \frac{\partial^3 V(\mathbf{x}, \mathbf{y})}{\partial x_i \partial x_k \partial x_l}\{\hat{\rho}(\mathbf{y}, t)\hat{\rho}(\mathbf{x}, t)\} d^3y \quad (8.72)$$

It is obvious that this hierarchy of operator balance equations continues ad infinitum. As we go to higher-rank tensor operators, more microscopic information is exposed through the higher-order derivatives, the complete hierarchy being equivalent to the equation of motion (8.61) for the nonlocal operator $\hat{\rho}(\mathbf{x}', \mathbf{x}'', t)$. It is interesting to note that the coupling of the nth equation to the $(n+1)$st equation occurs via the divergence term on the left, whereas the right-hand side of the equation is completely determined by lower-order equations. This is in contrast to the hierarchy of equations for the reduced density matrices, eqn (8.19), and also to the BBGKY hierarchy, eqn (7.15), in classical statistical mechanics, which couple via their right-hand sides. We will continue this discussion in the next section after we have defined macroscopic quantities as statistical averages of our tensor operators.

8.4. Macroscopic Balance Equations from the Operator Hierarchy

In the previous section we derived a hierarchy of equations for tensor operators which we now want to make the starting point for a derivation of the macroscopic balance equations of nonequilibrium thermodynamics. We again have to introduce two concepts that are foreign to a microscopic quantum mechanical theory of a many-body system: (1) a macroscopic velocity field and (2) statistical averages. We could achieve this formally by simply recalling that according to eqn (8.57) the statistical average

$$\langle \hat{\rho}(\mathbf{x}', \mathbf{x}'', t) \rangle = \rho_1(\mathbf{x}'; \mathbf{x}''; t) \tag{8.73}$$

is the first-order reduced density matrix which via eqns (8.22), (8.29), and (8.32) can be used to introduce the local mass density $\rho(\mathbf{x}, t)$ and a velocity field $\mathbf{v}(\mathbf{x}, t)$.

In this section, however, we want to use an approach in which the process of statistical averaging is considered separately from the identification of a macroscopic velocity field. The latter we introduce at the operator level by transforming the field operators $\psi(\mathbf{x}, t)$ from a fluid at rest to a fluid in motion by a unitary transformation (Bogolyubov, 1962)

$$\psi(\mathbf{x}, t) \to \psi(\mathbf{x}, t)\tilde{\Phi}(\mathbf{x}) \tag{8.74}$$

determined by

$$-i\hbar \frac{\partial}{\partial \mathbf{x}} \tilde{\Phi}(\mathbf{x}) = m\mathbf{v}(\mathbf{x})\tilde{\Phi}(\mathbf{x})$$

with the formal solution

$$\tilde{\Phi}(\mathbf{x}) = \exp\left[\frac{im}{\hbar} \int_{\mathbf{x}_0}^{\mathbf{x}} v_i(\mathbf{x}') \, dx_i'\right]$$

and the initial condition $\tilde{\Phi}(\mathbf{x}_0) = 1$ at some reference point $\mathbf{x}_0$.‡

Using eqn (8.74) we can then transform the tensor operators introduced in the last section to a moving fluid. We obtain (the superscript 0 refers to a fluid at rest, and we also suppress the time dependence for brevity)

$$\hat{\rho}(\mathbf{x}) = \hat{\rho}^{(0)}(\mathbf{x})$$

$$\hat{j}_k(\mathbf{x}) = \hat{j}_k^{(0)}(\mathbf{x}) + \frac{m}{\hbar} \hat{\rho}(\mathbf{x}) v_k(\mathbf{x})$$

$$\hat{T}_{kl}(\mathbf{x}) = \hat{T}_{kl}^{(0)}(\mathbf{x}) - \left(\frac{m}{\hbar}\right)^2 v_k(\mathbf{x}) v_l(\mathbf{x}) \hat{\rho}(\mathbf{x}) - \frac{m}{\hbar}(v_k \hat{j}_l^{(0)} + v_l \hat{j}_k^{(0)})$$

$$\hat{F}_{ikl}(\mathbf{x}) = \hat{F}_{ikl}^{(0)}(\mathbf{x}) + \frac{m}{\hbar} v_k \hat{T}_{il}^{(0)}(\mathbf{x}) + \frac{m}{\hbar} v_i \hat{T}_{kl}^{(0)}(\mathbf{x}) + \frac{m}{\hbar} v_l \hat{T}_{ik}^{(0)}(x)$$

$$+ \frac{1}{12} \frac{m}{\hbar} \hat{\rho}(\mathbf{x}) \left(\frac{\partial^2 v_i}{\partial x_k \, \partial x_l} + \frac{\partial^2 v_k}{\partial x_i \, \partial x_l} + \frac{\partial^2 v_l}{\partial x_i \, \partial x_k}\right)$$

$$- \left(\frac{m}{\hbar}\right)^3 \hat{\rho}(\mathbf{x}) v_i v_k v_l - \left(\frac{m}{\hbar}\right)^2 (\hat{j}_k^{(0)} v_i v_l + \hat{j}_l^{(0)} v_i v_k + \hat{j}_i^{(0)} v_k v_l) \quad (8.75)$$

and similar reductions for higher-rank tensors. To proceed next with the definition of macroscopic quantities, we have to take the statistical averages of the local operators in eqn (8.75). This averaging procedure necessitates the assumption of local equilibrium in the fluid; i.e. around each point $(\mathbf{x}, t)$ in space and time we assume a macroscopic but still 'small' volume and time

‡ Uniqueness of $\tilde{\Phi}(\mathbf{x})$ demands only that $m/\hbar \int v_i(\mathbf{x}') \, dx_i' = 2\pi n$ along a closed path. In general, there will be no phase coherence in the field operator, i.e. $n = 0$, except in superfluids where the above relation is the condition for the quantization of vortex lines (Onsager, 1949; Feynman, 1955). The velocity field introduced in eqn (8.74) is irrotational, i.e. $\nabla \times \mathbf{v} = 0$. The generalization to rotational flow has been given by Pokrovsky and Sergeev (1973).

interval over which averaging is done with an equilibrium density matrix. The requirement of a long enough time interval Δt over which the local equilibrium density matrix can be assumed constant is usually met in systems in which two distinct time scales exist:‡ a short characteristic time over which local thermalization is achieved, which can be identified as the regression time τ_{regr} for fluctuations, and a long characteristic time τ_{ev} over which macroscopic changes evolve. For time intervals Δt bracketed by $\tau_{\text{regr}} \ll \Delta t \ll \tau_{\text{ev}}$, we can then assume that a constant equilibrium density matrix exists. As to the size of the 'small cells' over which this density matrix can be defined, we have argued in the introduction that it should be possible to reach neighboring 'cells' by local fluctuations.

To construct the local equilibrium density matrix explicitly, let us divide the volume V of our system in small cells of size ΔV_i such that $V = \sum_i \Delta V_i$. Let us next decompose the hamiltonian, eqns (8.1)–(8.3) according to§

$$\hat{H} = \sum_i \hat{H}_i + \sum_{i,j} \hat{V}_{ij} \qquad (8.76)$$

where

$$\hat{H}_i = \int_{\Delta V_i} d\mathbf{x} \hat{H}(\mathbf{x}) \qquad (8.77)$$

and

$$\hat{V}_{ij} = \tfrac{1}{2} \int_{\Delta V_i} d\mathbf{x} \int_{\Delta V_j} d\mathbf{y} \psi^\dagger(\mathbf{x}) \psi^\dagger(\mathbf{y}) V(\mathbf{x}, \mathbf{y}) \psi(\mathbf{y}) \psi(\mathbf{x}) \qquad (8.78)$$

This last term $\hat{V}_{ij}$ represents two-particle interactions, i.e. scattering from cell i to cell j. But this is at most important between neighboring cells through their common interface. However, surface effects are thermodynamically small, because they are not extensive and can be treated at the level of fluctuations. In lowest order we therefore will neglect the $\hat{V}_{ij}$ terms in eqn (8.76) and observe that

$$[\hat{H}_i, \hat{H}_j] = 0 \qquad (8.79)$$

‡ See also similar arguments put forward in the derivation of the Onsager reciprocity relations in Section 3.2.
§ Our construction of the local equilibrium density matrix is quite similar to the one presented by Mori (1958).

because field operators at different space points commute (or anticommute) with each other. We next assume that cell i is in local equilibrium at an inverse temperature $\beta_i = (k_B T_i)^{-1}$ and a chemical potential μ_i and write the density operator for the system as

$$\hat{\rho} = \exp\left[-\sum_i \beta_i(\hat{H}_i - \mu_i \hat{N}_i)\right] \bigg/ Tr\left\{\exp\left[-\sum_i \beta_i(\hat{H}_i - \mu_i \hat{N}_i)\right]\right\}$$
$$= \prod_i \hat{\rho}_i \qquad (8.80)$$

where

$$\hat{\rho}_i = \exp\left[-\beta_i(\hat{H}_i - \mu_i \hat{N}_i)\right]/Tr\{\exp\left[-\beta_i(\hat{H}_i - \mu_i \hat{N}_i)\right]\} \quad (8.81)$$

is the density operator for strict local equilibrium in cell i.

$$\hat{N}_i = \int_{\Delta\Omega_i} \psi^\dagger(\mathbf{x})\psi(\mathbf{x}) \, d\mathbf{x} \qquad (8.82)$$

is the number operator in cell i. The second line of eqn (8.80) follows because of eqn (8.79).‡

Equation (8.80) is the exact quantum analogue of the lowest-order, i.e. strict local equilibrium, approximation to the single-particle distribution function $f_1^{(0)}(\mathbf{r}, \mathbf{p}, t)$ in the Boltzmann theory of dilute gases. [See the discussion around eqn (7.166).] As has been done there, we can now show that averaging the balance eqns (8.65), (8.66), and (8.70), after the transformations (8.75) have been performed, with the density operator (8.80) will result in the set (7.175) of macroscopic balance equations in which all dissipative effects are absent. To include the latter, one must include the interactions (8.78) between neighboring local equilibrium cells at the level of fluctuations. Zubarev (1974) has done this quite explicitly by constructing a nonequilibrium statistical operator in close analogy to the hydrodynamic approximation to the single-particle distribution function in the Boltzmann gas. This was discussed in great length in Section 7.7, particularly with regard to eqns (7.166) and (7.196). Zubarev succeeds in deriving the linear phenomenological laws of viscosity, diffusion, and thermal conduction and deriving microscopic expressions for the transport coefficients which are identical to those derived in

‡ A derivation based on a more rigorous variational principle has been presented by Zubarev (1974).

linear response theory. As we will deal with the latter at great length in Chapter 9, we prefer here to proceed in a rather phenomenological way, as suggested by Bogolyubov (1970) and used by Galasiewicz (1970).

We denote the average with the (unspecified) local statistical operator in the comoving Lagrange frame by pointed brackets ⟨- - - -⟩. We also have to remember that we must multiply an nth-rank tensor by $\hbar^n$ to obtain a macroscopic quantity from the operators (8.79) because, according to the correspondence principle, it is $[i\hbar(\partial/\partial\xi_k)]$ that corresponds to an observable quantity (momentum). In this way, we get the macroscopic mass density

$$\rho(\mathbf{x}, t) = m\langle\hat{\rho}(\mathbf{x}, t)\rangle \tag{8.83}$$

the macroscopic mass current density

$$\mathbf{j}(\mathbf{x}, t) = \hbar\langle\hat{\mathbf{j}}(\mathbf{x}, t)\rangle = \rho(\mathbf{x}, t)\mathbf{v}(\mathbf{x}, t) \tag{8.84}$$

(the term $\hbar\langle\hat{\mathbf{j}}^{(0)}(\mathbf{x}, t)\rangle = 0$ vanishes, because there is no current in a fluid at rest) the macroscopic kinetic energy-stress tensor (dropping the time dependence)

$$T_{kl}(\mathbf{x}) = -\frac{\hbar^2}{2m}\langle\hat{T}_{kl}(\mathbf{x})\rangle = T_{kl}^{(0)}(\mathbf{x}) + \tfrac{1}{2}\rho(\mathbf{x})v_k v_l \tag{8.85}$$

and a macroscopic energy-flux tensor

$$\begin{aligned}F_{ikl}(\mathbf{x}) = -\frac{\hbar^3}{2m^2}\langle\hat{F}_{ikl}(\mathbf{x})\rangle &= F_{ikl}^{(0)}(\mathbf{x}) + v_i T_{kl}^{(0)}(\mathbf{x}) \\ &+ v_k T_{il}^{(0)}(\mathbf{x}) + v_l T_{ik}^{(0)}(\mathbf{x}) + \tfrac{1}{2}\rho(\mathbf{x})v_i v_k v_l \\ &- \frac{\hbar^2}{24m^2}\rho(\mathbf{x})\left(\frac{\partial^2 v_k}{\partial x_i\,\partial x_l} + \frac{\partial^2 v_i}{\partial x_k\,\partial x_l} + \frac{\partial^2 v_l}{\partial x_i\,\partial x_k}\right)\end{aligned} \tag{8.86}$$

$T_{kl}^{(0)}(\mathbf{x}, t)$ is, similar to eqn (8.38), the kinetic energy contribution to the stress or pressure tensor, and $F_{ikl}^{(0)}(\mathbf{x}, t)$ is an internal energy-flux tensor.

With these expressions, we obtain the macroscopic continuity equation from eqn (8.64):

$$\frac{\partial\rho(\mathbf{x})}{\partial t} + \frac{\partial(\rho v_k)}{\partial x_k} = 0 \tag{8.87}$$

Equation (8.66) yields the macroscopic equation of motion or momentum balance equation

$$\frac{\partial(\rho v_k)}{\partial t} + 2\frac{\partial}{\partial x_l}[T^{(0)}_{kl}(\mathbf{x}) + \tfrac{1}{2}\rho v_k v_l]$$

$$= -\frac{1}{m}\rho(\mathbf{x})\frac{\partial \Phi(\mathbf{x})}{\partial x_k} - \int \frac{\partial V(\mathbf{x},\mathbf{y})}{\partial x_k}\langle\{\hat{\rho}(\mathbf{x})\hat{\rho}(\mathbf{y})\}\rangle\, d^3\mathbf{y} \quad (8.88)$$

involving the density-density or two-body correlation function $\langle\{\hat{\rho}\hat{\rho}\}\rangle$ which, according to eqn (8.62), is actually identical to the second-order reduced density matrix $\rho_2(\mathbf{x},\mathbf{y};\mathbf{x},\mathbf{y};t)$ in eqn (8.35).

To obtain a balance equation for the energy of the system, we have essentially three different approaches. We can start from the momentum balance (8.88), the hamiltonian density eqns (8.1)–(8.3), or eqn (8.70) for the kinetic energy tensor. The last option is unique to this derivation of balance equations and will allow us to separate out a balance for the kinetic pressure.‡

To obtain a balance equation for the kinetic energy, we multiply the momentum balance equation (8.88) by the velocity $\mathbf{v}$ and get, after some algebra,

$$\frac{\partial}{\partial t}\left(\frac{1}{2}\rho v^2\right) + 2v_k\frac{\partial T^{(0)}_{kl}}{\partial x_l} + \frac{\partial}{\partial x_l}\left(\frac{1}{2}v^2 v_l\right)$$

$$= -\frac{1}{m}\rho v_k\frac{\partial \Phi}{\partial x_k} - v_k \int \frac{\partial V(\mathbf{x},\mathbf{y})}{\partial x_k}\langle\{\hat{\rho}(\mathbf{x})\hat{\rho}(\mathbf{y})\}\rangle\, d^3\mathbf{y} \quad (8.89)$$

or, in tensor notation,

$$\frac{\partial}{\partial t}\left(\frac{1}{2}\rho v_k v_l\right) + v_k\frac{\partial T^{(0)}_{li}}{\partial x_i} + v_l\frac{\partial T^{(0)}_{ki}}{\partial x_i} + \frac{\partial}{\partial x_i}\left(\frac{1}{2}\rho v_i v_k v_l\right)$$

$$= -\frac{\rho}{2m}v_l\frac{\partial \Phi}{\partial x_k} - \frac{\rho}{2m}v_k\frac{\partial \Phi}{\partial x_l} - \frac{1}{2}\int \langle\{\hat{\rho}(\mathbf{x})\hat{\rho}(\mathbf{y})\}\rangle$$

$$\times\left[v_l(\mathbf{x})\frac{\partial V(\mathbf{x},\mathbf{y})}{\partial x_k} + v_k(\mathbf{x})\frac{\partial V(\mathbf{x},\mathbf{y})}{\partial x_l}\right]d^3\mathbf{y} \quad (8.90)$$

‡ The analogous procedure in classical kinetic theory is contained in the thirteen moment equations of Grad (1949) developed for the Boltzmann equation. As we have seen in eqn (7.140) there is no potential contribution to the pressure in a Boltzmann gas so that its total pressure is equal to the kinetic pressure. In the general case discussed here the total pressure has both kinetic and potential contributions. We derive in the following a balance equation of the kinetic pressure only which in a gas should be the dominant contribution to the total pressure.

Let us next take thermodynamic averages in eqn (8.70) for the tensor $\hat{T}_{kl}(\mathbf{x})$ and make use of eqns (8.85). We derive

$$\frac{\partial}{\partial t}(T_{kl}^{(0)} + \tfrac{1}{2}\rho v_k v_l) + \frac{\partial}{\partial x_i}\bigg[F_{ikl}^{(0)} + v_i T_{kl}^{(0)} + v_l T_{ki}^{(0)} + v_k T_{il}^{(0)}$$
$$- \frac{\hbar^2}{24m}\rho\bigg(\frac{\partial^2 v_i}{\partial x_k \partial x_l} + \frac{\partial^2 v_k}{\partial x_i \partial x_l} + \frac{\partial^2 v_l}{\partial x_i \partial x_k}\bigg) + \tfrac{1}{2}\rho v_i v_k v_l\bigg]$$
$$= -\frac{\rho}{2m}v_l\frac{\partial \Phi}{\partial x_k} - \frac{\rho}{2m}v_k\frac{\partial \Phi}{\partial x_l} + \frac{\hbar}{2m}\int \frac{\partial V(\mathbf{x},\mathbf{y})}{\partial x_k}\langle\{\hat{\rho}(\mathbf{y})\hat{j}_l(\mathbf{x})\}\rangle\, d^3\mathbf{y}$$
$$+ \frac{\hbar}{2m}\int \frac{\partial V(\mathbf{x},\mathbf{y})}{\partial x_l}\langle\{\hat{\rho}(\mathbf{y})\hat{j}_k(\mathbf{x})\}\rangle\, d^3\mathbf{y} \tag{8.91}$$

Subtracting eqn (8.90) from (8.91), we obtain

$$\frac{\partial T_{kl}^{(0)}(\mathbf{x})}{\partial t} + \frac{\partial}{\partial x_i}[F_{ikl}^{(0)} + v_i T_{kl}^{(0)}(\mathbf{x})]$$
$$= \frac{\hbar}{2m}\int \frac{\partial V(\mathbf{x},\mathbf{y})}{\partial x_l}\langle\{\hat{\rho}(\mathbf{y})\hat{j}_k^{(0)}(\mathbf{y})\}\rangle\, d^3\mathbf{y}$$
$$+ \frac{\hbar}{2m}\int \frac{\partial V(\mathbf{x},\mathbf{y})}{\partial x_k}\langle\{\hat{\rho}(\mathbf{y})\hat{j}_l^{(0)}(\mathbf{x})\}\rangle\, d^3\mathbf{y}$$
$$- T_{ki}^{(0)}\frac{\partial v_l}{\partial x_i} - T_{li}^{(0)}\frac{\partial v_k}{\partial x_i} \tag{8.92}$$

This is a balance equation for the kinetic pressure which we will now investigate further.‡

We are now again at the stage where we must find suitable approximations to the correlation functions appearing in the integrals over the two-body interactions on the right of the balance equations (8.88)–(8.92) to reduce them to the phenomenological equations of nonequilibrium thermodynamics. There is no need here to consider eqn (8.88) as we have studied

---

‡ The supplementary balance equations for the hamiltonian energy density

$$\hat{E}(\mathbf{x},t) = -\frac{\hbar^2}{2m}\hat{T}_{ll}(\mathbf{x},t) + \Phi(\mathbf{x})\hat{\rho}(\mathbf{x},t) + \tfrac{1}{2}\int V(\mathbf{x},\mathbf{y})\psi^\dagger(\mathbf{x},t)\hat{\rho}(\mathbf{y},t)\psi(\mathbf{x},t)\, d^3\mathbf{y}$$

and the internal energy density

$$\rho(\mathbf{x},t)u(\mathbf{x},t) = \langle\hat{E}(\mathbf{x},t)\rangle - \tfrac{1}{2}\rho v^2 - \Phi(\mathbf{x})\rho(\mathbf{x},t)$$

are given by Kreuzer (1975).

its equivalent, namely eqn (8.40), to derive the Navier–Stokes equation. Instead, we now turn to a similar analysis of the balance equation (8.92) for the kinetic pressure. We discuss first the trivial case of an ideal gas. Obviously, we have [compare this to the classical treatment in eqn (9.62)]

$$T_{kl}^{(0)}(\mathbf{x}, t) = \tfrac{1}{3} p \, \delta_{kl} \tag{9.93}$$

where p is the hydrostatic pressure. Thus, eqn (8.92) reduces to a continuity equation for p, namely

$$\frac{\partial p}{\partial t} + \boldsymbol{\nabla} \cdot (p\mathbf{v}) = 0 \tag{8.94}$$

and the off-diagonal elements yield a relation

$$\frac{1}{2} \left(\frac{\partial v_l}{\partial x_k} + \frac{\partial v_k}{\partial x_l} \right) = \Lambda_{kl} = 0 \tag{8.95}$$

confirming that in an ideal fluid the rate-of-strain tensor Λ_{kl}, as defined in eqn (2.47), vanishes.

In a real fluid, however, the rate-of-strain tensor will contribute to the pressure balance, as will locally generated pressure flow through the term $(\partial/\partial x_i) F_{ikl}^{(0)}(\mathbf{x})$ in eqn (8.92). Moreover, the interactions will, through correlations, introduce local sinks and sources for the kinetic part of the pressure. To understand the nature of these sinks and sources, let us expand the density-current correlations for small deviations from equilibrium, the expansion parameter being again the local velocity field $\mathbf{v}(\mathbf{x})$. We can write

$$\langle \{ \hat{\rho}(\mathbf{y}) \hat{\jmath}_l^{(0)}(\mathbf{x}) \} \rangle = \langle \{ \hat{\rho}(\mathbf{x}+\mathbf{r}) \hat{\jmath}_l^{(0)}(\mathbf{x}) \} \rangle$$
$$= v_l \Xi_0(r) + \frac{\partial v_l}{\partial x_i} r_i \Xi_1(r) + \frac{\partial v_i}{\partial x_i} r_l \Xi_2(r) + \cdots \tag{8.96}$$

where $\Xi_i(r)$ are unspecified functions of the distance $r = |\mathbf{x} - \mathbf{y}|$. This expansion is certainly not complete but is sufficient for our purposes. Inserting (8.96) into (8.92), we obtain

$$\frac{\partial}{\partial t} T_{kl}^{(0)}(\mathbf{x}) + \frac{\partial}{\partial x_i} [F_{ikl}^{(0)}(\mathbf{x}) + v_i T_{kl}^{(0)}(\mathbf{x})]$$
$$= -2 G \Lambda_{kl} - B \frac{\partial v_k}{\partial x_k} \delta_{kl} - \left(T_{ki}^{(0)} \frac{\partial v_l}{\partial x_i} + T_{li}^{(0)} \frac{\partial v_k}{\partial x_i} \right) \tag{8.97}$$

where

$$G = -\frac{1}{2}\frac{\hbar}{2m}\int r^3 \frac{\partial V}{\partial r} \Xi_1(r)\, d^3\mathbf{r}$$

$$B = -\frac{2}{3}\frac{\hbar}{2m}\int r^3 \frac{\partial V}{\partial r} \Xi_2(r)\, d^3\mathbf{r} \qquad (8.98)$$

Comparing eqn (8.97) with the phenomenological equations (2.50) for the viscoelastic pressure tensor, we can identify G as the shear modulus and B as a relaxation bulk modulus which is equal to $B = (\zeta/\eta)G$, where ζ is the bulk viscosity and η the shear viscosity. Thus, the sinks for the kinetic presssure are due to the energy dissipation during the viscous deformation (change in volume) in the fluid. Again we find that ideal or incompressible fluids will satisfy a continuity equation for the pressure without any sinks.

9
Linear Response Theory

9.1. The Formalism of Linear Response Theory

9.1.1. Introductory Remarks

IN previous sections we have argued repeatedly that systems in local equilibrium respond to external stimuli through ever-present and all-pervasive spontaneous local fluctuations. This fact has been used in Chapter 1 to assess the range of the local equilibrium assumption in nonequilibrium systems by arguing as follows. We imagine our system to be divided into 'small' finite cells which are assumed to be in thermodynamic equilibrium over a sufficiently long time, long compared to the microscopic regression time for fluctuations to achieve equilibrium but short compared to the macroscopic evolution time of the system as a whole. It is evident then that spatially neighboring cells must also be close to each other in thermodynamic phase space, in fact so close that their thermodynamic states can be reached from each other by local fluctuations which are typically of the (relative) order $N^{-\frac{1}{2}}$, where N is the number of microscopic particles in a (finite) cell. This analysis was confirmed in Section 7.7 for a dilute gas by the explicit construction of the Chapman-Enskog solution to the Boltzmann equation in the hydrodynamic regime.

In Chapter 3 it was similarly argued in the 'proofs' of the Onsager reciprocity relations that under an external force local (time-reversible) fluctuations drive the system irreversibly to a steady state compatible with the constraints. As a very similar idea is exploited in this chapter, it might be useful to repeat the argument here in a slightly different language. We consider a system that in the distant past ($t \to -\infty$) was prepared in a state of (local, if necessary) equilibrium. There will exist spontaneous time-reversal invariant fluctuations of such a nature that the stability of the equilibrium state is guaranteed by the fact that the fluctuations lead to an excess entropy $\delta^2 S < 0$ (or $\delta^2 s < 0$ locally) that is always negative to ensure that equilibrium is characterized by a maximal entropy. Let us now, in the distant past, switch on adiabatically an external force. As a result, the system will no longer be in stable equilibrium. On the other hand, internal affairs in the system will not be much affected by a small force, and the only response will be through local equilibrium fluctuations that move the system to a new stable state compatible with the imposed constraints. We obviously must, first of all, estimate when an external force can be considered 'small.' In the case of a dilute gas we have been able to give an explicit criterion, namely that the correction $\phi^{(1)}$ in the Chapman-Enskog solution of the Boltzmann equation, eqn (7.166),

$$f_1 = f_1^{(0)}(1 + \phi^{(1)}) \tag{9.1}$$

has to be small compared to one over the relevant range of momenta. In the more general case considered here, we can argue that the energy transmitted to a 'particle' by the external force over a characteristic distance in the system has to be small compared to its average energy in local equilibrium.‡

This criterion is, in fact, very conservative in a situation where the external force is switched on adiabatically. Adiabatic here means slow on the time scale of the regression of local fluctuations. In this case, the system only has to adjust at any instant of time by an infinitesimal amount to the new external constraint. If we follow this time evolution of the system via the statistical operator, this implies that first-order time-dependent perturbation theory should be adequate. As we will see below, such an

---

‡ Consider, for example, a metal under an applied electric field. We can be sure that its electron system is not greatly disturbed if the energy gained by an electron in the electric field over a mean-free path l is small compared to the Fermi energy ε_F of the metal. At $T \approx 300$ K we have typically $l \sim 10^{-6} - 10^{-5}$ cm. With $\varepsilon_F \sim 1-10$ eV, we see that for electric fields E as high as 10^4 V/cm [corresponding to the highest current densities of 10^9 A/cm^2 in metals with conductivities σ of the order 10^5 mhos/cm at which Ohm's law is still valid (Guth and Meyerhoefer, 1940)], we still have

$$|eEl| \ll \varepsilon_F \tag{9.2}$$

where e is the electronic charge. In this context, the term 'particle' may also refer to an appropriate quasiparticle, e.g. to a phonon in a solid if thermal effects are studied.

approach leads automatically to constitutive relations that are linear in the applied external forces. It has therefore received the name *linear response theory*. The use of first-order time-dependent perturbation theory in the study of transport phenomena in many-body systems was first advocated by Callen and Welton (1951) and Green (1952). Most current accounts follow a paper of Kubo (1957), where a rather general framework was laid out for the calculation of the system's response to mechanical forces within a hamiltonian formalism. Of the many recent reviews, we mention Zwanzig (1965), Martin (1968), and Forster (1975).

The formalism of linear response theory is extremely simple and straightforward, so much so that in most accounts of it little effort is made to bring out the physics behind it. In addition to what was said above, we want to stress two more points. It should be a puzzle to the observant reader why first-order time-dependent perturbation theory should work for all times. Again life is made easy by the presence of spontaneous fluctuations! To see this let us assume that at discrete time intervals Δt (large compared to the regression time for fluctuations, i.e. to the time needed to establish local equilibrium), the external force is increased suddenly by small discrete amounts, indeed arbitrarily small due to the hypothesis of adiabatic switching. The system's response can then obviously be calculated in first-order time-dependent perturbation theory over a time interval Δt, during which the system will also undergo many fluctuations. These will, at the end of the interval Δt, have wiped out all dynamic memory of the system related to the small increase of the external field at the beginning of Δt, simply due to the fact that the achieved change in a property $\mathcal{P}(t)$ of the system has to be within the range of the spontaneous fluctuations. (See Fig. 9.1.) Thus the next small increase in the external force will again be in a system in local equilibrium and linearization or first-order perturbation theory is again possible. In this picture the random fluctuations are indeed leading to the molecular chaos assumed in Boltzmann's famous *Stosszahlansatz*.‡

---

‡ It is therefore of little surprise that the results of linear response theory correspond to those of the linearized Boltzmann equation in a dilute gas (Mori, 1958; Fujita, 1962; McLennan and Swenson, 1963).

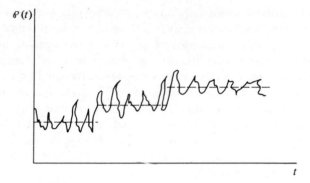

Fig. 9.1. Adiabatic switching of an external force implies that over times large compared to the regression time of fluctuations and small compared to macroscopic evolution times of the system, changes in properties $\mathcal{P}(t)$ must be within reach of equilibrium fluctuations.

With this physical picture linear response theory should yield acceptable results for slow processes, i.e. hydrodynamic problems. However, it turns out that the response functions are analytic in the frequency (introduced by a Fourier transform in time) through a set of Kramers-Kronig dispersion relations. Linear response theory thus provides its own analytic continuation and, due to the uniqueness of the latter, is valid for all ω (for fast processes).

Before we engage in the formalism, let us add a word of caution. Linear response theory, once restricted to systems as described above, makes no further assumptions except that the applied external force must be such that it can be incorporated as part of the total hamiltonian of the system. This, of course, means that the results of linear response theory are as general as can be, and are, as with most general results in physics, not directly useful for practical calculations. However, linear response theory is eminently suited to prove some general features of nonequilibrium physics like positivity of transport coefficients, validity of Onsager reciprocity relations, sum rules, dispersion relations, etc. To do practical calculations based on linear response theory, one can either employ equilibrium Green's function techniques, use kinetic equations, or resort to semiempirical models with experimental data as partial input. The latter approach is particularly rewarding since linear response theory expresses transport coefficients in terms of time-dependent correlation functions which are quite often available from experiments.

The rest of this section is devoted to setting up the formalism of linear response theory first for a system of classical particles and then for the quantum mechanical case. We next define response functions, establish dispersion relations for them, and arrive at the fluctuation-dissipation theorem. Then we concentrate on the hydrodynamics of a normal fluid, derive formulae for the transport coefficients of viscosity and heat conduction, and at last prove their positivity. As a practical application, we look at the ideal gas and then study electrical conductivity in a metal following Chester and Thellung (1959).

9.1.2. Classical Reponse Theory

We will now develop linear response theory for a system of classical particles. We assume that such a system has been isolated in the distant past, i.e. as $t \to -\infty$, and that it was at that stage controlled by a hamiltonian function $H_0 = H_0(\xi_1, \ldots, \xi_N)$, where $\xi_i = (\mathbf{r}_i, \mathbf{p}_i)$ are the six-dimensional vectors of space coordinates $\mathbf{r}_i$ and the momentum $\mathbf{p}_i$ of the ith particle. As time progresses an external force is switched on which adds a small time dependent perturbation

$$H_1(\xi_1, \ldots, \xi_N, t) = -F(t) A_1(\xi_1, \ldots, \xi_N) \tag{9.3}$$

to the total hamiltonian of the system

$$H = H_0 + H_1 \tag{9.4}$$

The factor $F(t)$ carries the explicit time dependence of the external perturbation.

One usually assumes that an external force acts on a many-body system by interacting with its constituents individually. We can therefore also write

$$H_1(\xi_1, \ldots, \xi_N, t) = -\sum_n \int \hat{A}_n(\mathbf{r}, \mathbf{p}) f_n(\mathbf{r}, t) \, d^3r \, d^3p \tag{9.5}$$

where the sum over n indicates the possibility that the system might be influenced by several external forces $f_n(\mathbf{r}, t)$ simultaneously which are, in addition, allowed to change locally throughout the system, and may also be time-dependent, e.g. oscillating. Adiabatic switching, if necessary, must also be included in $f_n(\mathbf{r}, t)$

through a factor $\exp(\varepsilon t)$ for $-\infty < t < 0$ with the implication that the limit $\varepsilon \to 0$ must be taken in the final result of the calculation. The factors $\hat{A}_n(\mathbf{r}, \mathbf{p})$ are single-particle operators in the classical phase space γ spanned by the position and momentum vectors $\mathbf{r}$ and $\mathbf{p}$, respectively (or some other set of canonical variables), of a single particle. Their $\mathbf{p}$ dependence can always be trivially integrated out in eqn (9.5), if the forces $f_n(\mathbf{r}, t)$ do not depend on the (canonical) momenta $\mathbf{p}$.

As an example let us construct the interaction hamiltonian for a metal subjected to an electric field $\mathbf{E}(\mathbf{r}, t)$. In the simplest model we can treat the electrons as a gas of free charged particles and can neglect the influence of the electric field on the positive charge background of the ions. The index n in eqn (9.5) then runs through the three (cartesian) components of $\mathbf{E}$ and

$$\mathbf{A}(\mathbf{r}, \mathbf{p}) = -e\mathbf{r}\hat{\rho}(\mathbf{r}, \mathbf{p}) \tag{9.6}$$

is the (vector) operator of the total dipole moment of a system of particles each with a charge $(-e)$. Here

$$\hat{\rho}(\mathbf{r}, \mathbf{p}) = \sum_{i=1}^{N} \delta(\mathbf{r} - \mathbf{r}_i)\delta(\mathbf{p} - \mathbf{p}_i) \tag{9.7}$$

is the single-particle density operator in γ space. For this example, eqn (9.5) then reads [remember that $\boldsymbol{\xi}_i = (\mathbf{r}_i, \mathbf{p}_i)$]

$$H_1(\boldsymbol{\xi}_1, \ldots, \boldsymbol{\xi}_N, t) = e \sum_{i=1}^{N} \mathbf{r}_i \cdot \mathbf{E}(\mathbf{r}_i, t) \tag{9.8}$$

We are, of course, only interested in a statistical description of the macroscopic response of the system to the external perturbation eqn (9.3), and therefore consider an ensemble of identical replicas of the physical system whose properties are controlled by the probability function $\rho_N(\boldsymbol{\xi}, \ldots, \boldsymbol{\xi}_N, t)$ in $6N$-dimensional Γ space, which is subject to Liouville's eqn (7.6),

$$\frac{\partial \rho_N}{\partial t} = \{H, \rho_N\} = \{H_0 + H_1, \rho_N\} \tag{9.9}$$

We denote by $\rho_N^{(0)}$ an equilibrium probability function in a system controlled by H_0. It satisfies the equation

$$0 = \frac{\partial}{\partial t} \rho_N^{(0)} = \{H_0, \rho_N^{(0)}\} = iL_0 \rho_N^{(0)} \tag{9.10}$$

where, by the last equality, we introduced the (differential) Liouville operator L_0 with the help of which we can rewrite eqn (9.9) formally as an integral equation, namely

$$\rho_N(\xi_1, \ldots, \xi_N, t) = \rho_N^{(0)}(\xi_1, \ldots, \xi_N) + \int_{-\infty}^{t} dt' e^{i(t-t')L_0}$$
$$\times \{\rho_N(\xi_1, \ldots, \xi_N, t'), A(\xi_1, \ldots, \xi_N)\} F(t') \quad (9.11)$$

as may be checked explicitly by direct differentiation. Assuming next that ρ_N does not differ much from $\rho_N^{(0)}$, we can hope that an iterative solution of eqn (9.11) exists and is given by

$$\rho_N(\xi_1, \ldots, \xi_N, t) = \rho_N^{(0)}(\xi_1, \ldots, \xi_N)$$
$$+ \sum_{k=1}^{\infty} \int_{-\infty}^{t} dt_1 \int_{-\infty}^{t_1} dt_2 \cdots \int_{-\infty}^{t_{k-1}} dt_k e^{i(t-t_1)L_0}$$
$$\times \{e^{i(t_1-t_2)L_0}\{\ldots \{e^{i(t_{k-1}-t_k)L_0}\rho_N^{(0)}(\xi_1, \ldots, \xi_N),$$
$$\times A(\xi_1, \ldots, \xi_N)\} \ldots \}\} F(t_1), \ldots, F(t_k)$$
$$(9.12)$$

Linear response theory results if only the lowest-order term is kept, in which case we have

$$\rho_N(\xi_1, \ldots, \xi_N, t) \approx \rho_N^{(0)}(\xi_1, \ldots, \xi_N) + \int_0^t dt_1 \, e^{i(t-t_1)L_0}$$
$$\times \{\rho_N^{(0)}(\xi_1, \ldots, \xi_N), A(\xi_1, \ldots, \xi_N)\} F(t_1) \, dt_1 \quad (9.13)$$

This explicit time-dependent probability density of the ensemble can be used to calculate the average change induced by H_1 in some dynamic variable $B = B(\xi_1, \ldots, \xi_N)$ according to

$$\bar{B}(t) = \int B(\xi_1, \ldots, \xi_N) \rho_N(\xi_1, \ldots, \xi_N, t) \, d\xi_1 \cdots d\xi_N$$
$$= \bar{B}_0 + \int d\xi_1 \cdots d\xi_N \int_{-\infty}^{t} dt_1 e^{i(t-t_1)L_0} \{\rho_N^{(0)}(\xi_1, \ldots, \xi_N),$$
$$A(\xi_1, \ldots, \xi_N)\} B(\xi_1, \ldots, \xi_N) F(t_1) \quad (9.14)$$

where the integration $d\xi_1 \cdots d\xi_N$ is over all the $6N$-dimensional phase space available to the ensemble and $\bar{B}_0 = \bar{B}(t = -\infty)$ is the ensemble average of B in the remote past when the system was prepared according to an equilibrium ensemble controlled by $\rho_N^{(0)}$.

By partial integration in eqn (9.14), we can shift the action of the Liouville operator L_0 from the Poisson bracket $\{\rho_N^{(0)}, A\}$ onto B and get

$$\bar{B}(t) = \bar{B}_0 + \int d\boldsymbol{\xi}_1 \cdots d\boldsymbol{\xi}_N \int_{-\infty}^{t} dt_1 F(t_1)$$
$$\times \{\rho_N^{(0)}(\boldsymbol{\xi}_1, \ldots, \boldsymbol{\xi}_N), A(\boldsymbol{\xi}_1, \ldots, \boldsymbol{\xi}_N)\} B(\boldsymbol{\xi}_1, \ldots, \boldsymbol{\xi}_N; t - t_1) \quad (9.15)$$

where

$$B(\boldsymbol{\xi}_1, \ldots, \boldsymbol{\xi}_N; t - t_1) = e^{-i(t-t_1)L_0} B(\boldsymbol{\xi}_1, \ldots, \boldsymbol{\xi}_N) \quad (9.16)$$

can be thought of as a time-dependent Heisenberg operator in classical phase space subject to the equation-of-motion [note the sign change from eqn (9.10)]

$$\dot{B} = \{B, H_0\} \quad (9.17)$$

Equation (9.15) states that the response in a property B of the system at time t due to an external perturbation H_1 switched on with the function $F(t)$ is a cumulative effect of pulses $F(t_1) dt_1$ during the past history of the system, i.e. for $-\infty < t_1 < t$. This point can be brought out more clearly by defining a response function

$$\phi_{BA}(t) = \int d\boldsymbol{\xi}_1 \cdots d\boldsymbol{\xi}_N \{\rho_N^{(0)}(\boldsymbol{\xi}_1, \ldots, \boldsymbol{\xi}_N), A(\boldsymbol{\xi}_1, \ldots, \boldsymbol{\xi}_N)\}$$
$$\times B(\boldsymbol{\xi}_1, \ldots, \boldsymbol{\xi}_N; t) \quad (9.18)$$

which measures the linear response in B at time t to a unit pulse in H_1, at time $t = 0$, calculated at each and every moment t as an average over the initial equilibrium distribution $\rho_N^{(0)}$. With this we can rewrite eqn (9.15) as

$$\bar{B}(t) = \bar{B}_0 + \int_{-\infty}^{t} \phi_{BA}(t - t_1) F(t_1) \, dt_1$$
$$= \bar{B}_0 + \int_{0}^{\infty} \phi_{BA}(t) F(t - \tau) \, d\tau \quad (9.19)$$

confirming the above interpretation.

With H_1 given by eqn (9.5), we can calculate from eqns (9.18) and (9.19) the local response in a property n represented by the operator $\hat{A}_n(\mathbf{r}, \mathbf{p})$ due to the force $f_m(\mathbf{r}, t)$ acting on property m. It is given by

$$\bar{A}_m(\mathbf{r}, t) = \int d\boldsymbol{\xi}_1 \cdots d\boldsymbol{\xi}_N \int d^3p \hat{A}_m(\mathbf{r}, \mathbf{p}) \hat{\rho}(t)$$

$$= \bar{A}_m^{(0)}(\mathbf{r}) + \sum_n \int_{-\infty}^t dt_1 \int d^3r_1 \phi_{mn}(\mathbf{r}, \mathbf{r}_1; t - t_1) f_n(\mathbf{r}_1, t_1) \quad (9.20)$$

where the response function is now given by

$$\phi_{mn}(\mathbf{r}, \mathbf{r}_1; \tau) = \int d\boldsymbol{\xi}_1 \cdots d\boldsymbol{\xi}_N \{\rho_N^{(0)}(\boldsymbol{\xi}_1, \ldots, \boldsymbol{\xi}_N), \hat{A}_n(\mathbf{r})\} \hat{A}_m(\mathbf{r}, \tau)$$

with (9.21)

$$\hat{A}_m(\mathbf{r}, \tau) = e^{-i\tau L_0} \hat{A}_m(\mathbf{r}) \quad (9.22)$$

and

$$\hat{A}_m(\mathbf{r}) = \int d^3p \hat{A}_m(\mathbf{r}, \mathbf{p}). \quad (9.23)$$

9.1.3. Quantum-Mechanical Response Theory

Analogous to the classical linear response theory of the previous section, we now want to calculate the response of a quantum-mechanical system. We assume that the latter is controlled by a hamiltonian operator

$$\hat{H} = \hat{H}_0 + \hat{H}_1(t) \quad (9.24)$$

where $\hat{H}_0$ is the hamiltonian of an isolated system and

$$\hat{H}_1(t) = -\sum_n \int d^3r \hat{A}_n(\mathbf{r}) f_n(\mathbf{r}, t) \quad (9.25)$$

describes the interaction with the external world in analogy to eqn (9.5). The sum over n indicates the possibility that the system might be influenced by several external forces $f_n(\mathbf{r}, t)$ which are, in addition, allowed to change locally throughout the system, and may also be time-dependent, e.g. oscillating. Adiabatic switching, if necessary, must also be done in $f_n(\mathbf{r}, \mathbf{t})$. The operator density $\hat{A}_n(\mathbf{r})$ is representative of that property of the microscopic constituents of the system on which the force $f_n(\mathbf{r}, t)$ acts. For the example of a free electron gas subjected to an electric field, the operators $\hat{A}_n(\mathbf{r})$ are again the three (cartesian) components of the local electric dipole moment, given by

$$\hat{\mathbf{A}}(\mathbf{r}) = -e\mathbf{r}\hat{\rho}(\mathbf{r}) \qquad (9.26)$$

with

$$\hat{\rho}(\mathbf{r}) = \psi^+(\mathbf{r})\psi(\mathbf{r}) \qquad (9.27)$$

being the quantum-mechanical density operator with $\psi(\mathbf{r})$ field operators in second quantization.

The statistical time evolution of the system controlled by $\hat{H}$ from eqn (9.24) follows from the Liouville–von Neumann equation (8.7)

$$i\hbar\dot{\hat{\rho}} = [\hat{H}, \hat{\rho}] \qquad (9.28)$$

for the density operator $\hat{\rho}$. To find a solution for this equation in time-dependent perturbation theory, we transform all operators to the interaction picture by the unitary transformation

$$_I\hat{O}(t) = e^{i\hat{H}_0 t/\hbar} \hat{O} e^{-i\hat{H}_0 t/\hbar} \qquad (9.29)$$

where $\hat{O}$ is any (time-independent) operator in the Schrödinger picture. Equation (9.28) then becomes

$$i\hbar \frac{\partial}{\partial t} {}_I\hat{\rho}(t) = [{}_I\hat{H}_1(t), {}_I\hat{\rho}(t)] \qquad (9.30)$$

or in integral form

$$_I\hat{\rho}(t) = {}_I\hat{\rho}(t = -\infty) + \frac{1}{i\hbar} \int_{-\infty}^t [{}_I\hat{H}(t'), {}_I\hat{\rho}(t')] \, dt' \qquad (9.31)$$

Observe that

$$_I\hat{\rho}(t = -\infty) = \hat{\rho}_0 \qquad (9.32)$$

is just the initial statistical operator describing the equilibrium of the system as prepared in the remote past. Transforming eqn (9.31) into the Schrödinger picture with the inverse of eqn (9.29), we obtain

$$\hat{\rho}(t) = \hat{\rho}_0 + \frac{1}{i\hbar} \int_{-\infty}^t e^{-i\hat{H}_0 t/\hbar} [{}_I\hat{H}_1(t'), {}_I\hat{\rho}(t')] e^{i\hat{H}_0 t/\hbar} \, dt' \qquad (9.33)$$

We next assume that $\hat{\rho}(t)$ never changes much from $\hat{\rho}_0$ in the course of the time evolution of the system so that an iterative solution of eqn (9.33) may be attempted to yield

$$\hat{\rho}(t) = \hat{\rho}_0 + \sum_{k=1}^{\infty} \left(\frac{1}{i\hbar}\right)^k \int_{-\infty}^{t} dt_1 \int_{-\infty}^{t_1} dt_2 \cdots \int_{-\infty}^{t_{k-1}} dt_k e^{-i\hat{H}_0 t/\hbar}$$
$$\times [_I\hat{H}_1(t_1), [_I\hat{H}_1(t_2) \cdots [_I\hat{H}_1(t_k), \hat{\rho}_0] \cdots] e^{i\hat{H}_0 t/\hbar} \quad (9.34)$$

The lowest-order term, which is also linear in $\hat{H}_1$, reads more explicitly

$$\hat{\rho}(t) \approx \hat{\rho}_0 + \frac{1}{i\hbar} \sum_n \int_{-\infty}^{t} dt_1 e^{-iH_0(t-t_1)/\hbar}$$
$$\times \int d^3 r_1 [\hat{A}_n(\mathbf{r}_1), \hat{\rho}_0] e^{i\hat{H}_0(t-t_1)/\hbar} f_n(\mathbf{r}_1, t_1) \quad (9.35)$$

This is the quantum analogue of eqn (9.13). With it we can follow the time evolution of any property of the system by calculating

$$\bar{A}_m(\mathbf{r}, t) = Tr\{\hat{A}_m(\mathbf{r})\hat{\rho}(t)\}$$
$$= \bar{A}_m^{(0)}(\mathbf{r}) - \frac{1}{i\hbar} \sum_n \int_{-\infty}^{t} dt_1 \int d^3 r_1 f_n(\mathbf{r}_1, t_1)$$
$$\times Tr\{\hat{A}_m(\mathbf{r}) e^{-i\hat{H}_0(t-t_1)/\hbar} [\hat{A}_n(\mathbf{r}), \hat{\rho}_0] e^{i\hat{H}_0(t-t_1)/\hbar}\} \quad (9.36)$$

In the last line, we used the fact that

$$Tr\hat{\rho}(t) = Tr(e^{iHt/\hbar}\hat{\rho}_0 e^{-iHt/\hbar}) = Tr\hat{\rho}_0 = 1 \quad (9.37)$$

due to the invariance of the trace under cyclic permutation of its arguments. We also define the equilibrium average

$$\bar{A}_m^{(0)}(\mathbf{r}) = Tr[\hat{A}(\mathbf{r})\hat{\rho}_0] \quad (9.38)$$

which in most cases will be independent of $\mathbf{r}$. Using the cyclic invariance of the trace once more in eqn (9.35) we obtain

$$\bar{A}_m(\mathbf{r}, t) = \bar{A}_m^{(0)}(\mathbf{r}) - \frac{1}{i\hbar} \sum_n \int_{-\infty}^{t} dt_1 \int d^3 r_1 f_n(r_1, t_1)$$
$$\times Tr\{\hat{A}_m(\mathbf{r}, t - t_1)[\hat{A}_n(\mathbf{r}_1), \hat{\rho}_0]\} \quad (9.39)$$

where

$$\hat{A}_m(\mathbf{r}, \tau) = e^{i\hat{H}_0 \tau/\hbar} \hat{A}_m(\mathbf{r}) e^{-i\hat{H}_0 \tau/\hbar} \quad (9.40)$$

in accordance with eqn (9.29).

Next we define response functions [in analogy to eqn (9.21)]

$$\phi_{mn}(\mathbf{r}, \mathbf{r}_1; t - t_1) = -\frac{1}{ih} Tr\{\hat{A}_m(\mathbf{r}, t - t_1)[\hat{A}_n(\mathbf{r}_1), \hat{\rho}_0]\} \quad (9.41)$$

measuring the response of the system in its property m at time t and at point $\mathbf{r}$ to a unit pulse in the external force acting in property n at time t and point $\mathbf{r}_1$ in the system. If the ansatz (9.25) is exhaustive, then all possible mechanical linear cross-coupling effects are taken care of by the response function ϕ_{mn}. With the help of the latter, we now write

$$\delta \bar{A}_m(\mathbf{r}, t) = \bar{A}_m(\mathbf{r}, t) - \bar{A}_m^{(0)}(\mathbf{r})$$

$$= \sum_n \int_{-\infty}^{t} dt_1 \int d^3 r_1 \phi_{mn}(\mathbf{r}, \mathbf{r}_1; t - t_1) f_n(\mathbf{r}_1, t_1) \quad (9.42)$$

This says again that in linear response theory the local response of a system to an external perturbation is the cumulative effect resulting from individual pulses of strength $f_n(\mathbf{r}_1, t_1) \, d^3 r_1 \, dt_1$ at positions $\mathbf{r}_1$ in the system during its complete prior history, i.e. for $-\infty < t_1 < t$.

Let us briefly study the situation where the external forces are switched on adiabatically in the remote past as usual, but are then suddenly switched off at time $t = 0$; i.e. we assume that

$$f_n(\mathbf{r}, t) = f_n(\mathbf{r}) e^{\varepsilon t} \qquad \text{for } -\infty < t < 0$$

and

$$f_n(\mathbf{r}, t) = 0 \qquad \text{for } t \geq 0 \quad (9.43)$$

At time $t = 0$, the system will then find itself in a nonequilibrium state appropriate to the external constraints $f_n(\mathbf{r})$, and starts to relax to equilibrium. From eqn (9.42) we can then write for $t > 0$

$$\delta \bar{A}_m(\mathbf{r}, t) = \sum_n \int \Phi_{mn}(\mathbf{r}, \mathbf{r}_1; t) f_n(\mathbf{r}_1) \, d^3 r_1 \quad (9.44)$$

where

$$\Phi_{mn}(\mathbf{r}, \mathbf{r}_1; t) = \lim_{\varepsilon \to 0} \int_0^{\infty} \phi_{mn}(\mathbf{r}, \mathbf{r}_1; t_1) e^{-\varepsilon t_1} \, dt_1 \quad (9.45)$$

is referred to as the relaxation function of the system (Kubo, 1957).

9.2. General Properties of Response Functions
9.2.1. Symmetries, Analyticity, and Dispersion Relations

Many experiments on nonequilibrium systems do not follow the time evolution of the latter directly but rather look at their frequency response. It is therefore worthwhile to consider the Fourier transforms of the response functions ϕ_{mn}. We first decompose the external fields‡

$$f_n(\mathbf{r}, t) = \int_{-\infty}^{+\infty} \frac{d\omega}{2\pi} \int \frac{d^3 k}{(2\pi)^3} e^{-i\omega t} e^{+i\mathbf{k}\cdot\mathbf{r}} f_n(\mathbf{k}, \omega) \qquad (9.46)$$

and the quantities

$$\delta \bar{A}_m(\mathbf{r}, t) = \int_{-\infty}^{+\infty} \frac{d\omega}{2\pi} \int \frac{d^3 k}{(2\pi)^3} e^{-i\omega t} e^{+i\mathbf{k}\cdot\mathbf{r}} \delta \bar{A}_m(\mathbf{k}, \omega) \qquad (9.47)$$

into their Fourier components and take the Fourier transform of eqn (9.42) to obtain

$$\delta(\bar{A}_m(\mathbf{k}, \omega) = \sum_n \chi_{mn}(\mathbf{k}, \omega) f_n(\mathbf{k}, \omega) \qquad (9.48)$$

where we identified

$$\chi_{mn}(\mathbf{k}, \omega) = \lim_{\varepsilon \to 0} \chi_{mn}(\mathbf{k}, \omega + i\varepsilon) \qquad (9.49)$$

and

$$\chi_{mn}(\mathbf{k}, z) = \int_0^\infty d(t - t_1) \int d^3(\mathbf{r} - \mathbf{r}_1)$$
$$\times e^{+iz(t-t_1)} e^{-i\mathbf{k}\cdot(\mathbf{r}-\mathbf{r}_1)} \phi_{mn}(\mathbf{r}, \mathbf{r}_1, t - t_1) \qquad (9.50)$$

with Im $z > 0$ to ensure convergence. In the last definition, averages in an equilibrium system can only depend on the relative separation $|\mathbf{r} - \mathbf{r}_1|$ due to the translational invariance of such a system.§

---

‡ We will sometimes use the same symbol for a function and its Fourier transform, distinguishing the two by their arguments $(\mathbf{r}, t)$ and $(\mathbf{k}, \omega)$, respectively.

§ The last statement is, of course, only true in fluids. In crystalline solids, the reduced symmetry under discrete translations must be employed. The existence of the Fourier transforms (9.50) is guaranteed in most systems by the fact that the cause-effect relation between two events m and n in the system should cease as they become infinitely separated in space and time, i.e. as $|t - t_1| \to \infty$ and $|\mathbf{r} - \mathbf{r}_1| \to \infty$. This may not always be true, particularly in systems with long-range forces or in systems close to critical points.

The response functions ϕ_{mn} possess a number of symmetry properties. We first rewrite eqn (9.41) as

$$\phi_{mn}(\mathbf{r}, \mathbf{r}_1; t - t_1) = \frac{i}{\hbar} Tr\{\hat{A}_m(\mathbf{r}, t)[\hat{A}_n(\mathbf{r}_1, t_1), \hat{\rho}_0]\}$$

$$= \frac{i}{\hbar} Tr\{[\hat{A}_m(\mathbf{r}, t), \hat{A}_n(\mathbf{r}_1, t_1)]\hat{\rho}_0\}$$

$$= \phi_{mn}(\mathbf{r}, t; \mathbf{r}_1, t_1) \quad (9.51)$$

and see immediately that

$$\phi_{mn}(\mathbf{r}, t; \mathbf{r}_1, t_1) = -\phi_{nm}(\mathbf{r}_1, t_1; \mathbf{r}, t) \quad (9.52)$$

due to the fact that ϕ_{mn} is a trace involving an (antisymmetric) commutator. Next observe that all operators $\hat{A}_m$ must be self-adjoint if they are representing physical observables. This implies that

$$\phi_{mn}(\mathbf{r}, t; \mathbf{r}_1, t_1) = \phi^*_{mn}(\mathbf{r}, t; \mathbf{r}_1, t_1) \quad (9.53)$$

is real. Last, we perform time reversal on eqn (9.50); i.e. we reverse the sign of time, momenta, angular momenta, spin, magnetic fields **B**, and other quantities that are odd under time reversal. We then obtain

$$\phi_{mn}(\mathbf{r}, t; \mathbf{r}_1, t_1; \mathbf{B}) = -\varepsilon_m^T \varepsilon_n^T \phi_{mn}(\mathbf{r}, -t; \mathbf{r}_1, -t_1; -\mathbf{B}) \quad (9.54)$$

where $\varepsilon_m^T = 1$ if property m does not change sign (i.e., is even) under time reversal, as it does not for mass and energy density, and $\varepsilon_m^T = -1$ if m changes sign (i.e., is odd) under time reversal. In fact,

$$\hat{A}_m(\mathbf{r}, -t) = \varepsilon_m^T \hat{A}_m(\mathbf{r}, t) \quad (9.55)$$

It will turn out to be useful to define a two-sided temporal Fourier transform of ϕ_{mn} according to

$$\phi_{mn}(\mathbf{r}, t; \mathbf{r}_1, t_1) = 2i \int_{-\infty}^{-\infty} \frac{d\omega}{2\pi} e^{-i\omega(t - t_1)}$$

$$\times \int \frac{d^3 k}{(2\pi)^3} e^{i\mathbf{k} \cdot (\mathbf{r} - \mathbf{r}_1)} \chi''_{mn}(\mathbf{k}, \omega) \quad (9.56)$$

in addition to the one-sided temporal Fourier transform $\chi_{mn}(\mathbf{k}, \omega)$, eqn (9.50). To see the significance of χ''_{mn}, let us insert eqn (9.56) into eqn (9.50) and perform the time integration. We obtain

$$\chi_{mn}(\mathbf{k}, z) = \frac{1}{\pi} \int_{-\infty}^{+\infty} \frac{d\omega}{\omega - z} \chi''_{mn}(\mathbf{k}, \omega) \qquad (9.57)$$

a relation which, for Im $z \neq 0$, allows us to continue $\chi_{mn}(\mathbf{k}, z)$ into the lower half of the z plane. Inverting the Fourier transforms in eqn (9.55), we can also show, using the symmetries (9.53) and (9.54), that

$$\chi''_{mn}(\mathbf{k}, \omega) = \frac{1}{2i}[\chi_{mn}(\mathbf{k}, \omega) - \varepsilon_m^T \varepsilon_n^T \chi_{mn}^*(\mathbf{k}, \omega)] \qquad (9.58)$$

This implies that for the diagonal elements of the response matrix

$$\chi''_{mm}(\mathbf{k}, \omega) = \text{Im } \chi_{mm}(\mathbf{k}, \omega) \qquad (9.59)$$

The symmetry relations (9.52), (9.53), and (9.54) can easily be translated into statements about the partial Fourier transforms

$$\chi''_{mn}(\mathbf{r}, \mathbf{r}_1; \omega) = \frac{1}{2i} \int_{-\infty}^{+\infty} d(t - t_1) e^{i\omega(t-t_1)} \phi_{mn}(\mathbf{r}, t; \mathbf{r}_1, t_1) \qquad (9.60)$$

From (9.52) we find

$$\chi''_{mn}(\mathbf{r}, \mathbf{r}_1; \omega) = -\chi''_{nm}(\mathbf{r}, \mathbf{r}_1; -\omega) \qquad (9.61)$$

From (9.53) we obtain

$$[\chi''_{mn}(\mathbf{r}, \mathbf{r}_1; \omega)]^* = -\chi''_{mn}(\mathbf{r}, \mathbf{r}_1; -\omega) \qquad (9.62)$$

Finally, from (9.54) we obtain

$$\chi''_{mn}(\mathbf{r}, \mathbf{r}_1; \omega, \mathbf{B}) = -\varepsilon_m^T \varepsilon_m^T \chi''_{mn}(\mathbf{r}, \mathbf{r}_1; -\omega, -\mathbf{B}) \qquad (9.63)$$

Additional symmetries are easily derived in isotropic systems which are also invariant under rotations and reflections. The latter imply that

$$\phi_{mn}(\mathbf{r}, t; \mathbf{r}_1, t_1) = \varepsilon_m^P \varepsilon_n^P \phi_{mn}(-\mathbf{r}, t; -\mathbf{r}_1, t_1) \qquad (9.64)$$

where ε_m^P is the signature of $\hat{A}_m(\mathbf{r}, t)$ under a parity transformation defined by

$$\hat{A}_m(\mathbf{r}, t) = \varepsilon_m^P \hat{A}_m(-\mathbf{r}, t) \qquad (9.65)$$

e.g. $\varepsilon_m^P = 1$ for mass and energy densities and spin, and $\varepsilon_m^P = -1$

for momentum. The resulting symmetries in $\chi''_{mn}(\mathbf{k}, \omega)$ are summarized as follows

$$\begin{aligned}
\chi''_{mn}(\mathbf{k}, \omega) &= \varepsilon^T_m \varepsilon^T_n \varepsilon^P_m \varepsilon^P_n [\chi''_{mn}(\mathbf{k}, \omega)]^* \\
&= \varepsilon^T_m \varepsilon^T_n \varepsilon^P_m \varepsilon^P_n \chi''_{mn}(\mathbf{k}, \omega) \\
&= \varepsilon^P_m \varepsilon^P_n \chi''_{mn}(-k, \omega) \\
&= -\varepsilon^T_m \varepsilon^T_n \chi''_{mn}(\mathbf{k}, -\omega)
\end{aligned} \quad (9.66)$$

i.e. χ''_{mn} is either real and symmetric in the indices m and n or purely imaginary and antisymmetric.

We next turn to the derivation of dispersion relations. Adding to eqn (9.57) its complex conjugate we obtain, using the first relations (9.66),

$$\chi_{mn}(\mathbf{k}, z) + \chi^*_{mn}(\mathbf{k}, z) = \frac{1}{\pi} \int_{-\infty}^{+\infty} d\omega' \chi''_{mn}(\mathbf{k}, \omega') \\
\times \left(\frac{1}{\omega' - z} + \frac{\varepsilon^T_m \varepsilon^T_n \varepsilon^P_m \varepsilon^P_n}{\omega' - z^*} \right) \quad (9.67)$$

Next we concentrate on the diagonal elements of the response matrix and let z approach the real axis ω from above. We obtain

$$\begin{aligned}
\operatorname{Re} \chi_{mm}(\mathbf{k}, \omega) &= \lim_{\varepsilon \to 0} \frac{1}{2} [\chi_{mm}(\mathbf{k}, \omega + i\varepsilon) + \chi^*_{mm}(\mathbf{k}, \omega - i\varepsilon)] \\
&= \frac{1}{2\pi} \lim_{\varepsilon \to 0} \int_{-\infty}^{+\infty} d\omega' \chi''_{mm}(\mathbf{k}, \omega') \\
&\quad \times \left(\frac{1}{\omega' - \omega - i\varepsilon} + \frac{1}{\omega' - \omega + i\varepsilon} \right) \\
&= \frac{1}{\pi} \fint_{-\infty}^{+\infty} \frac{d\omega'}{\omega' - \omega} \chi''_{mm}(\mathbf{k}, \omega')
\end{aligned} \quad (9.68)$$

where the last integral is a Cauchy principal value. Remembering from eqn (9.59) that $\chi''_{mm}(\mathbf{k}, \omega)$ is the imaginary part of $\chi_{mm}(\mathbf{k}, \omega)$, we see that eqn (9.68) is a dispersion relation

$$\operatorname{Re} \chi_{mm}(k, \omega) = \frac{1}{\pi} \fint_{-\infty}^{\infty} \frac{d\omega'}{\omega' - \omega} \operatorname{Im} \chi_{mm}(\mathbf{k}, \omega') \quad (9.69)$$

for the diagonal elements of the Fourier-transformed response matrix.

9.2.2. Sum Rules and Fluctuation-Dissipation Theorem

Sum rules are constraints imposed on the response functions by the dynamics of the system. For their derivation we start from the definition (9.51) and take the jth derivative with respect to time t, i.e.

$$\left(i\frac{\partial}{\partial t}\right)^j\left[\frac{1}{2i}\phi_{mn}(\mathbf{r}, t; \mathbf{r}_1, t_1)\right]$$

$$= \frac{1}{2\hbar} Tr\left\{\left[\left(i\frac{\partial}{\partial t}\right)^j \hat{A}_m(\mathbf{r}, t), \hat{A}_n(\mathbf{r}_1, t_1)\right]\hat{\rho}_0\right\} \quad (9.70)$$

Next we eliminate the time derivative on the left by employing the relation (9.56) to obtain

$$\left(i\frac{\partial}{\partial t}\right)^j\left[\frac{1}{2i}\phi_{mn}(\mathbf{r}, t; \mathbf{r}_1, t_1)\right]$$

$$\times \int \frac{d^3k}{(2\pi)^3} e^{i\mathbf{k}\cdot(\mathbf{r}-\mathbf{r}_1)} \int_{-\infty}^{+\infty} \frac{d\omega}{2\pi} \omega^j e^{-i\omega(t-t_1)} \chi''_{mn}(\mathbf{k}, \omega) \quad (9.71)$$

On the right-hand side of (9.70), we use Heisenberg's equation of motion to replace the time derivative by repeated equal-time commutation relations

$$\left(i\frac{\partial}{\partial t}\right)^j A_m(\mathbf{r}, t) = \left(i\frac{\partial}{\partial t}\right)^{j-1}\frac{1}{\hbar}[\hat{A}_m(\mathbf{r}, t), \hat{H}_0]$$

$$= \hbar^{-j}[\cdots[[\hat{A}_m(\mathbf{r}, t), \underbrace{H_0], H_0], \ldots, H_0]}_{j \text{ times}} \quad (9.72)$$

Equation (9.70) then reads for $t_1 = t$ after multiplication with $\exp[-i\mathbf{k}\cdot(\mathbf{r}-\mathbf{r}_1)]$ and integration over $d^3(\mathbf{r}-\mathbf{r}_1)$

$$\int \frac{d\omega}{\pi} \omega^j \chi''_{mn}(\mathbf{k}, \omega) = \hbar^{-j-1} \int d^3(\mathbf{r}-\mathbf{r}_1) e^{-i\mathbf{k}\cdot(\mathbf{r}-\mathbf{r}_1)}$$

$$\times Tr\{[\cdots[[\hat{A}_m(\mathbf{r}, t), \hat{H}_0], \hat{H}_0], \ldots, \hat{H}_0], \hat{A}_n(\mathbf{r}_1, t)]\hat{\rho}_0\} \quad (9.73)$$

For $j = 1$, this simplifies to

$$\frac{1}{\pi}\int_{-\infty}^{+\infty} d\omega\,\omega\chi''_{mn}(\mathbf{k}, \omega) = \hbar^{-2}\int d^3(\mathbf{r}-\mathbf{r}_1) e^{-i\mathbf{k}\cdot(\mathbf{r}-\mathbf{r}_1)}$$

$$\times Tr\{[[\hat{A}_m(\mathbf{r}, t), \hat{H}_0], \hat{A}_n(\mathbf{r}_1, t)]\hat{\rho}_0\} \quad (9.74)$$

Let us, as an example, work out this sum rule for the case where both operators $\hat{A}_m$ and $\hat{A}_n$ refer to the local particle density, i.e.

$$\hat{A}_m(\mathbf{r}, t) = \hat{\rho}(\mathbf{r}, t) = \psi^+(\mathbf{r}, t)\psi(\mathbf{r}, t)$$
$$\hat{A}_n(\mathbf{r}_1, t) = \hat{\rho}(\mathbf{r}_1, t) = \psi^+(\mathbf{r}_1, t)\psi(\mathbf{r}_1, t) \qquad (9.75)$$

where ψ^+ and ψ are field operators (see the beginning of Chapter 8). We know from eqn (8.66) that

$$i\hbar\partial_t\psi^+(\mathbf{r}, t)\psi(\mathbf{r}, t) = [\psi^+\psi, H_0]$$
$$= -i\frac{\hbar^2}{m}\boldsymbol{\nabla}\cdot\hat{\mathbf{j}}(\mathbf{r}, t) \qquad (9.76)$$

where, eqn (8.65),

$$\hat{\mathbf{j}}(\mathbf{r}, t) = \frac{1}{2i}[\psi^+(\mathbf{r}, t)\boldsymbol{\nabla}\psi(\mathbf{r}, t) - (\boldsymbol{\nabla}\psi^+(\mathbf{r}, t))\psi(\mathbf{r}, t)) \qquad (9.77)$$

Equation (9.74) then reads

$$\frac{1}{\pi}\int_{-\infty}^{\infty} d\omega\,\omega\chi''_{\rho\rho}(\mathbf{k}, \omega) = -\frac{i}{m}\int d^3(\mathbf{r}-\mathbf{r}_1)e^{-i\mathbf{k}\cdot(\mathbf{r}-\mathbf{r}_1)}$$
$$\times Tr\{[\boldsymbol{\nabla}\cdot\hat{\mathbf{j}}(\mathbf{r}, t), \hat{\rho}(\mathbf{r}_1, t)]\hat{\rho}_0\}$$
$$= \frac{1}{m}\mathbf{k}\cdot\int d^3(\mathbf{r}-\mathbf{r}_1)e^{-i\mathbf{k}\cdot(\mathbf{r}-\mathbf{r}_1)}$$
$$\times Tr\{[\hat{\mathbf{j}}(\mathbf{r}, t), \hat{\rho}(\mathbf{r}_1, t)]\hat{\rho}_0\} \qquad (9.78)$$

where the last line follows by partial integration. Next observe that

$$i[\hat{\mathbf{j}}(\mathbf{r}, t), \hat{\rho}(\mathbf{r}_1, t)] = \boldsymbol{\nabla}[\psi^+(\mathbf{r}_1, t)\psi(\mathbf{r}, t)\delta(\mathbf{r}-\mathbf{r}_1)]$$
$$- \delta(\mathbf{r}-\mathbf{r}_1)\boldsymbol{\nabla}\hat{\rho}(\mathbf{r}, t) \qquad (9.79)$$

which, used in eqn (9.78), yields simply

$$\int_{-\infty}^{+\infty}\frac{d\omega}{\pi}\,\omega\chi''_{\rho\rho}(\mathbf{k}, \omega) = \frac{1}{m}\left(\frac{N}{V}\right)k^2 \qquad (9.80)$$

where

$$\frac{N}{V} = Tr\{\psi^+(\mathbf{r}, t)\psi(\mathbf{r}, t)\hat{\rho}_0\} \qquad (9.81)$$

is the time- and space-independent equilibrium particle density in a system of N particles enclosed in a volume V. The sum rule, eqn (9.80), is an exact constraint on the density-density response function within the range of linear response theory, i.e. for systems close to equilibrium, and can, among others, also be used to normalize experimental data on $\chi''_{\rho\rho}(\mathbf{k}, \omega)$.

To make the aforementioned connection with experiments more apparent, let us define correlation functions

$$S_{mn}(\mathbf{r}, t; \mathbf{r}_1, t_1) = \langle \hat{A}_m(\mathbf{r}, t)\hat{A}_n(\mathbf{r}_1, t_1) \rangle - \langle \hat{A}_m(\mathbf{r}, t) \rangle\langle \hat{A}_n(\mathbf{r}_1, t_1) \rangle \tag{9.82}$$

where $\langle \cdots \rangle = Tr(\ldots, \hat{\rho}_0)$ indicates an equilibrium average. S_{mn} is a measure for the correlated probability of finding property m at space-time point $(\mathbf{r}, t)$ provided that n is given at point $(\mathbf{r}_1, t_1)$, and vice versa. Subtracting the product of the space- and time-dependent equilibrium values

$$\langle \hat{A}_m(\mathbf{r}, t) \rangle\langle \hat{A}_n(\mathbf{r}_1, t_1) \rangle = \bar{A}_m^{(0)} \bar{A}_n^{(0)} \tag{9.83}$$

guarantees that $S_{mn} \to 0$ as $(t - t_1) \to \infty$ and/or $|\mathbf{r} - \mathbf{r}_1| \to \infty$, an asymptotic behavior that will be welcome when taking Fourier transforms. Obviously, we also have

$$S_{mn}(\mathbf{r}, t; \mathbf{r}_1, t_1) = \langle [\hat{A}_m(\mathbf{r}, t) - \bar{A}_m^{(0)}][\hat{A}_n(\mathbf{r}_1, t_1) - \bar{A}_n^{(0)}] \rangle \tag{9.84}$$

and the connection with ϕ_{mn} is made by

$$-i\hbar\phi_{mn}(\mathbf{r}, t; \mathbf{r}_1, t_1) = S_{mn}(\mathbf{r}, t; \mathbf{r}_1, t_1) - S_{mn}(\mathbf{r}_1, t_1; \mathbf{r}, t) \tag{9.85}$$

Let us next perform the equilibrium averages explicitly in a canonical ensemble in which we set

$$\hat{\rho}_0 = \frac{e^{-\beta H_0}}{Tr e^{-\beta H_0}} \tag{9.86}$$

We first evaluate

$$\frac{Tr[\hat{A}_n(\mathbf{r}_1, t_1)\hat{A}_m(\mathbf{r}, t)e^{-\beta H_0}]}{Tr e^{-\beta H_0}}$$

$$= \frac{Tr[\hat{A}_n(\mathbf{r}_1, t_1)e^{-\beta H_0} e^{iH_0(t - i\beta\hbar)/\hbar}\hat{A}_m(\mathbf{r})e^{-iH_0(t - i\beta\hbar)/\hbar}]}{Tr e^{-\beta H_0}}$$

$$= \frac{Tr[\hat{A}_m(\mathbf{r}, t - i\beta\hbar)\hat{A}_n(\mathbf{r}_1, t_1)e^{-\beta H_0}]}{Tr e^{-\beta H_0}} \tag{9.87}$$

where property (9.40) and the cyclic invariance of the trace has been used repeatedly.

This gives in eqn (9.85)

$$
\begin{aligned}
-i\hbar\phi_{mn}(\mathbf{r}, t; \mathbf{r}_1, t_1) &= S_{mn}(\mathbf{r}, t; \mathbf{r}_1, t_1) \\
&\quad - S_{mn}(\mathbf{r}, t - i\beta\hbar; \mathbf{r}_1, t_1) \\
&= (1 - e^{-i\beta\hbar\partial/\partial t})S_{mn}(\mathbf{r}, t; \mathbf{r}_1, t_1)
\end{aligned} \quad (9.88)
$$

Due to time translational invariance the left- and right-hand sides of this equation can only depend on $(t - t_1)$. If we therefore take a two-sided Fourier transform with respect to $(t - t_1)$, we obtain with eqn (9.60)

$$2\hbar\chi''_{mn}(\mathbf{r}, \mathbf{r}_1; \omega) = (1 - e^{-\hbar\omega\beta})S_{mn}(\mathbf{r}, \mathbf{r}_1; \omega) \quad (9.89)$$

where

$$S_{mn}(\mathbf{r}, \mathbf{r}_1; \omega) = \int_{-\infty}^{+\infty} d(t - t_1)e^{i\omega(t - t_1)}S_{mn}(\mathbf{r}, t; \mathbf{r}_1, t_1) \quad (9.90)$$

If the system is, in addition, invariant under spatial translations, we can take Fourier transforms with respect to $(\mathbf{r} - \mathbf{r}_1)$ and obtain with eqn (9.56)

$$2\hbar\chi''_{mn}(\mathbf{k}, \omega) = (1 - e^{-\hbar\omega\beta})S_{mn}(\mathbf{k}, \omega) \quad (9.91)$$

where

$$S_{mn}(\mathbf{k}, \omega) = \int d^3(\mathbf{r} - \mathbf{r}_1)e^{-i\mathbf{k}\cdot(\mathbf{r} - \mathbf{r}_1)}S_{mn}(\mathbf{r}, \mathbf{r}_1; \omega) \quad (9.92)$$

Equations (9.89) and (9.91) are two versions of the famous fluctuation-dissipation theorem, in this general form attributable to Kubo (1957). To understand its meaning and significance, let us first observe that the correlation function S_{mn} is defined with reference to the equilibrium state of the system and contains statistical information about spontaneous equilibrium fluctuations as indicated in eqn (9.84).

The fact that χ''_{mn} has to do with energy dissipation in a nonequilibrium system was already hinted at in eqn (9.59), which states that for the diagonal elements of the response function χ''_{mn} is nothing but the imaginary part of the (complex) admittance χ_{mn} of the system. To prove our point in general, we calculate explicitly the energy dissipation as the rate of change of the total

LINEAR RESPONSE THEORY   255

energy from eqn (9.24)

$$\frac{dE}{dt} = \frac{d}{dt} Tr(\hat{H}\hat{\rho}_t) = Tr\left(\frac{\partial \hat{H}}{\partial t}\hat{\rho}_t\right) + Tr[\hat{H}\dot{\hat{\rho}}(t)] \quad (9.93)$$

The last term vanishes because, with eqn (9.28), we obtain using the cyclic invariance of the trace

$$Tr[\hat{H}\dot{\hat{\rho}}(t)] = \frac{1}{i\hbar} Tr\{\hat{H}[\hat{H}, \hat{\rho}(t)]\}$$

$$= \frac{1}{i\hbar} Tr\{[\hat{H}, \hat{H}]\hat{\rho}(t)\} = 0 \quad (9.94)$$

Assuming the explicit time dependence of $\hat{H}_1(t)$ according to eqn (9.25) and using eqns (9.36) and (9.42), we obtain

$$\frac{dE}{dt} = -\sum_m \int d^3r \frac{\partial f_m(\mathbf{r}, t)}{\partial t} Tr[\hat{A}_m(\mathbf{r})\hat{\rho}(t)]$$

$$= -\sum_m \int d^3r \frac{\partial f_m(\mathbf{r}, t)}{\partial t} \left[\bar{A}_m^{(0)} + \sum_n \int_{-\infty}^t dt_1 \, d^3r_1\right.$$

$$\left. \times \phi_{mn}(\mathbf{r}, \mathbf{r}_1, t - t_1) f_n(\mathbf{r}_1, t_1)\right] \quad (9.95)$$

Next we integrate this over all times to get the total energy dissipated by the system

$$E_{\text{diss}} = \int_{-\infty}^{+\infty} dt \frac{dE}{dt} = -\sum_m \bar{A}_m^{(0)}$$

$$\times \int d^3r [f_m(\mathbf{r}, t = +\infty) - f_m(\mathbf{r}, t = -\infty)]$$

$$-2i \sum_{n,m} \int_{-\infty}^{\infty} dt \int_{-\infty}^t dt_1$$

$$\times \int d^3r \, d^3r_1 \frac{\partial f_m(\mathbf{r}, t)}{\partial t} f_n(\mathbf{r}_1, t_1)$$

$$\times \int_{+\infty}^{-\infty} \frac{d\omega}{2\pi} e^{-i\omega(t-t_1)}$$

$$\times \int \frac{d^3k}{(2\pi)^3} e^{i\mathbf{k}\cdot(\mathbf{r}-\mathbf{r}_1)} \chi''_{mm}(\mathbf{k}, \omega) \quad (9.96)$$

The first term can be dropped if we assume that the external forces are switched on and off adiabatically. In the remaining second term we insert the Fourier decomposition of the external forces f_m and f_n from eqn (9.46) and perform all time and space integrations. The result is

$$E_{\text{diss}} = \int_{-\infty}^{+\infty} \frac{d\omega}{2\pi} \int \frac{d^3k}{(2\pi)^3} \sum_{n,m} f_m^*(\mathbf{k}, \omega)\omega\chi''_{mn}(\mathbf{k}, \omega)f_n(\mathbf{k}, \omega) \tag{9.97}$$

where we also used the fact that $f_m(-\mathbf{k}, -\omega) = f_m^*(\mathbf{k}, \omega)$ if $f_m(\mathbf{r}, t)$ is real. Equation (9.97) shows clearly that χ''_{mn} controls the energy dissipation in the system and, as a by-product of our effort, suggests that

$$\sum_{n,m} f_m^*(\mathbf{k}, \omega)\omega\chi''_{mn}(\mathbf{k}, \omega)f_n(\mathbf{k}, \omega) \geq 0 \tag{9.98}$$

must be a positive semidefinite quadratic form to ensure that $E_{\text{diss}} > 0$ is satisfied for any choice of forces f_m and f_n; i.e. that irreversible effects during the systems time evolution do, indeed, dissipate energy. This implies for the matrix $\omega\chi''_{mn}(\mathbf{k}, \omega)$

$$\omega\chi''_{mm}(\mathbf{k}, \omega) > 0 \tag{9.99}$$

and

$$\chi''_{mm}(\mathbf{k}, \omega)\chi''_{nn}(\mathbf{k}, \omega) > \tfrac{1}{4}[\chi''_{mn}(\mathbf{k}, \omega) + \chi''_{nm}(\mathbf{k}, \omega)]^2 \tag{9.100}$$

for all ω and $\mathbf{k}$.

We can now grasp the central importance of the fluctuation-dissipation theorem, eqns (9.89) or (9.91). The right-hand side involves a correlation function, a quantity that results from and is a measure of spontaneous fluctuations in equilibrium, i.e. of ever-present thermal or statistical noise. The response function on the left-hand side incorporates the mechanical, i.e. dynamical, response of a system that has been removed from equilibrium by the imposition of external forces or constraints. The fluctuation-dissipation theorem then says that nonequilibrium transport processes, linear in the external forces, in systems close to equilibrium are related to and can, indeed, be calculated from equilibrium fluctuations.

For the study of relaxation phenomena, i.e. for external forces with a time dependence like eqn (9.43), we introduced in eqns

LINEAR RESPONSE THEORY

(9.44) and (9.45) a relaxation function $\Phi_{mn}(\mathbf{r}, \mathbf{r}_1; t)$. If we introduce its spatial Fourier transform in a translational invariant system by

$$\Phi_{mn}(\mathbf{k}, t) = \int d^3(\mathbf{r} - \mathbf{r}_1) \exp[-i\mathbf{k} \cdot (\mathbf{r} - \mathbf{r}_1)] \phi_{mn}(\mathbf{r}, \mathbf{r}_1, t) \quad (9.101)$$

we find its relation with the response functions, eqns (9.50), to be given by

$$\chi_{mn}(\mathbf{k}, \omega) = \Phi_{mn}(\mathbf{k}, t=0) + i\omega \int_0^\infty dt\, e^{i\omega t} \Phi_{mn}(\mathbf{k}, t) \quad (9.102)$$

This can be seen by first noting that for forces $f_n(\mathbf{r}, t)$ adiabatically switched on for $-\infty < t < 0$ and suddenly switched off at time $t = 0$ according to eqns (9.43), we find

$$f_n(\mathbf{k}, \omega) = \int_{-\infty}^\infty dt\, e^{i\omega t} \int d^3 k\, e^{-i\mathbf{k}\cdot\mathbf{r}} f_n(\mathbf{r}, t)$$

$$= \frac{1}{i\omega + \varepsilon} f_n(\mathbf{k}) \quad (9.103)$$

where

$$f_n(\mathbf{k}) = \int d^3 k\, e^{-i\mathbf{k}\cdot\mathbf{r}} f_n(\mathbf{r}) \quad (9.104)$$

Next, taking spatial and temporal Fourier transforms of eqn (9.42), we obtain

$$\delta \bar{A}_m(\mathbf{k}, \omega) = \int_{-\infty}^{+\infty} dt\, e^{i\omega t} \int d^3 k\, e^{-i\mathbf{k}\cdot\mathbf{r}} \delta \bar{A}_m(\mathbf{r}, t)$$

$$= \sum_n f_n(\mathbf{k}) \left[\int_{-\infty}^0 dt\, e^{i\omega t} \int_{-\infty}^t dt_1\, \phi_{mn}(\mathbf{k}; t-t_1) e^{\varepsilon t_1} \right.$$

$$\left. + \int_0^\infty dt\, e^{i\omega t} \int_{-\infty}^t dt_1\, \phi_{mn}(\mathbf{k}; t-t_1) e^{\varepsilon t_1} \right]$$

$$= \sum_n f_n(\mathbf{k}) \left[\frac{1}{i\omega + \varepsilon} \Phi(\mathbf{k}, t=0) + \int_0^\infty dt\, e^{i\omega t} \Phi_{mn}(\mathbf{k}; t) \right]$$

$$(9.105)$$

Combined with eqns (9.45) and (9.48) this proves eqn (9.102).

For the relaxation phenomena just considered, it is sometimes advantageous to formulate linear response as an initial value problem. Let us first observe that at time $t = 0$, we obtain from eqn (9.42), also using eqn (9.104)

$$\delta \bar{A}_m(\mathbf{k}, t=0) = \sum_n \chi_{mn}(\mathbf{k}, \omega=0) f_n(\mathbf{k}) \quad (9.106)$$

Let us next define a one-sided Fourier transform over positive times only as

$$\delta \bar{A}_m^{(+)}(\mathbf{k}, z) = \int_0^\infty dt e^{izt} \delta \bar{A}_m(\mathbf{k}, t) = \delta \bar{A}_m(\mathbf{k}, z) - \int_{-\infty}^0 dt e^{izt} \delta \bar{A}_m(\mathbf{k}, t)$$

$$= \frac{1}{iz} \sum_m \chi_{mn}(\mathbf{k}, z) f_n(\mathbf{k}) - \int_{-\infty}^0 dt e^{izt} \sum_n f_n(\mathbf{k}) \int_{-\infty}^0 dt_1 \phi_{mn}(\mathbf{k}; t-t_1) e^{\varepsilon t} \quad (9.107)$$

$\delta \bar{A}_m(\mathbf{k}, z)$ in the second line is the two-sided Fourier transform, which in the last line is given with eqn (9.48), using again (9.104). In the last term we simply inserted eqn (9.42) after spatial Fourier transforming. Performing the trivial time integrations and inserting the formal inverse of eqn (9.106), namely

$$f_n(\mathbf{k}) = \sum_l \chi_{nl}^{-1}(\mathbf{k}, \omega=0) \delta \bar{A}_l(\mathbf{k}, t=0) \quad (9.108)$$

we obtain

$$\delta \bar{A}_m^{(+)}(\mathbf{k}, z) = \frac{1}{iz} \sum_n \left[\sum_l \chi_{ml}(\mathbf{k}, z) \chi_{ln}^{-1}(\mathbf{k}, \omega=0) - \delta_{mn} \right] \delta \bar{A}_n(\mathbf{k}, t=0) \quad (9.109)$$

This relation allows us to calculate the relaxation of the system with the help of the response functions and the initial nonequilibrium data $\delta \bar{A}_n(\mathbf{k}, t=0)$ as prepared by the action of the adiabatically switched forces $f_n(\mathbf{r})$ over times $-\infty < t < 0$.

9.2.3. Current Response and Onsager Reciprocity Relations
Let us assume that the equilibrium properties of our system can be described with the density operator of a canonical ensemble, i.e.

$$\hat{\rho}_0 = \frac{e^{-\beta \hat{H}_0}}{Tr(e^{-\beta \hat{H}_0})} \quad (9.110)$$

Observe that for any operator $\hat{A}$

$$e^{\beta H_0}[\hat{A}, e^{-\beta H_0}] = e^{\beta H_0} \hat{A} e^{-\beta H_0} - \hat{A} = \int_0^\beta e^{\lambda H_0}[H_0, \hat{A}] e^{-\lambda H_0} d\lambda \quad (9.111)$$

as one can check by differentiating the last equality. Using Heisenberg's equation of motion

$$i\hbar \dot{\hat{A}} = [\hat{A}, \hat{H}_0] \tag{9.112}$$

in (9.111), we obtain

$$[\hat{A}, e^{-\beta H_0}] = \frac{\hbar}{i} e^{-\beta H_0} \int_0^\beta e^{\lambda H_0} \dot{\hat{A}} e^{-\lambda H_0} \, d\lambda$$

$$= -i\hbar e^{-\beta H_0} \int_0^\beta \dot{\hat{A}}(-i\hbar\lambda) \, d\lambda \tag{9.113}$$

With this relation we can write the response function (9.41) as

$$\phi_{mn}(\mathbf{r}, \mathbf{r}_1; t - t_1) = \int_0^\beta Tr[A_m(\mathbf{r}, t - t_1)\hat{\rho}_0 \dot{\hat{A}}_n(\mathbf{r}_1, -i\hbar\lambda)] \, d\lambda \tag{9.114}$$

For those operators $\hat{A}_m$ that correspond to extensive thermodynamic variables (electric and magnetic dipole moments, for example), we must obviously ask what currents

$$Y_m(\mathbf{r}, t) = \langle \dot{\hat{A}}_m(\mathbf{r}, t) \rangle = \langle \hat{J}_m(\mathbf{r}, t) \rangle \tag{9.115}$$

are generated by the relevant forces in H_1. From (9.42) and (9.115), we find that

$$Y_m(\mathbf{r}, t) = \sum_n \int_{-\infty}^t dt_1 \int d^3 r_1 \tilde{\Phi}_{mn}(\mathbf{r}, \mathbf{r}_1, t - t_1) f_n(\mathbf{r}_1, t_1)$$

with
$$\tag{9.116}$$

$$\tilde{\phi}_{mn}(\mathbf{r}, \mathbf{r}_1, t - t_1) = \int_0^\beta Tr[\dot{\hat{A}}_m(\mathbf{r}, t - t_1)\hat{\rho}_0 \dot{\hat{A}}_n(\mathbf{r}_1, -i\hbar\lambda)] \, d\lambda$$

$$= \int_0^\beta Tr[\hat{J}_m(\mathbf{r}, t - t_1)\hat{\rho}_0 \hat{J}_n(\mathbf{r}_1, -i\hbar\lambda)] \, d\lambda \tag{9.117}$$

a formula first obtained by Kubo (1957) expressing the response functions in terms of current-current correlation functions.

Taking spatial and temporal Fourier transforms on (9.117) according to eqns (9.46), (9.47), and (9.50), we obtain

$$Y_m(\mathbf{k}, \omega) = \sum_n \tilde{\chi}_{mn}(\mathbf{k}, \omega) f_n(\mathbf{k}, \omega) \tag{9.118}$$

where

$$\tilde{\chi}_{mn}(\mathbf{k}, \omega) = \int_0^\infty d(t-t_1) \int d^3(\mathbf{r}-\mathbf{r}_1) \\ \times e^{i\omega(t-t_1)} e^{-i\mathbf{k}\cdot(\mathbf{r}-\mathbf{r}_1)} \tilde{\phi}_{mn}(\mathbf{r}, \mathbf{r}_1, t-t_1) \quad (9.119)$$

The limit

$$\lim_{\mathbf{k}\to 0} \lim_{\omega \to 0} \tilde{\chi}_{mn}(\mathbf{k}, \omega) = L_{mn} \quad (9.120)$$

is nothing but the linear transport coefficient in a uniform and static external field.

Let us next calculate the complex conjugate of eqn (9.117), namely

$$\tilde{\phi}^*_{mn}(\mathbf{r}, \mathbf{r}_1, \tau) = \int_0^\beta d\lambda \, Tr[\dot{\hat{A}}_n^+(\mathbf{r}_1, i\hbar\lambda)\hat{\rho}_0 \dot{\hat{A}}_m^+(\mathbf{r}, \tau)] \quad (9.121)$$

Checking in Heisenberg's equation of motion (9.112) that $\dot{\hat{A}}_n^+ = \dot{\hat{A}}_n$ for a hermitian operator and shifting the evolution operators from $\dot{\hat{A}}_m$ onto $\dot{\hat{A}}_n$, we obtain

$$\tilde{\phi}^*_{mn}(\mathbf{r}, \mathbf{r}_1, \tau)$$
$$= \int_0^\beta d\lambda \frac{Tr[e^{-i(\tau-i\hbar\lambda)H_0}\dot{\hat{A}}_n(\mathbf{r}, 0)e^{i(\tau-i\hbar\lambda)H_0}e^{-\beta H_0}\dot{\hat{A}}_m(\mathbf{r}, 0)e^{-\beta H_0}e^{\beta H_0}]}{Tr(e^{-\beta H_0})}$$
$$= \int_0^\beta d\lambda \, Tr[e^{-i[\tau-i\hbar(\lambda-\beta)]H_0}\dot{\hat{A}}_n(\mathbf{r}_1, 0)e^{i[\tau-i\hbar(\lambda-\beta)]H_0}\dot{\hat{A}}_m(\mathbf{r}, 0)\hat{\rho}_0]$$
$$\quad (9.122)$$

Next change integration variables $\lambda' = \beta - \lambda$ and shift the thermal evolution operators onto $\dot{\hat{A}}_m(\mathbf{r}, 0)$ to get

$$\tilde{\phi}^*_{mn}(\mathbf{r}, \mathbf{r}_1, \tau) = \int_0^\beta d\lambda' \, Tr[\dot{\hat{A}}_m(\mathbf{r}, i\hbar\lambda')\hat{\rho}_0 \dot{\hat{A}}_n(\mathbf{r}_1, \tau)]$$
$$= \tilde{\Phi}_{nm}(\mathbf{r}_1, \mathbf{r}, \tau) \quad (9.123)$$

Similarly, we can show that

$$\phi_{mn}(\mathbf{r}, \mathbf{r}_1, -\tau) = \phi_{mn}(\mathbf{r}, \mathbf{r}_1, \tau) \quad (9.124)$$

and in particular that

$$\tilde{\chi}^*_{mn}(-\mathbf{k}, -\omega) = \tilde{\chi}_{nm}(\mathbf{k}, \omega) \quad (9.125)$$

implying that

$$\text{Re } \tilde{\chi}_{mn}(-\mathbf{k}, -\omega) = \text{Re } \tilde{\chi}_{nm}(\mathbf{k}, \omega)$$
$$\text{Im } \tilde{\chi}_{mn}(-\mathbf{k}, -\omega) = -\text{Im } \tilde{\chi}_{nm}(\mathbf{k}, \omega) \qquad (9.126)$$

This shows that the linear transport coefficients L_{mn}, eqn (9.120), in a static external field are real and symmetric, i.e.

$$L_{nm} = L_{mn} \qquad (9.127)$$

This constitutes another proof of Onsager's reciprocity relations for cross effects in the response of a system to mechanical perturbations that can be incorporated into a hamiltonian formalism. All important thermal effects are not covered by this proof. We should also observe that the transport coefficients calculated by linear response theory are adiabatic ones rather than isothermal as measured in most experiments. The difference between the two can, however, be assumed small as long as the external perturbation only affects a part of the total system whose partial heat capacity is small compared to the whole (Kubo, 1957).

9.2.4. Density-Density Correlation Functions from Scattering Experiments

In this section we briefly consider how correlation functions can be determined from scattering experiments. We envisage a beam of monoenergetic test particles, for our purposes most likely electrons, photons, or neutrons, hitting a fluid target of volume V and being scattered as a result of the interaction of the test particles with the constituents of the fluid. Such a scattering process will probe the microscopic structure of the fluid if the momentum transfer from the beam to the target

$$\mathbf{k} = \mathbf{k}_i - \mathbf{k}_f \qquad (9.128)$$

i.e. the difference between the initial momentum $\mathbf{k}_i$ of an incident test particle and its final momentum $\mathbf{k}_f$ when leaving the fluid volume after the completed scattering event, is of the order of the inverse separation distance between fluid particles; i.e.

$$|\mathbf{k}| \sim \frac{2\pi}{d} \qquad (9.129)$$

where d is typically a few angstroms in a liquid. The scattering

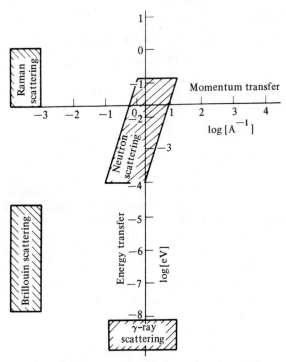

Fig. 9.2. Regions in the energy-momentum transfer plane accessible to various scattering processes to measure the molecular correlation functions in fluids. After Egelstaff (1967).

process itself must be inelastic in nature, accompanied by an energy transfer

$$\hbar\omega = \varepsilon_f - \varepsilon_i \qquad (9.130)$$

where ε_i and ε_f are initial and final energy of the scattering probes, of about 1 eV, corresponding to the depth of the intermolecular potential in the liquid.

Suitable probes for such momentum and energy transfers are thermal neutrons of a de Broglie wavelength of some 1–10 Å. The suitability of other test particles for probing molecular correlation functions is summarized in Fig. 9.2, which also shows their range of applicability in the momentum transfer–energy transfer plane.

From now on we will concentrate on neutron scattering.‡ Neutrons have the great advantage, from a theoretical point of view, that their interaction with an atom can for our purposes be approximated by a pointlike δ potential in its nucleus;§ i.e. we can write

$$V(\mathbf{r}) = \frac{2\pi\hbar^2}{m} c_0 \delta(\mathbf{r}) \quad (9.131)$$

where $\mathbf{r}$ is the relative separation of the neutron from the nucleus and $m = m_n A/(A+1)$ is the reduced mass of the neutron-nucleus system with m_n the neutron mass and A the mass number of the nucleus. c_0 is a characteristic length parameter of the system, more specifically the s-wave scattering length. Let us first calculate the cross section for neutron scattering off a single atom. We have to find a solution of the Schrödinger equation

$$(\nabla^2 + k^2)\psi^{(+)}(\mathbf{k}_i, \mathbf{r}) = \frac{2m}{\hbar^2} V(\mathbf{r})\psi^{(+)}(\mathbf{k}_i, \mathbf{r}) \quad (9.132)$$

which asymptotically for large $|\mathbf{r}| \to \infty$ contains in addition to the incoming plane wave, outgoing spherical waves (Newton, 1966)

$$\psi^{(+)}(\mathbf{k}_i, \mathbf{r}) \sim \psi_0(\mathbf{k}_i, \mathbf{r}) + A(\mathbf{k}_i, \mathbf{k}_f) \frac{e^{ikr}}{r} \quad (9.133)$$

where

$$\psi_0(\mathbf{k}_i, \mathbf{r}) = \frac{\sqrt{mk}}{(2\pi)^{\frac{3}{2}}\hbar} e^{i\mathbf{k}_i \cdot \mathbf{r}} \quad (9.134)$$

and the scattering amplitude is given by

$$A(\mathbf{k}_i, \mathbf{k}_f) = -\frac{(2\pi)^2}{k} \int \psi_0(\mathbf{k}_i, \mathbf{r}) V(\mathbf{r}) \psi^{(+)}(\mathbf{k}_f, \mathbf{r}) \, d^3r \quad (9.135)$$

Here $|\mathbf{k}_i| = |\mathbf{k}_f| = k$ and $\mathbf{k}_f = k\mathbf{r}/r$ is the momentum of the scattered neutron. For a zero-range potential, eqn (9.131), only s-wave

‡ Further details on other scattering techniques are discussed in reviews by Egelstaff (1967), Martin (1968), Chen (1971), McIntyre and Sengers (1968), and Enderby (1968).

§ This can be looked upon as the lowest-order pseudopotential for a hard-core interaction. See Huang and Yang (1957).

scattering is effective and we find

$$A(\mathbf{k}_i, \mathbf{k}_f) = -\frac{c_0}{1 - ic_0 k} \qquad (9.136)$$

which, in the very low energy region, can be approximated by

$$A(\mathbf{k}_i, \mathbf{k}_f) \approx -c_0 \qquad (9.137)$$

a result that is identical to the Born approximation in eqn (9.132). The differential cross section for scattering into a solid angle $d\Omega_f$ is then given by

$$\frac{d\sigma}{d\Omega_f} = |A(\mathbf{k}_i, \mathbf{k}_f)|^2 = c_0^2 \qquad (9.138)$$

which is simply the cross section for scattering off a hard sphere of radius c_0. The scattering length c_0 has to be determined experimentally from low energy neutron-nucleus scattering.

In case the nuclei have spin I, neutron scattering in s waves proceeds in the two eigenstates $(I \pm \frac{1}{2})$, for which we generally find different scattering lengths c_{0+} and c_{0-}. The cross section for scattering off a single nucleus then reads with the inclusion of the proper statistical weights in the two spin channels

$$\frac{d\sigma}{d\Omega_f} = \frac{I+1}{2I+1} c_{0+}^2 + \frac{I}{2I+1} c_{0-}^2 \qquad (9.139)$$

which is sometimes decomposed as

$$\frac{d\sigma}{d\Omega_f} = \left(\frac{d\sigma}{d\Omega_f}\right)_{\text{coh}} + \left(\frac{d\sigma}{d\Omega_f}\right)_{\text{incoh}} \qquad (9.140)$$

where the coherent cross section is given by

$$\frac{d\sigma}{d\Omega_f} = c_{\text{coh}}^2 = \left(\frac{I+1}{2I+1} c_{0+} + \frac{I}{2I+1} c_{0-}\right)^2 \qquad (9.141)$$

It includes interference effects between the two spin channels. The incoherent part $(d\sigma/d\Omega_f)_{\text{incoh}}$ exhausts the remainder of the cross section.

So far we have only dealt with elastic neutron scattering off a single atom. However, when a beam of neutrons impinges on a fluid, we must also take into account that the atoms in the fluid are moving around and, in particular, will recoil in a collision with a neutron, the transferred energy being rapidly distributed

LINEAR RESPONSE THEORY

among the many degrees of freedom of the fluid by the interatomic interaction. We are therefore generally concerned with inelastic neutron scattering in a potential

$$V_{\text{fluid}} = \frac{2\pi\hbar^2}{m_n} \sum_{i=1}^{N} b_i \delta[\mathbf{r} - \mathbf{r}_i(t_1)] \quad (9.142)$$

where $\mathbf{r}_i(t)$ is the position of the ith atom in the fluid at some time t_1.

The probability for a scattering event $\mathbf{k}_i \to \mathbf{k}_f$ to occur in unit time is then given by Fermi's golden rule as

$$P_{i \to f} = \frac{2\pi}{\hbar} |A(\mathbf{k}_i, \mathbf{k}_f)|^2 \, \delta(\varepsilon_f - \varepsilon_i - \hbar\omega) \quad (9.143)$$

and the total probability is

$$\sum_f P_{i \to f} = \frac{2\pi}{\hbar} \sum_{\varepsilon_f} \frac{1}{V} \sum_{\mathbf{k}_f} |A(\mathbf{k}_i, \mathbf{k}_f)|^2 \, \delta(\varepsilon_f - \varepsilon_i - \hbar\omega)$$

$$= \frac{2\pi}{\hbar} \sum_{\varepsilon_f} \int \frac{d^3 k_f}{(2\pi)^3} \frac{1}{2\pi\hbar} \int_{-\infty}^{+\infty} d(t - t_1) e^{i\omega(t-t_1)} e^{-i(\varepsilon_f - \varepsilon_i)(t-t_1)/\hbar}$$

$$\times \langle \mathbf{k}_f \varepsilon_f | V_{\text{fluid}} | \mathbf{k}_i \varepsilon_i \rangle \langle \mathbf{k}_i \varepsilon_i | V_{\text{fluid}} | \mathbf{k}_f \varepsilon_f \rangle \quad (9.144)$$

where in the last expression, we have taken the large volume limit and expressed the energy δ function by its Fourier transform. Next observe that with the help of the density operator

$$\hat{\rho}(\mathbf{r}, t) = \sum_m \delta[\mathbf{r} - \mathbf{r}_m(t)] \quad (9.145)$$

we can write, assuming that $b = b_m$ is independent of position,

$$e^{-i(\varepsilon_f - \varepsilon_i)t/\hbar} \langle \mathbf{k}_i \varepsilon_i | V_{\text{fluid}} | \mathbf{k}_f \varepsilon_f \rangle = \int d^3 r \, e^{-i(\mathbf{k}_i - \mathbf{k}_f) \cdot \mathbf{r}}$$

$$\times \langle \varepsilon_i | e^{i\varepsilon_i t/\hbar} b \hat{\rho}(\mathbf{r}, t) e^{-i\varepsilon_f t/\hbar} | \varepsilon_f \rangle$$

$$= \int d^3 r \, e^{i(\mathbf{k}_i - \mathbf{k}_f) \cdot \mathbf{r}} \langle \varepsilon_i | b \hat{\rho}(\mathbf{r}, t) | \varepsilon_f \rangle \quad (9.146)$$

Also recalling that the total cross section is given by the total transition probability, eqn (9.144), divided by the flux $\hbar |\mathbf{k}_i|/m_n$ of neutrons in the incoming beam, we get

$$\sigma = \left\langle \frac{m_n}{\hbar |\mathbf{k}_i|} \sum_f P_{i \to f} \right\rangle$$

$$= \frac{\hbar}{2m_n} \frac{1}{|\mathbf{k}_i|} \int \frac{d^3k_f}{(2\pi)^3} \int d^3r \, d^3r_1 e^{-i\mathbf{k}\cdot(\mathbf{r}-\mathbf{r}_1)}$$

$$\times \int d(t-t_1) e^{i\omega(t-t_1)} \langle b\hat{\rho}(\mathbf{r}, t) b^+ \hat{\rho}(\mathbf{r}_1, t_1) \rangle \qquad (9.147)$$

where the pointed brackets $\langle \cdots \rangle$ refer to an average over the initial equilibrium ensemble of the fluid. The differential cross section for scattering into a solid angle $d\Omega_f$ and energy interval $d\varepsilon_f$ is obtained by noting that

$$d^3k_f = k_f^2 \, dk_f \, d\Omega_f = \frac{m_n |\mathbf{k}_f|}{\hbar^2} d\varepsilon_f \, d\Omega_f \qquad (9.148)$$

and thus

$$\frac{d^2\sigma}{d\Omega_f \, d\varepsilon_f} = \frac{1}{2\pi\hbar} \frac{|\mathbf{k}_f|}{|\mathbf{k}_i|} \int d^3r \, d^3r_1 e^{i-\mathbf{k}_1(\mathbf{r}-\mathbf{r}_1)} \int d(t-t_1)$$

$$\times e^{i\omega(t-t_1)} \langle b\hat{\rho}(\mathbf{r}, t) b^+ \hat{\rho}(\mathbf{r}_1, t_1) \rangle \qquad (9.149)$$

At this stage we have to take proper account of the fact that the nuclei may carry spins I. We therefore write

$$b_m = b_0 + b_1 \hat{\sigma}_N \cdot \hat{I}_m \qquad (9.150)$$

where $\hat{\sigma}_N$ is the spin one-half operator for a neutron and I_m is the nuclear spin operator of the mth atom.

Assuming that the nuclear spins are uncorrelated, we then obtain

$$\langle b\hat{\rho}(\mathbf{r}, t) b^+ \hat{\rho}(\mathbf{r}_1, t_1) \rangle = |b_1|^2 \langle \hat{\rho}(\mathbf{r}, t) \hat{\rho}(\mathbf{r}_1, t_1) \rangle$$

$$+ |b_2|^2 I(I+1) \left\langle \sum_\alpha \delta[\mathbf{r} - \mathbf{r}_\alpha(t)] \delta[\mathbf{r}_1 - \mathbf{r}_\alpha(t_1)] \right\rangle \qquad (9.151)$$

Let us next, in accordance with eqn (9.82), introduce a density-density correlation function

$$S_{\rho\rho}(\mathbf{r}, t; \mathbf{r}_1, t_1) = \langle \hat{\rho}(\mathbf{r}, t) \hat{\rho}(\mathbf{r}_1, t_1) \rangle$$

$$- \langle \hat{\rho}(\mathbf{r}, t) \rangle \langle \hat{\rho}(\mathbf{r}_1, t_1) \rangle \qquad (9.152)$$

LINEAR RESPONSE THEORY

and also an auto-correlation function

$$S_{\text{self}}(\mathbf{r}, t; \mathbf{r}_1, t) = \left\langle \sum_m \delta[\mathbf{r} - \mathbf{r}_m(t)] \delta[\mathbf{r}_1 - \mathbf{r}_m(t_1)] \right\rangle \quad (9.153)$$

We can then write the differential cross section, eqn (9.149), as the sum of the coherent cross section

$$\frac{d^2\sigma_{\text{coh}}}{d\Omega_f d\varepsilon_f} = \frac{|\mathbf{k}_f|}{|\mathbf{k}_i|} \frac{1}{2\pi\hbar} |\langle b \rangle|^2 \int_{-\infty}^{+\infty} dt$$
$$\times \int d^3r \, d^3r_1 \, e^{i\omega(t-t_1)} e^{-i\mathbf{k}\cdot(\mathbf{r}-\mathbf{r}_1)} \left[S_{\rho\rho}(\mathbf{r}, t; \mathbf{r}_1, t_1) + \left(\frac{N}{V}\right)^2 \right] \quad (9.154)$$

and the incoherent cross section

$$\frac{d^2\sigma_{\text{incoh}}}{d\Omega_f d\varepsilon_f} = \frac{|\mathbf{k}_f|}{|\mathbf{k}_i|} \frac{1}{2\pi\hbar} (\langle b^+ b \rangle - \langle b \rangle^2) \int_{-\infty}^{+\infty} dt \int d^3r \, d^3r_1$$
$$\times e^{i\omega(t-t_1)} e^{-i\mathbf{k}\cdot(\mathbf{r}-\mathbf{r}_1)} S_{\text{self}}(\mathbf{r}, t; \mathbf{r}_1, t_1) \quad (9.155)$$

Using eqns (9.90) and (9.92) we can also get the coherent cross section per unit volume as

$$\frac{d^2\sigma_{\text{coh}}}{V d\Omega_f d\varepsilon_f} = \frac{|\mathbf{k}_f|}{|\mathbf{k}_i|} \frac{1}{2\pi\hbar} |\langle b \rangle|^2 \left[S_{\rho\rho}(\mathbf{k}, \omega) + (2\pi)^4 \left(\frac{N}{V}\right)^2 \delta(\mathbf{k}) \delta(\omega) \right] \quad (9.156)$$

These last three equations demonstrate that neutron scattering experiments can be used to determine density-density correlation functions in a fluid. Some typical results are given in Fig. 9.3a and 9.3b in liquid argon and in Fig. 9.4 in liquid lead.

We finally mention that the static structure factor

$$\left(\frac{N}{V}\right) S(\mathbf{k}) = \int_{-\infty}^{+\infty} S_{\rho\rho}(\mathbf{k}, \omega) \, d\omega \quad (9.157)$$

is the key function in the calculation of equilibrium properties of a fluid (Barker and Henderson, 1976) and is, in turn, given in terms of the (equilibrium) radial distribution function $g(r)$, defined by the equilibrium two-particle distribution function

$$f_2(\mathbf{r}_1, \mathbf{r}_2, t) = \left(\frac{N}{V}\right)^2 g(|\mathbf{r}_1 - \mathbf{r}_2|) \quad (9.158)$$

by

$$S(|\mathbf{k}|) = 1 + \left(\frac{N}{V}\right) \int e^{i\mathbf{k}\cdot\mathbf{r}} g(|\mathbf{r}|) \, d^3r \quad (9.159)$$

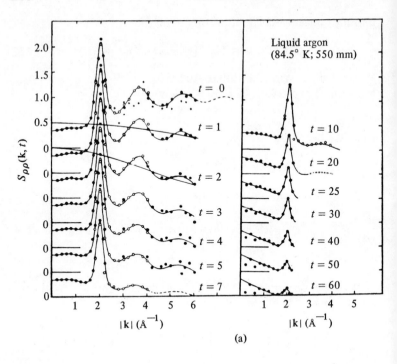

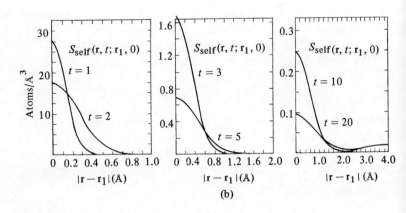

Fig. 9.3. Structure functions in liquid argon as determined by neutron scattering: (a) $S_{\rho\rho}(k, t) = S_{nn}(k, t)$ for different times in units of 10^{-23} sec. (b) $S_{\text{self}}(\mathbf{r}, t)$. After Martin (1968).

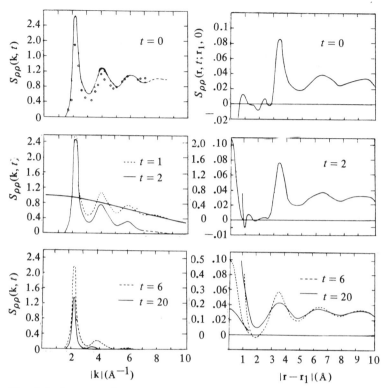

Fig. 9.4. Structure functions in liquid lead as determined by neutron scattering. After Martin (1968).

The connection between $S(|\mathbf{k}|)$ and $S(\mathbf{k}, \omega)$ is schematically indicated in Fig. 9.5, with the radial distribution function in liquid argon given as an example in Fig. 9.6.

9.3. Hydrodynamic Fluctuations and Transport Coefficients

9.3.1. Linearized Hydrodynamics

In this section we will see what predictive power linear response theory and the correlation function formalism have for the transport properties of a normal fluid, i.e. viscosity and thermal conduction. We must remind ourselves that linear response

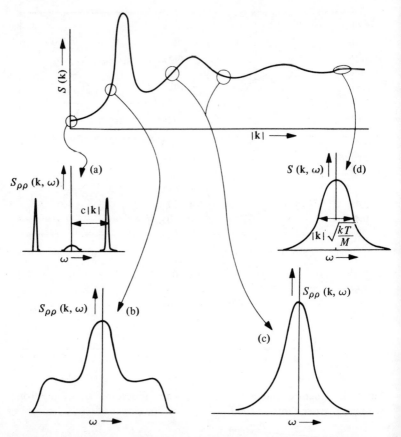

Fig. 9.5. Qualitative behavior of $S_{\rho\rho}(\mathbf{k}, \omega)$ for several values of $\mathbf{k}$. The upper curve is the function of $S(\mathbf{k})$, and the remaining curves show the spectral shape at fixed values of $\mathbf{k}$ marked by circles on $S(\mathbf{k})$. At low and high values of $\mathbf{k}$, the width can be calculated from simple considerations. From Egelstaff (1967).

theory, as outlined in previous sections of this chapter, can only deal with transport processes provoked by external forces $f_n(\mathbf{r}, t)$ acting on internal variables $A_m(\mathbf{r}, \mathbf{p})$ of the system that can be included as terms, eqn (9.4),

$$H_1(\boldsymbol{\xi}_1, \ldots, \boldsymbol{\xi}_N, t) = -\sum_n \int A_n(\mathbf{r}, \mathbf{p}) f_n(\mathbf{r}, t)\, d^3r\, d^3p \quad (9.160)$$

in the total hamiltonian of the system. All thermal and viscous processes, in which we are now interested, are therefore excluded so far. Schemes have been proposed according to which thermal effects can be incorporated into a hamiltonian formalism by a 'gravitational' potential and viscous effects by a vector potential (Luttinger, 1964; Picman, 1967). Other, more straightforward methods have been suggested by Kubo, Yokota, and Nakajima (1957), Mori (1958), Fujita (1962), and others; for a review, see Zwanzig (1965). For a linear transport theory of a normal fluid, we follow here the approach of Kadanoff and Martin (1963). (See also Galasiewicz, 1970, for an account of Bogolyubov's approach.)

A normal one-component fluid in local equilibrium in the absence of an external force is, according to Chapter 2, subject to

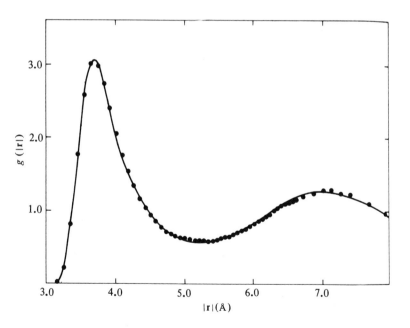

Fig. 9.6. Radial distribution function of liquid argon at 85 K (Yarnell et al., 1973). The solid curve is from a neutron diffraction experiment and the circles from Monte Carlo calculation. From Barker and Henderson (1979).

a set of general balance equations for the mass density $\rho(\mathbf{r}, t)$, the momentum density $\rho\mathbf{v}$, the energy density ρe, and the entropy density ρs. They are [eqns (2.6), (2.13), (2.24), and (2.61)]

$$\frac{\partial \rho}{\partial t}+\boldsymbol{\nabla}\cdot(\rho\mathbf{v})=0 \qquad (9.161a)$$

$$\frac{\partial(\rho\mathbf{v})}{\partial t}+\boldsymbol{\nabla}\cdot(\mathbf{P}+\rho\mathbf{v}\mathbf{v})=0 \qquad (9.161b)$$

$$\frac{\partial(\rho e)}{\partial t}+\boldsymbol{\nabla}\cdot\mathbf{j}_e=0 \qquad (9.161c)$$

$$\frac{\partial(\rho s)}{\partial t}+\boldsymbol{\nabla}\cdot\mathbf{j}_s=\sigma_s \qquad (9.161d)$$

As argued in Chapters 2 and 3, these balance equations have to be supplemented by a set of constitutive relations to achieve closure. For systems close to equilibrium, we can most certainly assume that these constitutive laws are linear; i.e. we write eqn (2.48) for a newtonian fluid

$$\mathbf{P}(\mathbf{r}, t)=p(\mathbf{r}, t)\mathbf{I}-2\eta\boldsymbol{\Lambda}-(\zeta+\tfrac{4}{3}\eta)(Tr\boldsymbol{\Lambda})\mathbf{I} \qquad (9.162)$$

where $p(\mathbf{r}, t)$ is the local hydrostatic pressure and

$$\Lambda_{\alpha\beta}=\frac{1}{2}\left(\frac{\partial v_\alpha}{\partial r_\beta}+\frac{\partial v_\beta}{\partial r_\alpha}\right) \qquad (9.163)$$

is the rate of strain tensor; η and ζ are the coefficients of shear and bulk viscosity, respectively. From eqn (2.26) we obtain

$$\mathbf{j}_e(\mathbf{r}, t)=pe\mathbf{v}+\mathbf{P}\cdot\mathbf{v}+\mathbf{j}_q \qquad (9.164)$$

with the heat current given by Fourier's law,

$$\mathbf{j}_q=-\lambda\boldsymbol{\nabla}T \qquad (9.165)$$

We also have, from eqns (2.69) and (2.70),

$$\mathbf{j}_s(\mathbf{r}, t)=\rho s\mathbf{v}+\frac{\mathbf{j}_q}{T} \qquad (9.166)$$

and

$$\sigma_s(\mathbf{r}, t)=\mathbf{j}_q\cdot\boldsymbol{\nabla}\left(\frac{1}{T}\right)-\frac{1}{T}(\mathbf{P}-p\mathbf{I}):(\boldsymbol{\nabla}\mathbf{v}) \qquad (9.167)$$

As we are interested here in the linear response of the fluid only, we further simplify the above equations by dropping all terms nonlinear in **v** from eqns (9.161a)–(9.161d). We then obtain the linearized hydrodynamic equations

$$\frac{\partial \rho}{\partial t} + \boldsymbol{\nabla} \cdot (\rho \mathbf{v}) = 0 \qquad (9.168a)$$

$$\frac{\partial (\rho \mathbf{v})}{\partial t} + \boldsymbol{\nabla} p - \boldsymbol{\nabla} \cdot (2\eta \boldsymbol{\Lambda} + (\zeta - \tfrac{2}{3})(Tr\boldsymbol{\Lambda})\mathbf{I}) = 0 \qquad (9.168b)$$

$$\frac{\partial (\rho e)}{\partial t} + \boldsymbol{\nabla} \cdot [(\rho e + p)\mathbf{v}] - \boldsymbol{\nabla} \cdot (\lambda \boldsymbol{\nabla} T) = 0 \qquad (9.168c)$$

If, close to equilibrium, the fluid flow is not too agitated and the temperature variations not too great, we can treat the transport coefficients as constant and state independent.

We now try to find a solution of eqns (9.168a)–(9.168c) in the form of the initial value problem given in eqn (9.109). As we are ultimately interested in microscopic expressions for the transport coefficients η, ζ, and λ (and not in the solution of a hydrodynamic boundary value problem), we choose the simplest possible system, namely an infinite one, in which case we do not have to worry about boundary conditions. We split the momentum vector $\rho \mathbf{v}$ into a longitudinal and a transverse component according to

$$\rho \mathbf{v}(\mathbf{r}, t) = \rho \mathbf{v}_l(\mathbf{r}, t) + \rho \mathbf{v}_t(\mathbf{r}, t) \qquad (9.169)$$

determined such that

$$\boldsymbol{\nabla} \cdot (\rho \mathbf{v}_t) = 0 \qquad (9.170)$$

and

$$\boldsymbol{\nabla} \cdot (\rho \mathbf{v}_t) = 0 \qquad (9.171)$$

In a linearized theory we can set $\rho \mathbf{v} \approx \rho_0 \mathbf{v}$, where ρ_0 is the equilibrium density in the system. From eqn (9.168b), we then find that $\rho_0 \mathbf{v}_t$ satisfies a diffusion equation, namely

$$\left(\frac{\partial}{\partial t} - \nu \nabla^2\right) \rho_0 \mathbf{v}_t(\mathbf{r}, t) = 0 \qquad (9.172)$$

where $\nu = \eta/\rho_0$ is the kinematic viscosity. We readily find the solution of eqn (9.172) to be

$$\mathbf{v}_t(\mathbf{r}, t) = \int d^3 r' \frac{e^{-(\mathbf{r}-\mathbf{r}')^2/4\nu t}}{(4\pi \nu t)^{\frac{3}{2}}} \mathbf{v}_t(\mathbf{r}', 0) \qquad (9.173)$$

in terms of some initial conditions at time $t=0$. Introducing a spatial Fourier transform

$$\mathbf{v}_t(\mathbf{k}, t) = \int d^3 r e^{-i\mathbf{k}\cdot\mathbf{r}} \mathbf{v}_t(\mathbf{r}, t) \tag{9.174}$$

we obtain

$$\mathbf{v}_t(\mathbf{k}, t) = e^{-t/\tau_m} \mathbf{v}_t(\mathbf{k}, t=0) \tag{9.175}$$

where we introduced a diffusion lifetime

$$\tau_m = \frac{1}{\nu k^2} \tag{9.176}$$

Observe immediately that $\tau_m \to \infty$ as $k \to 0$, i.e. long wavelength 'hydrodynamic' disturbances live longer. Finally we take a one-sided temporal Fourier transform for Im $z > 0$

$$\mathbf{v}_t^{(+)}(\mathbf{k}, z) = \int_0^\infty dt e^{izt} \mathbf{v}_t(\mathbf{k}, t) \tag{9.177}$$

and find from eqn (9.175)

$$\mathbf{v}_t^{(+)}(\mathbf{k}, z) = \frac{i}{z + i\nu k^2} \mathbf{v}_t(\mathbf{k}, t=0) \tag{9.178}$$

a solution of eqn (9.172) in the form of the initial value problem, eqn (9.109). Note that diffusion shows up as a pole in the lower half-plane.

Let us next perform a similar analysis on the rest of eqns (9.168). Linearized around the equilibrium state with density ρ_0, energy density e_0, and hydrostatic pressure p_0, they read

$$\frac{\partial \rho}{\partial t} + \rho_0 \boldsymbol{\nabla} \cdot \mathbf{v}_l = 0 \tag{9.179a}$$

$$\frac{\partial \mathbf{v}_l}{\partial t} + \frac{1}{\rho_0} \boldsymbol{\nabla} p - \frac{1}{\rho_0} (\zeta + \tfrac{4}{3}\eta) \boldsymbol{\nabla}(\boldsymbol{\nabla} \cdot \mathbf{v}_l) = 0 \tag{9.179b}$$

$$\frac{\partial \rho_0 e}{\partial t} + (\rho_0 e_0 + p_0) \boldsymbol{\nabla} \cdot \mathbf{v}_l - \lambda \nabla^2 T = 0 \tag{9.179c}$$

Eliminating the $\boldsymbol{\nabla} \cdot \mathbf{v}_l$ term in the energy balance with the help of the continuity eqns (9.179a), we obtain

$$\frac{\partial}{\partial t} \left(\rho_0 e - \frac{\rho_0 e_0 + p_0}{\rho_0} \rho \right) - \lambda \nabla^2 T = 0 \tag{9.180}$$

LINEAR RESPONSE THEORY

Recall now that below eqn (2.28) we introduced the heat per unit mass q by the equation

$$\frac{\partial(\rho q)}{\partial t} + \boldsymbol{\nabla}\cdot(\rho q\mathbf{v}) + \boldsymbol{\nabla}\cdot\mathbf{j}_q = 0 \qquad (9.181)$$

Observe that $(q\mathbf{v})$ is a term of second order in the deviations from equilibrium, so that the middle term in eqn (9.181) should be dropped in a linearized theory, leading to the identification

$$\rho_0 q(\mathbf{r}, t) = \rho_0 e(\mathbf{r}, t) - \frac{\rho_0 e_0 + p_0}{\rho_0}\rho(\mathbf{r}, t) \qquad (9.182)$$

which, in turn, makes eqn (9.180), divided by T_0, identical to the linearized entropy balance, eqn (2.61), namely

$$\frac{\partial \rho_0 s}{dt} + \boldsymbol{\nabla}\cdot\frac{\mathbf{j}_q}{T_0} = 0 \qquad (9.183)$$

As a last manipulation we write

$$\boldsymbol{\nabla} p(\mathbf{r}, t) = \left(\frac{\partial p_0}{\partial \rho_0}\right)_{s_0} \boldsymbol{\nabla}\rho(r, t) + \left(\frac{\partial p_0}{\partial \rho_0 s_0}\right)_{\rho_0} \boldsymbol{\nabla}\rho_0 s \qquad (9.184a)$$

$$\boldsymbol{\nabla} T(\mathbf{r}, t) = \left(\frac{\partial T_0}{\partial \rho_0}\right)_{s_0} \boldsymbol{\nabla}\rho + \left(\frac{\partial T_0}{\partial \rho_0 s_0}\right)_{\rho_0} \boldsymbol{\nabla}\rho_0 s \qquad (9.184b)$$

and observe that

$$T_0\left(\frac{\partial \rho_0 s_0}{\partial T_0}\right)_{\rho_0} = \rho_0 c_v \qquad (9.185)$$

where c_V is the specific heat per unit mass at constant volume V, and

$$\left.\frac{\partial p_0}{\partial \rho_0}\right|_{s_0} = c^2 \qquad (9.186)$$

is the square of the adiabatic sound velocity. Equation (9.179) now reads

$$\frac{\partial \rho}{\partial t} + \rho_0 \boldsymbol{\nabla}\cdot\mathbf{v}_l = 0 \qquad (9.187a)$$

$$\left[\frac{\partial}{\partial t} - \frac{1}{\rho_0}(\zeta + \tfrac{4}{3}\eta)\nabla^2\right]\mathbf{v}_l + \frac{1}{\rho_0}c^2\boldsymbol{\nabla}\rho + \left(\frac{\partial p_0}{\partial s_0}\right)_{\rho_0}\boldsymbol{\nabla} s = 0 \qquad (9.187b)$$

$$\left(\frac{\partial}{\partial t} - \frac{\lambda}{\rho_0 c_v}\nabla^2\right)s - \frac{\lambda}{\rho_0 T_0}\left(\frac{\partial T_0}{\partial \rho_0}\right)_{s_0}\nabla^2\rho = 0 \qquad (9.187c)$$

Introducing Fourier transforms similar to eqn (9.158) we can solve the set of coupled equations (9.168) as an initial value problem which reads in matrix form

$$
\begin{bmatrix}
z & -\rho_0 \mathbf{k} & 0 \\
-\dfrac{1}{\rho_0} c^2 \mathbf{k} & z + i\dfrac{i}{\rho_0}\left(\zeta + \dfrac{4}{3}\eta\right)k^2 & -\left(\dfrac{\partial \rho_0}{\partial s_0}\right)_s \mathbf{k} \\
i\dfrac{\lambda}{\rho_0 T_0}\left(\dfrac{\partial T_0}{\partial \rho_0}\right)_{s_0} k^2 & 0 & z + ik^2 \dfrac{\lambda}{\rho_0 c_v}
\end{bmatrix}
\begin{bmatrix}
\rho^{(+)}(\mathbf{k}, z) \\
\mathbf{v}_l^{(+)}(\mathbf{k}, z) \\
s^{(+)}(\mathbf{k}, z)
\end{bmatrix}
$$

$$
= i \begin{bmatrix} \rho(\mathbf{k}, t=0) \\ \mathbf{v}_l(k, t=0) \\ s(\mathbf{k}, t=0) \end{bmatrix} \quad (9.188)
$$

Inverting this matrix equation, we see that the solution of our linearized hydrodynamic initial problem is given in complete analogy to eqn (9.109). Putting the determinant of the above matrix equal to zero, we find that these solutions exhibit three poles in the lower half of the z plane which, for small k^2, are given by

$$z_1 = -i\kappa k^2 \quad (9.189)$$

$$z_{2,3} = \pm c\,|\mathbf{k}| - \frac{i}{2}\Gamma \quad (9.190)$$

where

$$\kappa = \frac{\lambda}{(\rho_0 c_p)} \quad (9.191)$$

is the thermal diffusivity, introduced in Table 6.1, and

$$\Gamma = \kappa\left(\frac{c_p}{c_V} - 1\right) + \left(\frac{\zeta}{\rho_0} + \tfrac{4}{3}\nu\right) \quad (9.192)$$

From our experience with the transverse momentum eqn (9.178), we can identify the pole z_1 with a diffusive heat mode with a lifetime

$$\tau_{\text{heat}} = \frac{1}{\kappa k^2} \quad (9.193)$$

The poles $z_{2,3}$ obviously allow the propagation of sound waves through their linear k dependence. They are, however, damped and have a lifetime

$$\tau_{\text{sound}} = \frac{2}{\Gamma k^2} \quad (9.194)$$

9.3.2. Connection with Linear Response Theory

By a formal comparison of the hydrodynamic solutions (9.188) with the general result (9.109), we are now in a position to write explicit expressions for various correlation functions. Let us start with the transverse momentum eqn (9.178). It does not couple to any of the other hydrodynamic eqns (9.188). A transverse velocity field would be set up by a force $\mathbf{F}_t$ for which $\mathbf{\nabla}\cdot\mathbf{F}_t = 0$ but $\mathbf{\nabla}\times\mathbf{F}_t \neq 0$. Such a force can obviously not be conservative. For the purpose of this derivation we can imagine that in a thought experiment we give an infinitesimally small electric charge to each fluid element and apply a weak magnetic fluid. The resulting Lorentz force will then be transverse and create transverse momentum correlations. This comment implies that it will be very hard and, indeed, impossible in most fluids to measure such correlations. The only correlations involving the transverse momentum density are therefore with itself, and can be defined by eqn (9.61).

$$\mathbf{S}_{tt}(\mathbf{r}, t; \mathbf{r}_1, t_1) = \hbar^2 \langle \hat{\mathbf{j}}(\mathbf{r}, t)\hat{\mathbf{j}}(\mathbf{r}_1, t_1)\rangle \quad (9.195)$$

where, from eqns (8.27) and (8.76), the current operator is given by

$$\hat{\mathbf{j}}(\mathbf{r}, t) = \frac{1}{2i}[\psi^+(\mathbf{r}, t)\mathbf{\nabla}\psi(\mathbf{r}, t) - (\mathbf{\nabla}\psi^+(\mathbf{r}, t))\psi(\mathbf{r}, t)] \quad (9.196)$$

where ψ^+ and ψ are field operators. The unnecessary complications of second quantization can be avoided by defining a momentum operator in phase space

$$\hat{\mathbf{p}}(\mathbf{r}, t) = \sum_{i=1}^{N} \mathbf{p}_i \delta[\mathbf{r} - \mathbf{r}_i(\mathbf{r}, t)] \quad (9.197)$$

In this case, $\mathbf{S}_{tt}$ is given by

$$\mathbf{S}_{tt}(\mathbf{r}, t; \mathbf{r}_1, t_1) = \left(\frac{N}{V}\right)^2 \langle \hat{\mathbf{p}}(\mathbf{r}, t)\hat{\mathbf{p}}(\mathbf{r}_1, t_1)\rangle \quad (9.198)$$

Observe further that the tensor $\mathbf{S}_{tt}$ has to be diagonal due to the isotropy of space in a normal fluid in equilibrium. Thus we can concentrate our attention on the scalar

$$S_t(\mathbf{r}, t; \mathbf{r}_1, t_1) = \tfrac{1}{3} Tr \mathbf{S}_{tt}(\mathbf{r}, t; \mathbf{r}_1, t_1) \quad (9.199)$$

Its Fourier transform, eqn (9.50), defines by eqn (9.56) a quantity $\chi_t''(\mathbf{k}, \omega)$ which, in turn, determines through eqn (9.57) $\chi_t(\mathbf{k}, \omega)$. Observe that (9.59) implies

$$\chi_t''(\mathbf{k}, \omega) = \operatorname{Im} \chi_t(\mathbf{k}, \omega) \qquad (9.200)$$

The comparison of eqns (9.109) and (9.178) then furnishes the result

$$\frac{1}{i\omega}\left[\frac{\chi_t(\mathbf{k}, \omega)}{\chi_t(\mathbf{k}, \omega = 0)} - 1\right] = \frac{i}{\omega + i\nu k^2} \qquad (9.201)$$

or

$$\chi_t(\mathbf{k}, \omega) = \frac{i\nu k^2}{\omega + i\nu k^2} \chi_t(\mathbf{k}, \omega = 0) \qquad (9.202)$$

To determine $\chi_t(\mathbf{k}, \omega = 0)$, we remind ourselves that hydrodynamical considerations can only be invoked, as done above, for small $\mathbf{k}$. If we therefore take the limit $\mathbf{k} \to 0$ in eqn (9.102), keeping ω fixed, we see that $\chi_t(\mathbf{k}, \omega = 0)$ must be independent of $\mathbf{k}$ and, from dimensional arguments, equal to ρ_0. We therefore have

$$\chi_t(\mathbf{k}, \omega) = \frac{i\nu k^2}{\omega + i\nu k^2} \rho_0 \qquad (9.203)$$

and

$$\chi_t''(\mathbf{k}, \omega) = \operatorname{Im} \chi_t(\mathbf{k}, \omega) = \frac{\nu k^2 \omega}{\omega^2 + i\nu^2 k^4} \rho_0 \qquad (9.204)$$

We have seen in eqn (9.99) that

$$\omega \chi_t''(\mathbf{k}, \omega) > 0 \qquad (9.205)$$

and can thus conclude that the kinematic viscosity has to be positive. It is in fact given by

$$\eta = \rho_0 \nu = \lim_{\omega \to 0} \lim_{\mathbf{k} \to 0} \frac{\omega}{k^2} \chi_t''(\mathbf{k}, \omega) \qquad (9.206)$$

Observe that the order of the two limits is crucial to obtain a nonzero shear viscosity η.

By an analogous analysis of eqn (9.188), one finds (see Forster,

1975) for small **k**, i.e. for $(\kappa k^2)^2 \ll c^2 k^2$

$$\frac{1}{\omega} \chi''_{\rho\rho}(\mathbf{k}, \omega) = \rho_0 \left(\frac{\partial \rho}{\partial p}\right)_{T_0} \quad (9.207a)$$

$$\frac{1}{\omega} \chi''_{ss}(\mathbf{k}, \omega) = \frac{\rho_0 c_p}{T_0} \frac{k^2 \kappa}{\omega^2 + (\kappa k^2)^2} \quad (9.207b)$$

$$\frac{1}{\omega} \chi''_{\rho s}(\mathbf{k}, \omega) = \left(\frac{\partial \rho}{\partial T}\right)_{p_0} \quad (9.207c)$$

where the normalizations were fixed through the relations

$$\lim_{\mathbf{k} \to 0} \chi_{\rho\rho}(\mathbf{k}, \omega = 0) = \rho_0 \left(\frac{\partial \rho}{\partial p}\right)_{T_0} \quad (9.208a)$$

$$\lim_{\mathbf{k} \to 0} \chi_{ss}(\mathbf{k}, \omega = 0) = \frac{\rho_0 c_p}{T_0} \quad (9.208b)$$

$$\lim_{\mathbf{k} \to 0} \chi_t(\mathbf{k}, \omega = 0) = \left(\frac{\partial \rho}{\partial T}\right)_{p_0} \quad (9.208c)$$

$$\lim_{\mathbf{k} \to 0} \chi_t(\mathbf{k}, \omega = 0) = \rho_0 \quad (9.208d)$$

Zero frequency cross correlations between vector and scalar quantities vanish due to time-reversal symmetry. Let us also note that with the help of the continuity equation for the longitudinal momentum $\mathbf{p}_l$

$$\frac{\partial}{\partial t} \rho + \boldsymbol{\nabla} \cdot \mathbf{p}_l = 0 \quad (9.209)$$

one sees immediately that

$$k^2 \chi''_l(\mathbf{k}, \omega) = k^2 \chi_{\mathbf{p}_l \mathbf{p}_l}(\mathbf{k}, \omega) = \omega^2 \chi''_{\rho\rho}(\mathbf{k}, \omega) \quad (9.210)$$

and, for any A,

$$\omega \chi''_{\rho A}(\mathbf{k}, \omega) = |\mathbf{k}| \chi''_{p_l A}(\mathbf{k}, \omega) \quad (9.211)$$

From eqn (9.188) we can now give microscopic expressions for the remaining transport coefficients, namely

$$\zeta + \tfrac{4}{3}\eta = \lim_{\omega \to 0} \lim_{\mathbf{k} \to 0} \frac{\omega^3}{k^4} \chi_{\rho\rho}(\mathbf{k}, \omega) \quad (9.212)$$

$$= \lim_{\omega \to 0} \lim_{\mathbf{k} \to 0} \frac{\omega}{k^2} \chi''_l(\mathbf{k}, \omega)$$

and
$$\kappa = T_0 \lim_{\omega \to 0} \lim_{\mathbf{k} \to 0} \frac{\omega}{k^2} \chi''_{ss}(\mathbf{k}, \omega) \tag{9.213}$$

It is particularly interesting to see that the diffusion of thermal energy, controlled by κ, is determined by the equilibrium fluctuations of the entropy of the system. The explicit calculation of transport coefficients according to any of the formulas (9.206), (9.212), and (9.213) or of the response functions $\chi_{mn}(\mathbf{k}, \omega)$ in eqn (9.208) obviously hinges on the possibility of evaluating the correlation functions on the right-hand side of these equations. This, in turn, demands the calculation of the linear response to an external perturbation, eqn (9.41). Attempts at this problem have been reviewed by Zwanzig (1965). The calculation of correlation functions is very simple for an ideal gas, as we will show in Section 9.4. In a dilute interacting gas, the resulting transport coefficients for viscosity and thermal conductivity have been shown to agree with those we calculated in Section 7.7 from the Boltzmann equation (Mori, 1958; Fujita, 1962; McLennan and Swenson, 1963). A survey of these attempts has been given by Ernst et al. (1969) who also critically assessed the possibility of extending time-correlation techniques to moderately dense gases. Let us finally mention that a simple, phenomenological ansatz for the structure of the correlation functions in (9.206) has been proposed by Forster et al. (1968) with reasonable success in the calculation of the shear viscosity.

9.4. Practical Results

9.4.1. The Ideal Gas

In this section we look at specific physical systems and evaluate the relevant predictions of linear response theory. We will examine the ideal gas and calculate the electrical conductivity in a metal. Starting with the ideal gas and calculating the density-density correlation function, eqn (9.152),

$$S_{\rho\rho}(\mathbf{r}_1, t_1; \mathbf{r}_2, t_2) = \langle \hat{\rho}(\mathbf{r}_1, t_1) \hat{\rho}(\mathbf{r}_2, t_2) \rangle - \langle \hat{\rho}(\mathbf{r}_1, t_1) \rangle \langle \hat{\rho}(\mathbf{r}_2, t_2) \rangle \tag{9.214}$$

The density operator is defined in a system of classical particles by eqn (9.145)

$$\hat{\rho}(\mathbf{r}, t) = \sum_m \delta[\mathbf{r} - \mathbf{r}_m(t)] \tag{9.215}$$

where $\mathbf{r}_m(t)$ is the position of the mth particle at time t. Its quantum-mechanical analogue is eqn (9.75)

$$\hat{\rho}(\mathbf{r}, t) = \psi^\dagger(\mathbf{r}, t)\psi(\mathbf{r}, t) \tag{9.216}$$

where $\psi(\mathbf{r}, t)$ is a field operator in second quantization (see also Section 8.3). We proceed with the quantum-mechanical calculation and expand $\psi(\mathbf{r}, t)$ in terms of free particle wavefunctions, getting

$$\psi(\mathbf{r}, t) = V^{-\frac{1}{2}} \sum_{\mathbf{k}} e^{i\mathbf{k}\cdot\mathbf{r}} a_{\mathbf{k}}(t) = V^{-\frac{1}{2}} \sum_{\mathbf{k}} e^{i\mathbf{k}\cdot\mathbf{r}} e^{-i\varepsilon(\mathbf{k})t/\hbar} a_{\mathbf{k}} \tag{9.217}$$

where $a_\mathbf{k}$ is a Schrödinger annihilation operator of a particle in momentum state $\mathbf{k}$ satisfying commutation relations

$$[a_\mathbf{k}, a^\dagger_{\mathbf{k}'}]_- = \delta_{\mathbf{k}\mathbf{k}'} \tag{9.218}$$

for bosons and anticommutation relations

$$[a_\mathbf{k}, a^\dagger_{\mathbf{k}'}]_+ = \delta_{\mathbf{k}\mathbf{k}'} \tag{9.219}$$

for fermions. $\varepsilon(\mathbf{k}) = (\hbar k)^2/2m$ is the energy of a particle of mass m. Inserting this into eqn (9.214), we obtain

$$S_{\rho\rho}(\mathbf{r}_1, t_1; \mathbf{r}_2, t_2) = V^{-2} \sum_{\mathbf{k}_1, \ldots, \mathbf{k}_4} e^{-i(\mathbf{k}_1-\mathbf{k}_2)\cdot\mathbf{r}_1} e^{i(\varepsilon_1-\varepsilon_2)t_1/\hbar}$$
$$\times e^{-i(\mathbf{k}_3-\mathbf{k}_4)\cdot\mathbf{r}_2} e^{i(\varepsilon_3-\varepsilon_4)t/\hbar} [\langle a_1^\dagger a_2 a_3^\dagger a_4\rangle - \langle a_1^\dagger a_2\rangle\langle a_3^\dagger a_4\rangle] \tag{9.220}$$

Here $\varepsilon_i = \varepsilon(\mathbf{k}_i)$ and $a_i = a_{\mathbf{k}_i}$ for brevity. The pointed brackets $\langle \cdots \rangle$ refer to thermal equilibrium averages in the grand canonical ensemble with

$$H_0 = \sum_\mathbf{k} \varepsilon(\mathbf{k}) a^\dagger_\mathbf{k} a_\mathbf{k} \tag{9.221}$$

i.e.

$$\langle \cdots \rangle = Tr(\cdots \hat{\rho}_0) \tag{9.222}$$

where the statistical operator is given by

$$\hat{\rho}_0 = \frac{\exp[-\beta(\hat{H}_0 - \mu\hat{N})]}{Tr\{\exp[-\beta(\hat{H}_0 - \mu\hat{N})]\}} \tag{9.223}$$

where μ is the chemical potential per particle and

$$\hat{N} = \sum_\mathbf{k} a^\dagger_\mathbf{k} a_\mathbf{k} \tag{9.224}$$

is the number operator. We first evaluate

$$\begin{aligned}\langle a_i^\dagger a_j\rangle &= Tr(a_i^\dagger a_j \hat{\rho}_0) \\ &= Tr(\delta_{ij} \pm a_j a_i^\dagger)\hat{\rho}_0\end{aligned} \quad (9.225)$$

where we used the commutation and anticommutation relations, eqns (9.218) and (9.219), respectively; the upper sign refers to bosons and the lower one to fermions. Using the cyclic invariance of the trace we get

$$\begin{aligned}Tr(a_j a_i^\dagger e^{-\beta(\hat{H}_0-\mu\hat{N})}) &= Tr(e^{-\beta(\hat{H}_0-\mu\hat{N})}a_j e^{\beta(\hat{H}_0-\mu\hat{N})}e^{-\beta(\hat{H}_0-\mu\hat{N})}a_i^\dagger) \\ &= e^{\beta(\varepsilon_j-\mu)}Tr(a_j e^{-\beta(\hat{H}_0-\mu\hat{N})}a_i^\dagger) \\ &= e^{\beta(\varepsilon_j-\mu)}Tr(a_i^\dagger a_j e^{-\beta(\hat{H}_0-\mu\hat{N})})\end{aligned} \quad (9.226)$$

where we also used the relation

$$e^{-\beta H_0}a_\mathbf{k}e^{+\beta H_0} = e^{\beta\varepsilon(\mathbf{k})}a_\mathbf{k} \quad (9.227)$$

Thus we get the well-known result

$$\langle a_\mathbf{k}^\dagger a_{\mathbf{k}'}\rangle = n(\mathbf{k})\delta_{\mathbf{k}\mathbf{k}'} \quad (9.228)$$

with

$$n(\mathbf{k}) = (e^{\beta[\varepsilon(\mathbf{k})-\mu]} \mp 1)^{-1} \quad (9.229)$$

the single-particle occupation functions for free bosons (upper sign) and fermions (lower sign). Similarly, we obtain‡

$$\begin{aligned}\langle a_1^\dagger a_2 a_3^\dagger a_4\rangle &= \delta_{23}\langle a_1^\dagger a_2\rangle \pm \langle a_1^\dagger a_3^\dagger a_2 a_4\rangle \\ &= e^{\beta(\varepsilon_2-\mu)}n(\mathbf{k}_1)n(\mathbf{k}_2)\delta_{23}\delta_{14} + n(\mathbf{k}_1)n(\mathbf{k}_3)\delta_{12}\delta_{34}\end{aligned} \quad (9.230)$$

where we also used the fact that

$$e^{-\beta\hat{H}_0}a_\mathbf{k}^\dagger e^{\beta\hat{H}_0} = e^{-\beta\varepsilon(\mathbf{k})}a_\mathbf{k}^\dagger \quad (9.231)$$

Thus we find

$$S_{\rho\rho}(\mathbf{r}_1,t_1;\mathbf{r}_2,t_2) = V^{-2}\sum_{\mathbf{k}_1,\mathbf{k}_2} e^{-i(\mathbf{k}_1-\mathbf{k}_2)\cdot(\mathbf{r}_1-\mathbf{r}_2)}e^{i(\varepsilon_1-\varepsilon_2)(t_1-t_2)/\hbar} \\ \times e^{\beta(\varepsilon_2-\mu)}n(\mathbf{k}_1)n(\mathbf{k}_2) = S_{\rho\rho}(\mathbf{r}_1-\mathbf{r}_2;t_1-t_2) \quad (9.232)$$

‡ This is a simple example of the contraction theorem by Bloch and De Dominicis (1958).

LINEAR RESPONSE THEORY

Replacing summations by integrations in the large volume limit according to

$$V^{-1}\sum_{\mathbf{k}} \xrightarrow[V\to\infty]{} \frac{1}{(2\pi)^3}\int d^3k \qquad (9.233)$$

we find for an ideal gas obeying classical Maxwell-Boltzmann statistics

$$n(\mathbf{k}) = e^{-\beta[\varepsilon(\mathbf{k})-\mu]} \qquad (9.234)$$

with

$$e^{\beta\mu} = \frac{N}{V}\frac{(2\pi\hbar)^3}{(2\pi m k_B T)^{\frac{3}{2}}} \qquad (9.235)$$

that

$$\begin{aligned}
S_{\rho\rho}&(\mathbf{r}_1-\mathbf{r}_2; t_1-t_2) \\
&= \left(\frac{e^{\beta\mu}}{(2\pi)^3}\int d^3k_1 e^{-i\mathbf{k}_1\cdot(\mathbf{r}_1-\mathbf{r}_2)} e^{i\varepsilon(\mathbf{k}_1)(t_1-t_2)/\hbar} e^{-\beta\varepsilon(\mathbf{k}_1)}\right) \\
&\quad\times \left(\frac{1}{(2\pi)^3}\int d^3k_2 e^{i\mathbf{k}_2\cdot(\mathbf{r}_1-\mathbf{r}_2)} e^{-i\varepsilon(\mathbf{k}_2)(t_1-t_2)/\hbar}\right) \\
&= \frac{N}{V}(m/(2\pi(t_1-t_2)(k_B T(t_1-t_2)+i\hbar)))^{\frac{3}{2}} \\
&\quad\times \exp\left[-\frac{(\mathbf{r}_1-\mathbf{r}_2)^2}{2(t_1-t_2)^2}\frac{mk_B T(t_1-t_2)^2 - i\hbar m(t_1-t_2)}{\hbar^2 + [k_B T(t_1-t_2)]^2}\right] \qquad (9.236)
\end{aligned}$$

Similar expressions can be derived for weakly and strongly degenerate quantum gases. In the classical limit, $\hbar \to 0$, we find from eqn (9.236)

$$S_{\rho\rho}(\mathbf{r}_1-\mathbf{r}_2; t_1-t_2) = \frac{N}{V}[\pi^2 v_0(t_1-t_2)]^{-3}$$
$$\times \exp\left[-\left(\frac{\mathbf{r}_1-\mathbf{r}_2}{v_0(t_1-t_2)}\right)^2\right] \qquad (9.237)$$

where $v_0^2 = 2k_B T/m$ is the average thermal velocity. Thus the density correlation function in a gas can be calculated classically for times

$$|t_1-t_2| \gg \frac{\hbar}{k_B T} \qquad (9.238)$$

i.e. if the Heisenberg uncertainty relation is well satisfied. Whereas $h/k_B T \sim 2 \cdot 10^{-14}$ sec at room temperature, i.e. of the order of the interaction time τ_0, introduced in Section 7.1, we infer from eqn (9.238) that, as temperature is lowered, a quantum-mechanical calculation of $S_{\rho\rho}$ becomes necessary for longer times, as expected. (For a clarifying discussion of approximate classical correlation functions, see Egelstaff, 1967.)

Let us next calculate the spatial Fourier transform of the classical density-density correlation function getting

$$S_{\rho\rho}(\mathbf{k}; t_1 - t_2) = \int d^3(\mathbf{r}_1 - \mathbf{r}_2) e^{i\mathbf{k}\cdot(\mathbf{r}_1 - \mathbf{r}_2)} S_{\rho\rho}(\mathbf{r}_1 - \mathbf{r}_2; t_1 - t_2)$$

$$= \frac{N}{V} \exp\left[-\tfrac{1}{4} v_0^2 (t_1 - t_2)^2 k^2\right] \quad (9.239)$$

An additional temporal Fourier transform yields

$$S_{\rho\rho}(\mathbf{k}, \omega) = \int_{-\infty}^{\infty} d(t_1 - t_2) e^{i\omega(t_1 - t_2)} S_{\rho\rho}(\mathbf{k}; t_1 - t_2)$$

$$= \frac{2\pi^{\frac{1}{2}}}{v_0 |\mathbf{k}|} \frac{N}{V} \exp\left(-\frac{\omega^2}{v_0^2 k^2}\right) \quad (9.240)$$

a simple Gaussian of width $v_0 |\mathbf{k}|$. From eqn (9.91) we can then determine the absorptive part of the response function in the classical limit

$$\chi''_{\rho\rho}(\mathbf{k}, \omega) = \frac{1}{2\hbar}(1 - e^{-\hbar\omega\beta}) S_{\rho\rho}(\mathbf{k}, \omega)$$

$$\approx \tfrac{1}{2} \beta\omega S_{\rho\rho}(\mathbf{k}, \omega)$$

$$= \pi^{\frac{1}{2}} \frac{\beta\omega}{v_0 |\mathbf{k}|} \frac{N}{V} \exp\left(-\frac{\omega^2}{v_0^2 k^2}\right) \quad (9.241)$$

Note that there is still an additional temperature dependence through $v_0^2 = 2k_B T/m$.

It remains to calculate $\chi_{\rho\rho}(\mathbf{k}, \omega)$ from eqn (9.57), i.e.

$$\chi_{\rho\rho}(\mathbf{k}, z) = \frac{1}{\pi} \int_{-\infty}^{+\infty} \frac{d\omega}{\omega - z} \chi''_{\rho\rho}(\mathbf{k}, \omega)$$

$$= \frac{\beta}{\pi} \frac{N}{V} \frac{\sqrt{\pi}}{v_0 |\mathbf{k}|} \int_{-\infty}^{+\infty} \frac{\omega \, d\omega}{\omega - z} e^{-\omega^2/v_0^2 k^2}$$

$$= \beta \frac{N}{V} + i\sqrt{\pi} \frac{N}{V} \beta \frac{z}{v_0 k} e^{-(z/v_0 k)^2} \operatorname{erf} c\left(-i \frac{z}{v_0 k}\right) \quad (9.242)$$

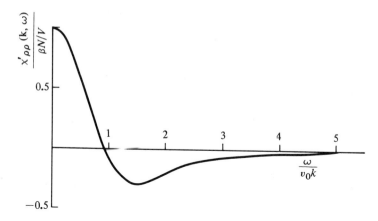

Fig. 9.7. Real part $\chi'_{\rho\rho}$ of the response function $\chi_{\rho\rho}$ versus ω.

where

$$\operatorname{erf} c(x) = 1 - \operatorname{erf}(x) \quad (9.243)$$

and

$$\operatorname{erf}(x) = \frac{2}{\sqrt{\pi}} \int_0^x e^{-t^2} dt \quad (9.244)$$

is the error function (see Abramowitz and Stegun, 1966). For the real part of $\chi_{\rho\rho}$, we obtain

$$\chi'_{\rho\rho}(\mathbf{k}, \omega) = \operatorname{Re} \chi_{\rho\rho}(\mathbf{k}, \omega)$$
$$= \beta \frac{N}{V}\left(1 - 2\frac{\omega}{v_0 k} \exp\left[-\left(\frac{\omega}{v_0 k}\right)^2\right] \int_0^{\omega/v_0 k} e^{x^2}\, dx\right) \quad (9.245)$$

which is the derivative of Dawson's integral and sketched in Fig. 9.7.

The ideal gas has been invented as a model for the calculation of the equilibrium properties of a very dilute gas in which the interparticle potential energy can be neglected in comparison with the kinetic energy. However, as a model for such a gas in nonequilibrium it is very limited, because although the interparticle interactions still do not play any role energy-wise, they are essential to retain collisions for (1) achieving randomization in the system and for (2) ultimately redistributing the initial nonequilibrium excess energy. Indeed, we have seen in Section 7.5 that a meaningful model of such a nonequilibrium gas must at least be controlled by a Boltzmann equation where the collision integral is given in the Grad limit.

The calculations of this paragraph of the linear response of an ideal gas are therefore only meaningful during the very early time evolution of a gas after some external constraint was suddenly switched on, for times $t < \tau_0$, where $\tau_0 = r/v_0$ is the interaction time for two atoms of average speed v_0 to pass through the range r of their mutual interaction. In this early statistical regime of the time evolution of a system, we do not expect any hydrodynamic behavior yet and should therefore not be surprised to find that from eqn (9.212)

$$\zeta + \tfrac{4}{3}\eta = \lim_{\omega \to 0} \lim_{|\mathbf{k}| \to 0} \frac{\omega^3}{k^4} \chi''_{\rho\rho}(\mathbf{k}, \omega) = 0 \qquad (9.246)$$

i.e. that transport coefficients of shear and bulk viscosity are both zero, indeed, the only meaningful value in an ideal gas. Calculating the entropy-entropy correlation, we would also find from eqn (9.212) that the thermal conductivity λ vanishes.

9.4.2. Electrical Conductivity of Metals

In this section we present an explicit calculation of the electrical conductivity of metals in a simple model and give a derivation of the law of Wiedemann and Franz, following the work by Chester and Thellung (1959, 1961). We start by specifying the hamiltonian part $H_1(t)$ in eqn (9.5). The external force in our case is an electric field $\mathbf{E}(t)$ that is adiabatically switched on in the remote part, i.e. as $t \to -\infty$, and which we assume to be uniform throughout the metal. It acts on the total electric dipole moment of the system and gives rise to the interaction term

$$H_1(t) = \sum_{i=1}^{N} e\mathbf{r}_i \cdot \mathbf{E}(t) \qquad (9.247)$$

where e is the electric charge of a particle and $\mathbf{r}_i(t)$ is its position at time t. The electric field $\mathbf{E}$ will produce a current $\mathbf{J}_e(t)$ which we calculate using linear response theory. The microscopic current is simply given by

$$\hat{\mathbf{J}}_e = \sum_{i=1}^{N} e\dot{\mathbf{r}}_i(t) \qquad (9.248)$$

and equation (9.115) then reads

$$\mathbf{J}_e(t) = \langle \hat{\mathbf{J}}_e(t) \rangle = \langle \hat{\mathbf{J}}_e(t=-\infty) \rangle$$
$$- \frac{1}{i\hbar} \int_{-\infty}^{t} dt' Tr \left\{ \mathbf{J}_e(t-t') \left[\sum_{i=1}^{N} e\mathbf{r}_i, \hat{\rho}_0 \right] \right\} \mathbf{E}(t') \quad (9.249)$$

where $\langle \cdots \rangle$ denotes the average over an equilibrium ensemble appropriate for the initial equilibrium state of the system in the absence of the electric field in the remote part, at which time there is, of course, also no current flowing. This implies that the first term in eqn (9.249) is zero. The general result, eqn (9.49), yields the frequency dependent conductivity tensor

$$\boldsymbol{\sigma}(\omega) = -\frac{1}{i\hbar} \int_0^{\infty} dt e^{-i\omega t} Tr \left\{ \hat{\mathbf{J}}_e(t) \left[\sum_{i=1}^{N} e\mathbf{r}_i, \boldsymbol{\rho}_0 \right] \right\} \quad (9.250)$$

If the equilibrium average $\langle \cdots \rangle$ is performed in a canonical ensemble

$$\hat{\rho}_0 = e^{-\beta H_0} \quad (9.251)$$

this can be rewritten according to eqn (9.116) as (Nakano, 1956)

$$\sigma_{\mu\nu}(\omega) = \int_0^{\infty} dt e^{-i\omega t} \int_0^{\beta} d\lambda Tr[\hat{\rho}_0 \hat{J}_{e\nu}(-i\hbar\lambda) \hat{J}_{e\mu}(t)] \quad (9.252)$$

To evaluate this expression, the hamiltonian H_0 of the unperturbed system has to be specified. We want to assume with Chester and Thellung (1959) that for the present purpose the metal is adequately described by a gas of free electrons moving in the periodic potential of the lattice. To obtain a finite conductivity we assume that static impurities are embedded randomly in the metal that act as elastic scattering centers for the electrons. We thus can write H_0 as a sum of single-particle hamiltonians

$$H_0 = \sum_{i=1}^{N} H_0^{(i)} \quad (9.253)$$

where each $H_0^{(i)}$ has the structure

$$H_0^{(i)} = H_e + \lambda V_e \quad (9.254)$$

Here $H_e = p^2/2m$ is the kinetic energy of a single (free) electron and λV_e is the scattering potential for one electron due to all impurities in the metal.

In a free electron gas we can also write, according to eqn (9.248), the total current operator $\hat{\mathbf{J}}_e(t)$ as a sum of single-particle current operators $\hat{\mathbf{j}}(t)$. In this case, the static conductivity tensor reduces to (Chester and Thellung, 1961)

$$\boldsymbol{\sigma} = -Tr\left[\frac{\partial f}{\partial H_0^{(i)}} \int_0^\infty dt\, \hat{\mathbf{j}}(t)\hat{\mathbf{j}}(0)\right]$$

$$= -\tfrac{1}{2} Tr\left\{\frac{\partial f}{\partial H_0^{(i)}} \int_0^\infty dt [\hat{\mathbf{j}}(t)\hat{\mathbf{j}}(0) + \hat{\mathbf{j}}(0)\hat{\mathbf{j}}(t)]\right\} \quad (9.255)$$

where

$$f^{-1}(H_0^{(i)}) = e^{\beta(H_0^{(i)} - \mu)} + 1 \quad (9.256)$$

The trace in eqn (9.255) is next evaluated in a basis in which H_e is diagonal; let us label its eigenstates by the wavevector $\mathbf{k}$ and call the eigenvalues $\varepsilon_{\mathbf{k}}$ (including the diagonal matrix elements of λV). We then obtain

$$\boldsymbol{\sigma} = -\sum_{\mathbf{k},\mathbf{k}'} \left\langle \mathbf{k} \left| \frac{\partial f}{\partial H_0^{(i)}} \right| \mathbf{k}' \right\rangle \int_0^\infty dt \langle \mathbf{k}' |\hat{\mathbf{j}}(t)| \mathbf{k}\rangle\langle \mathbf{k} |\hat{\mathbf{j}}(0)| \mathbf{k}'\rangle \quad (9.257)$$

where we used the fact that $\hat{\mathbf{j}}(0)$ is diagonal in this basis (in second quantization it is just $\sum \mathbf{p} a_{\mathbf{p}}^\dagger a_{\mathbf{p}}$).

We must realize that neither the term $(\partial f/\partial H_0)$ nor $\langle \mathbf{k}'|\hat{\mathbf{j}}(t')|\mathbf{k}\rangle$ can be evaluated exactly for an arbitrary scattering potential λV_e, produced by a collection of randomly distributed static impurities. We therefore want to make the further assumption on the system that λV is somehow small compared to H_e, allowing us to calculate $\boldsymbol{\sigma}$ in perturbation theory. Obviously,

$$\left\langle \mathbf{k} \left| \frac{\partial f}{\partial H} \right| \mathbf{k}' \right\rangle = \left\langle \mathbf{k} \left| \frac{\partial f}{\partial H} \right| \mathbf{k} \right\rangle \delta_{\mathbf{k}\mathbf{k}'} \quad (9.258)$$

has a nonvanishing zero-order term whereas

$$\int_0^\infty dt' \langle \mathbf{k} |\hat{\mathbf{j}}(t')| \mathbf{k}\rangle \quad (9.259)$$

will turn out to be proportional to λ^{-2} in lowest order. To see this we must assume that the scattering of the electrons by the random static impurities is a Markov process and can be

LINEAR RESPONSE THEORY

described by a master equation‡ (Pauli, 1928; van Hove, 1955, 1957)

$$\frac{dP_t(\mathbf{k}', \mathbf{k})}{dt} = 2\pi\lambda^2 \int d\mathbf{k}'' \langle \mathbf{k}' | V_e | \mathbf{k}'' \rangle \delta(\varepsilon_k - \varepsilon_{k''}) \langle \mathbf{k}'' | V_e | \mathbf{k}' \rangle P_t(\mathbf{k}'', \mathbf{k})$$

$$- 2\pi\lambda^2 \int d^3k'' |\langle \mathbf{k}' | V_e | \mathbf{k}'' \rangle|^2 \delta(\varepsilon_k - \varepsilon_{k''}) P_t(\mathbf{k}', \mathbf{k}) \quad (9.260)$$

Here $P_t(\mathbf{k}', \mathbf{k})$ is the conditional probability that an electron with momentum $\mathbf{k}$ at time $t = 0$ will scatter elastically in a time t to a momentum $\mathbf{k}'$ with $|\mathbf{k}| = |\mathbf{k}'|$.

With the initial conditions

$$P_{t=0}(\mathbf{k}', \mathbf{k}) = \delta_{\mathbf{k}\mathbf{k}'} \quad (9.261)$$

we can find a solution of (9.260) in the form

$$P_t(\mathbf{k}', \mathbf{k}) = e^{-2\lambda^2 t \Gamma(\mathbf{k})} \left[\delta_{\mathbf{k}', \mathbf{k}} + \sum_{n=1}^{\infty} \frac{(2\pi\lambda^2 t)^n}{n!} \right.$$

$$\left. \times \int d^3k_{n-1}, \ldots, d^3k_1 |\langle \mathbf{k}' | V_e | \mathbf{k}_{n-1} \rangle|^2 \cdots |\langle \mathbf{k}_1 | V_e | \mathbf{k} \rangle|^2 \right] \quad (9.262)$$

where

$$\Gamma(\mathbf{k}) = \pi \int d^3k_1 |\langle \mathbf{k} | V_e | \mathbf{k}_1 \rangle|^2 \delta(\varepsilon_k - \varepsilon_{k_1}) \quad (9.263)$$

assuming that the electronic single-particle energies $\varepsilon_\mathbf{k} = \varepsilon(|\mathbf{k}|)$ depend on the magnitude of the wavevector $\mathbf{k}$ only, as they do in a parabolic band model with $\varepsilon_\mathbf{k} = (\hbar\mathbf{k})^2/2m$. The matrix element of the current is then given by

$$\langle \mathbf{k} | \hat{\mathbf{j}}(t) | \mathbf{k} \rangle = \int d\mathbf{k}' P_t(\mathbf{k}', \mathbf{k}) \delta(\varepsilon_k - \varepsilon_{k'}) \langle \mathbf{k}' | \hat{\mathbf{j}}(0) | \mathbf{k}' \rangle \quad (9.264)$$

as the sum total of the transfers of current from states $\mathbf{k}'$ to $\mathbf{k}$ (with $|\mathbf{k}'| = |\mathbf{k}|$ because we are only considering elastic collisions), weighted with the probability $P_t(\mathbf{k}', \mathbf{k})$ that such a scattering event indeed takes place. Inserting eqn (9.264) into (9.257), we must

‡ The derivation of this equation (van Hove, 1955, 1957) involves a number of assumptions. Here, we mention only the weak coupling limit; i.e. in the derivation of the master equation it is assumed that as $t \to \infty$ the potential $\lambda \to 0$ is switched off but $\lambda^2 t = \text{const}$ is kept. Hence the appearance of a factor λ^2 in the master equation. A discussion of the weak coupling limit will be given in Chapter 10.

evaluate terms like

$$\int d^3k' \langle \mathbf{k} | V_e | \mathbf{k}' \rangle \delta(\varepsilon_k - \varepsilon_{k'}) \langle \mathbf{k}' | V_e | k \rangle \langle \mathbf{k}' | \mathbf{j}(0) | \mathbf{k}' \rangle \quad (9.265)$$

To do this we assume that the impurity potential is spherically symmetric and, we obtain

$$\langle \mathbf{k} | V_e | \mathbf{k}' \rangle = V_e(|\mathbf{k} - \mathbf{k}'|) = V_e(|\mathbf{k}|, \cos\theta) \quad (9.266)$$

where $\cos\theta = \mathbf{k}\cdot\mathbf{k}'/|\mathbf{k}|^2$ is the angle between $\mathbf{k}$ and $\mathbf{k}'$. Equation (9.265) then reads

$$\pi \int_{-\pi}^{\pi} \rho(\varepsilon) \cos\theta V_e^2(|\mathbf{k}|, \cos\theta) \sin\theta \, d\theta$$

$$\times \langle \mathbf{k} | \mathbf{j}(0) | \mathbf{k} \rangle = \Gamma_1(\varepsilon) \langle \mathbf{k} | \mathbf{j}(0) | \mathbf{k} \rangle \quad (9.267)$$

where the density of states $\rho(\varepsilon)$ has been defined by

$$d^3k = \rho(\varepsilon) \sin\theta \, d\theta \, d\phi \, d\varepsilon \quad (9.268)$$

This gives then

$$\langle \mathbf{k} | \hat{\mathbf{j}}(t) | \mathbf{k} \rangle = \langle \mathbf{k} | \mathbf{j}(0) | \mathbf{k} \rangle e^{-2\lambda^2 t(\Gamma - \Gamma_1)} \quad (9.269)$$

and

$$\int_0^\infty dt \langle \mathbf{k} | \mathbf{j}(t) | \mathbf{k} \rangle = \langle \mathbf{k} | \mathbf{j}(0) | \mathbf{k} \rangle \tau(\varepsilon_k) \quad (9.270)$$

where we defined a collision time

$$\tau^{-1}(\varepsilon_k) = 2\lambda^2 [\Gamma(\varepsilon_k) - \Gamma_1(\varepsilon_k)]$$

$$= 4\pi\lambda^2 \rho(\varepsilon_k) \int_{-\pi}^{\pi} (1 - \cos\theta) V_e^2(\varepsilon_k, \cos\theta) \sin\theta \, d\theta \quad (9.271)$$

Substituted back into eqn (9.257), we obtain for the conductivity tensor

$$\boldsymbol{\sigma} = -\sum_k \frac{\partial f}{\partial \varepsilon_k} \langle \mathbf{k} | \mathbf{j}(0) | \mathbf{k} \rangle \langle \mathbf{k} | \mathbf{j}(0) | \mathbf{k} \rangle \tau(\varepsilon_k) \quad (9.272)$$

which agrees with similar calculations starting from the quantum analogue of the Boltzmann equation (see Ziman, 1969, and Eliashberg, 1961).

A few comments are in order. The aim of this exercise was to show that a detailed computation of transport coefficients as given by linear response theory necessarily involves the use of additional information on the microscopic dynamics of the unperturbed model. The particular model used here, as done by Chester and Thellung (1959), is based on van Hove's form of the master equation. Within this model, the Nakano (1956) expression for the conductivity tensor can be reduced to the standard form obtained from kinetic theory, i.e. a Boltzmann equation. This might lend support to the master equation (9.260). Chester and Thellung (1959) also generalized their analysis to include asymmetric impurity scatterers and electron-phonon interaction, again recovering standard results. More importantly, they looked at the higher-order terms in λ in the expansion of the collision time (9.271) and found that the resulting corrections to the conductivity tensor (9.272) are small as long as

$$\frac{\hbar}{\tau} \ll \varepsilon_F \tag{9.273}$$

where ε_F is the Fermi energy of the electronic system and τ is the electronic collision time, i.e. a suitable average of eqn (9.271) or $\tau(\varepsilon_F)$. This is quite an important result in light of the fact that the kinetic theory approach via a Boltzmann equation necessitates the assumption that $\hbar/\tau \ll k_B T$, a condition that is seldom satisfied in metals and definitely never satisfied at low temperatures where the collision time becomes independent of temperature.‡

We finally want to show that the above theory of electrical conductivity lends itself readily to a derivation of the law of Wiedemann and Franz (1853). This law states that if λ is the thermal conductivity of a metal (neglecting any lattice contribution) and σ the electrical conductivity at temperature T, then the Lorenz number is given by

$$L = \frac{\lambda}{\sigma T} = \frac{1}{3} \pi^2 \left(\frac{k_B}{e}\right)^2 \tag{9.274}$$

universally for all metals.

We consider an isotropic thermocouple (see Section 3.5) in which a combined electric field $\mathbf{E}$ and a temperature gradient ∇T produce an electric current $\mathbf{J}_e$ and a heat current $\mathbf{J}_q$ according to

---

‡ See also Peierls (1934, 1955) who quotes an earlier argument by Landau in favor of condition (9.273).

eqns (3.107), i.e.

$$\mathbf{J}_e = eS_{11}\left(e\mathbf{E} + \frac{1}{T}\boldsymbol{\nabla}\frac{\mu}{T}\right) + eS_{12}\frac{1}{T}\boldsymbol{\nabla}T$$

$$\mathbf{J}_q = -S_{21}\left(e\mathbf{E} + \frac{1}{T}\boldsymbol{\nabla}\frac{\mu}{T}\right) - S_{22}\frac{1}{T}\boldsymbol{\nabla}T \qquad (9.275)$$

Thus the electrical conductivity is

$$\sigma = e^2 S_{11} \qquad (9.276)$$

and since the heat current is measured when $\mathbf{J}_e = 0$, we have for the thermal conductivity

$$\lambda = \frac{S_{11}S_{22} - S_{12}S_{21}}{TS_{11}} \qquad (9.277)$$

and

$$L = \frac{\lambda}{\sigma T} = \frac{S_{11}S_{22} - S_{12}S_{21}}{(eTS_{11})^2} \qquad (9.278)$$

The linear transport coefficients are given by eqns (9.252)

$$S_{ij} = \int_0^\infty dt \int_0^\beta d\lambda \langle S_i(0) S_j(t+i\lambda)\rangle \qquad i,j = 1,2 \qquad (9.279)$$

where S_1 is a component of the total current operator of the system and S_2 is the corresponding component of the total energy current operator. Let us now again assume that the hamiltonian H_0 of the unperturbed system is a sum of single-particle hamiltonians as in eqn (9.253) and again assume that $\hbar/\tau \ll \varepsilon_F$, which justifies use of the van Hove model. In analogy to eqn (9.257) for the conductivity tensor, we then obtain

$$S_{ij} = -\sum_{\mathbf{k},\mathbf{k}'} \frac{\partial f}{\partial \varepsilon_k} \langle \mathbf{k}|\hat{j}_i|\mathbf{k}\rangle\langle \mathbf{k}'|\hat{j}_j|\mathbf{k}'\rangle \delta(\varepsilon_k - \varepsilon_{k'}) \int_0^\infty P_t(\mathbf{k}',\mathbf{k})\,dt \qquad (9.280)$$

where $P_t(\mathbf{k}',\mathbf{k})$ again satisfies the master equation (9.260).

Chester and Thellung (1961) next define a generating function

$$\mathscr{L}(s) = -\sum_\varepsilon \frac{\partial f}{\partial \varepsilon} e^{-s\varepsilon} Q(\varepsilon) \qquad (9.281)$$

with

$$Q(\varepsilon) = \sum_{\mathbf{k},\mathbf{k}'} \delta_{\varepsilon\varepsilon_k}\delta_{\varepsilon\varepsilon_{k'}}\langle\mathbf{k}|\hat{j}_1|\mathbf{k}\rangle\langle\mathbf{k}'|\hat{j}_1|\mathbf{k}'\rangle \int_0^\infty P_t(\mathbf{k}',\mathbf{k})\,dt \qquad (9.282)$$

LINEAR RESPONSE THEORY

in terms of which we get

$$S_{11} = \mathcal{L}(0)$$

$$S_{12} = S_{21} = -\left.\frac{\partial \mathcal{L}(s)}{\partial s}\right|_{s=0}$$

$$S_{22} = \left.\frac{\partial^2 \mathcal{L}(s)}{\partial s^2}\right|_{s=0} \quad (9.283)$$

and

$$L = \frac{1}{e^2 T^2} \left.\frac{\partial^2 \ln \mathcal{L}(s)}{\partial s^2}\right|_{s=0} \quad (9.284)$$

Replacing the sum in eqn (9.281) by an integral we obtain

$$\mathcal{L}(s) = -\int_0^\infty G(\varepsilon) e^{-s\varepsilon} \frac{\partial f}{\partial \varepsilon} d\varepsilon \quad (9.285)$$

where

$$G(\varepsilon) = \rho(\varepsilon) Q(\varepsilon) \quad (9.286)$$

with $\rho(\varepsilon)$ the density of states.

If the electrons form a strongly degenerate Fermi-Dirac gas then $\partial f/\partial \varepsilon$ is a sharply peaked function at the Fermi surface and $\mathcal{L}(s)$ can be expanded in powers of $(k_B T/\varepsilon_F)$ yielding to second order

$$\mathcal{L}(s) = G(\varepsilon_F) e^{-s\varepsilon_F} + \frac{\pi^2}{6} (k_B T)^2 e^{-s\varepsilon_F}$$
$$\times \left[\frac{\partial^2 G(\varepsilon)}{\partial \varepsilon^2} - 2s \frac{\partial G(\varepsilon)}{\partial \varepsilon} + s^2 G(\varepsilon)\right]_{\varepsilon = \varepsilon_F} \quad (9.287)$$

from which we find with eqn (9.284)

$$L = \frac{\pi^2}{3}\left(\frac{k_B}{e}\right)^2 \quad (9.288)$$

i.e. the universal Wiedemann-Franz law which is well satisfied by most metals at temperatures well above the Debye temperature, where the difference in the collisional processes responsible for thermal and electrical conductivity becomes negligible (see Kittel, 1976).

In closing this chapter let us mention severe criticism voiced by van Kampen (1971) against the whole concept of linear response theory. His very intriguing argument starts with the statement that linear response theory assumes that the linearity of the macroscopic response means that the equation of motion for the density operator $\hat{\rho}(t)$ should be solved to first-order in the external fields. The equation for $\hat{\rho}$, however, is equivalent to the Schrödinger equation for the whole many-body system and contains therefore all the details of the microscopic motion of all individual particles. His point is then that linearity of the microscopic motion is entirely different from macroscopic linearity. To demonstrate his point, van Kampen considers conduction in a conductor. He supposes that the electrons move freely apart from occasional collisions with impurities. An external field $\mathbf{E}$ then has the effect of shifting their paths during the time t between collisions by an amount $\frac{1}{2}t^2$ ($e\,|\mathbf{E}|/m$). In order that this effect be linear in $|\mathbf{E}|$, one must have $\frac{1}{2}t^2(e\,|\mathbf{E}|/m) \ll d$, where d is a measure for the diameter of the impurity. Van Kampen then estimates that $|\mathbf{E}| < 10^{-18}$ V/cm for a linearized theory to hold, in striking contrast to the range over which Ohm's law, indeed, holds. The above argument certainly holds for an ideal gas of noninteracting electrons scattering off random impurities. But such a model for a metal is at best acceptable (1) for the calculation of some equilibrium properties like specific heat, Pauli spin paramagnetism, etc., and (2) for the calculation of initial transient phenomena after the electric field has been switched on, i.e. for times short compared to the collision time. It must, however, be recalled that in most metals at room temperatures electron-electron and electron-phonon collision times are of the order of 10^{-13} to 10^{-12} sec as compared to impurity scattering times of some 10^{-9} to 10^{-10} sec in relatively pure metals. The electron system has therefore a chance to randomize between impurity scattering events, justifying, as argued in the introduction to this chapter, either (1) linear time-dependent perturbation theory or (2) truncation of the BBGKY hierarchy with a *Stoszzahlansatz* to get a Boltzmann-type kinetic equation (see Section 7.5). Linearization of the equations of motion with respect to external forces over times large compared to two-body collision times calculates the change in a true many-body wavefunction. In cases where a single-particle picture can be constructed for the underlying many-body system, this implies that in linear response theory the effective single-particle mean-field potential (e.g. Hartree-Fock potential) is changed primarily, with a resultant change in the wavefunctions of the new quasiparticles, a highly nonlinear response for the original electrons. To apply arguments based on classical particle trajectories to a highly degenerate Fermi-Dirac gas of electrons can, indeed, be very misleading.

10
Master Equations

10.1. Introduction

IN CHAPTER 7 we looked at the statistical foundations of nonequilibrium thermodynamics within the framework of classical kinetic theory. We derived from the Liouville equation, cast in the equivalent form of the BBGKY hierarchy and without any further assumptions, the balance equations for certain 'mechanical' quantities—the macroscopic local mass density $\rho(\mathbf{r}, t)$, the momentum density $\rho\mathbf{v}$, the kinetic energy density $\frac{1}{2}\rho v^2$, and the internal energy density ρu. Assuming further that externally imposed forces are not too strong so as to keep the system locally close to equilibrium, we also derived the entropy balance and the linear laws of chapter 3 connecting thermodynamic forces and fluxes. Transport coefficients like thermal conductivity and viscosity were introduced formally but could only be calculated explicitly for a moderately dense gas for which Boltzmann's kinetic equation was derived.

In Chapter 9 we saw that for systems which respond linearly to an external force, switched on sufficiently slowly so that the internal dynamics of the system can keep it locally close to equilibrium, linear response theory provides general formulas for transport coefficients. The latter can, however, only be evaluated if kinetic equations describe the system's statistical time evolution in the absence of the external forces. An example was furnished in Section 9.4.2, where we calculated the electrical resistivity of metals. To generalize Boltzmann's kinetic equation to arbitrary systems is, of course, the task of nonequilibrium statistical mechanics. One of the early attempts for quantum systems was made by Pauli (1928) who derived a master equation, i.e. a linear, Markoffian rate equation for the probability distribution of the system.‡ Pauli's derivation was improved and generalized by van Hove (1955, 1961) who also derived the master equation used in Section 9.4.2 for the electron-impurity model of a metal. Extending these ideas, Prigogine and his collaborators (1961 and later) arrived at an exact master equation for an arbitrary system. Similar equations were derived at about the same time by Nakajima (1958), Zwanzig (1960), Prigogine and Résibois (1961), Montroll (1962), and Swenson (1962). Their equivalence was shown by Zwanzig (1964). These developments will be sketched in this chapter. We start with an introduction of Pauli's ideas about the role of a master equation in quantum systems. We then present in some detail van Hove's generalization thereof. In the last section we outline Prigogine's approach to nonequilibrium statistical mechanics and apply the generalized master equation to a study of hydrodynamic modes in a fluid.

10.2. Pauli's Master Equation

We consider a quantum-mechanical many-body system controlled by a hamiltonian

$$H = H_0 + \lambda \phi \tag{10.1}$$

and intend to study its statistical time evolution as a result of transitions between eigenstates of H_0 caused by the interaction ϕ. Because we are not interested in all the dynamical details of this system, we introduce a coarse-graining in its phase space (Γ space) by lumping a small number G_n of eigenstates labeled by $n_1, n_2, \ldots$, into a group labeled n. Denoting by W_{nm} the transition probability per unit time between the groups of states n and m, one hopes to establish an equation governing the rate of change of the probability $P_n(t)$ of finding the system in state n at time t, namely

$$\frac{dP_n(t)}{dt} = \sum_m [W_{nm} P_m(t) - W_{mn} P_n(t)] \tag{10.2}$$

‡ The use of master equations in phenomenological theories has been reviewed by Oppenheim, Shuler, and Weiss (1977). The role of master equations in the theory of fluctuations and the connection with the Fokker–Planck equation has been discussed extensively by van Kampen (1961, 1965, 1969). See also the criticism by Razavy (1976).

This is Pauli's master equation. If the index n can be treated as a continuous variable, say, α it is sometimes written as

$$\frac{\partial}{\partial t} P(\alpha, t) = \int [W(\alpha, \alpha')P(\alpha', t) - W(\alpha', \alpha)P(\alpha, t)] \, d\alpha' \tag{10.3}$$

Pauli (1928) has already pointed out that such an equation can only hold if the interaction ϕ, causing the transitions between the eigenstates of H_0 of an otherwise arbitrary system, is so small that the assumption of molecular chaos as applied to the microscopic motion resulting from H_0 is not invalidated. This implies, in particular, that the microscopic processes entering the transition probabilities W_{nm} or $W(\alpha, \alpha')$, respectively, can be assumed to be Markoffian, i.e. not to depend on the past history of the system.

We now proceed with the derivation of Pauli's master equation. Let us denote the eigenstates of H_0 by $|\nu_i\rangle$ such that

$$H_0 |\nu_i\rangle = \varepsilon_{\nu_i} |\nu_i\rangle \tag{10.4}$$

and let us expand the time-dependent solution $|\psi(t)\rangle$ of

$$(H_0 + \lambda \phi) |\psi(t)\rangle = i\hbar \frac{\partial}{\partial t} |\psi(t)\rangle \tag{10.5}$$

as

$$|\psi(t)\rangle = \sum_{\nu_i} \gamma_{\nu_i}(t) \exp\left(-\frac{i\varepsilon_{\nu_i} t}{\hbar}\right) |\nu_i\rangle \tag{10.6}$$

where the expansion coefficients must satisfy the set of differential equations

$$i\hbar \frac{d}{dt} \gamma_{\nu_i}(t) = \sum_{\mu_i} \langle \nu_i | \phi | \mu_i \rangle \exp\left[\frac{i(\varepsilon_{\nu_i} - \varepsilon_{\mu_i})t}{\hbar}\right] \gamma_{\mu_i}(t) \tag{10.7}$$

The probability of finding the system at time t in the eigenstate $|\nu_i\rangle$ of H_0 is then given by

$$P_{\nu_i}(t) = |\gamma_{\nu_i}(t)|^2 \tag{10.8}$$

To derive the master equation (10.2), Pauli (1928) solved (10.7) in first-order time-dependent perturbation theory. The result is that, if the system is initially in state $|\nu_j\rangle$, i.e. if

$$|\psi(t=0)\rangle = |\nu_j\rangle \qquad \gamma_{\nu_i}(t=0) = \delta_{\nu_i \nu_j} \tag{10.9}$$

then
$$\gamma_{\nu_i}(t) = -\sum_{\nu_j} \langle \nu_i|\phi|\nu_i\rangle \frac{\exp[i(\varepsilon_{\nu_i} - \varepsilon_{\nu_j})t/\hbar]}{\varepsilon_{\nu_i} - \varepsilon_{\nu_i}} \gamma_{\nu_j}(0) \quad (10.10)$$

and the probability of finding the system after a time t in state $|\nu_i\rangle$ is given by

$$P_{\nu_i}(t) = |\gamma_{\nu_i}(t)|^2 = 2\sum_{\nu_j} |\langle\nu_i|\phi|\nu_i\rangle|^2 \frac{1 - \cos(\varepsilon_{\nu_i} - \varepsilon_{\nu_j})t/\hbar}{(\varepsilon_{\nu_i} - \varepsilon_{\nu_i})^2} \quad (10.11)$$

which for times

$$t \gg \frac{\hbar}{\Delta E} \qquad \Delta E = \varepsilon_{\nu_i} - \varepsilon_{\nu_i} \quad (10.12)$$

yields

$$P_{\nu_i}(t) = \frac{2\pi}{\hbar} \sum_{\nu_j} |\langle\nu_i|\phi|\nu_i\rangle|^2 \delta((\varepsilon_{\nu_i} - \varepsilon_{\nu_j})t \quad (10.13)$$

The δ function enforces energy conservation. From the fact that $P_{\nu_i}(t)$ is proportional to t, we can identify the coefficient of proportionality as the transition probability

$$W_{\nu_i\nu_i} = \frac{2\pi}{\hbar}|\langle\nu_i|\phi|\nu_j\rangle|^2 \delta(\varepsilon_{\nu_i} - \varepsilon_{\nu_i}) \quad (10.14)$$

in eqn (10.2) whose symmetry

$$W_{\nu_i\nu_i} = W_{\nu_j\nu_i} \quad (10.15)$$

follows from the hermicicity of ϕ.

The above derivation suffers from the undesirable restriction that at time $t = 0$ the system must be in an eigenstate $|\nu_i\rangle$ of H_0 or, more generally, that at time $t = 0$ the density matrix of the system must be diagonal in the representation in which H_0 is diagonal, any phase relations in the off-diagonal matrix elements being discarded. Moreover, eqn (10.13) can only be valid for small times for which the $P_{\nu_i}(t)$ have not changed appreciably. Moreover, in order to establish (10.2) for all times, i.e. including the approach to equilibrium, one has to discard any build-up of phase relations by invoking a repeated random phase approximation at a series of times at microscopically small intervals.

Another undesirable feature is the fact that eqn (10.13) refers to pure quantum states $|\nu_i\rangle$ rather than coarse-grained groups of states for which we intended to derive the master equation.

In his analysis of the problem of deriving the master equation from quantum mechanics, van Kampen (1954, 1956) observes that by writing a master equation we only intend to get statements on macroscopically observable properties of (statistically) large systems, whereas the solution (10.6) still contains the full microscopic information and the exorbitantly complex time evolution of an N-body system, where N is a number of the order of 10^{23}. In trying to derive a master equation from quantum mechanics we must therefore first construct, following Pauli (1928), a suitable coarse-graining of phase space in such a way that the quantities of the statistical theory are only those that can be measured macroscopically, i.e. in particular, simultaneously. But this implies that our first task must be to construct out of generally noncommuting (microscopic) operators a set of commuting macroscopic operators.

We start with the energy operator $\hat{H}$ and note that any macroscopic measurement of the energy of a large system will occur with an inaccurracy ΔE which, according to Heisenberg's uncertainty relation, is bounded from below by

$$\Delta E \gg \delta E = \frac{\hbar}{\tau} \tag{10.16}$$

where τ is the time required to perform the energy measurement. But an N-body system has a large number of degrees of freedom and will therefore have a large number of energy eigenvalues within an interval ΔE. We are thus led to divide the energy spectrum into cells $E^{(1)}$, $E^{(2)}$, ... in such a way that a measurement of energy can only indicate that the system has an energy within a cell $E^{(n)}$. To treat all energy eigenvalues within one cell as equal, we construct from the microscopic energy operator

$$\hat{H} = \sum_i E_{n_i} |n_i\rangle\langle n_i| \tag{10.17}$$

a macroscopic energy operator (von Neumann, 1929; Watanabe, 1935; Pauli and Fierz, 1937; van Kampen, 1954; see Jancel, 1969, for a review)

$$\hat{\mathcal{H}} = \sum_n E^{(n)} |n\rangle\langle n| \tag{10.18}$$

where

$$|n\rangle\langle n| = \sum_{i=1}^{I_n} |n_i\rangle\langle n_i| \qquad (10.19)$$

with the sum over i running over all those exact eigenstates of $\hat{H}$ whose energy eigenvalues lie within a range ΔE of $E^{(n)}$. If the E_{n_i} were taken as nondegenerate (e.g. by double-counting, if necessary) then $E^{(n)}$ will be I_n times degenerate.

Suppose next that $\hat{A}$ is a microscopic operator that does not commute with $\hat{H}$, i.e. whose observable is not simultaneously measurable with the energy. Heisenberg's uncertainty principle then demands that

$$\delta E \, \delta A \geq \tfrac{1}{2} |\overline{[\hat{H}, \hat{A}]_-}| \qquad (10.20)$$

where

$$(\delta A)^2 = \overline{(\hat{A} - \bar{\hat{A}})^2} \qquad (10.21)$$
$$(\delta E)^2 = \overline{(\hat{H} - \bar{\hat{H}})^2}$$

Here $\bar{\hat{A}} = Tr(\hat{A}\hat{\rho})$ denotes a quantum-mechanical average with $\hat{\rho}$ being the density matrix for pure states. In a macroscopic measurement, the product of the measurement inaccuracies ΔE and ΔA exceeds their quantum-mechanical uncertainties δE and δA, i.e.

$$\Delta E \, \Delta A \gg \delta E \, \delta A \qquad (10.22)$$

Let us next evaluate the right-hand side of (10.20) in the representation $|n_i\rangle$ in which H is diagonal. We obtain

$$([\hat{A}, \hat{H}]_-)_{ij} = (E_i - E_j)A_{ij} \qquad (10.23)$$

and find from (10.22) by order of magnitude

$$(E_i - E_j)A_{ij} \sim \delta E \, \delta A \ll \Delta E \, \Delta A \qquad (10.24)$$

Next we choose two states i and j such that $(E_i - E_j) \gtrsim \Delta E$ and obtain

$$A_{ij} \sim \frac{\delta E}{\Delta E} \delta A \ll \Delta A \qquad (10.25)$$

Thus matrix elements A_{ij} connecting states i and j whose energy difference is of the order of the measurement inaccuracy ΔE are much smaller than the inaccuracy ΔA and can therefore be neglected. The matrix A_{ij} of the operator $\hat{A}$ is therefore reduced to a band of nonzero elements along the diagonal, with the width being approximately I_n in the nth energy cell. However, there will still be matrix elements A_{ij} connecting states in neighboring energy cells. We now make the crucial assumption that such matrix elements can be neglected. A justification must be that for times long compared to microscopic times, all real transitions must be energy conserving and virtual transitions become less important. As a result the matrix A_{ij} is reduced to a set of submatrices along the diagonal each corresponding to a single energy cell $E^{(n)}$.

We denote the operator constructed in this way from $\hat{A}$ by $\hat{\mathcal{A}}$ and observe immediately that

$$[\hat{\mathcal{A}}, \hat{\mathcal{H}}]_- = 0 \tag{10.26}$$

Within each cell $E^{(n)}$, we can next diagonalize $\hat{\mathcal{A}}$ to obtain eigenvalues $A_\alpha^{(n)}$. Because we can measure the property A only within an accuracy ΔA, we set all eigenvalues equal within a range ΔA, thus effecting a coarse-graining of A which is, of course, finer than the previous coarse-graining of the energy. This completes the construction of the commuting macroscopic operators $\hat{\mathcal{H}}$ and $\hat{\mathcal{A}}$. The procedure can formally be extended to further operators $\hat{B}, \hat{C}, \ldots$ chosen in decreasing order of measurement inaccuracy $\Delta B, \Delta C, \ldots$. With each additional operator the crucial neglect of matrix elements connecting (diagonal) cells of the previously constructed macroscopic operators becomes more and more questionable. The result, however, is a coarse-graining of phase space into macroscopically observable cells; let us label them with an index n. The diagonalization of all macroscopic operators has been achieved by a sequence of orthogonal transformations, leading from the eigenfunctions of $\hat{H}$, i.e.

$$H|n_i\rangle = E_{n_i}|n_i\rangle \tag{10.27}$$

to a system $|n, i\rangle$ such that for any macroscopic operator

$$\hat{\mathcal{A}}|n, i\rangle = A_n|n, i\rangle \tag{10.28}$$

within a cell n. Let us then expand an arbitrary vector $|\psi\rangle$ as

$$|\psi(t)\rangle = \sum_n \sum_{i=1}^{I_n} b_{ni}(t)|n, i\rangle \tag{10.29}$$

The expectation value of an operator $\hat{A}$ is then given by

$$\langle \psi(t)|\hat{A}|\psi(t)\rangle = \sum_n A_n \sum_i |b_{ni}(t)|^2$$

$$= \sum_n A_n P_n(t) \qquad (10.30)$$

where

$$P_n(t) = \sum_{i=1}^{I_n} |b_{ni}(t)|^2 \qquad (10.31)$$

is the probability of finding the system in the nth cell. To find the time evolution of $P_n(t)$ we observe that the coefficients $b_{ni}(t)$ must satisfy eqns (10.7) and are therefore given by

$$b_{ni}(t) = \sum_{n',i'} \langle n, i| U(t) |n', i'\rangle b_{n'i'}(0) \qquad (10.32)$$

with

$$\langle n, i| U(t) |n', i'\rangle = \sum_{n_i} \langle n, i | n_i\rangle e^{-iE_{n_i}t/\hbar} \langle n_i | n', i'\rangle \qquad (10.33)$$

so that

$$P_n(t) = \sum_{n',i',n'',i''} \left[\sum_i \langle n, i| U(t) |n', i'\rangle \right. \\ \left. \times \langle n, i| U(t) |n'', i''\rangle \right] b_{n'i'}(0) b^*_{n''i''}(0) \qquad (10.34)$$

To determine $P_n(t)$ we obviously need more than just $P_n(0)$, namely all $b_{ni}(0)$, or the complete information on the microstate of the system. However, we observe that in the sums over n', i', n'' and i'' in eqn (10.34), the diagonal terms with $n' = n''$ and $i' = i''$ are all real and nonnegative whereas all off-diagonal elements are complex. We will now assume that the latter will, in general, be randomly distributed in the complex number plane so that they will practically cancel each other, leaving only the diagonal terms as significant contributions. We can then write

$$P_n(t) = \sum_{n',i'} \left[\sum_i |\langle n, i| U(t) |n', i'\rangle|^2 \right] |b_{n'i'}(0)|^2 \qquad (10.35)$$

Still we have not determined $P_n(t)$ in terms of $P_n(0)$. We have to invoke initial chaos once more, namely by demanding that in the above sum we can replace

$$|b_{ni}(0)|^2 \to \frac{\sum_i |b_{ni}(0)|^2}{I_n} = \frac{P_n(0)}{I_n} \tag{10.36}$$

This finally yields

$$P_n(t) = \sum_n T_{nn'}(t) P_{n'}(0) \tag{10.37}$$

with

$$T_{nn'}(t) = I_n^{-1} \sum_{i,i'} |\langle n, i| U(t) |n', i'\rangle|^2 \tag{10.38}$$

Observe that

$$T_{nn'}(t) I_{n'} = T_{n'n}(t) I_n \tag{10.39}$$

and

$$\sum_n T_{nn'}(t) = 1 \tag{10.40}$$

We can therefore interpret the matrix elements $T_{nn'}(t)$ as transition probabilities.

To proceed with the derivation of the master equation, van Kampen (1956) next assumes that the state reached by our system after a time τ is such that the assumptions of molecular chaos invoked to go from (10.34) to (10.37) are valid. Under such conditions, we can write

$$P_n(\tau + t) = \sum_{n'} T_{nn'}(t) P_{n'}(\tau)$$

$$= \sum_{n',n''} T_{nn'}(t) T_{n'n''}(\tau) P_{n''}(0) \tag{10.41}$$

and

$$P_n(m\tau) = \sum_{n'} [\mathbf{T}^m(\tau)]_{nn'} P_{n'}(0) \tag{10.42}$$

for nonnegative m, where $\mathbf{T}^m(\tau)$ is the mth power of the matrix $\mathbf{T}(\tau)$. Equations (10.41) and (10.42) can only be true if the time evolution of our system over a time τ is such that the macroscopically observable properties, and thus also $P_n(t)$, change little whereas the microstate of the system changes so drastically that it appears to the macroscopic observer as chaotic as the initial state. The time τ therefore must be larger or of the order of the microscopic interaction time τ_0; τ can be chosen as the minimum time required according to Heisenberg's uncertainty principle to measure the energy of the system to an accuracy ΔE, i.e.

$$\tau \sim \frac{\hbar}{\Delta E} \tag{10.43}$$

It should be obvious that the repeated random phase approximation invoked at times $(m\tau)$ is equivalent to Boltzmann's *Stosszahlansatz*. It implies that we restrict our attention to the time evolution of the system on a coarse-grained time scale.

To obtain a differential equation from (10.42), define

$$T_{nn'}(\tau) = \delta_{nn'} + \tau W_{nn'} \tag{10.44}$$

and obtain

$$\frac{P_n(t+\tau) - P_n(t)}{\tau} = \sum_{n'} W_{nn'} P_{n'}(t) \tag{10.45}$$

Equations (10.39) and (10.40) imply that

$$\sum_n W_{nn'} = 0$$

$$W_{nn} = -\sum_{n' \neq n} W_{n'n}$$

$$W_{nn'} I_{n'} = W_{n'n} I_n \tag{10.46}$$

With the understanding that $P_n(t)$ changes slowly over times τ, we can replace (10.45) by a differential equation, also invoking relations (10.46), i.e.

$$\frac{d}{dt} P_n(t) = \sum_{n' \neq n} [W_{nn'} P_{n'}(t) - W_{n'n} P_n(t)] \tag{10.47}$$

which is Pauli's master equation. Van Kampen's derivation of the master equation highlights quite explicitly the inherent difficulties of nonequilibrium statistical mechanics. To derive a kinetic equation, we must postulate a number of (justifiable) mathematical assumptions, unfortunately in most cases without being able to give the explicit criteria on the microscopic dynamics of the system for such assumptions to be valid. The attitude is that because many large systems evolve smoothly on a macroscopic time scale, microscopic details are most likely not important, and must therefore be suppressed. In van Kampen's derivation of the master equation, this is done by invoking a repeated random phase approximation, i.e. by neglecting or suppressing any dynamical build-up of phases as time evolves. As the derivation shows these are obviously sufficient conditions for the validity of the master equation. A satisfactory theory, however, should establish the necessary conditions as well. An attempt in this direction has been made by van Hove (1955), whose derivation of the master equation we will study next.

10.3. Van Hove's Master Equation

In his attempt to derive Pauli's master equation without invoking a repeated random phase approximation, van Hove (1955, 1957, 1959) obtained a generalized master equation.

We again suppose that the hamiltonian of the system

$$\hat{H} = \hat{H}_0 + \lambda \hat{V} \qquad (10.48)$$

can be split into two terms and assume that we know the eigenstates of H_0 exactly. We denote them this time by $|\alpha\rangle$ for two reasons: one is that van Hove only considers operators diagonal in the basis $|\alpha\rangle$ (which we might identify as macroscopic operators) and, secondly, his theory deals with infinite systems in which limit, at least part of the spectrum $|\alpha\rangle$, has to be continuous. We therefore assume the normalization $\langle \alpha | \alpha' \rangle = \delta(\alpha - \alpha')$, with the right-hand side denoting a product of Dirac delta functions and Kronecker symbols for the continuous and discrete quantum numbers, respectively. Van Hove (1957) showed that for operators $\hat{A}_j$ that are diagonal in the basis $|\alpha\rangle$, one obtains

$$\langle \alpha | \hat{V}\hat{A}_1 \hat{V}, \ldots, \hat{A}_n \hat{V} | \alpha' \rangle = \delta(\alpha - \alpha') F_1(\alpha) + F_2(\alpha, \alpha') \qquad (10.49)$$

The singular term $\delta(\alpha - \alpha') F_1(\alpha)$ is called the diagonal part of the matrix element. It is, of course, absent in $\langle \alpha | \hat{V} | \alpha' \rangle$. The

wavefunction of the system evolves according to
$$|\psi(t)\rangle = U(t)|\psi(0)\rangle \tag{10.50}$$
where
$$|\psi(0)\rangle = \int c(\alpha)|\alpha\rangle\, d\alpha \tag{10.51}$$
and
$$U(t) = \exp[-i(\hat{H}_0 + \lambda \hat{V})t] \tag{10.52}$$

An operator $\hat{A}$; initially diagonal in the $|\alpha\rangle$ representation, so that
$$\hat{A}|\alpha\rangle = A(\alpha)|\alpha\rangle \tag{10.53}$$
with $A(\alpha)$ a smooth function, evolves in time according to
$$\langle\psi(t)|\hat{A}|\psi(t)\rangle = \int A(\alpha)P(\alpha, t)\, d\alpha \tag{10.54}$$
where $P(\alpha, t)$ is the (coarse-grained) probability of finding the system in state $|\alpha\rangle$ at time t. With eqns (10.51) and (10.52), this gives

$$\langle\psi(t)|\hat{A}|\psi(t)\rangle = \langle\psi(0)|U(-t)\hat{A}U(t)|\psi(0)\rangle$$
$$= \int A(\alpha'')\, d\alpha'' \int \tilde{W}(\alpha'', \alpha; t)\, d\alpha\, |c(\alpha)|^2$$
$$+ \int A(\alpha'')\, d\alpha'' \int I(\alpha'', \alpha, \alpha'; t)\, d\alpha\, d\alpha'\, c^*(\alpha)c(\alpha') \tag{10.55}$$

where $I(\alpha'', \alpha, \alpha'; t)$ has no δ singularity. Comparing eqns (10.54) and (10.55), we obtain

$$P(\alpha, t) = \int \tilde{W}(\alpha', \alpha; t)|c(\alpha')|^2\, d\alpha'$$
$$+ \int I(\alpha, \alpha', \alpha''; t)c^*(\alpha')c(\alpha'')\, d\alpha'\, d\alpha'' \tag{10.56}$$

For an initial state $|\psi(0)\rangle$ with random phases, the second term on the right is again negligible and we obtain
$$P(\alpha, t) = \int \tilde{W}(\alpha, \alpha'; t)|c(\alpha')|^2\, d\alpha' \tag{10.57}$$

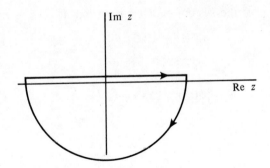

Fig. 10.1. Contour C in the definition of R(z), eqn (10.60).

Thus $\tilde{W}(\alpha, \alpha'; t)$ is the transition probability from α' to α in time t. It has been shown by van Hove (1957) that the energy-dependent partial transition probability $P_E(\alpha, \alpha'; t)$ defined by

$$\tilde{W}(\alpha, \alpha'; t) = \int_{-\infty}^{\infty} P_E(\alpha, \alpha', t)\, dE \qquad (10.58)$$

satisfies a generalized master equation

$$\frac{d}{dt} P_E(\alpha, \alpha_0; t) = \delta(\alpha - \alpha_0) f_E(\alpha; t)$$

$$- 2\pi\lambda^2 \int_0^t dt' \int d\alpha' w_E(\alpha, \alpha'; t-t') P_E(\alpha', \alpha_0; t')$$

$$- 2\pi\lambda^2 \int_0^t dt' \int d\alpha' w_E(\alpha', \alpha; t-t') P_E(\alpha, \alpha_0; t') \quad (10.59)$$

To derive the latter, one expresses the time evolution operator $U(t)$, eqn (10.52), in terms of its resolvent

$$U(t) = \frac{1}{2\pi i} \oint dz\, e^{-izt} R(z) \qquad (10.60)$$

$$\hat{R}(z) = (\hat{H}_0 + \lambda \hat{V} - z)^{-1}$$

with the contour C defined in Fig. 10.1.

We define a function $X_{zz'}(\alpha', \alpha)$ for a diagonal operator $\hat{A}$ by

$$[\hat{R}(z)\hat{A}\hat{R}(z')]_d |\alpha\rangle = |\alpha\rangle \int A(\alpha')\, d\alpha' X_{zz'}(\alpha', \alpha) \qquad (10.61)$$

where the suffix d indicates the diagonal part of the matrix in brackets. We then get

$$\tilde{W}(\alpha', \alpha; t) = -\frac{1}{(2\pi)^2} \oint_C dz \oint_C dz' e^{i(z-z')t} X_{zz'}(\alpha', \alpha) \quad (10.62)$$

$X_{zz'}(\alpha', \alpha)$ can be shown to satisfy the series

$$X_{zz'}(\alpha, \alpha_0) = D_z(\alpha) D_{z'}(\alpha) \delta(\alpha - \alpha_0)$$
$$+ \lambda^2 D_z(\alpha) D_{z'}(\alpha) \bigg[W_{zz'}(\alpha, \alpha_0) + \lambda^2 \int d\alpha_1 W_{zz'}(\alpha, \alpha_1)$$
$$\times D_z(\alpha_1) D_{z'}(\alpha_1) W_{zz'}(\alpha_1, \alpha_0) + \cdots \bigg] D_z(\alpha_0) D_{z'}(\alpha_0) \quad (10.63)$$

where

$$D_z(\alpha_1) = \langle \alpha_1 | \hat{D}(z) | \alpha_1 \rangle$$
$$\hat{D}(z) = (H_0 - \lambda^2 \hat{g}(z) - z)^{-1} \quad (10.64)$$

is the diagonal part of the resolvent $\hat{R}(z)$, with $\hat{g}(z)$ satisfying

$$\hat{g}(z) = \{\hat{V}\hat{D}(z)\hat{V} - \lambda \hat{V}\hat{D}(z)\hat{V}\hat{D}(z)\hat{V} + \cdots\}_{id} \quad (10.65)$$

where the suffix id stands for the irreducible diagonal part and implies that all intermediate states in the operator products must be taken differently from each other and from the initial state. $W_{zz'}(\alpha, \alpha_0)$ in eqn (10.63) is then defined by

$$\{(V - \lambda \hat{V}\hat{D}(z)\hat{V} + \cdots)\hat{A}(\hat{V} - \lambda \hat{V}\hat{D}(z')\hat{V} + \cdots)\}_{id} |\alpha\rangle$$
$$= |\alpha\rangle \int d\alpha' A(\alpha') W_{zz'}(\alpha', \alpha) \quad (10.66)$$

One can check that

$$g_z(\alpha) - g_{z'}(\alpha) = -i \int d\alpha' \tilde{W}_{zz'}(\alpha', \alpha) \quad (10.67)$$

with

$$\tilde{W}_{zz'}(\alpha', \alpha) = i[D_z(\alpha') - D_{z'}(\alpha')] W_{zz'}(\alpha', \alpha) \quad (10.68)$$

With eqn (10.64) we find from eqn (10.67) that

$$(z - z') D_z(\alpha) D_{z'}(\alpha) = D_z(\alpha) - D_{z'}(\alpha)$$
$$- i\lambda^2 \int d\alpha' \tilde{W}_{zz'}(\alpha', \alpha) D_z(\alpha) D_{z'}(\alpha) \quad (10.69)$$

and furthermore that

$$(z - z')X_{zz'}(\alpha, \alpha_0) = [D_z(\alpha) - D_{z'}(\alpha)] \delta(\alpha - \alpha_0)$$
$$- i\lambda^2 \int \tilde{W}_{zz'}(\alpha, \alpha') \, d\alpha' X_{zz'}(\alpha', \alpha_0)$$
$$- i\lambda^2 \int d\alpha' \tilde{W}_{zz'}(\alpha', \alpha) X_{zz'}(\alpha, \alpha_0) \quad (10.70)$$

If we then identify the partial transition probability at energy E by

$$P_E(\alpha, \alpha_0; t) = (2\pi^2)^{-1} \frac{t}{|t|} \oint_C dz \, e^{2izt} X_{E+z, E-z}(\alpha, \alpha_0) \quad (10.71)$$

and define

$$w_E(\alpha', \alpha; t) = (2\pi^2)^{-1} \oint_C dz \, e^{2izt} \tilde{W}_{E+z, E-z}(\alpha', \alpha)$$

$$f_E(\alpha; t) = i(2\pi^2)^{-1} \frac{t}{|t|} \oint_C dz \, e^{2izt} [D_{E+z}(\alpha) - D_{E-z}(\alpha)] \quad (10.72)$$

we find that $P_E(\alpha, \alpha_0; t)$ satisfies the generalized master equation (10.59) with the initial conditions

$$P_E(\alpha, \alpha_0; t = 0) = 0 \quad (10.73)$$

The above derivation and thus van Hove's generalized master equation do not hold in any isolated system but only if very special conditions are met. Foremost, the entire analysis is based on the existence of a special orthonormal set of states α which are eigenfunctions of H_0 for an infinite system; i.e. they represent free quasiparticles or plane-wave excitations. On this basis, matrix elements (10.49) are supposed to exhibit diagonal singularities. Next, phase-dependent terms are dropped on the grounds that they will vanish for macroscopic times if the initial amplitudes $c(\alpha)$ have incoherent phases. It is hoped that systems dominated by H_0, such as moderately dense gases and nearly harmonic solids, are described by the generalized master equation (10.59). However, its validity must be doubted for liquids (van Hove, 1959) in which the concept of quasi-particles is questionable and a description in terms of single-particle properties is at least not exhaustive.

In contrast to Pauli's master equation (10.2), the generalized master equation (10.59) is inhomogeneous due to the presence of the term $f_E(\alpha; t)$ and nonmarkoffian because $P_E(\alpha, \alpha_0; t)$ depends, via the time integrations on the right-hand side of (10.59), on the whole previous history of the system. Van Hove (1957) has shown that his master equation reduces to Pauli's in the weak-coupling limit, i.e. $\lambda \to 0$ when $t \to \infty$. To perform this limit we first observe that for $t \gg \tau_0$, $f_E(\alpha; t)$ and $w_E(\alpha, \alpha'; t)$ tend to zero uniformly for small λ. But because the integrals in eqn (10.59) will be small for $t \lesssim \tau_0$, we obtain for the initial time evolution

$$\frac{d}{dt} P_E(\alpha, \alpha_0; t) \approx f_E(\alpha; t) \delta(\alpha - \alpha_0) \qquad (10.74)$$

neglecting terms of order λ^2. For times $t \gg \tau_0$ on the other hand, the inhomogeneous term becomes negligible and

$$\frac{d}{dt} P_E \sim \frac{P_E}{\tau_1} \sim \lambda^2 w_E P_E \qquad (10.75)$$

defining a (large) time constant τ_1 controlling the slow, macroscopic evolution of the system. If this is so, we can write for $t \sim \tau_1$

$$\int_0^t dt' w_E(\alpha, \alpha'; t-t') P_E(\alpha', \alpha_0; t')$$

$$= \int_0^t dt_1 w_E(\alpha, \alpha'; t_1) P_E(\alpha', \alpha_0; t - t_1)$$

$$\approx P_E(\alpha', \alpha_0; t) \int_0^\infty dt_1 w_E(\alpha, \alpha'; t_1) \quad (10.76)$$

because the effective range of integration is controlled by w_E, which vanishes after a very short time $\tau_0 \ll \tau_1$ over which P_E has not changed much. Equation (10.75) is correct to order $\tau_0/\tau_1 \sim \lambda^2$. For consistency we therefore must expand f_E and w_E to order λ and obtain for $P(\alpha; t)$, eqn (10.57),

$$\frac{dP}{dt}(\alpha; t) = 2\pi\lambda^2 \int d\alpha' \, \delta[\varepsilon(\alpha) - \varepsilon(\alpha')] \bar{W}^{(1)}(\alpha, \alpha') P(\alpha'; t)$$

$$- 2\pi\lambda^2 \int d\alpha' \, \delta[\varepsilon(\alpha') - \varepsilon(\alpha)] \bar{W}^{(1)}(\alpha', \alpha) P(\alpha; t) \quad (10.77)$$

where $\varepsilon(\alpha)$ are the eigenvalues of $\hat{H}_0$ for the state $|\alpha\rangle$, and $\bar{W}^{(1)}$ is a matrix element of first order in λ.

Equation (10.77) is, of course, Pauli's markoffian master equation. According to this derivation, it is valid in an infinite system (continuous spectrum of $\hat{H}_0$; i.e. the thermodynamic limit is taken in van Hove's theory right at the beginning, for times $t \sim \tau_1 \sim \lambda^{-2}$). If we want it to be valid for all times, we demand that

$$\lambda \to 0$$
$$t \to \infty$$

such that

$$\lambda^2 t = \text{const} \qquad (10.78)$$

This mathematical limit might be interpreted physically as follows. According to eqn (10.77) the system will evolve from an initial nonequilibrium state due to the presence of a weak perturbation $\lambda \hat{V}$. As the system settles for long times into a new stationary state, the effect of $\lambda \hat{V}$ will be less and less apparent and can therefore be discarded eventually. If this cannot be done, then λ and τ_1 will remain finite and a repeated random phase approximation has to be invoked at time intervals $\tau_1, 2\tau_1, 3\tau_1, \ldots$, in the spirit of van Kampen's derivation of Pauli's master equation presented in the previous section.

Applying the above theory to a system consisting of free electrons scattering off randomly distributed, static, elastic scattering centers, van Hove and Verboven (1961), and Janner, van Hove, and Verboven (1962) derived a simplified master equation that serves as the starting point for the calculation of the electrical conductivity of the system presented in Section 9.4.2.

10.4. Prigogine's Approach to Nonequilibrium Statistical Mechanics

Since the mid-1950s Prigogine and his collaborators, the socalled Brussels school, have been engaged in setting up a general theory of nonequilibrium statistical mechanics.‡ The scope of this approach is very wide and ambitious, and we unfortunately have to restrict ourselves to an outline of the general principles, i.e. the derivation of a generalized master equation, and to a short discussion of hydrodynamic modes.

‡ A detailed exposition can be found in a monograph by Prigogine (1962) and in many conference reports. We rely mainly on Prigogine (1968, 1973). See also Balescu (1975).

The starting point for the description of the time evolution of a large N-body system is, of course, again the Liouville equation (7.6) if the particle dynamics are approximately classical or the von Neumann equation (8.7) if the system shows explicit quantum behavior. Both descriptions can be unified through the Liouville–von Neumann equation

$$i\frac{\partial \rho}{\partial t} = \hat{L}\rho \qquad (10.79)$$

where

$$\hat{L}\rho = -i\{H, \rho\} \qquad (10.80)$$

is the Poisson bracket in the classical case, with ρ being the N-body distribution function in Γ-space, and

$$\hat{L}\rho = [H, \rho] \qquad (10.81)$$

is the commutator in the quantum-mechanical case with ρ the density matrix. In both cases $\hat{L}$ is a hermitian operator (or better a superoperator because it acts on the operator ρ), i.e. $\hat{L}^+ = \hat{L}$. Equation (10.79) is obviously Lt-invariant; i.e. it remains unchanged under the simultaneous transformations

$$\hat{L} \to -\hat{L}$$
$$t \to -t \qquad (10.82)$$

The formal solution of eqn (10.80) reads

$$\rho(t) = e^{-i\hat{L}t}\rho(0) \qquad (10.83)$$

which we can express in terms of the resolvent of the evolution operator $\exp(-i\hat{L}t)$ as [see eqn (10.60)]

$$\rho(t) = \frac{1}{2\pi i}\oint_C dz\, e^{-izt}\frac{1}{\hat{L}-z}\rho(0) \qquad (10.84)$$

where the contour C is again chosen as in Fig. 10.1. This solution is retarded, appropriate for an initial value problem in which the time evolution of a system proceeds casually from a set of initial data. The Lt-invariance of eqn (10.80) is reflected in the fact that we can easily construct an advanced solution by choosing the contour C in eqn (10.84) to run parallel but below the real z axis and closing in the upper half of the z plane. The latter solution corresponds to a final value problem in which the system evolves causally toward a specified set of final data. Such a situation is usually not encountered in physics but is certainly possible.

The Brussels formalism can be most elegantly based on the decomposition of $\rho(t)$ into a sum of two components $\rho_0(t)$ and $\rho_c(t)$. Their choice depends, of course, on the particular physical system under study; e.g. in a homogeneous classical gas, $\rho_0(t)$ may be chosen to be the N-body momentum distribution function, in which case $\rho_c(t)$ will describe spatial correlations. In a quantum system one might choose $\rho_0(t)$ to be the diagonal part of the density matrix, with $\rho_c(t)$ its off-diagonal elements. Let us then perform such a decomposition of $\rho(t)$ using a pair of projection operators (Zwanzig, 1964) $\hat{P}$ and $\hat{Q}$ such that

$$\hat{P} + \hat{Q} = 1$$
$$\hat{P} = \hat{P}^\dagger = \hat{P}^2$$
$$\rho_0 = \hat{P}\rho$$
$$\rho_c = \hat{Q}\rho \quad (10.85)$$

To derive an equation for ρ_0, we write the resolvent as

$$\frac{1}{\hat{L}-z} = [\hat{P} + \hat{\zeta}(z)]\frac{1}{\hat{P}\hat{L}\hat{P} + \hat{\psi}(z) - z}$$
$$\times [\hat{P} + \hat{\mathscr{D}}(z)] + \frac{1}{\hat{Q}\hat{L}\hat{Q} - z}\hat{Q} \quad (10.86)$$

where we introduce the collision operator

$$\hat{\psi}(z) = -\hat{P}\hat{L}\hat{Q}\frac{1}{\hat{Q}\hat{L}\hat{Q}-z}\hat{Q}\hat{L}\hat{P} \quad (10.87)$$

the destruction operator

$$\hat{\mathscr{D}}(z) = -\hat{P}\hat{L}\hat{Q}\frac{1}{\hat{Q}\hat{L}\hat{Q}-z} \quad (10.88)$$

and the creation operator

$$\hat{\zeta}(z) = -\frac{1}{\hat{Q}\hat{L}\hat{Q}-z}\hat{Q}\hat{L}\hat{P} \quad (10.89)$$

Inserting eqn (10.86) into eqn (10.84), we obtain the Brussels master equation (sometimes also referred to as the Prigogine–Résibois master equation because of its precursor derived by

Prigogine and Résibois, 1961; see also Prigogine, 1962):

$$i\frac{\partial}{\partial t}\rho_0(t) + \hat{P}\hat{L}\hat{P}\rho_0(t)$$
$$= \int_0^t d\tau \hat{g}(t-\tau)\rho_0(\tau) + \frac{1}{2\pi i}\int_C dz e^{-izt}\hat{\mathcal{D}}(z)\rho_c(0) \quad (10.90)$$

where

$$\hat{g}(t) = \frac{1}{2\pi i}\int_C dz e^{-izt}\hat{\psi}(z) \quad (10.91)$$

It must be stressed that this master equation is still exact. No approximations have been introduced in its derivation from the Liouville–von Neumann equation. Several of its features are noteworthy. The last term in eqn (10.90) describes the influence of initial data about $\rho_c(0) = \hat{Q}\rho(0)$ at time $t = 0$ on the subsequent time evolution of the system. It can be assumed that in most systems their effect should disappear as time goes on, thus the name destruction operator for $\mathcal{D}(z)$. The time integration in the collision term on the right-hand side of eqn (10.90) indicates that the value of $\rho_0(t)$ at time t will depend on the past history of the system; the master equation is nonlocal in time, i.e. nonmarkoffian. This, of course, does not imply that ρ_0 depends on the complete history of the past evolution; rather the kernel $\hat{g}(t-\tau)$ will for most systems be such that it vanishes for large values of $(t-\tau)$. Let us then write the collision term in eqn (10.90) as

$$\int_0^t d\tau \hat{g}(t-\tau)\rho_0(\tau) = \int_0^t d\tau' \hat{g}(\tau')\rho_0(t-\tau') \quad (10.92)$$

and assume that $\hat{g}(\tau')$ is only nonzero for times $\tau' \lesssim \tau_0$. Assuming further that τ_0 is microscopic, i.e. so short that $\rho_0(t-\tau')$ does not change appreciably over times τ', we can write for times $t \gg \tau_0$

$$\int_0^t d\tau' \hat{g}(\tau')\rho_0(t-\tau') \approx \int_0^t d\tau' \hat{g}(\tau')\rho_0(t)$$
$$\approx \int_0^\infty d\tau' \hat{g}(\tau')\rho_0(t) \quad (10.93)$$

Also neglecting the destruction term for large times, we then obtain from (10.90)

$$i\frac{\partial \rho_0(t)}{\partial t} = \hat{\psi}(+i0)\rho_0(t)$$

$$\hat{\psi}(+i0) = \lim_{z \to +i\varepsilon} \hat{\psi}(z) \qquad (10.94)$$

This is a local kinetic equation of the Boltzmann type. To examine it further, note that from the definition of the collision operator eqn (10.87), it follows that

$$\hat{\psi}(x+iy) = -\hat{P}\hat{L}\hat{Q}\frac{\hat{Q}\hat{L}\hat{Q}-x}{(\hat{Q}\hat{L}\hat{Q}-x)^2+y^2}\hat{Q}\hat{L}\hat{P}$$

$$-i\hat{P}\hat{L}\hat{Q}\frac{y}{(\hat{Q}\hat{L}\hat{Q}-x)^2+y^2}\hat{Q}\hat{L}\hat{P} \qquad (10.95)$$

The second term has a definite sign, namely

$$-y\hat{P}\hat{L}\hat{Q}\frac{1}{(\hat{Q}\hat{L}\hat{Q}-x)^2+y^2}\hat{Q}\hat{L}\hat{P} < 0 \quad \text{for } y > 0$$

$$> 0 \quad \text{for } y < 0 \qquad (10.96)$$

and it vanishes in the limit $y \to 0$ if the spectrum of $\hat{L}$ is discrete, i.e. in a finite system. However, if we perform the thermodynamic limit $N \to \infty$ and $V \to \infty$ with $N/V = \text{const}$ before taking the limit $z \to +i0$, part of the spectrum of $\hat{L}$ will become continuous and eqn (10.96) may not vanish. Then we have

$$\hat{\psi}(+i0) \neq \hat{\psi}(-i0) \qquad (10.97)$$

which is called the dissipativity condition because the real part of $\hat{\psi}$ will lead to oscillatory solutions of (10.94), whereas an imaginary contribution in $\hat{\psi}(+i0)$ can lead to damped or growing solutions. In the general framework outlined so far this is, of course, all speculation. General proofs of such behavior are also missing, but these ideas have been tested in several models (Grecos and Mareschal, 1976; Grecos and Prigogine, 1972; Résibois and Mareschal, 1978).

The above framework has been applied to several explicit physical systems, e.g. in spin systems by Borckmans and Walgraef (1967), in open systems by Walgraef (1974) and Lugiato and Milani (1976), and in fluids; see Résibois and De Leener (1977) for a review. Recently, it has also been extended to a so-called theory of subdynamics (Prigogine *et al.*, 1973) for the study of ergodicity and the origin of irreversibility in statistical mechanics. These topics, however, go far beyond the scope of this book, and we want to be content here to outline the application of the Brussels master equation to the theory of hydrodynamic modes in a fluid.‡ In this endeavor, we will recover many results that we know from linear response theory, in particular those in Section 9.3, which will allow us to estimate the relative worth of the two approaches.

We will now be dealing with a classical fluid. Its hamiltonian is given by

$$H(\mathbf{r}_1, \ldots, \mathbf{r}_N, \mathbf{p}_1, \ldots, \mathbf{p}_N) = \sum_{k=1}^{N} \frac{\mathbf{p}_k^2}{2m} + \sum_{k<l} V(|\mathbf{r}_k - \mathbf{r}_l|) \tag{10.98}$$

where $\mathbf{p}_k$ is the momentum and $\mathbf{r}_k$ the position of the kth particle. The Liouville operator is then given by

$$L = -i \sum_{k=1}^{N} \mathbf{v}_k \cdot \frac{\partial}{\partial \mathbf{r}_k} + \frac{i}{2m} \sum_{k \neq l} \frac{\partial V(|\mathbf{r}_k - \mathbf{r}_l|)}{\partial \mathbf{r}_k} \cdot \left(\frac{\partial}{\partial \mathbf{v}_k} - \frac{\partial}{\partial \mathbf{v}_l} \right) \tag{10.99}$$

where $\mathbf{v}_k = \mathbf{p}_k/m$ is the velocity of the kth particle. We know that in the hydrodynamic regime a fluid must be in local equilibrium. For simplicity let us assume that the fluid is, indeed, only slightly perturbed from a state of global (canonical) equilibrium which is given by

$$\rho_{eq} = e^{-H/k_B T}/Z$$

$$Z = \int d\mathbf{p}_1, \ldots, d\mathbf{p}_N \, d\mathbf{r}_1, \ldots, d\mathbf{r}_N e^{-H/k_B T} \tag{10.100}$$

For the one-particle velocity distribution function we write§

$$f(\mathbf{r}, \mathbf{v}, t) = \frac{N}{V} f^{(0)}(\mathbf{v})[1 + \phi(\mathbf{r}, \mathbf{v}, t)] \tag{10.101}$$

---

‡ We follow closely a set of lectures by Grecos (1977) and Prigogine and Grecos (1978). See also Résibois (1970) and Résibois and De Leener (1977).

§ This ansatz is in obvious analogy to what we have done to find the normal (i.e. hydrodynamic) solutions of the Boltzmann equation in Section 7.7. But note that there we were dealing with momentum rather than velocity distribution functions. See eqns (7.152) and (7.166).

where

$$f^{(0)}(\mathbf{v}) = \left(\frac{m}{2\pi k_B T}\right)^{\frac{3}{2}} e^{-mv^2/2k_B T} \tag{10.102}$$

is the Maxwell–Boltzmann velocity distribution function for a fluid in equilibrium. The perturbation $\phi(\mathbf{r}, \mathbf{v}, t)$ must satisfy $\phi \ll 1$ over the relevant range of velocities (see the discussion toward the end of Section 7.7).

The starting point in the Brussels theory of nonequilibrium statistical mechanics is the definition of an appropriate projection operator. For the present discussion of hydrodynamic modes in a fluid, the projection

$$\langle \mathbf{r}, \mathbf{v} | \hat{P}_\mathbf{q} | \mathbf{r}', \mathbf{v}' \rangle = \sum_{k=1}^{N} (\rho_{eq}/f^{(0)}(\mathbf{v}_k))$$
$$\times e^{i\mathbf{q}\cdot(\mathbf{r}_k-\mathbf{r}'_k)}[\delta(\mathbf{v}_k - \mathbf{v}'_k) - \frac{N}{V}f^{(0)}(\mathbf{v}'_k)(\delta_{\mathbf{q},0} + C_\mathbf{q}) \tag{10.103}$$

is appropriate where $C_\mathbf{q}$ is the Fourier transform of the direct (equilibrium) correlation function. (We omit a detailed definition of $C_\mathbf{q}$ as it will not be used below. For a discussion, see Egelstaff, 1967). With (10.103) we can construct explicitly the Brussels master equation (10.90) for $\rho_0 = \hat{P}_\mathbf{q}\rho$, which we integrate immediately over the positions of all particles and the velocities of all but one to obtain an equation for the Fourier transform of the single-particle distribution function (10.101)

$$f_\mathbf{q}(\mathbf{v}, t) = \int e^{i\mathbf{q}\cdot\mathbf{r}} f(\mathbf{r}, \mathbf{v}, t)\, d\mathbf{r} \tag{10.104}$$

which, after linearization, reads

$$\frac{\partial}{\partial t} f_q + i\mathbf{q}\cdot\mathbf{v} f_\mathbf{q} - i\mathbf{q}\cdot\mathbf{v} C_\mathbf{q} \frac{N}{V} f^{(0)}(\mathbf{v}) \int d\mathbf{v}' f_\mathbf{q}(\mathbf{v}', t)$$
$$= -i \int_0^t d\tau \int d\mathbf{v}'\, g_\mathbf{q}(\mathbf{v}, \mathbf{v}', t-\tau) f_\mathbf{q}(\mathbf{v}, \tau) + F_\mathbf{q}(t) \tag{10.105}$$

It is left to the reader to evaluate the kernel $g_\mathbf{q}$ and the source term $F_\mathbf{q}$ explicitly. Hydrodynamic modes must be solutions of (10.105) for long times showing an exponential time dependence like $\exp(-iz_q t)$. Dropping F_q for long times, expressing $g_\mathbf{q}$ in terms of the collision operator (10.94) and denoting the hydrodynamic or normal solutions of (10.105) by $\tilde{f}_q$ we obtain ($q = |\mathbf{q}|$)

$$\hat{M}_q(z_q)\tilde{f}_q = \mathbf{q} \cdot \mathbf{v}\tilde{f}_q - \frac{N}{V}\mathbf{q} \cdot \mathbf{v}C_\mathbf{q}f^{(0)}(\mathbf{v})\int d\mathbf{v}'\tilde{f}_q(\mathbf{v}')$$

$$+ \int d\mathbf{v}'\psi_q(\mathbf{v}, \mathbf{v}'; z_q)\tilde{f}_q(\mathbf{v}') = z_q\tilde{f}_q \quad (10.106)$$

as the eigenvalue problem for the hydrodynamic modes in a fluid where

$$\int d\mathbf{v}'\psi_q(\mathbf{v}, \mathbf{v}'; z_q)f_q(\mathbf{v}')$$

$$= -\int d\mathbf{v}_2, \ldots, d\mathbf{v}_N\, d\mathbf{r}_1, \ldots, d\mathbf{r}_N e^{i\mathbf{q}\cdot\mathbf{r}_1}\psi_\mathbf{q}(z_\mathbf{q})P_\mathbf{q}\rho \quad (10.107)$$

To solve for the latter, we note that the asymptotic scattering operator $\hat{\psi}_0(+i0) \equiv \hat{M}_0(+i0)$ must admit five collisional invariants (see Section 7.6)

$$\psi_0(+i0)|\alpha\rangle = 0 \quad \alpha = 1, \ldots, 5 \quad (10.108)$$

where

$$|1\rangle = 1$$

$$|\alpha\rangle = v_\alpha\sqrt{\frac{m}{k_BT}} \quad \alpha = 2, 3, 4$$

$$|5\rangle = \sqrt{\frac{2}{3}}\left(\frac{mv^2}{2\pi k_BT} - \frac{3}{2}\right) \quad (10.109)$$

We therefore project (10.106) into the five-dimensional space generated by the vectors $\{|\alpha\rangle\}$ by applying a projection $\hat{P}$ and $\hat{Q} = \hat{P}_q - \hat{P}$ to obtain

$$\hat{M}u_q \equiv \left[\hat{P}\hat{M}_q\hat{P} - \hat{P}\hat{M}_q\hat{Q}\frac{1}{\hat{Q}\hat{M}_q\hat{Q} - z_q}\hat{Q}\hat{M}_q\hat{P}\right]u_q = z_qu_q$$

$$(10.110)$$

where

$$\hat{P}f_q = u_q$$

$$\hat{Q}f_q = -\frac{1}{\hat{Q}\hat{M}_q\hat{Q} - z_q}\hat{Q}\hat{M}_q\hat{P}f_q \quad (10.111)$$

Next we expand the matrix elements $M_{\alpha\beta} = \langle \alpha | \hat{M}_{\mathbf{q}} | \beta \rangle$ up to second order in q and z_q and get

$$M_{\alpha\alpha} \approx \frac{\partial M_{\alpha\alpha}}{\partial q^2} q^2 \qquad \alpha = 2, 3, 4$$

$$M_{55} \approx \frac{\partial M_{55}}{\partial z_q} z_q + \frac{\partial^2 M_{55}}{\partial z_q^2} z_q^2 + \frac{\partial^2 M_{55}}{\partial q^2} q^2$$

$$M_{12} \approx \frac{\partial M_{12}}{\partial q} q$$

$$M_{21} = \frac{\partial M_{21}}{\partial q} q$$

$$M_{52} = \frac{\partial M_{52}}{\partial q} q + \frac{\partial^2 M_{52}}{\partial q \, \partial z_q} q z_q$$

$$M_{25} = \frac{\partial M_{25}}{\partial q} q + \frac{\partial^2 M_{25}}{\partial q \, \partial z_q} q z_q \qquad (10.112)$$

the derivatives being taken at $|\mathbf{q}| = z_q = 0$. But for hydrodynamic modes, we must also demand that

$$z_q \approx A \, |\mathbf{q}| + B \, |\mathbf{q}|^2 \qquad (10.113)$$

and can calculate A and B for the five modes from

$$\det |M_{\alpha\beta}(|\mathbf{q}|, z)| = 0 \qquad (10.114)$$

we obtain

$$\left.\begin{array}{l} z_1 = c \, |\mathbf{q}| - i\Gamma \, |\mathbf{q}|^2 \\ z_2 = -c \, |\mathbf{q}| - i\Gamma \, |\mathbf{q}|^2 \end{array}\right\} \quad \text{sound modes}$$
$$z_3 = z_4 = -i\nu \, |\mathbf{q}|^2 \qquad \text{shear mode}$$
$$z_5 = -i\kappa \, |\mathbf{q}|^2 \qquad \text{thermal mode} \qquad (10.115)$$

The sound modes are, of course, sound waves propagating with the speed of sound c and damped with a damping constant Γ. The shear and thermal modes are diffusive, with ν being the kinematic shear viscosity and κ the thermal diffusivity (see Table

6.1). From eqn (10.113) we obtain

$$c^2 = -\frac{\partial M_{12}}{\partial |\mathbf{q}|}\frac{\partial M_{21}}{\partial |\mathbf{q}|} - \frac{\partial M_{25}}{\partial |\mathbf{q}|}\frac{\partial M_{52}}{\partial |\mathbf{q}|}\left(1 - \frac{\partial M_{55}}{\partial z}\right)^{-1}$$

$$\nu = \frac{\partial^2 M_{33}}{\partial |\mathbf{q}|^2} = \frac{\partial^2 M_{44}}{\partial |\mathbf{q}|^2}$$

$$\kappa = \frac{\partial^2 M_{55}}{\partial |\mathbf{q}|^2}\frac{\partial M_{12}}{\partial |\mathbf{q}|}\frac{\partial M_{21}}{\partial |\mathbf{q}|} c^{-2}\left(1 - \frac{\partial M_{55}}{\partial z}\right)^{-1} \qquad (10.116)$$

expressing the linear transport coefficients in terms of derivatives of the collision operator. It can be shown (Résibois, 1972) that they are identical to those obtained in linear response theory (see Section 9.3). However, the theory presented here, although less straightforward than linear response theory, is obviously capable of an extension to systems far away from global equilibrium as it is, indeed, not a perturbation theory but the starting point of a rigorous theory of nonequilibrium statistical mechanics.‡

---

‡ For further elaboration of the Brussels theory, we refer the reader to Prigogine (1962), Balescu (1975), and Résibois and De Leener (1977).

11
Irreversibility and the Approach to Equilibrium

11.1. Defining the Problem

IN 1897 it appeared to Planck that it was the principal task of theoretical physics to explain irreversible processes in terms of conservative interactions. Although we have mentioned irreversibility and its sources repeatedly in this book, the concept is so fundamental to thermodynamics (and so fascinating to the mind) that we must devote a chapter to its study. We begin by a careful statement of the problem.

At the level of a macroscopic description within the framework of nonequilibrium thermodynamics we call, according to Planck‡ a process irreversible if 'it can in no way be completely reversed..., all other processes (are called) reversible. That a process may be irreversible, it is not sufficient that it cannot be directly reversed. This is the case with many mechanical processes which are not irreversible. The full requirement is, that it be impossible, even with the assistance of all agents in nature, to restore everywhere the exact initial state when the process has once taken place... the generation of heat by friction, the expansion of a gas without the performance of external work and the absorption of external heat, the conduction of heat, etc., are irreversible processes.'

To put these ideas into more straightforward language, we recall that in taking a system from an equilibrium state 1 to an equilibrium state 2 its change in entropy satisfies the inequality

$$\Delta S_{12} \geq 0 \tag{11.1}$$

‡ This description of an irreversible process can be found in Planck's *Treatise on Thermodynamics*, 3rd ed., 1945, but can be traced back to the first edition of 1897.

with $\Delta S_{12} = 0$ if the process is reversible, and $\Delta S_{12} > 0$ if it is irreversible. Note that the initial and final states have to be equilibrium states because otherwise the concept of entropy is meaningless. If the two states are infinitesimally close to each other in thermodynamic phase space, then (11.1) reduces to its differential form

$$dS = dS_e + dS_i \geq 0 \qquad (11.2)$$

where dS_i is the change in entropy due to internal rearrangement of the system and dS_e is added to it from the surroundings. For open systems in local equilibrium, we have seen in Section 2.3 that the local entropy density must satisfy a balance equation

$$\frac{\partial \rho s}{\partial t} + \nabla \cdot \mathbf{j}_s = \sigma_s \qquad (11.3)$$

with $\sigma_s = 0$ if all processes considered are reversible and $\sigma_s > 0$ for irreversible processes. Equation (11.3) is obviously not invariant under reversal of time, because changing $t \to -t$ and reversing the direction of $\mathbf{j}_s$ would imply that σ_s has to change sign in order to keep eqn (11.3) form invariant in contradiction to the second law of thermodynamics which demands that $\sigma_s \geq 0$. We can therefore state an additional criterion for a process to be irreversible, namely that it is not invariant under the reversal of the direction of time; i.e. an irreversible cannot be made to run backwards by any means whatsoever.

It appears that all processes occurring in nature are irreversible, with reversible processes being a mere abstraction that, however, is of great theoretical importance for setting up equilibrium thermodynamics and can, indeed, also be realized quite accurately in experiments. The fact that irreversible processes are not invariant under the reversal of the direction of time brings us back to the opening statement of this chapter. We are faced with the contradiction that irreversible processes certainly do occur in a thermodynamically isolated system and must therefore be part of a macroscopic, i.e. thermodynamic, theory, whereas the same system, described at the microscopic level, is controlled, if isolated mechanically, by Hamilton's or Heisenberg's equations of motion for a collection of N interacting particles, both equations being invariant under the reversal of time.‡ This means that in a system of classical particles changing $t \to -t$ and reversing momenta $\mathbf{p} \to -\mathbf{p}$ or particle velocities $\mathbf{v} \to -\mathbf{v}$ results in another solution of Hamilton's equations and thus is also a possible motion of the system as long as the forces are conservative. The last proviso must be made to ensure microscopic energy conservation, implying that the sum of kinetic and potential energy of the N particles comprising the (isolated) system must remain constant, whereas at the macroscopic level we must include heat as a possible form of energy to ensure energy conservation. Connected with the phenomenon of heat, however, is the occurrence of irreversible processes.

Efforts to explain this apparent paradox between microscopic reversibility and macroscopic irreversibility and to understand the phenomenon 'heat' began in the mid-nineteenth century with the creation of kinetic theory and culminated in Boltzmann's work, particularly his $\mathcal{H}$ theorem. The gist of the kinetic argument is that any statement on the macroscopic behavior of a large system is, by necessity, a probabilistic one. To extract a macroscopic description from the microscopic theory a deliberate falsification of the dynamics of an interacting many-body system has to be made; the most successful one so far is, of course, Boltzmann's *Stosszahlansatz*.

Before we go into a detailed analysis of this idea and the developments since Boltzmann, we will briefly mention an attempt by Planck to explain irreversibility as a consequence of radiation damping. These ideas were put forward in a series of six papers (nine, including two precursors and a review article) between 1895 and 1901. All are entitled 'On Irreversible Radiation Processes.§ It appeared to Planck that the only irreversible process in nature that results directly from conservative forces is the radiation damping of an initially excited oscillator. After arguing about acoustic oscillators, Planck concentrates on an electromagnetic oscillator in a cavity enclosed by perfectly

‡ There is a slight difference between thermodynamical and mechanical isolation of a system. In the first case, we demand that the average energy remains constant with thermodynamic fluctuations around this value still possible, whereas in the second case we must insist that energy is strictly conserved.

§ A modern revival of these ideas may be seen in the absorber theory of radiation (Wheeler and Feynman, 1945), as reviewed by Pegg (1975).

reflecting walls.‡ He notices that an oscillator experiences damping due to two sources: (1) energy loss due to Joule's heating within the oscillator (e.g. a dipole antenna), mainly important at low frequencies, and (2) radiation of energy, dominating at high frequencies. The irreversibility of the radiation process due to Joule's heating Planck terms consumptive damping or consumptive irreversibility. It is obviously a secondary effect and should be traced back to the origin of ohmic resistance of a metal in the dynamics of electrons. Consumptive irreversibility always appears if the effects of a part of the system (e.g. a heat bath) are collectively described by energy consuming processes. Consumptive irreversibility is of utmost importance in open systems. In contrast, Planck terms the damping of his electromagnetic oscillator due to radiation conservative irreversibility as arising from conservative interactions alone. To explain conservative irreversibility is the fundamental task of theoretical physics, according to Planck.

The sequence of Planck's papers on irreversible radiation processes, although ultimately of little importance for the understanding of the origin of irreversibility, of course led to the creation of quantum theory in 1900. Apart from this they are, combined with Boltzmann's criticism, a fascinating part of the history of nonequilibrium physics which we will briefly relate here. In the first two papers of the series, Planck (1896, 1897) shows that a small electromagnetic dipole oscillator (without ohmic resistance) within a large container with perfectly reflecting walls acts (1) to make an incident plane electromagnetic wave more isotropic by emitting secondary waves into all directions,§ (2) to make an incident spherical wave with intensity fluctuations smoother through the time delay in the emission process, and (3) to cause a frequency redistribution. Planck interprets these three effects as showing that the system is approaching an

‡ In 1898 Planck identified this oscillator with atoms or molecules inside the cavity. His classical calculation of radiation damping is, of course, no longer acceptable but should be redone quantum-mechanically. For historical reasons we want to point out that Planck as early as 1899 identified a new universal constant, called b then, and calculated it to be $b = 6.885 \times 10^{-27}$ erg sec. He identified it in 1901 with h, which he had introduced in the same year.

§ This idea has been revived by Prigogine (1967) in his statistical theory of potential scattering.

equilibrium state with a distribution of the electromagnetic energy uniform in space and an appropriate frequency spectrum. Boltzmann (1897a) argued immediately that Planck's interpretation of his results was totally misleading because Maxwell's equations are time-reversal invariant, implying that a reversal of time, magnetic fields, and polarization at a given time will make the system retrace its previous evolution.‡ Moreover, Boltzmann points out, if the oscillator is free from ohmic losses, it is simply part of the ideally reflecting walls and nothing happens but specular reflection at its surfaces. In a second note, Boltzmann (1897b) underlines the usefulness of calculating the entropy of the radiation oscillator§ and stresses the importance in taking the thermodynamic limit on any mechanical or electromagnetic system before one can hope to recover thermodynamics. In his third paper of 1897, Planck (1897c) points out that his approximate solution is only valid if (1) the linear size of the oscillator is small compared to the resonance wavelength, (2) this wavelength is small compared to the cavity, and (3) the damping constant is small. This last statement implies that Planck's approximate solution must break down for large times. He therefore suggests decreasing the damping constant as time goes on; this is the precursor of the weak-coupling limit so important in the perturbation theories of the 1950s and 1960s (see Chapter 10). To avoid reflection at the boundaries, Planck (1901) suggests moving them to cosmic distances. One more important lesson was learned from Planck's papers, namely that his theory of radiation damping will not work for special, i.e. resonant, initial conditions. This argument occurs sixty years later in van Hove's theory (van Hove, 1958).

‡ But notice the significant difference in the complexity of the 'initial' conditions that have to be specified for the direct and the time-reversed evolution. In the first case we must specify an incoming plane wave, whereas in the time-reversed version we must give all the details of the scattered waves. To see that this difference in the complexity of initial data, already discussed by Einstein (1909) in the context of the undulatory theory of light, is not a trivial one, consider the evolution following the dropping of a stone into a lake and its time reversal after a finite time has evolved. See also Popper (1956, 1957) for this and similar examples.

§ Planck's efforts in this direction eventually led him to the discovery of the law of blackbody radiation and thus to the creation of quantum mechanics (Planck, 1900 and 1901).

In order not to leave an incorrect impression, we must stress that Planck's papers on irreversible radiation processes, although fascinating in their pursuit of an idea and very educating if taken together with Boltzmann's criticism, have been of little consequence to nonequilibrium statistical mechanics except for the introduction of a few new concepts.‡ With hindsight we must say that Planck's law of blackbody radiation, which he found in his desire to explain conservative irreversiblity, turns out to be a key example for consumptive irreversibility of an open system, caused by the emission of radiation through a pinhole in the cavity walls into outer space as demonstrated in the calculation of the electromagnetic entropy (Planck, 1901).

In the remainder of this chapter we will explore the mechanical theories of large interacting systems to find an explanation of thermodynamic irreversibility. In the next section we take another look at Boltzmann's theory and study his $\mathcal{H}$ theorem. Discussing the Loschmidt and Zermélo 'paradoxes,' we are led to an investigation of the thermodynamic limit, which we will undertake in Section 11.3 in some simple mechanical models, namely harmonically coupled oscillator systems. Section 11.4 will deal with the inclusion of irreversibility in ensemble theories and the connection with information theory. This is followed in Section 11.5 with a brief discussion of generalized $\mathcal{H}$ theorems.

11.2. Irreversibility in the Boltzmann Equation

11.2.1. The $\mathcal{H}$ Theorem

We have seen in Section 7.5 that the Boltzmann equation can be derived from the first member of the BBGKY hierarchy at the expense of an additional hypothesis, namely Boltzmann's *Stosszahlansatz*.§ It states that in a dilute gas in which triple

---

‡ In his popular lectures, Planck (1960) says that 'it was L. Boltzmann who, by the introduction of the atomic theory, explained the meaning of the second law and at the same time all irreversible processes.'

§ Boltzmann (1895a) had to defend his theory of gases against the accusation that the *Stosszahlansatz*, not being a mechanical model but a hypothesis based on probability theory, rendered it unacceptable as a 'true physical theory.' The pragmatic philosopher Boltzmann came to the fore in his reply when he first quoted Hertz (1892) as saying, 'The rigour of science requires, that we distinguish well the undraped figure of nature itself from the gay-coloured vesture with which

(*Continued overleaf*)

encounters can be neglected, two gas particles will enter a collision without any dynamical memory of previous encounters. They will, however, leave each other's interaction sphere with the full dynamical information of a two-body collision, which is then used in the reduction of the collision integral to the well-known Boltzmann form (see Section 7.5 for details). The *Stosszahlansatz* makes an explicit distinction between the events 'before a collision' and 'after a collision' and thus introduces an 'arrow of time' (Eddington, 1929).‡ Whereas the exact equations of motion of an N-body system, Hamilton's equations or Liouville's equation, are invariant under time-reversal, i.e. remain unchanged under the transformations

$$t \to t' = -t$$
$$\mathbf{v} \to \mathbf{v}' = -\mathbf{v} \quad (\text{or } \mathbf{p} \to -\mathbf{p}) \qquad (11.4)$$

this is no longer true for Boltzmann's equation, which reads after time reversal

$$\left[\frac{\partial}{\partial t'} + \frac{\mathbf{p}'}{m} \cdot \mathbf{\nabla}_r + m\mathbf{F}(\mathbf{r}) \cdot \mathbf{\nabla}_{p'}\right] f_1 = -\left(\frac{\partial f_1}{\partial t}\right)_{\text{coll}} \qquad (11.5)$$

with the sign of the collision integral changed. Thus if $f_1(\mathbf{r}, \mathbf{p}, t)$ is a solution of the Boltzmann equation (7.120), then $f_1(\mathbf{r}, -\mathbf{p}, -t)$ is not.

Irreversibility of the evolution of macroscopic systems expresses itself in two effects: (1) the dissipation of energy in an irreversible process and (2) in an isolated system, the eventual approach to equilibrium if given enough time. Equilibrium is a very special stationary state of the system, namely one that, being compatible with external constraints, is independent of the initial state from which the approach to equilibrium started and thus is independent of any portion of the previous evolution of the system before the time when it settled in equilibrium.

‡ A good study of the 'arrow of time' in the Kac ring model has been given by Coopersmith and Mandeville (1974a and b). Less exact contemplations on time can be found in a series of books by Fraser *et al.* (1972, 1975, 1978) and many other books of a similar nature. See also Gold (1967).

---

we clothe it at our pleasure.' Boltzmann (1895a) continues, 'But I think the predilection for nudity would be carried too far if we were to forego every hypothesis. Only we must not demand too much from hypotheses.... Every hypothesis must derive indubitable results from mechanically well-defined assumptions by mathematically correct methods.

We have seen in Section 7.7 that the normal solutions of the Boltzmann equation lead to energy-dissipating effects like heat conduction and viscosity. Let us also recall that in Section 7.6 we introduced a $\mathscr{H}$ quantity (adding a subscript B for Boltzmann)

$$\mathscr{H}_B(t) = \frac{N}{V} \int f_1(\mathbf{r}, \mathbf{p}, t) \ln\left[\frac{N}{V} f_1(\mathbf{r}, \mathbf{p}, t)\right] d^3r\, d^3p \tag{11.6}$$

in an attempt to find a balance equation for entropy. We had seen then that in a system isolated from the outside world at time $t = 0$ we obtain

$$\frac{d\mathscr{H}_B(t)}{dt} = \int \sigma_h(\mathbf{r}, t)\, d^3r \leq 0 \tag{11.7}$$

where σ_h is given in eqn (7.146). Moreover, $d\mathscr{H}_B(t)/dt = 0$ if and only if

$$f_1(\mathbf{r}, \mathbf{p}, t) = (2\pi m k_B T)^{-\frac{3}{2}} \exp\left(-\frac{(\mathbf{p} - m\mathbf{v})^2}{2m k_B T}\right) \tag{11.8}$$

is the equilibrium distribution function. This last fact should, indeed, be very surprising since the Boltzmann equation is a partial differential equation and as such poses an initial value problem. On the other hand, the nonlinearity in the collision integral has such a structure that in an isolated system it determines uniquely the final state of the system as well.

We still must clarify why we can take the (monotonically decaying) $\mathscr{H}_B$ quantity as a measure for the irreversibility of the time evolution. We notice first that if f_1 in eqn (11.6) is the equilibrium distribution function (11.8), then [see eqn (7.153)]

$$-k_B \mathscr{H}_B = S + \text{const} \tag{11.9}$$

Moreover, if the system is in a nonequilibrium state but still in local equilibrium, i.e. described by the normal solutions of the Boltzmann equation, then, with the help of eqns (7.202) and (7.203) we can identify $-k_B h(\mathbf{r}, t)$ as the local entropy density in the system. But for such systems the entropy balance equation is the appropriate differential statement of the second law of thermodynamics and $\mathscr{H}_B(t)$, indeed, contains global information on the irreversible behavior of the system. We must stress, as we did in the discussion of eqn (7.156), that it is generally inappropriate to call $-k_B h(\mathbf{r}, t)$ the local entropy density. It is simply a time-dependent quantity that in the final stages of the approach to equilibrium, namely after the conditions of local equilibrium have emerged, goes over into the local entropy density.

11.2.2. The Loschmidt Paradox: Time Reversal

When Boltzmann presented his kinetic equation and his $\mathcal{H}$ theorem in 1872, he immediately encountered strong criticism and opposition to his ideas. Two of the most important objections to his $\mathcal{H}$ theorem were in the form of 'paradoxes.' From our derivation of the Boltzmann equation, it will be obvious that both 'paradoxes' miss the essential idea in Boltzmann's kinetic theory, but still their discussion is fruitful.

The first paradox, attributable to Loschmidt (1876),‡ may be stated as follows. The laws of mechanics, which govern any isolated system, are time-reversal invariant in the sense that instantaneously reversing the velocities of all the particles results in the system retracing its motion back to its initial state. Thus, for any evolution of a system toward equilibrium, there is another possible motion which takes the system away from equilibrium. There then seems to be a direct contradiction between the microscopic reversibility of the dynamics of the system and the irreversibility contained in the Boltzmann equation.§

The consequences of velocity inversion for the time evolution of a system can be illustrated dramatically in computer simulations (Orban and Bellemans, 1967) and in spin-echo experiments

‡ The Loschmidt paradox was formulated precisely by Boltzmann himself in his rebuttal (Boltzmann, 1877).

§ The velocity inversion paradox was picked up later by Culverwell (1894), Burbury (1894), and Bryan (1894). In one of his rebuttals Boltzmann (1895a) says: 'It can never be proved from the equations of motion alone, that the minimum function $\mathcal{H}$ must always decrease. It can only be deduced from the laws of probability, that if the initial state is not specially arranged for a certain purpose, but haphazard governs freely, the probability that $\mathcal{H}$ decreases is always greater than that it increases.... If in (a given) state $\mathcal{H}$ is greater than $\mathcal{H}_{min}$ it will not be certain, but very probable, that $\mathcal{H}$ decreases and finally reaches not exactly but very nearly the value $\mathcal{H}_{min}$, and the same is true at all subsequent instants of time. If in an intermediate state we reverse all velocities, we get an exceptional case where $\mathcal{H}$ increases for a certain time and then decreases again. But the existence of such cases does not disprove our theorem. On the contrary, the theory of probability itself shows that the probability of such cases is not mathematically zero, only extremely small.'

(Hahn, 1950; Rhim, Pines, and Waugh, 1971). Using computer simulation method of molecular dynamics (for a review, see Alder, 1973), Orban and Bellemans followed the exact time evolution of a two-dimensional dilute gas of 100 hard disks in a periodic square box at a density of 0.04 of that of close packing. Their initial conditions at time $t = 0$ are that the molecules are on the vertices of a regular square network in the box with the absolute magnitude of their velocities all the same but with the initial direction of motion chosen at random. As time proceeds, collisions between the disks and with the walls tend to establish a maxwellian velocity distribution (Alder and Wainwright, 1960). This time evolution is globally followed by calculating $\mathcal{H}(t)$, eqn (11.6). It is found that $\mathcal{H}(t)$ reaches a stationary minimum (maximal entropy) with small fluctuations after about 200 collisions (Fig. 11.1). In a second and third run, respectively, the computer is stopped after 50 or 100 collisions and all velocities are reversed in sign. With this new set of initial conditions, the system retraces its previous evolution with $\mathcal{H}(t)$ increasing until the original state at time $t = 0$ is reached after which the system evolves as if the velocity inversion had not occurred.

This computer experiment shows that a dilute system with random initial conditions evolves according to Boltzmann's $\mathcal{H}$ theorem, i.e. it shows kinetic or Boltzmann behavior. If, however, we specify special initial conditions such as the ones generated by velocity inversion in which a great amount of dynamic information is included, then antikinetic or anti-Boltzmann situations can arise in which $\mathcal{H}(t)$ increases over a finite period of time. But note again the vast difference in complexity between the two sets of initial conditions. It is very striking that small errors introduced into the initial data of the velocity inversion experiment will prevent the system from retracing its history completely to the extent that if the errors become too large, the antikinetic behavior is suppressed more or less completely (Fig. 11.1).

Antikinetic behavior can be seen in spin-echo experiments, the main features of which we will outline briefly following Rhim *et al.* (1971). They show that by applying a suitable sequence of strong rf fields, a system of dipolar-coupled nuclear spins can be made to behave as though the sign of the dipolar hamiltonian representing the dipole-dipole coupling in a rigid lattice of like spins had been reversed and the system appears to develop

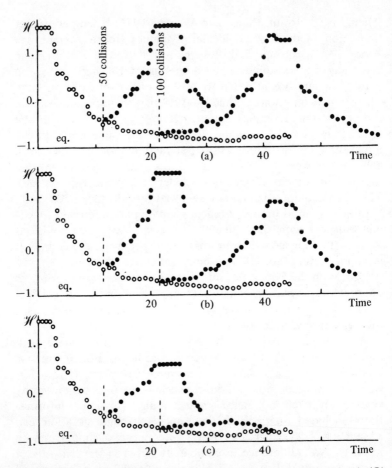

Fig. 11.1. Plot of $\mathcal{H}$ with time (in arbitrary units) showing the kinetic (○) and the antikinetic (●) evolutions for velocity inversions taking place at 50 or 100 collisions, with random errors 10^{-8}, 10^{-5}, and 10^{-2}, respectively. After Orban and Bellemans (1967).

backward in time. Let us then for times $t < 0$ polarize the spins in a strong Zeeman field $H_0 = \omega_0/\gamma$ in the z direction, applied sufficiently long so that the magnetization in the z direction reaches its equilibrium value appropriate for the lattice temperature T_s (Fig. 11.2a). The following discussion will take place in a reference frame rotating at frequency ω_0. At time $t = 0$, an rf

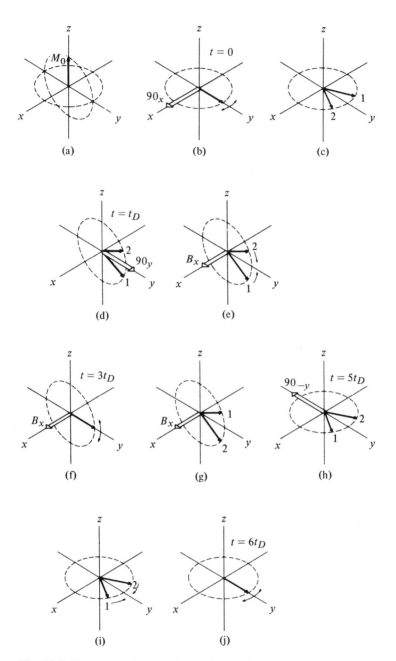

Fig. 11.2. Rotating-reference-frame description of the spin-echo experiment. After Rhim, Pines, and Waugh (1971).

pulse is applied along the x direction such that the magnetization nutates into the xy plane (Fig. 11.2b). The magnetization will start to decay with 'fast' spins processing clockwise and 'slow' spins anticlockwise about the z axis (Fig. 11.2c). After a time t_2, say, the magnetization will have decayed to zero, but this does not imply that the system is in complete equilibrium. Because for the latter to have occurred, we must wait a much longer time $t_1 \gg t_2$ for the spin and lattice degrees of freedom to interact, with the result that the spins are completely randomized.‡ If we therefore apply the right sequence of rf pulses to the system within a transient time $t \ll t_1$, we will be able to recover some or even most of the dynamic information of the nonequilibrium state. To do this, we apply at time $t = t_D$ an rf pulse in the y direction to rotate the already dephased spins into the yz plane (Fig. 11.2d). Another rf burst applied along the x axis will make the spins precess about the x axis with 'slow' and 'fast' spins having interchanged their sense of precession (Fig. 11.2e). As a result, at time $t = 3t_D$ they actually realign along the y axis (Fig. 11.2f). The spins now dephase about the x axis (Fig. 11.2g) and at $t = 5t_D$ are finally brought back into the xy plane by an rf pulse along the $-y$ axis (Fig. 11.2h). This is a situation similar to Fig. 11.2c, except that 'fast' and 'slow' spins are interchanged (Fig. 11.2i) so that they can actually rephase to form the echo (Fig. 11.2j) of the initial state. The experimental observation of this time reversal is shown in Fig. 11.3 for the spin echo in the ^{19}F transient magnetization in CaF_2.§

We should finally note the difference between the velocity inversion computer experiment and the time reversal spin-echo experiment. In the former we must employ a Maxwell demon that has control over all microscopic dynamical variables and can thus change the sign of all velocities. The resulting motion is antikinetic. In the spin-echo experiment, on the other hand, one uses a Loschmidt demon (Rhim et al., 1971) that, through a sequence of rf pulses, manages to reverse the apparent sign of the

‡ A review of the establishment of thermal equilibrium in paramagnetic crystals has been given by Gill (1975).

§ Note that the whole spin-echo sequence is a transient phenomenon that takes place over times of the order of the (magnetic) interaction time before any kinetic or hydrodynamic behavior sets in. We will return to transient phenomenon in Chapter 12.

IRREVERSIBILITY AND THE APPROACH TO EQUILIBRIUM

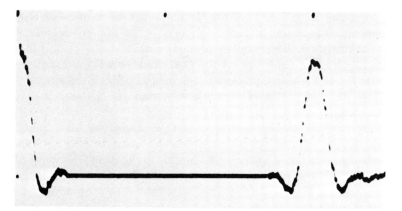

Fig. 11.3. Oscilloscope trace of the ^{19}F transient magnetization in CaF$_2$. Following the initial free induction decay, a time-reversing burst is applied at the noise-free section of the trace and an echo then appears, $t_D = 88$ μsec. Total length of the pulse sequence of Fig. 11.2 is $t_B = 350$ μsec. After Rhim, Pines, and Waugh (1971).

dipole-dipole interaction as viewed from the rotating frame of reference, thus forcing the system to retrace its history.

As implied by the choice of terminology, antikinetic or anti-Boltzmann behavior in a velocity inversion or time reversal experiment cannot be covered by solutions of the Boltzmann equation.‡ The reason is as follows: In order to perform velocity inversion or time reversal at a given time exactly, we must know the complete dynamics of the system necessitating a full solution of the BBGKY hierarchy of the N-body system. In an experiment, this is obviously done. But in the kinetic theory of Boltzmann, we truncate this hierarchy by invoking a *Stosszahlansatz* and thus approximate or, better, ignore all higher correlations in the system. An initial state prepared by velocity inversion in a real system, however, entails very intricate correlations that have been built up dynamically. Thus it cannot be accommodated by the

‡ Anitkinetic behavior has already been discussed by Boltzmann (1895). He says: 'Finally there is the difference between the ordinary cases, where $\mathcal{H}$ decreases or is near to its minimum value, and the very rare cases, where $\mathcal{H}$ is far from the minimum value and still increasing. In the last cases, $\mathcal{H}$ will reach, probably in a very short time, a maximum value. Then it will decrease from that value to the well-known minimum value.'

Boltzmann equation. The latter assumes that the initial state and, indeed, the complete time evolution is, to a high degree of approximation, described by a single-particle distribution function only. The resolution of the Loschmidt paradox is therefore that the Boltzmann equation, being an approximate kinetic equation and not the exact equation of motion of a dilute gas, is restricted to such systems in which the initial state is chaotic, i.e. shows no dynamical correlations. One might call such initial conditions 'natural' (Planck, 1898) in the sense that they are the ones to be found in systems which prior to the commencement of a controlled experiment (to be described as an initial value problem in kinetic theory) have been left alone long enough to settle in a stationary state or even in equilibrium compatible with the external constraints at that time.

11.2.3. The Zermélo Paradox: Poincaré Recurrence

A second objection against Boltzmann's kinetic theory was raised by Zermélo (1896) in the form of a recurrence paradox. Zermélo argued that there must be something fundamentally wrong with Boltzmann's equation because its irreversibility as expressed in the $\mathcal{H}$ theorem is in direct contradiction to Poincaré's recurrence theorem (Poincaré, 1892). The latter states that any isolated, finite, conservative system will, in a finite time, come arbitrarily close to its initial configuration. Thus, Zermélo argues $\mathcal{H}(t)$ cannot decrease monotonically but must eventually reverse its slope and approach $\mathcal{H}(0)$ in a finite time, the Poincaré recurrence time τ_P.

To find a satisfactory resolution of Zermélo's paradox, we must first inquire about the typical magnitude of τ_P.‡ Hemmer, Maximon, and Wergeland (1958) found an estimate for the linear chain of N classical mass points of mass m each coupled together with nearest neighbor harmonic forces (see Section 11.3). They first showed that the eigenvector of the jth normal mode

$$Z_j = p_j + im\omega_j q_j = a_j e^{i\omega_j t} \qquad (11.10)$$

rotates in the complex plane with the eigenfrequency ω_j of the jth mode. q_j is the jth normal coordinate and p_j its conjugate momentum. If we then demand that a specified state

$$Z_j(t_0) = |Z_j(t_0)| \, e^{i\phi_j^0} \qquad (11.11)$$

‡ Boltzmann (1896) gives a rather inconclusive estimate for a gas.

is recurring at time $t = \tau_p$ if in

$$Z_j(t) = |Z_j(t)| e^{i\phi_j} \tag{11.12}$$

the angular accuracy $|\phi_j - \phi_j^{(0)}| \leq \Delta\phi_j$ for $j = 1, \ldots, N$, then

$$\tau_P = \frac{\prod_{j=1}^{N-1} \frac{2\pi}{\Delta\phi_i}}{\sum_{j=1}^{N-1} \frac{\omega_j}{\Delta\phi_j}} \tag{11.13}$$

As an example take a linear chain of $N = 10$ atoms with the harmonic force constant such that the highest eigenfrequency is $\omega_0 \doteq 10 \text{ sec}^{-1}$. Setting $\Delta\phi = \pi/100$ we find approximately

$$\tau_P = 10^{10} \text{ years} \tag{11.14}$$

i.e. about the age of the universe.

A somewhat more general estimate for harmonically coupled oscillators in n dimensions was obtained by Mazur and Montroll (1960). They showed that for a given fluctuation, i.e. a relative deviation from the statistical mean of order $N^{-\frac{1}{2}}$, to recur takes a Poincaré time τ_P of the order of the mean period of normal modes. Thus in a Debye solid τ_P is of the order of the inverse Debye frequency, i.e. typically $\tau_P \sim 10^{-13} - 10^{-12}$ sec. It is noteworthy that this time is independent of the size of the system even though the magnitude of the (recurring) fluctuations, of course, is. For true nonequilibrium states, independent of N and outside the noise level, to reoccur, Mazur and Montroll (1960) found that

$$\tau_P \sim C^N \qquad C > 1 \tag{11.15}$$

Note that $\tau_P \to \infty$ as $N \to \infty$. There can, of course, be no doubt that Poincaré's theorem is right and that an isolated, finite, conservative system will, indeed return arbitrarily close to a given state in a finite time τ_p. However, Boltzmann (1896, 1897) argues far from being in contradiction to the kinetic theory, Poincaré's theorem clarifies the role and justifies the methods of probability theory used in kinetic theory. Two cases must be clearly distinguished. In the first situation we consider an isolated, finite (con-

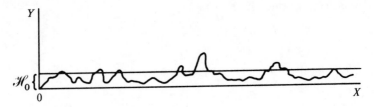

Fig. 11.4. The fluctuating $\mathcal{H}$ quantity for a finite system in equilibrium. Reproduced from Boltzmann's (1897) paper.

servative) system, a gas, say, a long time after it has been isolated. A macroscopic observation will find the system in equilibrium, i.e. in a state of maximal entropy. A microscopic probe will then find that the gas molecules have, on the average, velocities according to the Maxwell–Boltzmann distribution, eqn (7.152). However, this distribution is not completely stationary but small fluctuations of relative order $N^{-\frac{1}{2}}$ occur. They also manifest themselves in the fact that the $\mathcal{H}$ quantity is not constant and equal to $(-S_{\max}/k_B T)$ but fluctuates randomly as depicted schematically in Fig. 11.4, which is taken from Boltzmann (1897).

On the other hand, to describe the time evolution of an isolated, conservative system from an initial (chaotic) nonequilibrium state towards equilibrium, we are interested in changes in $\mathcal{H}_B$ that are large compared to the size of its equilibrium fluctuations. The latter, however, are a manifestation of the finiteness of the system and can be suppressed by taking the thermodynamic limit, i.e. by letting the volume V of the container and the number N of particles in it go to infinity, keeping the density (N/V) constant. Thus, for Boltzmann, expediency demands that to study time-dependent phenomena in equilibrium systems, i.e. fluctuations, we must consider a finite system but examine the equations of motion for very large times, whereas the study of the time evolution of nonequilibrium systems necessitates taking the thermodynamic limit first for finite times. The resulting model system is, however, no longer finite, and thus Poincaré's theorem does not apply and Zermélo's objection is pointless.

Let us summarize this discussion. All physical systems that are amenable to direct observation are finite.‡ Their complete time evolution is extremely complicated, particularly in the initial stages developing from a nonequilibrium state, and also after each Poincaré cycle because all the initial complications must occur in reversed order. Boltzmann's kinetic theory simplifies the initial stage by invoking the *Stosszahlansatz* (similar hypotheses are made in more general kinetic theories) which deliberately falsifies the dynamics over times of the order of the two-body interaction time. It must be stressed that this is achieved without introducing any errors in the predicted evolution over times of the order of the collision time and in the subsequent hydrodynamic evolution. This falsification of the dynamics must in a finite system result in drastic errors after a Poincaré time τ_P. But because for large systems τ_P is much larger than the time over which we observe a system and indeed, for most systems is even larger than the age of the universe, we should simply suppress the Poincaré recurrence in our theory. This is done by invoking the thermodynamic limit, with the resulting equations and their solutions being much simpler. The role of the thermodynamic limit is therefore to construct a simpler, asymptotic statistical theory for infinite systems in which the Poincaré recurrence is suppressed.

This discussion naturally leads us to the question of why irreversibility is connected, at least in that part of the universe accessible to observation, with a decrease of the $\mathcal{H}$ quantity or, more generally, why the second law of thermodynamics holds. We will quote Boltzmann (1895): 'I will conclude this paper with an idea of my old assistant, Dr. Schuetz. We assume that the whole universe is, and rests for ever, in thermal equilibrium (characterized by the 3K background radiation?). The probability that one (only one) part of the universe is in a certain state, is the smaller the further this state is from thermal equilibrium; but this probability is greater, the greater the universe itself is. If we assume the universe great enough we can make the probability of one relatively small part being in any given state (however far from the state of thermal equilibrium), as great as we please. We can also make the probability great that, though the whole universe is in thermal equilibrium, our world is in its present

---

‡ With this statement we exclude the total universe as a thermodynamic system and are discharged from the duty to discuss any cosmological speculations.

state. It may be said that the world is so far from thermal equilibrium that we cannot imagine the improbability of such a state. But can we imagine, on the other side, how small a part of the whole universe this world is? Assuming the universe great enough, the probability that such a small part of it as our world should be in its present state, is no longer small. If this assumption were correct, our world would return more and more to thermal equilibrium; but because the whole universe is so great, it might be probable that at some future time some other world might deviate as far from thermal equilibrium as our world does at present. Then the aforementioned $\mathscr{H}$ curve would form a representation of what takes place in the universe. The summits of the curve would represent the worlds where visible motion and life exist.' We add the prosaic comment that quantification of these ideas is difficult.

We conclude this part of the discussion of irreversibility by saying that in an isolated, finite system irreversible behavior is observed, apart from possible antikinetic transient, over times short compared to the Poincaré recurrence time as the most probable evolution of a large system. We note that no matter how great care is taken, isolation of a system from the rest of the world is never complete (Blatt, 1959). But then exchange of energy with the surroundings of the system is possible. Our system becomes at least marginally open and thus consumptive by being coupled to the practically infinite heat supply of the sun for the case of thermodynamic systems in our solar system. But any open system brought in thermal contact with an equilibrium heat reservoir with a substantially bigger heat capacity will, if otherwise left alone, irreversibly approach a state characterized by the equilibrium properties of the reservoir.

We close this paragraph with another quotation from Boltzmann (1896): 'All objections raised against the mechanistic viewpoint of nature are therefore pointless and are based on errors. Those, however, who cannot overcome the difficulties inherent in getting a clear understanding of the gas kinetic theory, should follow Mr. Zermélo's advice and give it up altogether.'‡

‡ An interesting collection of quotations on the current divergent views of the origin of irreversibility and time's arrow has been given by P. T. Landsberg in Fraser et al. (1972).

11.3. Irreversibility in Systems of Coupled Oscillators

11.3.1. The Thermodynamic Limit in the Classical Harmonic Chain

In the previous section, we gave estimates of Poincaré recurrence times in n-dimensional systems of harmonically coupled mass points. Apart from being the standard models for crystalline solids, these systems have been of great use in statistical mechanics both for the study of the ergodic problem‡ and for the demonstration of non-Boltzmann behavior in systems out of equilibrium (see Prigogine, 1967).

We consider a (one-dimensional) chain of N classical mass points of mass m, distributed in equilibrium at regular intervals a distance a apart. Calling the nonequilibrium deviation of the nth mass point from its equilibrium position ξ_n and assuming only nearest-neighbor harmonic interactions for $|\xi_n| \ll a$, the equations of motion are

$$m\frac{d^2\xi_n}{dt^2} = m\ddot{\xi}_n = f(\xi_{n+1} - \xi_n) - f(\xi_n - \xi_{n-1})$$

$$n = 1, 2, \ldots, N \quad (11.16)$$

They can be solved with arbitrary boundary conditions as a finite Fourier series

$$\xi_n(t) = \frac{2}{N+1} \sum_{n'=1}^{N} \sin\left(\frac{n'n}{N+1}\pi\right) \left\{ \cos\left[\nu t \sin\left(\frac{n'}{N+1}\frac{\pi}{2}\right)\right] \right.$$
$$\times \sum_{\kappa=1}^{N} \xi_\kappa(0) \sin\left(\frac{\kappa n}{N+1}\pi\right) + \frac{1}{\nu \sin\frac{n'\pi}{2(N+1)}}$$
$$\left. \times \sin\left[\nu t \sin\left(\frac{n'}{N+1}\frac{\pi}{2}\right)\right] \sum_{\kappa=1}^{N} \dot{\xi}_\kappa(0) \sin\left(\frac{\kappa n}{N+1}\pi\right) \right\} \quad (11.17)$$

Here, $\xi_\kappa(0)$ and $\dot{\xi}_\kappa(0)$ are the initial displacement and velocity of the κth particle at time $t = 0$ and $\nu^2 = 4f/m$. Note that the frequencies $\nu \sin[n'\pi/2(N+1)]$ are incommensurate. The system can therefore never return completely to its initial state but can only come arbitrarily close to it if we wait long enough. We depict the solution (11.17) in Fig. 11.5 for a chain of $N = 11$

‡ For a recent contribution, see Vigfússon (1976). Earlier work is quoted there.

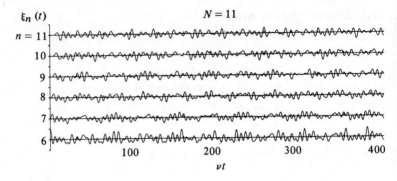

Fig. 11.5. Time-dependent displacement of the nth atom in a harmonic chain of $N = 11$ atoms in which the center atom ($n = 6$) has been displaced at time $t = 0$ by one length unit.

atoms with the initial conditions that only the atom in the middle of the chain is displaced at $t = 0$ by an amount

$$\xi_{(N+1)/2}(t = 0) = \varepsilon$$
$$\xi_n(t = 0) = 0 \quad \text{for all other } n$$
$$\dot{\xi}_n(t = 0) = 0 \quad \text{for all } n \quad (11.18)$$

This atom will respond with damped oscillations, giving its energy to its neighbors until the echo signal returns in a time $t_r = 2n/\nu$ after reflection at the ends of the chain. From then on, the response looks quite chaotic although it is, of course, completely causal. We know from the estimate of Hemmer, Maximom, and Wergeland (1958) that for

$$\omega_0 = 2\pi\nu = 10 \text{ sec}^{-1} \quad (11.19)$$

it takes the system about 10^{10} years to return (approximately) to the initial state. To exhibit the initial oscillatory decay of a finite harmonic chain more clearly, we show in Fig. 11.6 the response of a chain of 101 atoms for the initial conditions (11.8). The initially excited center atom again shows damped oscillations which could be fitted very well by a Bessel function of zero order (see 11.23) until the signal returns after reflection at the ends of the chain. Moreover, the nth atom responds significantly only after a time $t \sim 2[n - (N+1)/2]/\nu$ needed by the excitation to travel a distance $d = [n - (N+1)/2]a$ at the speed of sound $a\nu/2$.

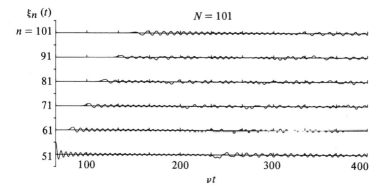

Fig. 11.6. Time-dependent displacement of the nth atom in a harmonic chain of $N = 101$ atoms in which the center atom ($n = 51$) has been displaced at time $t = 0$ by one length unit.

To avoid the echo and the Poincaré recurrence, we will now take the thermodynamic limit; i.e., we let the length of the chain L and the number N of mass points go to infinity with the spacing $a = L/N$ kept constant. Schrödinger (1914) studied this system in his attempt to understand the wavelike motion of a continuous elastic medium in terms of its atomic dynamics. Rather than solving this problem by Fourier methods, Schrödinger introduced new variables (the index n now runs more conveniently from $n = -\infty$ to $n = +\infty$)

$$x_{2n} = \sqrt{m}\, \frac{d\xi_n}{dt}$$

and

$$x_{2n+1} = \sqrt{f}(\xi_n - \xi_{n+1}) \qquad (11.20)$$

and finds the equation of motion

$$\frac{dx_n}{dt} = \frac{-\nu}{2}(x_{n+1} - x_{n-1}) \qquad (11.21)$$

where $\nu = 2\sqrt{f/m}$. These are recognized as the recurrence relations for Bessel functions. The general solution is given by

$$x_n(t) = \sum_{k=-\infty}^{\infty} x_k^0 J_{n-k}(\nu t) \qquad n = -\infty, \ldots, \infty \qquad (11.22)$$

with x_k^0 being the predetermined initial conditions. Let us specialize them according to (11.18). The system's response is then given by

$$\xi_n(t) = \varepsilon J_{2n}(\nu t)$$

$$\frac{d\xi_n(t)}{dt} = \frac{\varepsilon}{2}[J_{2n-1}(\nu t) - J_{2n+1}(\nu t)] \qquad (11.23)$$

A few observations are in order. First, all mass points start to move in the positive direction as soon as the mass at the origin is released. However, the nth point undergoes appreciable motion only after a time $t = 2n/\nu$ in which the original perturbation has travelled the distance a at the speed of sound $a\nu/2$. Secondly, the oscillations at a fixed mass point will decay for large times like

$$\xi_n(t) \xrightarrow[t \to \infty]{} \varepsilon \sqrt{\frac{2}{\pi \nu t}} \cos\left(\nu t - \frac{4n+1}{4}\pi\right) \qquad (11.24)$$

In Schrödinger's words, this is due to the 'damping influence of the neighbors... without any dissipative forces.' Also, notice that the maximum of the oscillations remains at the site of the original perturbation, which is similar to the propagation of a temperature perturbation in a linear chain but is quite unlike the normal propagation of sound. Finally, the linear chain of elastically coupled mass points will exhibit a wavelike motion similar to sound propagation in a continuous elastic medium, provided the set of initial conditions ξ_n^0 and $\dot{\xi}_n^0$ closely approximate continuous functions which vary slowly over distances large compared to the lattice constant a.

Let us stress that this model of a harmonic chain is completely deterministic. No statistical statements are made anywhere. It is a classical many-body system at temperature $T = 0$ and irreversibility—implying in this system simply the absence of a Poincaré recurrence— is due to the thermodynamic limit. A finite amount of energy, fed initially into the system by displacing the atom at the origin from its equilibrium position, is spread among infinitely many degrees of freedom as time goes on. Note that the equations of motion (11.16) are invariant under the transformation $t \to t' = -t$ even for the solution (11.23) of the infinite system, because

$$\xi_n(t') = \varepsilon J_{2n}(-\nu t) = \varepsilon J_{2n}(\nu t) = \xi_n(t)$$
$$\frac{d\xi_n(t')}{dt'} = \frac{d\xi_n(t)}{dt} \qquad (11.25)$$

so that the system shows the same damped oscillations as $t \to -\infty$ as it does as $t \to +\infty$.

Recently, a quantum-mechanical version of the infinite harmonic chain has been advanced by Razavy (1979). To quantize the classical system given by (11.16), we write its hamiltonian

$$H = \sum_{n=-\infty}^{+\infty} \frac{1}{2m} p_n^2 + \frac{1}{2} \sum_{n,k=-\infty}^{+\infty} \tfrac{1}{4} m\nu^2 (\xi_n - \xi_k)^2 \, \delta_{n,k+1} \qquad (11.26)$$

where p_j is the canonical momentum of the jth particle. The total energy of the classical system

$$E_c = \sum_{n=-\infty}^{+\infty} \tfrac{1}{2} m\dot{\xi}_n^2 + \frac{1}{2} \sum_{n,k=-\infty}^{\infty} \tfrac{1}{4} m\nu^2 (\xi_n - \xi_k)^2 \, \delta_{n,k+1} \qquad (11.27)$$

is a constant of the motion, and for the initial data (11.19) is given by

$$E_c = \tfrac{1}{2} m\nu^2 [\xi_0(t=0)]^2 \qquad (11.28)$$

Let us next observe that for the motion of the nth particle the terms ξ_{n-1} and ξ_{n+1} in (11.16) should be treated as time-dependent terms, given by (11.23), i.e.

$$\xi_{n\pm 1}(t) = \xi_0(t=0) J_{2n\pm 2}(\nu t) \qquad (11.29)$$

Denoting the canonical coordinate of the nth particle by q_n, its time-dependent hamiltonian reads

$$H_N = \frac{1}{2m} p_n^2 + \tfrac{1}{8} m\nu^2 [(q_n - \xi_{n-1})^2 + (q_n - \xi_{n+1})^2] \qquad (11.30)$$

for which the time-dependent Schrödinger equation reads

$$-\frac{\hbar^2}{2m} \frac{\partial^2 \psi_n}{\partial q_n^2} + \frac{1}{4} m\nu^2 \{q_n^2 - q_n[\xi_{n-1}(t) - \xi_{n+1}(t)] + \tfrac{1}{2}(\xi_{n+1}^2 - \xi_{n-1}^2)\} \psi_n$$
$$= i\hbar \frac{\partial}{\partial t} \psi_n \qquad (11.31)$$

To solve it we write

$$\psi_n(q_n, t) = \phi_n[q_n - \xi_n(t)] \exp\left[\frac{i}{\hbar}(mq_n\dot{\xi}_n(t) + C_n(t))\right] \quad (11.32)$$

where $C_n(t)$ is a solution of

$$\dot{C}_n(t) = \tfrac{1}{8}m\nu^2[2\xi_n^2(t) - \xi_{n+1}^2(t) - \xi_{n-1}^2(t)] - \tfrac{1}{2}m\dot{\xi}_n^2(t) - \varepsilon_k \quad (11.33)$$

and ϕ_n satisfies the wave equation for the harmonic oscillator

$$-\frac{\hbar^2}{2m}\frac{d^2\phi_n}{dy_n^2} + (\tfrac{1}{4}m\nu^2 y_n^2 - \varepsilon_n)\phi_n = 0 \quad (11.34)$$

where $y_n = q_n - \xi_n(t)$. The eigenvalues and normalized eigenfunctions are, with $\alpha^2 = m\nu/(2^{\frac{1}{2}}\hbar)$

$$\varepsilon_n^{(i_n)} = (i_n + \tfrac{1}{2})\hbar\nu/\sqrt{2}$$

$$\phi_n^{(i_n)} = \left(\frac{\alpha}{\pi^{\frac{1}{2}}2^{i_n}i_n!}\right)^{\frac{1}{2}} H_{i_n}(\alpha y_n) \exp(-\tfrac{1}{2}\alpha^2 y_n^2) \quad (11.35)$$

where $H_{i_n}(\alpha y_n)$ are hermitian polynomials. Since the total hamiltonian (11.26) is the sum of the single-particle hamiltonians H_n, eqn (11.30), we know that the total wavefunction is a product of single-particle wavefunctions. For the states (11.32), we calculate the expectation of the energy to be

$$\langle E_n \rangle = \int_{-\infty}^{\infty} \psi_n^{(i_n)*}\left(i\hbar\frac{\partial}{\partial t}\right)\psi_n^{(i_n)} dq_n$$
$$= \tfrac{1}{2}m\dot{\xi}_n^2(t) + \tfrac{1}{8}m\nu^2\{[\xi_n(t) - \xi_{n-1}(t)]^2 + [\xi_n(t) - \xi_{n+1}(t)]^2\} + \varepsilon_n^{(i_n)} \quad (11.36)$$

so that the energy of the nth particle is given by

$$e_n = \langle E_n \rangle - \tfrac{1}{8}m\nu^2[\xi_{n-1}^2(t) + \xi_{n+1}^2(t)] \quad (11.37)$$

The total energy of the quantal system is

$$\langle E \rangle = \sum_{k=-\infty}^{+\infty} \langle E_k \rangle = E_c + \sum_{k=-\infty}^{\infty} \varepsilon_n^{(i_n)} = \text{const} \quad (11.38)$$

From eqn (11.37) we find for the initially excited particle

$$e_0(t) = \tfrac{1}{4}m^2[\xi_0(t=0)]^2 \nu^2$$
$$\times \{J_0^2(\nu t) + 2[J_1^2(\nu t) - J_0(\nu t)J_2(\nu t)]\} + \varepsilon_0^{(i_0)} \quad (11.39)$$

so that its asymptotic energy loss is

$$\frac{de_0(t)}{dt} \xrightarrow[t \to \infty]{} m\nu^2[\xi_0(t=0)]^2 \frac{\cos 2\nu t}{2\pi t} \qquad (11.40)$$

That is, the energy transfer to the rest of the system shows damped oscillations. Again from eqn (11.39) it is evident that $e_0(-t) = e_0(t)$.

As the disturbance propagates through the system, each oscillator will have a nonzero probability of making a transition to an excited state. Because the Schrödinger equation (11.31) can be regarded as that of a forced harmonic oscillator with the time-dependent force given by

$$F(t) = \tfrac{1}{4}m\nu^2 \xi_0(t=0)[J_{2n-2}(\nu t) + J_{2n+2}(\nu t)] \qquad (11.41)$$

we find for the transition probability from the ith to the jth level in the nth oscillator (particle) at time t

$$p_n^{i \to j}(t) = \frac{i!}{j!}(\zeta_n)^{i-j} e^{-\zeta_n}[L_i^{j-i}(\zeta_n)]^2 \qquad j \geq i \qquad (11.42)$$

where L_i^{j-i} is the Laguerre polynomial and

$$\zeta_n(t) = (2^{\frac{1}{2}} m\hbar\nu)^{-1} \left\{ \int_0^t \tfrac{1}{4}m\nu^2 \xi_0(t=0)[J_{2n-2}(\nu t') \right.$$

$$\left. + J_{2n+2}(\nu t')] \cos\left(\frac{\nu t'}{2^{\frac{1}{2}}}\right) dt' \right\}^2 \qquad (11.43)$$

is the amount of energy (in units of $\hbar\nu/2^{\frac{1}{2}}$) stored at time t in a classical oscillator subject to a force $F(t)$. Observe that

$$\zeta_n(t) \xrightarrow[t \to \infty]{} 0 \qquad (11.44)$$

so that asymptotically there will be no transitions between different energy levels of this (or any other) oscillator.

Considering only the ground state, Razavy (1979) goes on to show that the probability density of the nth particle

$$\rho_n(q_n, t) = |\phi_n[q_n - \xi_n(t)]|^2$$

$$= \left(\frac{m\nu}{\pi\hbar 2^{\frac{1}{2}}}\right)^{\frac{1}{2}} \exp\left\{-\frac{m\nu}{2^{\frac{1}{2}}\hbar}[q_n - \xi_n(t)]^2\right\} \qquad (11.45)$$

satisfies a Fokker–Planck equation

$$\frac{\partial}{\partial t}\rho_n + \frac{\partial}{\partial q_n}(v_n\rho_n) - \frac{\hbar}{2m}\frac{\partial^2}{\partial q_n^2}\rho_n = 0 \qquad (11.46)$$

where

$$v_n(q_n, t) = -\frac{\nu}{2^{\frac{1}{2}}}[q_n - \xi_n(t)] + \dot{\xi}_n(t) \qquad (11.47)$$

The probability current density

$$j_n(q_n, t) = v_n\rho_n - \frac{\hbar}{2m}\frac{\partial \rho_n}{\partial q_n} = \dot{\xi}_n(t)\rho_n(q_n, t) \qquad (11.48)$$

turns out to be purely convective so that eqn (11.46) reduces to a simple continuity equation

$$\frac{\partial}{\partial t}\rho_n + \frac{\partial}{\partial q_n}j_n = 0 \qquad (11.49)$$

In closing, let us point out that by quantizing the classically known time-dependent (irreversible) motion of a subsystem (one atom in the chain) of a large (infinite) reversible quantum-mechanical system, Razavy (1979) has avoided all the difficulties usually encountered in the quantization of open systems (Razavy, 1977, 1978).

11.3.2. Time Evolution of a Reacting Two-Component Gas

In the previous section we looked at the time evolution of a finite and infinite harmonic chain, both classically and quantum mechanically, to shed more light on the thermodynamic limit as a means of ensuring irreversible behavior. We performed these calculations in chains at temperature $T = 0$. We now wish to study the time evolution in a more realistic system, namely a finite-temperature ideal gas mixture of two reacting components in which the two kinds of particles can be transformed into each other by an external field (Kreuzer and Nakamura, 1974). We will concentrate here on the following initial value problem: Assume that in the distant past a gas of component 1 was prepared in equilibrium at temperature T with mass density ρ_1^0. At time $t = 0$, switch on an external potential that can change molecules of kind 1 into kind 2 and vice versa. As a result, the density $\rho_1(t)$ will diminish at the expense of creating $\rho_2(t)$. It is this time evolution that we want to study in an exactly soluble model in which the external potential is taken to be spatially constant or diagonal in the momentum representation.

This two-component system in a volume L^3 (we reserve the symbol V for potential energy!) is described by a hamiltonian in creation and annihilation operator representation

$$H = H_0 + F(t)V$$

$$= \sum_{\mathbf{k}} (\varepsilon_{\mathbf{k}}^{(1)} a_{\mathbf{k}}^{(1)\dagger} a_{\mathbf{k}}^{(1)} + \varepsilon_{\mathbf{k}}^{(2)} a_{\mathbf{k}}^{(2)\dagger} a_{\mathbf{k}}^{(2)})$$

$$+ F(t)L^{-3} \sum_{\mathbf{k},\mathbf{k}'} V_{\mathbf{k}\mathbf{k}'}(a_{\mathbf{k}}^{(1)\dagger} a_{\mathbf{k}'}^{(2)} + a_{\mathbf{k}'}^{(2)\dagger} a_{\mathbf{k}}^{(1)}) \quad (11.50)$$

where the $\varepsilon_k^{(i)}$ are single-particle energies and $a_\mathbf{k}^{(i)}$ are annihilation operators for particles of kind i in momentum state $\mathbf{k}$. They obey either Fermi-Dirac or Bose-Einstein statistics. A physical situation where this hamiltonian is appropriate is a (gaseous) spin one-half system which, for times $t < 0$, was magnetized in a very strong external magnetic field that is removed at time $t = 0$ so that weak magnetic impurities, randomly distributed throughout the system, become effective scattering centers to readjust the magnetization (Abraham, Barouch, Gallavotti, and Martin-Löf, 1970). Another system is a gas of nitric oxide (NO) in which electromagnetic radiation is employed to excite the low-lying electronic excitation (creating particle 2) from NO molecules in their ground state (destroying particle 1). (See Kreuzer, 1976.)

Before we study the time evolution in the above model, we will derive the macroscopic balance equations for mass, momentum, and kinetic energy using the approach of Section 8.3. To this end we rewrite the hamiltonian (11.50) in coordinate space as

$$H = T_1 + T_2 + F(t)V$$

$$= -\frac{\hbar^2}{2m_1} \int \psi_1^\dagger(\mathbf{x}, t) \nabla^2 \psi_1(\mathbf{x}, t) \, d^3x$$

$$- \frac{\hbar^2}{2m_2} \int \psi_2^\dagger(\mathbf{x}, t) \nabla^2 \psi_2(\mathbf{x}, t) \, d^3x$$

$$+ F(t) \int d^3x V(\mathbf{x})[\psi_1^\dagger(\mathbf{x}, t)\psi_2(\mathbf{x}, t) + \psi_2^\dagger(\mathbf{x}, t)\psi_1(\mathbf{x}, t)] \quad (11.51)$$

where $\psi_i(\mathbf{x}, t)$ are field operators. The coupled Heisenberg equations of motion read ($\partial_t = \partial/\partial t$)

$$i\hbar\, \partial_t \psi_1(\mathbf{x}, t) = -\left(\frac{\hbar^2}{2m_1}\right)\nabla^2 \psi_1(\mathbf{x}, t) + F(t)V(\mathbf{x})\psi_2(\mathbf{x}, t) \quad (11.52a)$$

$$i\hbar\, \partial_t \psi_2(\mathbf{x}, t) = -\left(\frac{\hbar^2}{2m_2}\right)\nabla^2 \psi_2(\mathbf{x}, t) + F(t)V(\mathbf{x})\psi_2(\mathbf{x}, t) \quad (11.52b)$$

Using the retarded free Green Function

$$G_R(\mathbf{x}, t; \mathbf{x}', t) = \theta(t-t')\frac{1}{i\hbar}\left(\frac{im}{2\pi\hbar(t-t')}\right)^{\frac{3}{2}}\exp\left(\frac{im}{\hbar}\frac{(\mathbf{x}-\mathbf{x}')^2}{t-t'}\right)$$

(11.53)

we solve (11.52b) formally

$$\psi_2(\mathbf{x}, t) = \psi_2^{\text{in}}(\mathbf{x}, t) + \int dt' F(t') \int d^3x'\, G_R(\mathbf{x}, t; \mathbf{x}', t') V(\mathbf{x}')\psi_1(\mathbf{x}', t')$$

(11.54)

and substitute into eqn (11.52a)

$$i\hbar\, \partial_t \psi_1(\mathbf{x}, t) = -\left(\frac{\hbar^2}{2m_1}\right)\nabla^2 \psi_1(\mathbf{x}, t)$$

$$+ F(t)V(\mathbf{x}) \int dt' \int d^3x'\, G_R(\mathbf{x}, t; \mathbf{x}', t')$$

$$\times F(t')V(\mathbf{x}')\psi_1(\mathbf{x}', t') + F(t)V(\mathbf{x})\psi_2^{\text{in}}(\mathbf{x}, t) \quad (11.55)$$

The term $\psi_2^{\text{in}}(\mathbf{x}, t)$ reflects the presence of particles 2 at times $t \to -\infty$ and will be dropped hereafter; i.e., we assume that in the remote past only particles 1 were present which gradually, as the interaction is switched on, are transformed into particles 2 so that at large times $t \to +\infty$ a mixture of particles 1 and 2 is present.

From eqn (11.55) we obtain in the usual fashion (see Section 8.3) the equation of motion for the operator density

$$i\hbar\, \partial_t [\psi_1^\dagger(\mathbf{x}, t)\psi_1(\mathbf{x}, t)] = -\left(\frac{\hbar^2}{2m_1}\right)[\psi_1^\dagger(\mathbf{x}, t)\nabla^2\psi_1^\dagger(\mathbf{x}, t)$$
$$- \nabla^2\psi_1^\dagger(\mathbf{x}, t)\psi_1(\mathbf{x}, t)]$$
$$+ V(\mathbf{x})F(t)\int dt' F(t') \int d^3x'\, V(\mathbf{x}')$$
$$\times [G_R(\mathbf{x}, t; \mathbf{x}', t')\psi_1^\dagger(\mathbf{x}, t)\psi_1(\mathbf{x}', t')$$
$$- G_R^*(\mathbf{x}, t; \mathbf{x}', t')\psi_1^\dagger(\mathbf{x}', t')\psi_1(\mathbf{x}, t)] \quad (11.56)$$

Let us introduce the macroscopic mass density for particles 1

$$\rho_1(\mathbf{x}, t) = m_1 Tr[\psi_1^\dagger(\mathbf{x}, t)\psi_1(\mathbf{x}, t)\hat{\rho}] \quad (11.57)$$

and its nonlocal generalization

$$\langle\psi_1^\dagger(\mathbf{x}, t)\psi_1(\mathbf{x}', t')\rangle = Tr[\psi_1^\dagger(\mathbf{x}, t)\psi_1(\mathbf{x}', t')\hat{\rho}] \quad (11.58)$$

where $\hat{\rho}$ is the statistical operator (density matrix) that at this stage need not be specified. From eqn (11.57) we then obtain formally the equation of continuity (see Section 8.3)

$$\partial_t \rho_1(\mathbf{x}, t) + \boldsymbol{\nabla} \cdot \mathbf{j}_1(\mathbf{x}, t)$$
$$= F(t)\left(\frac{2m_1}{\hbar}\right)V(\mathbf{x})\int dt' F(t') \int d^3x' V(\mathbf{x}')$$
$$\times \text{Im}\,[G_R(\mathbf{x}, t; \mathbf{x}', t')\langle\psi_1^\dagger(\mathbf{x}, t)\psi_1(\mathbf{x}', t')\rangle] \quad (11.59)$$

where Im denotes the imaginary part and the mass current is given by

$$\mathbf{j}_1(\mathbf{x}, t) = \left(\frac{\hbar}{2i}\right)\langle\psi_1^\dagger(\mathbf{x}, t)\boldsymbol{\nabla}\psi_1(\mathbf{x}, t) - (\boldsymbol{\nabla}\psi_1^\dagger(\mathbf{x}, t))\psi(\mathbf{x}, t)\rangle \quad (11.60)$$

The sink on the right-hand side of eqn (11.59) reflects the fact that particles 1 are transformed into particles 2 by the field $V(\mathbf{x})$.‡ The corresponding mass balance for particles 2 would show a source term of similar structure. It should be noted that these production terms are strongly nonlocal in time depending on the whole previous history of the system starting from the time the interaction was switched on. This fact alone already suggests that the approach to equilibrium will be nonexponential, as will be shown explicitly below.

For completeness let us write the momentum balance for

---

‡ This procedure (commonly referred to as projection operator techniques—see also Section 10.4) of eliminating all degrees of freedom but those of immediate interest in a particular aspect of the problem is widely used in the theory of open systems. The result is usually a set of balance equations with explicit sinks and sources which are, in general, nonlocal in space and time, i.e., which show memory effects. The most common models are for lasers (see Haken, 1975, for a review) in which the infinitely many degrees of freedom of the photon field coupled to a set of two-level atoms is eliminated except for the resonant mode in order to arrive at macroscopic rate equations of the Langevin type. The precursor of these models is, of course, the Wigner–Weisskopf model (Weisskopf and Wigner, 1930). For a complete discussion, see Davidson and Kozak (1975).

component 1, suppressing the subscript 1 from now on and writing $\partial_l = \partial/\partial x_l$ and $\partial'_k = \partial/\partial x'_k$,

$$\partial_t \mathbf{j}_k(\mathbf{x}, t) - \left(\frac{\hbar}{m_1}\right) \partial_l T_{kl}(\mathbf{x}, t)$$
$$= F(t)\Big([\partial_k V(\mathbf{x})] \int d^3x' \, dt' F(t') V(\mathbf{x}')$$
$$\times \text{Re}\, [G_R(\mathbf{x}, t; \mathbf{x}', t)\langle \psi^\dagger(\mathbf{x}, t)\psi(\mathbf{x}', t')\rangle]$$
$$+ V(\mathbf{x}) \int d^3x' \, dt' F(t') V(\mathbf{x}')$$
$$\times \text{Re}\, [G_R(\mathbf{x}, t; \mathbf{x}', t')(\partial_k - \partial'_k)\langle \psi^\dagger(\mathbf{x}, t)\psi(\mathbf{x}', t')\rangle]\Big) \quad (11.61)$$

where the kinetic-energy tensor is given by eqn (8.68)

$$T_{kl}(\mathbf{x}, t) = \tfrac{1}{4}\langle [\partial_k \, \partial_l \psi^\dagger(\mathbf{x}, t)]\psi(\mathbf{x}, t) - [\partial_k \psi^\dagger(\mathbf{x}, t)] \partial_l \psi(\mathbf{x}, t)$$
$$- [\partial_l \psi^\dagger(\mathbf{x}, t)] \partial_k \psi(\mathbf{x}, t) + \psi^\dagger(\mathbf{x}, t) \partial_k \, \partial_l \psi(\mathbf{x}, t)\rangle \quad (11.62)$$

and is itself subject to the balance equation

$$\partial_t T_{kl}(\mathbf{x}, t) + \left(\frac{\hbar}{m_1}\right) \partial_t F_{ikl}(\mathbf{x}, t)$$
$$= -\left(\frac{1}{2\hbar}\right) F(t)[\partial_k \, \partial_l V(\mathbf{x})] \int d^3x' \, dt' F(t') V(x')$$
$$\times \text{Im}\, [G_R(\mathbf{x}, t; \mathbf{x}', t')\langle \psi^\dagger(\mathbf{x}, t)\psi(\mathbf{x}', t')\rangle]$$
$$- \left(\frac{1}{2\hbar}\right) F(t) V(\mathbf{x}) \int d^3x' \, dt' F(t') V(\mathbf{x}')$$
$$\times \text{Im}\, [G_R(\mathbf{x}, t; \mathbf{x}', t')(\partial_k - \partial'_k)(\partial_l - \partial'_l)\langle \psi^\dagger(\mathbf{x}, t)\psi(\mathbf{x}', t')\rangle]$$
$$\quad (11.63)$$

with F_{ikl} defined in eqn (8.70). We want to stress again the time nonlocality in the source terms of these balance equations. Further work with these equations at the macroscopic level would demand some approximations on these nonlocalities. We will dismiss this matter here and proceed with the detailed calculation of the time evolution of the system as controlled by the hamiltonian (11.50) for an interaction potential $L^{-3}V_{\mathbf{kk'}} = V_0 \, \delta_{\mathbf{kk'}}$. We compute the time-dependent mass density

$$\rho_i(t) = m_i \sum_{\mathbf{k}} n_{\mathbf{k}}^{(i)}(t) \quad (11.64)$$

where

$$n_{\mathbf{k}}^{(i)}(t) = \frac{Tr[a_{\mathbf{k}}^{(i)\dagger}(t) a_{\mathbf{k}}^{(i)}(t) e^{-\beta H_0}]}{Tr e^{-\beta H_0}} \quad (11.65)$$

Here $a_{\mathbf{k}}^{(i)}(t)$ are solutions of the Heisenberg equations of motion

$$i\hbar \, \partial_t a_{\mathbf{k}}^{(1)} = [a_{\mathbf{k}}^{(1)}, H] = \varepsilon_{\mathbf{k}}^{(1)} a_{\mathbf{k}}^{(1)} + F(t) V_0 a_{\mathbf{k}}^{(2)}$$
$$i\hbar \, \partial_t a_{\mathbf{k}}^{(2)} = [a_{\mathbf{k}}^{(2)}, H] = \varepsilon_{\mathbf{k}}^{(2)} a_{\mathbf{k}}^{(2)} + F(t) V_0 a_{\mathbf{k}}^{(1)} \quad (11.66)$$

and are formally given by

$$a_{\mathbf{k}}^{(i)}(t) = e^{iHt/\hbar} a_{\mathbf{k}}^{(i)}(0) e^{-iHt/\hbar} \quad (11.67)$$

To evaluate these expressions we diagonalize the hamiltonian (11.50) by an orthogonal transformation to quasiparticle operators

$$\alpha_{\mathbf{k}}^{(1)} = \cos \theta_{\mathbf{k}} a_{\mathbf{k}}^{(1)} + \sin \theta_{\mathbf{k}} a_{\mathbf{k}}^{(2)}$$
$$\alpha_{\mathbf{k}}^{(2)} = -\sin \theta_{\mathbf{k}} a_{\mathbf{k}}^{(1)} + \cos \theta_{\mathbf{k}} a_{\mathbf{k}}^{(2)} \quad (11.68)$$

Choosing

$$\tan 2\theta_{\mathbf{k}} = \frac{2V_0}{\varepsilon_{\mathbf{k}}^{(1)} - \varepsilon_{\mathbf{k}}^{(2)}} \quad (11.69)$$

transforms the hamiltonian (11.50) for $t > 0$ into

$$H = \sum_{\mathbf{k}} (\lambda_{\mathbf{k}}^{(1)} \alpha_{\mathbf{k}}^{(1)\dagger} \alpha_{\mathbf{k}}^{(1)} + \lambda_{\mathbf{k}}^{(2)} \alpha_{\mathbf{k}}^{(2)\dagger} \alpha_{\mathbf{k}}^{(2)}) \quad (11.70)$$

where the quasiparticle energies are given by

$$\lambda_{\mathbf{k}}^{(1)} = \tfrac{1}{2}(\varepsilon_{\mathbf{k}}^{(1)} + \varepsilon_{\mathbf{k}}^{(2)}) \mp [V_0^2 + \tfrac{1}{4}(\varepsilon_{\mathbf{k}}^{(1)} - \varepsilon_{\mathbf{k}}^{(2)})^2]^{\tfrac{1}{2}} \quad (11.71)$$

We then get

$$a_{\mathbf{k}}^{(1)}(t) = \cos \theta_{\mathbf{k}} e^{-i\lambda_{\mathbf{k}}^{(1)} t/\hbar} \alpha_{\mathbf{k}}^{(1)} - \sin \theta_{\mathbf{k}} e^{-i\lambda_{\mathbf{k}}^{(2)} t/\hbar} \alpha_{\mathbf{k}}^{(2)}$$
$$a_{\mathbf{k}}^{(2)}(t) = \sin \theta_{\mathbf{k}} e^{-i\lambda_{\mathbf{k}}^{(1)} t/\hbar} \alpha_{\mathbf{k}}^{(1)} + \cos \theta_{\mathbf{k}} e^{-i\lambda_{\mathbf{k}}^{(2)} t/\hbar} \alpha_{\mathbf{k}}^{(2)} \quad (11.72)$$

in which way we find, using the transformation (11.68) once more, the time evolution of $a_{\mathbf{k}}^{(i)}(t)$ in terms of $a_{\mathbf{k}}^{(i)}(t=0)$. But then notice that H_0 is diagonal in these latter operators so that the trace in (11.65) involves expressions

$$\frac{Tr[a_{\mathbf{k}}^{(i)\dagger}(t=0) a_{\mathbf{k}'}^{(j)}(t=0) e^{-\beta H_0}]}{Tr e^{-\beta H_0}} = \delta_{ij} \delta_{\mathbf{k}\mathbf{k}'} n_{\mathbf{k}}^{(i)}(t=0) \quad (11.73)$$

where

$$n_{\mathbf{k}}^{(i)}(t=0) = \frac{1}{\exp[-\beta_i(\varepsilon_{\mathbf{k}}^{(i)} - \mu^{(i)}) \pm 1]} \quad (11.74)$$

The plus (minus) sign must be taken if the particles obey Fermi-Dirac (Bose-Einstein) statistics. $T_i = (k_B \beta_i)^{-1}$ is the initial temperature of component i and $\mu^{(i)}$ is its (ideal gas) chemical potential.

To finally evaluate eqn (11.64) we take the thermodynamic limit in our system, i.e. we replace

$$\sum_{\mathbf{k}} \to \frac{L^3}{(2\pi)^3} \int d^3k \quad (11.75)$$

and find

$$\rho_1(t) = \rho_1(t=0) + \frac{m_1}{(2\pi)^2} \int_0^\infty k^2 \, dk \, \frac{V_0^2}{V_0^2 + \frac{1}{4}(\varepsilon_{\mathbf{k}}^{(1)} - \varepsilon_{\mathbf{k}}^{(2)})^2}$$
$$\times \{1 - \cos[(\lambda_{\mathbf{k}}^{(1)} - \lambda_{\mathbf{k}}^{(2)})t/\hbar]\}[n_{\mathbf{k}}^{(2)}(t=0) - n_{\mathbf{k}}^{(1)}(t=0)] \quad (11.76)$$

and a similar expression for the density $\rho_2(t)$ of the second component. This is the exact time evolution in a two-component system in which in an initial equilibrium system an external interaction is switched on at time $t=0$ that can transform the two species into each other.

We proceed with a detailed discussion of this result. In a first example, we assume that initially the two components of the gas mixture are at temperatures T_1 and T_2, respectively. Both are high enough to use Maxwell–Boltzmann statistics. Then, we have

$$n_k^{(1)}(0) = \exp\left(\beta_1 \mu^{(1)} - \beta_1 \frac{\hbar^2 k^2}{2m_1}\right)$$

and

$$n_k^{(2)}(0) = \exp\left[\beta_2 \mu^{(2)} - \beta_2\left(\frac{\hbar^2 k^2}{2m_2} + \varepsilon_2\right)\right] \quad (11.77)$$

where $\beta_i = 1/k_B T_i$. We have chosen $\varepsilon_{\mathbf{k}}^{(1)} = \hbar^2 k^2/2m_1$ and $\varepsilon_{\mathbf{k}}^{(2)} = \hbar^2 k^2/2m_2 + \varepsilon_2$ as those of free particles where ε_2 is the threshold energy for the formation of particles 2. The chemical potentials (per particles) are given by

$$\mu^{(i)} = -k_B T_i \ln\left(\frac{(2\pi m_i k_B T_i)^{\frac{3}{2}}}{\hbar^3} \frac{m_i}{\rho_i(0)}\right) \quad (11.78)$$

Introducing new variables

$$\mu = \frac{m_1 m_2}{m_2 - m_1} \qquad \delta = \frac{\mu}{m_1} \beta_1 V_0 \qquad \gamma = \frac{m_1 \beta_2}{m_2 \beta_1}$$

$$r = \frac{\varepsilon_2}{V_0} \qquad \tau = \frac{2 V_0 t}{\hbar} \qquad x^2 = \frac{\hbar^2 k^2}{2\mu V_0} \qquad (11.79)$$

we can rewrite eqn (11.76)

$$\frac{\rho_1(t)}{\rho_1(0) + \rho_2(0) m_1/m_2} = \frac{\rho_1(0)}{\rho_1(0) + \rho_2(0) m_1/m_2}$$

$$\times \left\{ 1 + \frac{2}{\sqrt{\pi}} \delta^{\frac{3}{2}} \int_0^\infty \frac{x^2}{1 + \frac{1}{4}(x^2 - r)^2} [1 - \cos \tau \sqrt{1 + \tfrac{1}{4}(x^2 - r)^2}] \right.$$

$$\left. \times \left(\gamma^{\frac{3}{2}} \frac{m_1}{m_2} \frac{\rho_2(0)}{\rho_1(0)} e^{-\gamma \delta x^2} - e^{-\delta x^2} \right) dx \right\} \qquad (11.80)$$

To obtain the general behavior for large times, we can apply Kelvin's stationary phase argument (see Sneddon, 1951; for a detailed discussion, see Kreuzer and Hiob, 1976). The endpoint contributions to the integral of eqn (11.80) are $0(\tau^{-\frac{3}{2}})$ since the integrand vanishes at $x = 0$ and $x = \infty$. Thus, to lowest order, we need only apply the stationary phase argument to the stationary point at $x = r^{\frac{1}{2}}$ and find, for large τ,

$$\frac{\rho_1(t)}{\rho_1(0) + \rho_2(0) \frac{m_1}{m_2}} = \frac{\rho_1(\infty)}{\rho_1(0) + \rho_2(0) \frac{m_1}{m_2}} - C \frac{\cos\left(\tau + \frac{\pi}{4}\right)}{\sqrt{\tau}}$$

$$(11.81)$$

where

$$C = \frac{\rho_1(0)}{\rho_1(0) + \rho_2(0) \frac{m_1}{m_2}} (8 r \delta^3)^{\frac{1}{2}} \left(\gamma^{\frac{3}{2}} \frac{\rho_2(0) m_1}{\rho_1(0) m_2} e^{-\gamma \delta r} - e^{-\delta r} \right)$$

$$(11.82)$$

Thus, oscillations with frequencies differing from 1 will die out faster than an oscillation with frequency $\omega = 1$. Therefore, asymptotically, the oscillations of a two channel system, with any initial conditions will always have a period $\pi(\hbar/V_0)$. Note that this infinite system of coupled oscillators shows a similar asymptotic behavior as the harmonic chain [see eqn (11.24)]. The fact that our system is now at a finite temperature only affects the amplitude C, whereas the frequency and the decay law are determined by the dynamics, i.e. by V_0.

For small τ, the behavior of eqn (11.80) can be quite different. The largest contribution to the time evolution then comes from the integration region where the nonoscillating factor of the integrand has its maximum. We define

$$g(x) = \tilde{g}_1(x) - \tilde{g}_2(x)$$

$$\tilde{g}_1(x) = \gamma^{\frac{3}{2}} \frac{\rho_2(0) m_1}{\rho_1(0) m_2} \frac{x^2}{1 + \frac{1}{4}(x^2 - r)^2} e^{-\gamma \delta x^2}$$

$$\tilde{g}_2(x) = \frac{x^2}{1 + \frac{1}{4}(x^2 - r)^2} e^{-\delta x^2} \qquad (11.83)$$

and set $\rho_2(0) = 0$. We have to look at two limiting cases. If δ is large, then the main contribution to the integral comes from the region around $x = \delta^{-\frac{1}{2}}$ where the factor

$$x^2 e^{-\delta x^2} \qquad (11.84)$$

attains its maximum value. Consequently, the oscillations of eqn (11.80) have frequency

$$\omega = \left[1 + \frac{1}{4} \left(\frac{1}{\delta} - r \right)^2 \right]^{\frac{1}{2}} \qquad (11.85)$$

for small times. These oscillations will decrease in amplitude, with the decrease being faster the further ω is from the asymptotic frequency $\omega = 1$.

On the other hand, if δ is small then $e^{-\delta x^2} \sim 1$ for x not too large, and the main contribution to the integral is from the region around $x = r^{\frac{1}{2}}$ where the factor

$$[1 + \frac{1}{4}(x^2 - r)^2]^{-1} \qquad (11.86)$$

attains its maximum value. Then, the oscillations of eqn (11.80) have frequency $\omega = 1$ for all times and will generally decay very slowly (like $t^{-\frac{1}{2}}$). This behavior is illustrated in Fig. 11.7.

Let us next study the limit $\delta \to 0$ in eqn (11.80) with $\rho_2(0)$ set equal to zero for simplicity. This corresponds to either the high temperature limit $T \to \infty$ or to the weak coupling limit $V_0 \to 0$. We find that the first minimum in the oscillations is not very

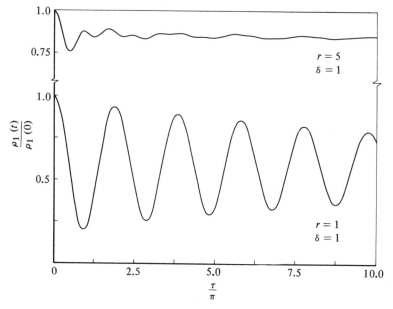

Fig. 11.7. Two-component system. Observe the fast decay of high-frequency oscillation in the upper graph and slow decay of the 'asymptotic' frequency ($\omega = 1$) oscillation in lower graph.

sensitive to the choice of r and occurs, for $r = 0$, at the first zero of the Bessel function $J_{\frac{1}{4}}(\tau)$, i.e. at $\tau \approx 2.405$. Thus, for finite temperatures, letting $\delta \to 0$ implies that $V_0 \to 0$ since $\delta = (\mu \beta_1 / m_1) V_0$ and then we have $r = (\varepsilon_0 / V_0) \to \infty$. Finally, since $\tau = 2t V_0 / \hbar = \text{const}$, we have $t \to \infty$, i.e. the weak coupling limit without any oscillations at finite times. However, the amplitude goes to zero concurrently. This indicates the limited value of the weak coupling limit as introduced formally and without justification in Section 10.3.

The above model of a reacting two-component ideal gas has been examined further by Kreuzer and Hiob (1976) who studied in detail the influence of various initial conditions. Looking at the importance of quantum statistics, they found that for a weakly degenerate Fermi–Dirac gas the oscillations in the response of the system to the external catalytic interaction are smaller. For a weakly degenerate Bose–Einstein gas, they are larger in amplitude than those for the classical Maxwell–Boltzmann statistics considered here. This can readily be understood as a consequence of the fact that the Pauli exclusion principle acts as a repulsion against further creation of particles, thus decreasing the effect of the external potential, whereas Bose–Einstein statistics favors a larger occupation in any given energy state, thus enhancing the transformation of particles in the external field. The generalization of this model to three- and four-component systems has also been given and examined by Kreuzer and Hiob (1976).

In closing, let us stress that systems of harmonically coupled oscillators will generally respond to an external perturbation in an oscillatory fashion; i.e., they will exhibit overstability in the sense of Eddington. This is due to the fact that no truly dissipative mechanisms are acting in the systems; irreversibility of the time evolution is simply due to the thermodynamic limit, i.e. due to the extension of the number of degrees of freedom of the system to infinity. One consequence of this overstability is the fact that Boltzmann's $\mathcal{H}$ quantity, as calculated for the subsystem consisting of particles 1 only, also shows damped oscillations (Kreuzer and Nakamura, 1974). In particular, it was found that for all times (1) $d\mathcal{H}(t)/dt \leq 0$, (2) $[\mathcal{H}_1(t) - \mathcal{H}_1(0)] \leq 0$, and (3) in the weak-coupling limit $\mathcal{H}_1(t)$ is a monotonic function. These findings are in agreement with all proven generalized $\mathcal{H}$ theorems (Jancel, 1969).

The lack of truly dissipative (many-body) effects in harmonically coupled oscillator systems has the further consequence that such systems will generally not approach a state of equilibrium as $t \to \infty$. Rather, their final state will usually depend on the initial state (Kreuzer and Hiob, 1976). Such models are therefore restricted to a study of transient effects in the earlier stages of the time response of a system to an external perturbation.

11.4. Irreversibility in an Ensemble

It should be obvious by now that the problem of irreversibility has many features. In particular, the thermodynamic limit can enter a statistical theory of nonequilibrium systems in various ways. So far we have taken the physical size of the system to infinity by performing the limits $N \to \infty$, $V \to \infty$ with $N/V = \text{const.}$

A second and more indirect way consists of constructing an ensemble theory explicitly. Recall that an ensemble is a collection of infinitely many replicates of the physical system under study. In such an ensemble theory, even a single-particle system will show (statistical) irreversibility. This has been demonstrated repeatedly in the following simple model:[‡] We consider a system consisting of just one classical point particle of mass m trapped inside a one-dimensional box of length L with perfectly reflecting walls. Of course, we are here not interested in the trajectory of a single such system, but rather in the statistical time evolution of an ensemble of such systems. At some time $t = 0$, we therefore specify statistical initial conditions through a distribution function $f(x, p, t = 0)$. The evolution of the system is then given through a time-dependent distribution function $f(x, p, t)$, which is a solution of the Liouville equation

$$\frac{\partial f}{\partial t} = \{H, f\} = -\frac{p}{m}\frac{\partial f}{\partial k} \quad (11.87)$$

subject to the prescribed initial and boundary conditions. The latter can easily be taken care of by replacing the original distribution function $f(x, p, 0)$ by an extended distribution function $\bar{f}(x, p, 0)$ satisfying periodic boundary conditions and defined as

$$\bar{f}(x, p, 0) = f(x, p, 0) \quad \text{for } 0 \leq x \leq L$$

and outside this interval

$$\bar{f}(-x, -p, 0) = \bar{f}(x, p, 0)$$

and

$$\bar{f}(x + L, p, 0) = f(x, p, 0) \quad (11.88)$$

In the new problem, the particles will move freely and the exact solution of the Liouville equation is

$$\bar{f}(x, p, t) = \bar{f}\left(x - \frac{pt}{m}, p, 0\right) \quad (11.89)$$

[‡] Our discussion follows the papers by Hobson (1966, 1968) and Hobson and Loomis (1968). See also Blatt (1959), Lee (1974), and Kreuzer and Teshima (1977). This simple model has been studied at numerous earlier occasions in the literature. No attempt is made here to cover its history. A nonequilibrium ensemble theory has also been formulated for the Jepsen (1965) model by Anstis, Green, and Hoffman (1973).

in terms of the initial conditions. To proceed with our analysis, we now assume that initially the position and velocity of the particle in the box were uncorrelated, i.e. that we can write

$$f(x, p, t = 0) = g(x, t = 0)h(p, t = 0) \tag{11.90}$$

We then find the distribution function to be

$$f(x, p, t) = h(p, 0)\left\{\frac{1}{L} + \sum_{n=1}^{\infty} a_n \cos\left[\frac{n\pi}{L}\left(x - \frac{pt}{m}\right)\right]\right\} \tag{11.91}$$

where

$$a_n = \frac{2}{L}\int_0^L g(x, 0) \cos\left(\frac{n\pi x}{L}\right) dx$$

Obviously, the momentum part of the initial distribution function is not affected by the free-particle Liouville equation because forces are required to mix momenta. However, strong correlations between position and momentum will build up as expected. Let us further simplify the problem by assuming that the momentum part of the initial distribution function was maxwellian

$$h(p, 0) = \left(\frac{\beta}{2\pi m}\right)^{\frac{1}{2}} \exp\left(-\frac{\beta p^2}{2m}\right) \tag{11.92}$$

and that the spatial distribution was given by

$$g(x, 0) = \frac{1}{L}\left[1 + \cos\left(\frac{2\pi x}{L}\right)\right] \tag{11.93}$$

With these choices, we obviously have

$$f(x, p, t) = h(p, 0)\left\{\frac{1}{L} + \frac{1}{L}\cos\left[\frac{2\pi}{L}\left(x - \frac{pt}{m}\right)\right]\right\} \tag{11.94}$$

It is instructive to calculate the reduced distribution functions

$$h(p, t) = \int_0^L f(x, p, t)\, dx = h(p, 0) \tag{11.95}$$

and

$$g(x, t) = \int_{-\infty}^{\infty} f(x, p, t)\, dp$$
$$= \frac{1}{L}\left[1 + \exp\left(-\frac{2\pi^2 t^2}{mL^2\beta}\right)\cos\left(\frac{2\pi}{L}x\right)\right] \tag{11.96}$$

Thus, the momentum distribution remains maxwellian and the position distribution relaxes monotonically to the final distribution

$$g(x, \infty) = \frac{1}{L} \qquad (11.97)$$

with a relaxation time $\tau_{\text{rel}} = (L/\pi)(\beta m/2)^{\frac{1}{2}}$. This relaxation is monotonic due to our particular choice of the initial momentum distribution. Also, observe that the final distribution is *not* the equilibrium state of the ensemble due to the strong correlations between momenta and positions. We can thus conclude that this simple model demonstrates that irreversibility can occur in the time evolution of an isolated, finite, microscopically reversible system if its initial state is only specified through a continuous statistical distribution function obeying Liouville's equation. However, for a system to relax to equilibrium, a mechanism is needed to overcome the build-up of correlations, most likely the residual interaction of the system with the rest of the world due to incomplete isolation (Blatt, 1959) or a proper coupling to a heat reservoir (Kreuzer and Teshima, 1977). Let us remark that the irreversibility in the above system is not in contradiction to the Poincaré recurrence theorem because the latter refers only to the motion of a single point in phase space and not to the evolution of a continuous distribution function.

We will use our results on this simple model to introduce some useful concepts from information theory (Hobson, 1966). The connection becomes clear with the observation that statistical mechanics is basically the study of mechanical systems about which not all possible information is available or desirable, while information theory is concerned with making the best possible predictions about large systems, consistent with given but limited information, without assuming anything more than the given information.

According to information theory, the quantity

$$\mathcal{H}_G(t) = \int \rho(\mathbf{x}_1, \ldots, \mathbf{x}_N, t)$$
$$\times \ln \rho(\mathbf{x}_1, \ldots, \mathbf{x}_N, t) \, d^6\mathbf{x}_1, \ldots, d^3\mathbf{x}_N \quad (11.98)$$

is a measure of the information about the system contained in ρ. In statistical mechanics, $\mathcal{H}_G(t)$ is known as the Gibbs fine-grained

$\mathcal{H}$ quantity. The distribution function ρ satisfies Liouville's equation

$$\frac{d\rho}{dt} = \frac{\partial \rho}{\partial t} - \{\rho, H\} = 0 \tag{11.99}$$

and is obviously a constant of motion. This implies that $\mathcal{H}_G(t)$ is a constant with the simple interpretation that, as long as the system is isolated and no new measurements are made, no information about the system is gained or lost.

In the example of a particle in a one-dimensional box, the qualitative behavior of the information function is clear. Initially, we have

$$h(p, 0) = \left(\frac{\beta}{2\pi m}\right)^{\frac{1}{2}} \exp - \left(\frac{\beta p^2}{2m}\right) \tag{11.100}$$

and

$$g(x, 0) = \frac{1}{L} \left[1 + \cos\left(\frac{2\pi x}{L}\right)\right] \tag{11.101}$$

If we define a momentum information function by

$$I_p(t) = \int_{-\infty}^{\infty} h(p, t) \ln h(p, t) \, dp \tag{11.102}$$

and a position information function by

$$I_x(t) = \int_0^L g(x, t) \ln g(x, t) \, dx \tag{11.103}$$

then the particular choice (11.100) for $h(p, 0)$ minimizes $I_p(t)$ but the information about the positions is not at a minimum since, for example, we know that there can be no particles at $x = L/2$ initially. As time progresses, $I_p(t)$ remains at its minimum value and the information $I_x(t)$ about position tends monotonically to its minimum value, corresponding to $g(x, \infty) = 1/L$. Since Liouville's theorem ensures that no information is lost, some other information must have been gained. Initially, there were no correlations between x and p but, as time progresses, correlations increase and all the lost information regarding position reappears as information about correlations between x and p. The fact that the final state still contains all the original information can be seen by considering the Gedanken experiment of Loschmidt.

Reversing all velocities will obviously cause the system to return to its initial state with all information intact. This implies that the system can never reach a true equilibrium in which all information about the initial conditions has been destroyed. However, if an observer is limited to certain types of measurements (e.g. he can measure $h(p, t)$ and $g(x, t)$ but not correlations), then the above system's behavior will appear to be a monotonic approach to equilibrium and the quasi-equilibrium distribution function $f(x, p, t)$, which includes all correlations, is indistinguishable from the true equilibrium distribution function. Irreversibility, then, is the 'flow' of information from relevant (because measurable) lower-order distribution functions to irrelevant (because unobtainable) information in higher-order correlations. Such an explanation of irreversibility is often justified by arguing that all physical measurements are actually averages over some finite time τ. Thus, $f(x, p, t)$ is not accessible to direct measurement but only such quantities as

$$f_{av}(x, p, t) = \frac{1}{\tau} \int_{t}^{t+\tau} f(x, p, t') \, dt' \qquad (11.104)$$

Since the correlations present in $f(x, p, t)$ oscillate rapidly with increasing p and t, such a physical averaging will smooth out and hide the effects of the correlations, rendering them inaccessible to measurements.

Starting from a nonequilibrium initial distribution, a system will rapidly evolve, by its own dynamics, to a quasi-equilibrium state in which the measured macroscopic parameters of the system have their equilibrium values (conservative irreversibility) but in which there remain some quantities which are from equilibrium. In practice, these far-from-equilibrium quantities will involve complex correlations and will seldom, if ever, be measured. Thus the distribution function of an isolated system will show irreversible evolution to an apparent equilibrium but, due to the inability of an isolated system to 'forget,' it is only through external interactions that true equilibrium can be reached (consumptive irreversibility). This will generally take longer than the time period for the attainment of the above quasi-equilibrium.

The manner in which a thermal interaction can allow a system to reach true equilibrium can easily be described in information-theoretical terms. It is consistent with the usual ideas of a thermal reservoir to assume that the interaction with the system does not change the properties of the reservoir in any measurable way.‡ Thus, information may be transferred to the reservoir, causing a net loss of information from the system. This may be regarded as a random thermal interaction which acts to destroy the fine-grained correlations present in an isolated system and allows it to evolve to a true equilibrium. It should be remembered that this interaction may be regarded as random only for times much less than the Poincaré recurrence time of the composite system; however, this time may be made as long as desired by considering a large thermal reservoir.

11.5. Generalized $\mathcal{H}$ Theorems

We have seen in Section 11.2 that in a dilute gas Boltzmann's $\mathcal{H}$ quantity, eqn (11.6) or (7.148),

$$\mathcal{H}_B(t) = \frac{N}{V} \oint f_1(\mathbf{r}, \mathbf{p}, t) \ln \frac{N}{V} f_1(\mathbf{r}, \mathbf{p}, t) \, d^3r \, d^3p \quad (11.105)$$

is a relevant measure for the irreversibility in the time evolution of the system. The usefulness of this definition hinges on the fact that within Boltzmann's kinetic theory only the single-particle distribution function is needed for a successful description of nonequilibrium as well as equilibrium phenomena.

In most systems, higher-order correlation functions will play a crucial role, and Boltzmann's $\mathcal{H}$ quantity seems to be of little value as it does not contain all the relevant information about the system. To remedy the situation, Gibbs defined what is now known as the fine-grained $\mathcal{H}$ quantity

$$\mathcal{H}_G(t) = \int \rho(\mathbf{r}_1, \mathbf{p}_1; \mathbf{r}_2, \mathbf{p}_2; \ldots; \mathbf{r}_N, \mathbf{p}_N; t)$$
$$\times \ln \rho(\mathbf{r}_1, \mathbf{p}_1; \ldots; \mathbf{r}_N, \mathbf{p}_N; t) \, d^3r_1 \, d^3p_1, \ldots, d^3r_N \, d^3p_N \quad (11.106)$$

where ρ is the N-body distribution function. Because ρ satisfies Liouville's equation (7.6), we can show easily that

$$\frac{d\mathcal{H}_G(t)}{dt} = 0 \quad (11.107)$$

‡ Walls (1970) has looked at the effects of a system on a reservoir using projection operator techniques.

implying that $\mathcal{H}_G(t)$ is a constant of the motion and can therefore not be used to characterize any aspect of the time evolution of a large system.

To generalize Boltzmann's $\mathcal{H}$ theorem to any system, the Ehrenfests (1911) suggested a procedure to make $\mathcal{H}_G$ time-dependent, namely coarse-graining. Let us divide phase space Γ into cells or grains of size $\Delta\Gamma$, determined by the resolution in a given experiment, and average the complete distribution function over each cell, yielding quantities

$$\bar{\rho}(\mathbf{X}, T) = (\Delta\Gamma)^{-1} \int_{\Delta\Gamma} \rho(\mathbf{X}', t) \, d^{6N}X' \qquad (11.108)$$

where $\Delta\Gamma$ includes the point $\mathbf{X}$.

Let us then define a coarse-grained $\mathcal{H}$ quantity by

$$\bar{\mathcal{H}}(t) = \int_\Gamma \rho(\mathbf{X}, t) \ln \bar{\rho}(\mathbf{X}, t) \, d^{6N}X \qquad (11.109)$$

and recall that, for any two positive functions $\rho_1(\mathbf{X}, t)$ and $\rho_2(\mathbf{X}, t)$ normalized such that

$$\int_\Gamma \rho_1(\mathbf{X}, t) \, d^{6N}X = \int_\Gamma \rho_2(\mathbf{X}', t) \, d^{6N}X' \qquad (11.110)$$

the Gibbs inequality

$$\int_\Gamma \rho_1 \ln \rho_1 \, d^{6N}X \geq \int_\Gamma \rho_1 \ln \rho_2 \, d^{6N}X \qquad (11.111)$$

holds. Applying this result to ρ and $\bar{\rho}$, we immediately find the inequality

$$\mathcal{H}_G(t) \geq \bar{\mathcal{H}}(t) \qquad (11.112)$$

If we choose an initial fine-grained distribution ρ which is constant over each of the small cells, then

$$\bar{\mathcal{H}}(0) = \mathcal{H}_G(0) \qquad (11.113)$$

Moreover, since $\mathcal{H}_G(t)$ is constant, $\mathcal{H}_G(t) = \mathcal{H}_G(0)$ and we find that

$$\bar{\mathcal{H}}(0) \geq \bar{\mathcal{H}}(t) \qquad (11.114)$$

which is a generalized $\mathcal{H}$ theorem valid in any system.

What is the origin of the decrease in $\bar{\mathcal{H}}(t)$? Suppose that the system at $t = 0$ is represented by phase points in the initial cell i with a constant density. At some later time, these will have moved out and will occupy a domain of the same volume but spread through several of the coarse-grained cells in Γ space. The total volume in phase space over which we then calculate the average of $\ln \bar{\rho}$ includes these partially filled cells and so is much larger than the corresponding volume for $\ln \rho$, leading to a decrease in $\bar{\mathcal{H}}(t)$ due to the loss of information in the coarse-graining process. Note that the amount of coarse-graining depends crucially on the accuracy with which relevant properties of the system are measured.

So far we have looked at classical systems. In general, the introduction of quantum mechanics does not significantly change the problem of irreversibility. Most of the difficulties of the classical theory have direct analogues in quantum mechanics. In particular, the quantum version of the Liouville equation (von Neumann equation) is time-reversal invariant and has an interpretation in terms of ensembles similar to that of its classical counterpart. The derivation of a particular generalized $\mathcal{H}$ theorem in quantum statistical mechanics, known as Klein's lemma (Klein, 1931; see also Jancel, 1969, for a general discussion of this topic) proceeds similarly to the derivation of master equations described in Chapter 10.

Let us assume that, in an isolated system with a hamiltonian H_0, an external interaction H'_t is switched on which causes transitions between eigenstates of H_0. We can write the wavefunction of the kth member of an ensemble of such systems as

$$\psi^k(\mathbf{x}_1, \ldots, \mathbf{x}_N, t_0) = \sum_n c_n^k(t_0) \phi_n(\mathbf{x}_1, \ldots, \mathbf{x}_n) \quad (11.115)$$

where the ϕ_n are a complete set of eigenstates of H_0. For times $t > t_0$, we then have

$$\psi^k(t) = \sum_{n,m} U_{nm}(t, t_0) c_m^k(t_0) \phi_n = \sum_n c_n^k(t) \phi_n \quad (11.116)$$

where U is the unitary time development operator. Thus

$$c_n^k(t) = \sum_m U_{nm}(t, t_0) c_m^k(t_0) \quad (11.117)$$

and

$$c_m^{k*}(t)c_n^k(t) = \sum_{i,j} U_{ni}(t, t_0)U_{mj}^*(t, t_0)c_i^k(t_0)c_j^{k*}(t_0) \quad (11.118)$$

Upon averaging over the $\mathcal{N}$ members of an ensemble, we introduce the density matrix

$$\rho_{nm}(t) = \frac{1}{\mathcal{N}} \sum_{k=1}^{\mathcal{N}} c_m^{k*}(t)c_n^k(t)$$

$$= \sum_{i,j} U_{ni} U_{mj}^* \frac{1}{\mathcal{N}} \sum_{k=1}^{\mathcal{N}} c_i^k(t_0)c_j^{k*}(t_0) \quad (11.119)$$

To simplify this expression, we assume that the phases associated with the wavefunctions of the members of the ensemble are initially random (Pauli, 1928), i.e.

$$\rho_{ij}(t_0) = \rho_{ii}(t_0)\,\delta_{ij} \quad (11.120)$$

This random phase assumption is an initial condition for ρ_{ij} and plays a role similar to the assumption that the fine-grained distribution is initially uniformly spread over several cells in Γ space in the discussion of classical coarse-graining. We then obviously have

$$\rho_{nn}(t) = \sum_i |U_{ni}(t, t_0)|^2\, \rho_{ii}(t_0) \quad (11.121)$$

and can introduce the notation

$$|U_{ni}(t, t_0)|^2 = T_{ni}(t_0, \Delta t) \quad (11.122)$$

where $\Delta t = t - t_0$ and T_{ni} is the probability of a transition from a state i to a state n in the time interval Δt. Also, note that the diagonal element $\rho_{nn}(t)$ is just the probability $P_n(t)$ that a system is in state n at time t. Thus, eqn (11.121) reads

$$P_n(t_0 + \Delta t) = \sum_i T_{ni}(t_0, \Delta t)P_i(t_0) \quad (11.123)$$

From the interpretation of the T_{ni} as transition probabilities, we should have

$$\sum_n T_{ni} = \sum_i T_{ni} = 1 \quad (11.124)$$

and

$$0 \le T_{ni} \le 1 \tag{11.125}$$

If the inverse of the matrix T_{ni} exists, we also have

$$P_i(t_0) = \sum_n T_{in}^{-1} P_n(t_0 + \Delta t) \tag{11.126}$$

However, since $\sum_n T_{in}^{-1} T_{nj} = \delta_{ij}$, it follows that not all of the elements T_{in}^{-1} can be between 0 and 1 unless $T_{nj} = \delta_{nj}$. Thus, if $T_{nj} \ne \delta_{nj}$, the elements of T^{-1} cannot be interpreted as transition probabilities and eqn (11.126) does not describe the reversed evolution to eqn (11.123). The origin of this irreversibility is the assumption that, at t_0, the density matrix is diagonal. Then, unless $|U_{ni}|^2 = \delta_{ni}$, for $t > t_0$, there will be some off-diagonal elements in $\rho_{ij}(t)$ which, however, are neglected when the random phase approximation is invoked. Thus, for $t > t_0$, no simple equation involving only the transition probabilities between states can be rigorously valid, as we have seen in Chapter 10.

Starting from eqn (11.123), it is possible to prove Klein's lemma. To do so, define

$$\mathcal{H}_p(t) = \sum_i P_i(t) \ln P_i(t) = \sum_i \rho_{ii}(t) \ln \rho_{ii}(t) \tag{11.127}$$

and

$$Q_{ji}(t_0, \Delta t) = \rho_{jj}(t_0)[\ln \rho_{jj}(t_0) - \ln \rho_{ii}(t_0 + \Delta t) - 1] \\ + \rho_{ii}(t_0 + \Delta t) \ge 0 \tag{11.128}$$

From the last inequality it follows that

$$\sum_{i,j} T_{ij}(t_0, \Delta t) Q_{ij}(t_0, \Delta t) = \sum_j \rho_{jj}(t_0) \ln \rho_{jj}(t_0) \\ - \sum_i \rho_{ii}(t_0 + \Delta t) \ln \rho_{ii}(t_0 + \Delta t) \ge 0 \tag{11.129}$$

since $T_{ij} \ge 0$, and where the fact that

$$\sum_j \rho_{jj}(t_0) = \sum_i \rho_{ii}(t_0 + \Delta t) \tag{11.130}$$

was used.

Thus, using the definition (11.127) of $\mathcal{H}_p(t)$ gives

$$\mathcal{H}_p(t_0 + \Delta t) \le \mathcal{H}_p(t_0) \tag{11.131}$$

This $\mathcal{H}$ theorem results from the spreading of the density matrix away from the diagonal form which it had at t_0. Even with this initial condition, this does not imply a monotonic decay of $\mathcal{H}_p(t)$, since, for $t > t_0$, $\rho_{ij}(t)$ is no longer diagonal and eqn (11.127) no longer holds. Only if the off-diagonal part of ρ_{ij} can be neglected at all times can it be shows that $\mathcal{H}_p(t)$ decays monotonically to an equilibrium value.

The relationship (11.131) is more fundamental since it assumes diagonality only at an initial instant. Although the weak nature of this $\mathcal{H}$ theorem and its derivation from an initial condition on ρ_{ij} are suggestive of the classical coarse-grained $\mathcal{H}$ theorem, no coarse-graining procedure was introduced in deriving (11.131). This $\mathcal{H}$ theorem has no classical analogue, a reflection of the fact that $\sum_i \rho_{ii} \ln \rho_{ii}$ is not the trace of any operator and hence is basis-dependent. Thus, except at the instant when ρ_{ij} is diagonal, it does not represent an observable and so this $\mathcal{H}$ theorem cannot provide an explanation of observed macroscopic irreversibility.

The situation is quite different once a master equation (10.2) has been established to describe the time evolution of a particular system properly. We can then show immediately that

$$\frac{d\mathcal{H}_p(t)}{dt} \leq 0 \tag{11.132}$$

In equilibrium, we have $d\mathcal{H}_p(t)/dt = 0$ and

$$W_{nm}P_n^{\text{eq}} = W_{mn}P_m^{\text{eq}} \tag{11.133}$$

This is the statement of detailed balance. More generalized $\mathcal{H}$ theorems have recently been derived by the Brussels school in their theory of subdynamics. We refer the reader to the original literature on the subject—e.g. Grecos and Prigogine (1978) and Prigogine, George, Henin, and Rosenfeld (1973).

12
Transient Effects in the Time Evolution of an Ideal Gas in an External Potential

12.1. Formulation of the Problem

WE have argued before that after an external perturbation has been switched on in a system, its time evolution proceeds generally in three stages that are distinguished from each other by characteristic time scales. In the first stage, which we called the transient or statistical regime, fast events take place on the time scale of the interaction time. Once the system has evolved to a certain complexity, smoother (average) features in the time evolution show up that vary on the scale of the collision time or, more generally, on the scale of a kinetic time. For this kinetic regime we have derived, with certain approximations, kinetic equations such as the Boltzmann equation for a dilute gas and master equations for more general systems. Except for the formal and exact master equation (10.90), these kinetic equations already exhibit the irreversibility so characteristic of the evolution of large systems. Normal solutions of these kinetic equations that vary smoothly over macroscopic times have then been obtained. These were used to derive the linear phenomenological laws of macroscopic nonequilibrium thermodynamics and to establish the validity of the local equilibrium hypothesis.

We will conclude this volume with a glimpse at the fascinating transient phenomena that take place more or less pronounced in every system in the initial stage of its time evolution after the external perturbation is switched on. Transient effects will generally be severely suppressed if the external perturbation is switched on over times large compared to τ_0 and will, of course, be absent if the switching process is adiabatic. We have already seen in Section 11.2 that transient phenomena can be the dominant feature for fast switching.

We will present in this chapter a rather complete discussion of the transient phenomena in the time evolution of an ideal gas which is subjected at some time, $t = 0$ say, to a static external potential (Kreuzer and Teshima, 1977; Kreuzer, 1978). It will become obvious that this very simple problem has a number of remarkable and interesting features which are not attainable by perturbation theory and can only be satisfactorily understood in an exact solution of the time evolution of the system. This, we must realize, cannot be achieved for all external potentials. However, for the class of separable potentials we will construct the explicit analytic solution to the problem and study it in great numerical detail, not only for weak potentials but also for cases in which the gas particles can be trapped in bound states or resonances.

To introduce the problem properly, we assume that, for times $t < 0$, we have a system of free particles of mass m (fermions or bosons), described by a hamiltonian in second quantized form

$$H_0 = \sum_{\mathbf{k}} \varepsilon_{\mathbf{k}} a_{\mathbf{k}}^\dagger a_{\mathbf{k}} \tag{12.1}$$

where the single-particle energies are $\varepsilon_{\mathbf{k}} = \hbar^2 \mathbf{k}^2 / 2m$ and $a_{\mathbf{k}}^\dagger$ and $a_{\mathbf{k}}$ are creation and annihilation operators of particles in momentum states $\mathbf{k}$. We further assume that the system is initially in equilibrium and that its statistical properties are described by the density operator of the canonical ensemble

$$\hat{\rho}_0 = e^{-\beta H_0} / Tr(e^{-\beta H_0}) \tag{12.2}$$

At time $t = 0$ we switch on a static external potential. The system will then evolve according to the new hamiltonian

$$H = H_0 + \theta(t) V^{-1} \sum_{\mathbf{k},\mathbf{k}'} V_{\mathbf{k}\mathbf{k}'} a_{\mathbf{k}}^\dagger a_{\mathbf{k}'} \tag{12.3}$$

where $\theta(t) = 1$, for $t > 0$, and zero otherwise. Here, V is the (large) volume in which our system is enclosed. To be specific, we want to calculate the time evolution of the local particle density

$$\rho(\mathbf{r}, t) = Tr[\psi^\dagger(\mathbf{r}, t) \psi(\mathbf{r}, t) \hat{\rho}_0] \tag{12.4}$$

where

$$\psi(\mathbf{r}, t) = V^{-\frac{1}{2}} \sum_{\mathbf{k}} e^{i\mathbf{k} \cdot \mathbf{r}} a_{\mathbf{k}}(t) \tag{12.5}$$

is a field operator defined in terms of the Heisenberg annihilation operators

$$a_{\mathbf{k}}(t) = e^{i(Ht/\hbar)} a_{\mathbf{k}}(0) e^{-i(Ht/\hbar)} \tag{12.6}$$

which satisfy the equation of motion

$$i\hbar \frac{\partial a_{\mathbf{k}}}{\partial t} = [a_{\mathbf{k}}, H] \tag{12.7}$$

According to eqn (12.4) we evaluate the time evolution at the operator level and only introduce statistics at the final stage by taking traces of the form

$$Tr(a_{\mathbf{k}}^{\dagger} a_{\mathbf{k}'} \hat{\rho}_0) = \delta_{\mathbf{k}\mathbf{k}'} \eta_{\mathbf{k}} \tag{12.8}$$

which are simply our initial data at time $t = 0$.

The methods developed in later sections will allow us to construct the exact time evolution in eqn (12.6) for external potentials that are separable. The simplest example, acting in s waves only, is given by

$$V_{\mathbf{k}\mathbf{k}'} = g v_{\mathbf{k}} v_{\mathbf{k}'} \tag{12.9}$$

The generalization to superpositions of terms like (12.9), also in higher partial waves, is straightforward.‡ The properties of a single particle in such a separable external potential are summarized in the appendix to this chapter, giving conditions for the occurrence of bound states and resonances.

Let us briefly outline the contents of this chapter. In Section 12.2, we will calculate the exact evolution of the particle creation and annihilation operators which will lead in a straightforward way in Section 12.3 to the evolution of the time-dependent local particle density (12.4). In Section 12.4, we will study explicit examples of weak attractive and repulsive potentials and look at the small and large time behavior. Next, in Section 12.5, we will examine the case of a strongly attractive potential that can develop a bound state. A detailed analysis is then made in Section 12.6 of the time evolution of a system with a strong

‡ Separable potentials as introduced by Wheeler (1936) have been used very successfully in potential scattering and nuclear physics to describe the nucleon-nucleon interaction as well as for the phenomenological shell model and optical model potentials.

potential which develops a resonance. A series of three-dimensional plots of the quantity $r^2[\rho(r,t)/\rho(t=0)-1]$ over the rt plane will be used to illustrate these transient phenomena.

12.2. Time Evolution of Operators $a_{\mathbf{k}}(t)$

As outlined in Section 12.1, we will calculate the time evolution of the particle creation and annihilation operators in order to compute the macroscopic density evolution. The annihilation operator ($\hbar$ set equal to 1)

$$a_{\mathbf{k}}(t) = e^{iHt} a_{\mathbf{k}}(0) e^{-iHt} \tag{12.10}$$

is subject to the equation of motion

$$i\frac{\partial a_{\mathbf{k}}}{\partial t} = \varepsilon_{\mathbf{k}} a_{\mathbf{k}} + \theta(t) V^{-1} \sum_{\mathbf{k}'} V_{\mathbf{k}\mathbf{k}'} a_{\mathbf{k}'} \tag{12.11}$$

which can be integrated formally to yield

$$a_{\mathbf{k}}(t) = a_{\mathbf{k}}^{in}(t) - i \int_0^t e^{-i\varepsilon_{\mathbf{k}}(t-t')} V^{-1} \sum_{\mathbf{k}'} V_{\mathbf{k}\mathbf{k}'} a_{\mathbf{k}'}(t')\, dt' \tag{12.12}$$

where the in-field is given by

$$a_{\mathbf{k}}^{in}(t) = e^{-i\varepsilon_{\mathbf{k}} t} a_{\mathbf{k}}(0) \tag{12.13}$$

Taking Laplace transforms of eqn (12.12), we obtain

$$A_{\mathbf{k}}(z) = L[a_{\mathbf{k}}(t)] = \int_0^\infty e^{-zt} a_{\mathbf{k}}(t)\, dt$$

$$= A_{\mathbf{k}}^{in}(z) - i\frac{1}{z+i\varepsilon_{\mathbf{k}}} V^{-1} \sum_{\mathbf{k}'} V_{\mathbf{k}\mathbf{k}'} A_{\mathbf{k}'}(z) \tag{12.14}$$

where

$$A_{\mathbf{k}}^{in}(z) = \frac{1}{z+i\varepsilon_k} a_{\mathbf{k}}(0) \tag{12.15}$$

Let us for the potential (12.9) define an operator

$$A(z) = V^{-1} \sum_{\mathbf{k}'} v_{\mathbf{k}'} A_{\mathbf{k}'}(z) \tag{12.16}$$

for which we find, multiplying eqn (12.14) by $V^{-1} v_{\mathbf{k}}$ and summing over **k**,

$$A(z) = [1 + iI(z)]^{-1} V^{-1} \sum_{\mathbf{p}} \frac{v_{\mathbf{p}}}{z + i\varepsilon_{\mathbf{p}}} a_{\mathbf{p}}(0) \qquad (12.17)$$

where

$$I(z) = gV^{-1} \sum_{\mathbf{k}} \frac{v_{\mathbf{k}}^2}{z + i\varepsilon_{\mathbf{k}}} \qquad (12.18)$$

This solves eqn (12.14), and we can next take inverse Laplace transforms to obtain the explicit time evolution

$$a_{\mathbf{k}}(t) = e^{-i\varepsilon_{\mathbf{k}} t} a_{\mathbf{k}}(0) + V^{-1} \sum_{\mathbf{p}} F_{\mathbf{kp}}(t) a_{\mathbf{p}}(0) \qquad (12.19)$$

with

$$F_{\mathbf{kp}}(t) = -igL^{-1} \left[\frac{v_{\mathbf{k}}}{z + i\varepsilon_{\mathbf{k}}} \frac{v_{\mathbf{p}}}{z + i\varepsilon_{\mathbf{p}}} \frac{1}{1 + iI(z)} \right] \qquad (12.20)$$

where

$$L^{-1}[f(z)] = \frac{1}{2\pi i} \int_{c-i\infty}^{c+i\infty} e^{zt} f(z) \, dz \qquad (12.21)$$

To carry the calculation further, we specify the potential form factor to be

$$v_{\mathbf{k}} = (k^2 + \gamma^2)^{-1} \qquad (12.22)$$

and take $\varepsilon_{\mathbf{k}} = k^2/2m$. Invoking here the thermodynamic limit by replacing sums $V^{-1} \sum_{\mathbf{k}}$ by integrals $(2\pi)^{-3} \int d^3k$ and thus introducing irreversibility, we find ($k = |\mathbf{k}|, p = |\mathbf{p}|$)

$$I(z) = \frac{g}{8\pi\gamma} \left(\sqrt{z} + \frac{e^{i\pi/4}\gamma}{\sqrt{2m}} \right)^{-2} \qquad (12.23)$$

and

$$F_{kp}(t) = \frac{g v_k v_p}{\varepsilon_p - \varepsilon_k} L^{-1} \left\{ \left(\frac{1}{z + i\varepsilon_p} - \frac{1}{z + i\varepsilon_k} \right) \right. \\ \left. \times \left[1 - \frac{ig}{8\pi\gamma} (\sqrt{z} - e^{i\pi/4}\alpha_1)^{-1} (\sqrt{z} - e^{i\pi/4}\alpha_2)^{-1} \right] \right\} \qquad (12.24)$$

where

$$\alpha_{1,2} = -\frac{\gamma}{\sqrt{2}m} \pm \left(-\frac{g}{8\pi\gamma}\right)^{\frac{1}{2}} \quad (12.25)$$

After further decomposition of eqn (12.25) into partial fractions, we finally obtain

$$\begin{aligned} F_{kp}(t) = &\frac{gv_k v_p}{\varepsilon_p - \varepsilon_k} \Bigg\{ e^{-i\varepsilon_p t} \left[1 - \frac{g}{8\pi\gamma} \frac{1}{(\alpha_1 + i\sqrt{\varepsilon_p})(\alpha_2 + i\sqrt{\varepsilon_p})}\right] \\ &- e^{i\varepsilon_k t}\left(1 - \frac{g}{8\pi\gamma} \frac{1}{(\alpha_1 + i\sqrt{\varepsilon_k})(\alpha_2 + i\sqrt{\varepsilon_k})}\right)\Bigg\} \\ &+ \frac{g^2 v_k v_p}{16\pi\gamma}\left(\frac{8\pi\gamma}{-g}\right)^{\frac{1}{2}}\left[\frac{1}{(\varepsilon_p + \alpha_1^2)(\varepsilon_k + \alpha_1^2)} e^{i\alpha_1^2 t} Erfc(-\alpha_1 e^{i\pi/4}\sqrt{t}) \right.\\ &\left.- \frac{\alpha_2}{(\varepsilon_p + \alpha_2^2)(\varepsilon_k + \alpha_2^2)} e^{i\alpha_2^2 t} Erfc(-\alpha_2 e^{i\pi/4}\sqrt{t})\right] \\ &- i\frac{g^2 v_k v_p}{4\pi}\frac{1}{\sqrt{2}m}\frac{1}{\varepsilon_k - \varepsilon_p}\left[\frac{\sqrt{\varepsilon_p}}{(\varepsilon_p + \alpha_1^2)(\varepsilon_p + \alpha_2^2)} e^{-i\varepsilon_p t} Erfc(\sqrt{-i\varepsilon_p}t)\right. \\ &\left.- \frac{\sqrt{\varepsilon_k}}{(\varepsilon_k + \alpha_1^2)(\varepsilon_k + \alpha_2^2)} e^{-i\varepsilon_k t} Erfc(\sqrt{-i\varepsilon_k}t)\right] \quad (12.26) \end{aligned}$$

where the error functions are

$$Erfc(x) = 1 - \frac{2}{\sqrt{\pi}}\int_0^x e^{-t^2}\,dt = 1 - Erf(x) \quad (12.27)$$

and the following inverse Laplace transform has been used

$$\frac{1}{2\pi i}\oint e^{zt}\frac{1}{z + i\varepsilon}\frac{1}{\sqrt{z} - \alpha e^{i\pi/4}}\,dz = \frac{e^{-i\pi/4}}{\varepsilon + \alpha^2}(i\sqrt{\varepsilon} - \alpha)e^{-i\varepsilon t}$$

$$+ \alpha e^{i\alpha^2 t} Erfc(-\alpha\sqrt{it}) - i\sqrt{\varepsilon}e^{-i\varepsilon t}Erfc(\sqrt{-i\varepsilon t}) \quad (12.28)$$

The class of potential form factors v_k for which the inverse Laplace transforms in $F_{kp}(t)$ can be explicitly evaluated is unfortunately rather restricted. However, we do not expect that the qualitative features of the evolution will depend crucially on the details of the potential but rather on qualitative features such as the existence of bound states and resonances.

12.3. Density Evolution

Before we continue to calculate the time evolution of the local macroscopic particle density given by eqn (12.4), recall the physical situation under study. We assume that, for times $t<0$, an ideal gas of particles of mass m is in equilibrium, uniformly distributed throughout a volume V. At time $t=0$, an external field is switched on and a redistribution of matter takes place, which we wish to follow as a function of space and time. To do this, we use eqn (12.19) and its conjugate in eqns (12.4) and (12.5). This gives us

$$\rho(\mathbf{r}, t) = \rho(t=0) + V^{-2} \sum_{\mathbf{k},\mathbf{k}'} [e^{i(\mathbf{k}-\mathbf{k}')\cdot\mathbf{r}} e^{i\varepsilon_{\mathbf{k}}t} n_{\mathbf{k}'} F_{\mathbf{k}\mathbf{k}'}(t) + c.c.]$$

$$+ V^{-3} \sum_{\mathbf{p}} \sum_{\mathbf{k},\mathbf{k}'} e^{i(\mathbf{k}-\mathbf{k}')\cdot\mathbf{r}} F^*_{\mathbf{k}'\mathbf{p}}(t) F_{\mathbf{k}\mathbf{p}}(t) n_{\mathbf{p}} \quad (12.29)$$

where we used the initial conditions of an ideal gas. Most of the following numerical work is done for classical Maxwell–Boltzmann statistics for which we have

$$\begin{aligned} n_{\mathbf{k}} &= e^{-\beta(\varepsilon_{\mathbf{k}}-\mu)} \\ &= \frac{e^{-\beta\varepsilon_{\mathbf{k}}} \rho(0)(2\pi\hbar)^3}{(2\pi m k T)^{\frac{3}{2}}} \end{aligned} \quad (12.30)$$

Note that, so far, all results are also true for Fermi–Dirac or Bose–Einstein statistics.

We now proceed with the evaluation of eqn (12.29) for the form factor $v_{\mathbf{k}}$ given in eqn (12.22) with the corresponding function $F_{kp}(t)$ given in eqn (12.26). Let us again take the thermodynamic limit in eqn (12.29) and introduce new variables which also show the dependence on Planck's constant $\hbar$

$$\begin{aligned} V_0 &= |g|\gamma^{-1} & \delta &= \frac{2V_0}{k_B T} \\ \sigma &= \frac{g}{|g|} & x &= k\lambda_0 \\ \lambda_0 &= \frac{\hbar}{(4mV_0)^{\frac{1}{2}}} & \tau &= \frac{2V_0 t}{\hbar} \\ \nu &= \lambda_0 \gamma & a_i &= \frac{\alpha_i}{\sqrt{2V_0}} \end{aligned} \quad (12.31)$$

Equation (12.30) then becomes
$$n_k = (2\pi)^3 \left(\frac{\delta}{\pi}\right)^{\frac{3}{2}} \lambda_0 \rho(0) e^{-\delta x^2} \tag{12.32}$$

and eqn (12.29) reads ($r = |\mathbf{r}|$)

$$\frac{\rho(r,t)}{\rho(0)} = 1 + \frac{\nu}{\pi}\left(\frac{\delta}{\pi}\right)^{\frac{3}{2}}\left(\frac{\lambda_0}{r}\right)^2 \int_0^\infty y \sin\left(\frac{yr}{\lambda_0}\right) e^{-\delta y^2}$$

$$\times \left[e^{iy^2\tau}\int_0^\infty x\, dx \sin\left(\frac{xr}{\lambda_0}\right) F(x,y,\tau) + \text{c.c.}\right] dy$$

$$+ (4\pi^3)^{-1}\left(\frac{\delta}{\pi}\right)^{\frac{3}{2}}\left(\frac{\lambda_0}{r}\right)^2 \nu^2$$

$$\times \int_0^\infty y^2 e^{-\delta y^2} \left|\int_0^\infty x \sin\left(\frac{xr}{\lambda_0}\right) F(x,y,\tau)\, dx\right|^2 dy \tag{12.33}$$

with

$$F(x,y,\tau) = 2\nu\lambda_0^{-3} F_{kp}(t) = \frac{1}{x^2+\nu^2}\frac{1}{y^2+\nu^2}$$

$$\times \left\{\frac{1}{y^2+x^2}\left[e^{-iy^2\tau}\left(1 - \frac{\sigma}{16\pi}\frac{1}{a_1+iy}\frac{1}{a_2+iy}\right)\right.\right.$$

$$\left.- e^{-ix^2\tau}\left(1 - \frac{\sigma}{16\pi}\frac{1}{a_1+ix}\frac{1}{a_2+ix}\right)\right]$$

$$+ \frac{1}{8\sqrt{-\sigma\pi}}\frac{a_1}{(x^2+a_2^2)(y^2+a_2^2)} e^{ia_1^2\tau}\text{Erfc}(-a_1 e^{i\pi/4}\sqrt{\tau})$$

$$- \frac{1}{8\sqrt{-\sigma\pi}}\frac{a_2}{(x^2+a_2^2)(y^2+a_2^2)} e^{ia_2^2\tau}\text{Erfc}(-a_2 e^{i\pi/4}\sqrt{\tau})$$

$$- i\frac{\nu}{8\pi}\frac{1}{x^2-y^2}\left[\frac{y}{(y^2+a_1^2)(y^2+a_2^2)} e^{iy^2\tau}\text{Erfc}(-e^{i\pi/4}y\sqrt{\tau})\right.$$

$$\left.\left.- \frac{x}{(x^2+a_1^2)(x^2+a_2^2)} e^{-ix^2\tau}\text{Erfc}(e^{-i\pi/4}x\sqrt{\tau})\right]\right\} \tag{12.34}$$

The x integration in eqn (12.33) can actually be carried out explicitly and yields the final result

$$\frac{\rho(r,t)}{\rho(0)} = 1 - \frac{\sigma}{4\pi\sqrt{\pi}}\lambda_0\gamma\left(\frac{\lambda_0}{r}\right)^2 \delta^{\frac{3}{2}}\int_0^\infty \frac{y}{y^2+\nu^2} e^{-\delta y^2}$$

$$\times \left(2\sin\left(\frac{yr}{\lambda_0}\right)\text{Re}[G] - \frac{\sigma}{16\pi}\frac{\nu y}{y^2+\nu^2}|G|^2\right) dy \tag{12.35}$$

where

$$G = e^{iy^2\tau} \sum_{k=1}^{5} A_k e^{ic_k^2\tau}\left\{e^{-r/\lambda}Erfc(e^{i\pi/4}c_k\sqrt{\tau}) \right.$$
$$\left. - e^{c_k r/\lambda_0}Erfc\left(e^{i\pi/4}c_k\sqrt{\tau} + e^{-i\pi/4}\frac{r}{(2\lambda_0\sqrt{\tau})}\right)\right\} \quad (12.36)$$

with

$$c_1 = \nu - \left(\frac{-\sigma}{16\pi}\right)^{\frac{1}{2}}$$
$$c_2 = \nu + \left(\frac{-\sigma}{16\pi}\right)^{\frac{1}{2}}$$
$$c_3 = -\nu$$
$$c_4 = iy$$
$$c_5 = -iy \quad (12.37)$$
$$A_1 = c_1(c_1+\nu)^{-1}(c_1^2+y^2)^{-1}$$
$$A_2 = c_2(c_2+\nu)^{-1}(c_2^2+y^2)^{-1}$$
$$A_3 = 4\nu^2(y^2+\nu^2)^{-1}\left(4\nu^2+\frac{\sigma}{16\pi}\right)^{-1}$$
$$A_4 = A_5^* = -(\nu-iy)(\nu+iy)^{-1}(c_1-iy)^{-1}(c_2-iy)^{-1}.$$

Equation (12.35) describes exactly the time evolution of the radial density of an ideal gas obeying Maxwell–Boltzmann statistics after an external potential of arbitrary strength is switched on at time $t = 0$. For comparison, we also list the analogous expression for an ideal Fermi–Dirac gas at zero temperature

$$\frac{\rho_{FD}(r,\tau)}{\rho(0)} = 1 - \lambda_0 \gamma \left(\frac{\lambda_0}{r}\right)^2 \frac{\sigma}{16\pi} \frac{3}{y_F^3} \int_0^{y_F} \frac{y}{y^2+\nu^2}$$
$$\times \left(2\sin\left(\frac{yr}{\lambda_0}\right)\text{Re}[G] - \frac{\sigma}{16\pi}\frac{y}{y^2+\nu^2}|G|^2\right) dy \quad (12.38)$$

where

$$y_F = \lambda_0 k_F = [6\pi^2 \lambda_0^3 \rho(0)]^{\frac{1}{3}} \quad (12.39)$$

is the dimensionless Fermi momentum. Dealing with quantum gases at finite temperatures does not present any difficulty as the

12.4. Weak Potentials

Let us first discuss the case in which, at time $t = 0$, a weak attractive potential with the form factor $v_k = 1/(k^2 + \gamma^2)$ is switched on, so weak that gas particles of mass m *cannot* be trapped into bound states. Figure 12.1 gives some exact radial density profiles, calculated from eqn (12.35) for some selected times. As expected, particles are attracted toward the potential center but, to satisfy overall mass conservation, a shell of decreased mass density develops further out followed by more small

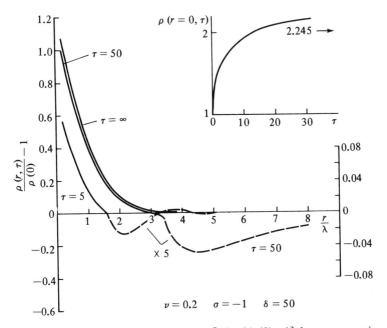

Fig. 12.1. Radial density distribution $[\rho(r, \tau)/\rho(0) - 1]$ for an attractive potential as a function of (r/λ) for various times. The dashed portions are magnified by a factor of 5 (right scale). The insert shows the time evolution at $r = 0$ with $\lim_{\tau \to \infty} \rho(r = 0, \tau)/\rho(0) = 2.245$.

density oscillations. For small τ we find

$$\frac{\rho(r=0,\tau)}{\rho(0)} \approx 1 - \frac{\sigma}{\pi}\nu\delta\sqrt{\frac{\tau}{2\pi}}[1 - \nu\sqrt{\pi}\delta e^{-\delta\nu^2}Erfc(\nu\sqrt{\delta})] + \cdots \tag{12.40}$$

which shows an immediate mass enhancement at $r=0$ with infinite slope and which is larger for a lower initial temperature, i.e. for larger δ. However, away from the origin, i.e. for $r \neq 0$, we find

$$\frac{\rho(r,\tau)}{\rho(0)} \approx 1 + A(r,\nu,\delta)\tau^2 + \cdots \tag{12.41}$$

for small τ, i.e. a quadratic time dependence.

As time increases, the depletion layer moves out and gets shallower because mass conservation demands that we have

$$\int_0^\infty r^2 \left[\frac{\rho(r,\tau)}{\rho(0)} - 1\right] dr = 0 \tag{12.42}$$

This has an interesting implication for the large time limit. Using the facts that

$$\lim_{z \to \infty} e^{z^2} Erfc(z) = 0 \quad \text{if } |\arg z| < \frac{3\pi}{4} \tag{12.43}$$

and that

$$Erfc(-z) = 2 - Erfc(z) \tag{12.44}$$

we find from eqn (12.36)

$$\lim_{\tau \to \infty} G = 2\frac{\nu - iy}{(\nu + iy)(c_1^2 + y^2)}(e^{iyr/\lambda_0} - e^{-r/\lambda})$$

$$+ 2\theta(-c_1)\frac{c_1}{(c_1+\nu)(c_1^2+y^2)} e^{i(y^2+c^2)\tau}(e^{-r\gamma} - e^{c_1 r/\lambda_0}) \tag{12.45}$$

Here, the last term is the boundstate contribution which, of course, is absent for weak or repulsive potentials. Also, in the last bracket observe the explicit radial damping over the range γ^{-1} of the potential and over the extent of the boundstate wavefunction. Inserting eqn (12.45) into eqn (12.35), we find for a weakly attractive potential that

$$\lim_{\tau \to \infty} \frac{\rho(r=\text{const},\tau)}{\rho(0)} > 1 \tag{12.46}$$

TRANSIENT EFFECTS IN THE TIME EVOLUTION OF AN IDEAL GAS 381

because the mass-conserving depletion layer will have moved past any finite r after a finite time and will be of vanishingly small depth due to the factor r^2. On the other hand, for any finite τ, however large, eqn (12.42) is indeed satisfied, as can be checked explicitly by a straightforward but very tedious integration. The insert in Fig. 12.1 shows the time dependence at $r = 0$ for a weakly attractive potential. The density increases monotonically

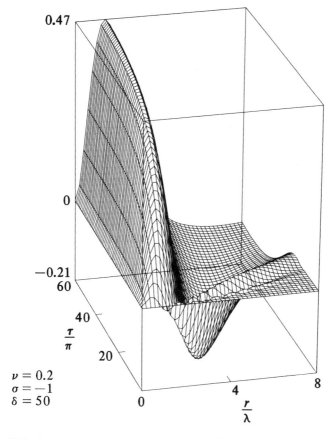

Fig. 12.2. Perspective view of the surface $(r/\lambda)^2[\rho(r, \tau)/\rho(0) - 1]$ over the $r\tau$ plane. Yaw $= -15°$, pitch $= 25°$ at distance of the observer equals ten times the diagonal of the cube.

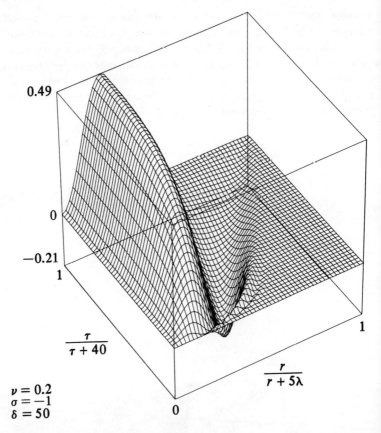

Fig. 12.3. Perspective view of the surface $(r/\lambda)^2[\rho(r,\tau)/\rho(0)-1]$ over the rescaled $r\tau$ plane.

but not exponentially. To give an overall picture of the density evolution, we have plotted in Fig. 12.2 a perspective view of the surface defined by

$$(r\gamma)^2\left[\frac{\rho(r,\tau)}{\rho(0)}-1\right] \qquad (12.47)$$

over the $r\tau$ plane. Close to the origin, a ridge of mass enhancement develops as a function of time followed, at larger values of r, by an initially quite deep depletion layer which propagates outward as a damped, wavelike disturbance and which is seen as a small ridge running diagonally across the density surface towards the right-hand back corner. To get a complete picture of the density evolution over the entire $r\tau$ plane, we have rescaled r into $r' = r/(r + r_0)$ and τ into $\tau' = \tau/(\tau + \tau_0)$ with appropriately chosen constants r_0 and τ_0. Such a plot is presented in Fig. 12.3, where we can now follow the density evolution from the initial uniform distribution at $\tau' = \tau = 0$ to the final stationary distribution in the external field at $\tau' = 1$ or $\tau = \infty$ for all values of the radial coordinate r.

To see the effect of the initial gas temperature, we have increased δ (decreased temperature) considerably in Fig. 12.4. More structure becomes visible due to the lack of temperature smearing. It should be noticed that the surface shown in Fig. 12.4 agrees within a few percent with the $\delta \to \infty$ limit of eqn (12.35), for which we can actually perform all integrations in closed form to yield

$$\lim_{\delta \to \infty} \frac{\rho(r,\tau)}{\rho(0)} = \left| 1 - \frac{\sigma}{16\pi} \frac{1}{r\gamma} \sum_{k=1}^{4} B_k e^{i\tilde{c}_k^2 \tau} \left[e^{-r\gamma} Erfc(e^{i\pi/4}\tilde{c}_k \sqrt{\tau}) \right. \right.$$
$$\left. \left. - e^{\tilde{c}_k r/\lambda_0} Erfc\left(e^{i\pi/4}\tilde{c}_k\sqrt{\tau} + e^{-i\pi/4}\frac{r}{2\lambda_0\sqrt{\tau}}\right) \right] \right|^2 \quad (12.48)$$

where

$$\tilde{c}_1 = \nu - \left(\frac{-\sigma}{16\pi}\right)^{\frac{1}{2}}$$

$$\tilde{c}_2 = \nu + \left(\frac{-\sigma}{16\pi}\right)^{\frac{1}{2}}$$

$$\tilde{c}_3 = -\nu$$

$$\tilde{c}_4 = 0$$

$$B_1 = \frac{1}{\tilde{c}_1(\tilde{c}_1 + \nu)} \quad (12.49)$$

$$B_2 = \frac{1}{\tilde{c}_2(\tilde{c}_2 + \nu)}$$

$$B_3 = \frac{4}{4\nu^2 + (\sigma/16\pi)}$$

$$B_4 = -\frac{2}{\nu^2 + (\sigma/16\pi)}$$

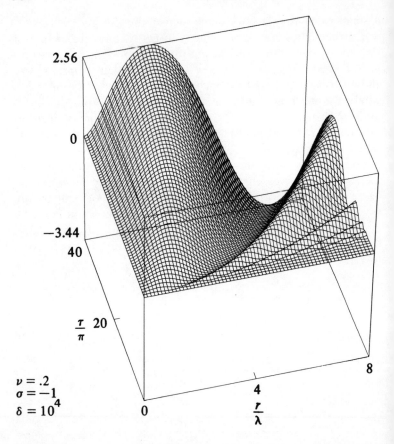

Fig.12.4. Perspective view of the surface $(r/\lambda)^2[\rho(r,\tau)/\rho(0)-1]$ over the $r\tau$ plane.

It must be realized that the limit $\delta \to \infty$ is unphysical since it is the zero temperature limit of a classical Maxwell–Boltzmann gas. It is, however, very useful, due to its simplicity, in getting suggestions, both analytically and numerically, for the analysis of the finite temperature case. As an example, let us look at the large time limit of eqn (12.48):

$$\lim_{\delta\to\infty}\frac{\rho(r,\tau\to\infty)}{\rho(0)} = 1 - \frac{\sigma}{16\pi\gamma r}$$
$$\times\left[\frac{1-e^{-r\gamma}}{\nu^2+(\sigma/16\pi)} + \theta(-\tilde{c}_1)\frac{e^{i\tilde{c}_1^2\tau}(e^{-r\gamma}-e^{\tilde{c}_1 r/\lambda_0})}{\tilde{c}_1(\tilde{c}_1+\nu)}\right]\Big|^2 \quad (12.50)$$

TRANSIENT EFFECTS IN THE TIME EVOLUTION OF AN IDEAL GAS   385

We see immediately that the statement in eqn (12.46) can, in general, be true only for a weakly attractive potential, i.e. with $\sigma = -1$ but without boundstate contributions.

It has been argued repeatedly that for a weak enough external potential, lowest- (i.e. second-) order perturbation theory should give a good approximation for times that are not too long. We will verify this explicitly by calculating the time evolution of the occupation function in second-order perturbation theory:

$$n_k(t) \approx n_k(0)\left[1 + 2V^2 \sum_{k'} V_{kk'}^2 \frac{\cos(\varepsilon_k - \varepsilon_{k'})t - 1}{(\varepsilon_k - \varepsilon_{k'})^2}\right]$$

$$+ 2V^2 \sum_{k'} V_{kk'}^2 \, n_{k'}(0) \frac{\cos(\varepsilon_k - \varepsilon_{k'})t - 1}{(\varepsilon_k - \varepsilon_{k'})^2} \quad (12.51)$$

We can extract the large time behavior from this expression by using the relation

$$\lim_{t \to \infty} \frac{\sin^2 \omega t}{\omega^2} = t\pi\delta(\omega) \quad (12.52)$$

This gives us

$$\lim_{t \to \infty} n_k(t) = n_k(0) + \frac{f_k^{(2)} g^2 t}{V} \quad (12.53)$$

where $f_k^{(2)}$ is some well-defined function of k. This result implies that for finite volume V, $n_k(t)$ increases without bound—an unacceptable result. To overcome this difficulty one argues (Planck, 1896; van Hove, 1955; Prigogine, 1967; and others) that to justify the use of perturbation theory for large times one must switch off the interaction, i.e. let $g \to 0$ as $t \to \infty$ such that $g^2 t = \text{const}$ in the weak-coupling limit. At least in the context of the present calculation, this is an empty statement as long as the constant $g^2 t$ is not fixed. Let us also mention here that Prigogine (1967) used potential scattering in second-order perturbation theory as an illustration of his general theory of nonequilibrium statistical mechanics. He summarizes his results by saying that 'in this potential scattering experiment we have a contribution to $\rho_0(t)$ coming from $\rho_0(0)$ which (1) is linearly growing in time as it should be for the scattering of the incoming beam; (2) introduces a collision operator whose only effect (within the Born approximation) is to increase the spherical symmetry of the initial distribution; and (3) yields an irreversible behavior, the stationary state being a spherically symmetric velocity distribution.' This interpretation has to be taken with some caution. For conclusion (1) we believe that Prigogine (1967) did not specify initial conditions, typical for a scattering experiment and, indeed, our result is derived for a gas initially in equilibrium, i.e. completely isotropic and homogeneous. This also makes conclusion (2) less convincing. This, as we have seen in Section 11.1, has already been advanced by Planck (1896).

Let us mention finally that the same large time behavior as eqn (12.53) appears in the exact result as well, namely

$$n_k(t) = Tr[a_k^\dagger(t)a_k(t)\rho_0]$$

$$= n_k(0) + V^{-1}[F_{kk}(t)e^{i\varepsilon_k t} + c.c.]n_k(0) + V^{-2}\sum_p F_{kp}^*(t)F_{kp}(t)n_p(0)$$

$$\underset{t\to\infty}{\approx} n_k(0) + \frac{f_k t}{V} \quad (12.54)$$

We have thus no room for plausibility arguments but must continue with a rigorous analysis which demands that we first take the limit $V \to \infty$, and the unphysical term proportional to t disappears. But this is not surprising because an external potential of finite range cannot and does not lead to a finite change in the occupation function $n_k(t)$ which, after all, is an average quantity over the whole infinite system. To obtain meaningful results, one has to study local quantities in this problem, such as the local distribution function introduced by Wigner (see Section 8.3)

$$f^w(\mathbf{x}, \mathbf{p}, t) = \frac{1}{(2\pi)^3}\int e^{i\mathbf{p}\cdot\boldsymbol{\xi}}Tr[\psi^\dagger(\mathbf{x}-\tfrac{1}{2}\boldsymbol{\xi})\psi(\mathbf{x}+\tfrac{1}{2}\boldsymbol{\xi})\hat{\rho}_t]\,d^3\boldsymbol{\xi}$$

(12.55)

or local macroscopic quantities as we have done above. If one is careful to do this, no difficulties occur in computation or interpretation.‡

12.5. Boundstates

As is pointed out in the appendix to this chapter, an attractive potential of sufficient strength will trap a particle of mass m in an s-wave boundstate. In our dimensionless units, this happens for

‡ To obtain nontrivial results one could also generalize the system to include a large number of randomly distributed, static scattering centers. This has been done for a two-component system by Kreuzer and Kurihara (1977).

$\sigma = -1$ if

$$\nu < \frac{1}{4\sqrt{\pi}} \qquad (12.56)$$

and the normalized boundstate energy is then

$$\frac{E_0^{bs}}{2V_0} = -\left(-\nu + \frac{1}{4\sqrt{\pi}}\right)^2 = -c_1^2 \qquad (12.57)$$

To get an idea of what features to expect in the density evolution in this case, let us first examine the limit $\delta \to \infty$, i.e. a situation in which temperature smearing is negligible. Let us recall eqn (12.48)

$$\frac{\rho(r, \tau \to \infty)}{\rho(0)} \approx \left|1 - \frac{\sigma}{16\pi\gamma r}\left[\frac{1 - e^{-r\gamma}}{\nu^2 + (\sigma/16\pi)}\right.\right.$$
$$\left.\left. + \theta(-\tilde{c}_1)\frac{e^{i\tilde{c}_1^2 \tau}(e^{-r\gamma} - e^{-\tilde{c}_1 r/\lambda_0})}{\tilde{c}_1(\tilde{c} + \nu)}\right]\right|^2 \qquad (12.58)$$

The first two time-independent terms are present for any potential. However, in the boundstate case we have an additional constant term as well as a term oscillating in time with a frequency $\tilde{c}_1^2$. This is illustrated in Fig. 12.5. The energy scale in this problem is set by the boundstate energy rather than by V_0 as it is in the absence of a boundstate.

For finite but large δ, we still expect some oscillations in the density evolution but they should be damped by temperature smearing. This is analytically achieved by the remaining integration in eqn (12.35) which involves the Boltzmann factor. This is demonstrated in the numerical example in Fig. 12.6 in which only δ has a different value from Fig. 12.5. Decreasing δ further leads to a significant decrease in the observed structure and to a substantial suppression of the oscillations.

In a Fermi–Dirac gas at zero temperature, damping of the oscillations is due to the remaining integration over the Fermi sphere in eqn (12.38). This damping is seen in Fig. 12.7, which again represents the evolution over the entire rescaled $r\tau$ plane. We have chosen the dimensionless Fermi momentum, eqn (12.39), to have the value $y_F = 0.1$ which leads to a damping of the temporal oscillations along the boundstate ridge similar to that resulting from a choice $\delta = 100$ in the Maxwell–Boltzmann case. Also, observe the diagonal disturbance of outgoing waves behind

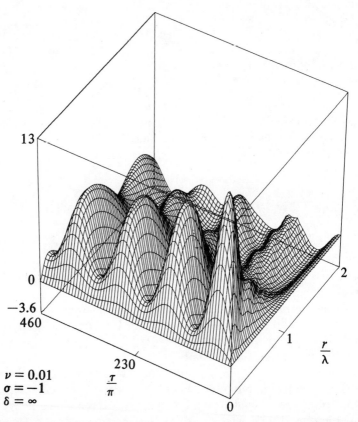

Fig. 12.5. Perspective views (above and right) of the surface $(r/\lambda)^2[\rho(r,\tau)/\rho(0)-1]$ over the $r\tau$ plane.

which pronounced radial oscillations build up into the final stationary distribution for $\tau \to \infty$, as can also be inferred from the explicit limit in eqn (12.48).

The case of a strong attractive potential serves very well to illustrate the transient nature of the calculations in this section. We have chosen this potential so strong that a boundstate for a particle of mass m develops at an energy $E_0 < 0$. What we have calculated here is the time evolution of an ideal gas in such a potential. Now all particles of an ideal gas have positive energies, i.e. $\varepsilon_\mathbf{k} = \hbar^2 k^2/2m > 0$. Moreover, no inelastic processes are in-

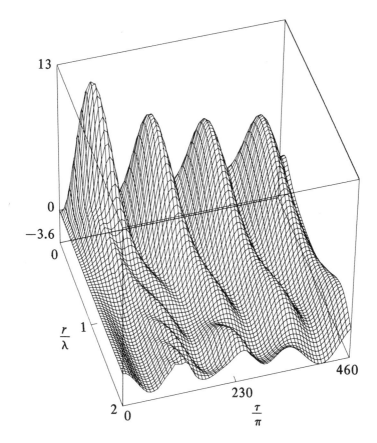

cluded in our model. It is therefore impossible for any gas particles to make the transition from the continuum into the boundstate. What we observe in Figs. 12.5–12.7 is only a readjustment of the wavefunctions in the presence of the external potential. In the limit $\tau \to \infty$, this adjustment is completed and a stationary mass distribution is achieved. However, to reach the equilibrium distribution with a thermal occupation of the boundstate, energy-dissipating mechanisms must be added to the model, a realistic one being simply the inclusion of two-body interactions. In this case two gas particles can collide within range of the external (attractive) potential with one particle losing enough energy to drop into the boundstate, the other particle

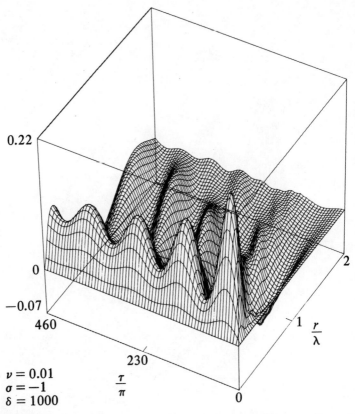

Fig. 12.6. Perspective views (above and right) of the $(r/\lambda)^2[\rho(r,\tau)/\rho(0)-1]$ over the $r\tau$ plane.

picking up this energy. The excess momentum is absorbed by the potential center which acts here as a third (infinitely heavy) scattering partner.

12.6. Resonances

In the appendix, we show that our separable potential will develop a resonance at the energy

$$\frac{E_R}{2V_0} = \frac{2^{-5}}{\pi - \nu^2} + \left[\frac{2^{-5}}{\pi(2^{-5}/\pi - 4\nu^2)}\right]^{\frac{1}{2}} \qquad (12.59)$$

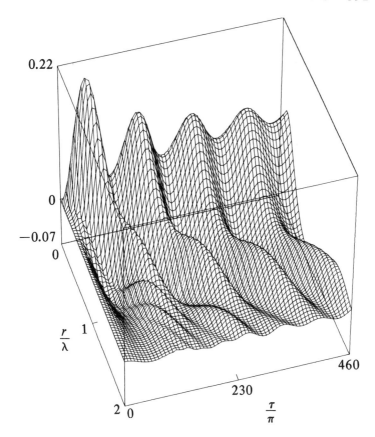

if $\sigma = +1$ and if

$$\nu < \frac{1}{8\sqrt{2\pi}} \quad (12.60)$$

The T matrix is then given by

$$T(x) = \frac{-\nu x/[8\pi(x^2+\nu^2)]}{x^2+\nu^2-1/(16\pi)(x^2-\nu^2-2i\nu)/(x^2+\nu^2)} \quad (12.61)$$

In our numerical example for the density evolution in a resonating potential, we again choose $\nu = 0.01$. This leads to a resonance at $E_R/2V_0 = 0.0196$. Approximating the T matrix by a Breit–Wigner form, this resonance will have a width $\Gamma/4V_0 = 3.96 \times 10^{-4}$. In Fig. 12.8 we show the density evolution at the

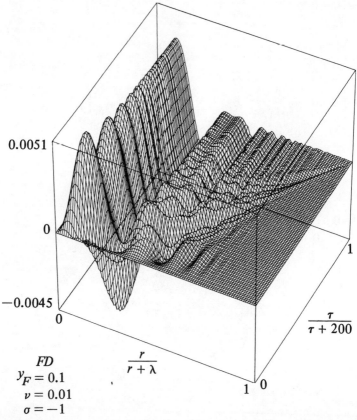

FD
$y_F = 0.1$
$\nu = 0.01$
$\sigma = -1$

Fig. 12.7. Perspective views (above and right) of the quantity $(r/\lambda)^2[\rho(r,\tau)/\rho(0)-1]$ in a Fermi–Dirac gas at $T=0$ plotted over the rescaled $r\tau$ plane.

origin of the potential. The dashed curves present the results for an attractive potential with a boundstate showing more or less undamped oscillations for $\sigma = \infty$, damped oscillations for $\delta = 10^3$, and a rather smooth time evolution for $\delta = 100$ due to overwhelming temperature smearing. In the resonance case a rather similar behavior occurs initially; however, the oscillations with periods of the order of the inverse resonance energy now sit on a sloping background that decays toward the stationary density with the lifetime $\tau_{\text{rel}} = \hbar/\Gamma$ of the resonance. For $\delta = \infty$, we see

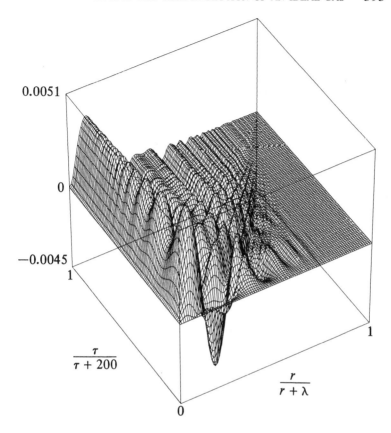

that the second minimum dips down to a strikingly low density of 2.6×10^{-5}. This feature is very sensitive to the choice of potential parameters. In particular, for $\delta = \infty$ we find from eqn (12.48) that $\rho(r=0, t)$ will dip down to zero at discrete times given by

$$\tau = (2n - \tfrac{1}{4})16\pi + 0\left(\frac{\ln n}{\sqrt{n}}\right) \qquad n = 1, 2, \ldots \quad (12.62)$$

if ν takes the value

$$\nu = \frac{\ln \tau\pi/16}{\tau\sqrt{\pi} + \pi\sqrt{\tau/2}} \quad (12.63)$$

For other values of ν we find that the density will go to zero at various radial distances from the potential center at different

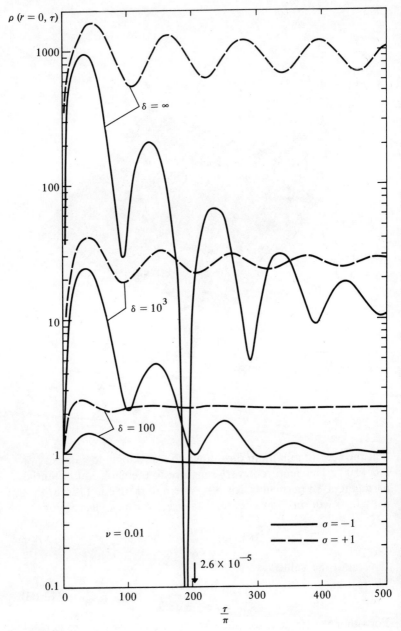

Fig. 12.8. Density at $r = 0$ as a function of τ for various δ and $\sigma = \pm 1$.

times. These dips are certainly surprising at first glance, but we must recall that they occur only in very strong potentials, indeed much stronger than is necessary for the occurrence of resonances. In this case, we find that the system's response to the external potential overshoots initially in the huge resonating mass enhancement peak, responding in turn with a complete mass depletion. Predictably, we then also find similar effects in very strong attractive potentials exhibiting boundstates. It is also not surprising that these complete depletion dips are readily filled in for finite δ.

In Fig. 12.9 we again show the density $(r\gamma)^2[\rho(r,\tau)/\rho(0)-1]$ over a finite portion of the $r\tau$ plane. The big peak near the origin of the $r\tau$ plane is the resonating mass enhancement. It is followed for both larger r and larger τ by secondary resonating peaks and further out by substantial density waves carrying the disturbance further out in space and time. Analytically, we can trace these waves back to the last term in eqn (12.48). The argument of the error function is, for $k=2$,

$$e^{i\pi/4}\sqrt{\tau}\left[\nu + i\left(\frac{1}{4\sqrt{\pi}} - \frac{r\gamma}{2\nu\tau}\right)\right] \tag{12.64}$$

giving roughly the wave velocity in the inner bracket. The first term, combined with the exponential prefactor in front of the error function, determines the wave envelope for fixed τ. This is fairly steep for smaller r and tails off for larger r. The waves will thus not only move out but the innermost one will disappear and feed into the next outer one as a result of dispersion.

In Fig. 12.10 we present a finite δ calculation for the same potential parameters as in Fig. 12.9. Some of the structure is smeared out and the series of waves combines into a single wave of rather complicated appearance. However, the resonance peaks are still followed by a trough of mass depletion at more or less constant r, reflecting the fact that this potential, though allowing a resonance, is basically repulsive. We complete our examples involving strong resonating potentials with a complete view in Fig. 12.11 of the density evolution over the rescaled $r\tau$ plane.

This concludes our discussion of transient effects in the initial time evolution of large systems, as illustrated here in an exactly solvable model. We hope that this quite detailed exposition serves both to give a glimpse at the fascinating aspects of trans-

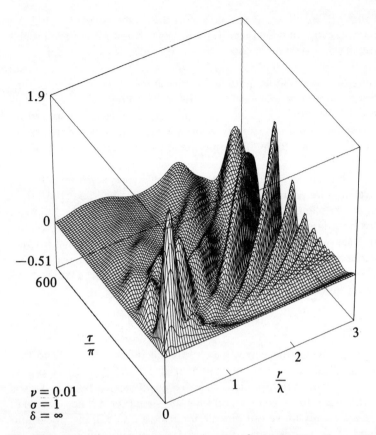

Fig. 12.9. Perspective view of the surface $(r/\lambda)^2[\rho(r,\tau)/\rho(0)-1]$ over the $r\tau$ plane.

ient (i.e. nonthermodynamic) effects and to stress the powerful role of an exactly solvable model in their study.

12.7. Appendix: A Single Particle in a Separable Potential

In this appendix we summarize the properties of a single particle in a static external potential consisting of a single separable term in each partial wave l, i.e. (see Newton, 1967)

$$V_{\mathbf{k}\mathbf{k}'} = \frac{g}{(2\pi)^3} \sum_{l=0}^{\infty} \sum_{m=0}^{l} v_l(\mathbf{k}) v_l^*(\mathbf{k}) Y_{lm}^*(\hat{\mathbf{k}}) Y_{lm}(\hat{\mathbf{k}}') \quad (12.65)$$

where $Y_{lm}(\hat{\mathbf{k}})$ is a spherical harmonic and $\hat{\mathbf{k}} = \mathbf{k}/k$. The outgoing stationary scattering solution of the nonrelativistic Schrödinger equation is given in momentum space by

$$\chi_{\mathbf{k}}(\mathbf{p}) = \sqrt{mk}\,\delta(\mathbf{p}-\mathbf{k}) + \frac{mg}{\pi^2} \lim_{\delta \to 0} \sum_{l,m_l} \frac{\sqrt{mk}}{k^2 - p^2 + i\delta^*}$$
$$\times \frac{v_l(p)v_l^*(k)}{1 - \frac{mg}{\pi^2}\int q^2 v_l^2(q)/(k^2 - q^2 + i\delta)\,dq} Y_{lm_l}(\hat{\mathbf{p}})Y_{lm_l}(\hat{\mathbf{k}}) \quad (12.66)$$

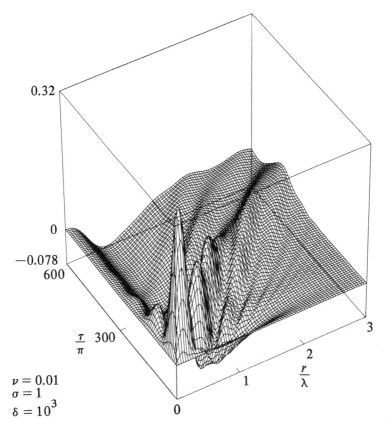

$\nu = 0.01$
$\sigma = 1$
$\delta = 10^3$

Fig. 12.10. Perspective view of the surface $(r/\lambda)^2[\rho(r,\tau)/\rho(0) - 1]$ over the $r\tau$ plane.

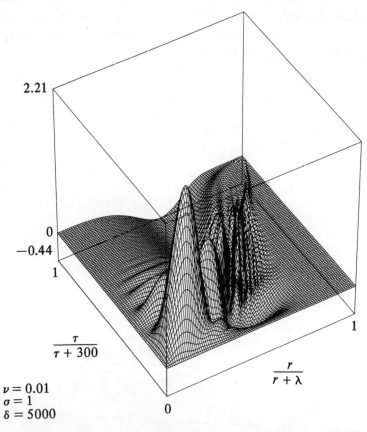

Fig. 12.11. Perspective views (above and right) of the quantity $(r/\lambda)^2[\rho(r,\tau)/\rho(0)-1]$ in a Fermi–Dirac gas at $T=0$ plotted over the rescaled $r\tau$ plane.

We define the on-shell, partial-wave T matrix by

$$\langle \mathbf{k}_1 | T | \mathbf{k}_2 \rangle = \sum_{l,m_l} T_l(k) Y^*_{lm_l}(\hat{\mathbf{k}}_1) Y_{lm_l}(\hat{\mathbf{k}}_2) \qquad (12.67)$$

where $|\mathbf{k}_1| = |\mathbf{k}_2| = k = (2mE)^{\frac{1}{2}}$ and where

$$T_l(k) = -\frac{\dfrac{g}{2\pi} mk v_l(k) v_l^*(k)}{1 - \dfrac{g}{2\pi^2}\int p^2 v_l(p) v_l^*(p)/(E - p/2m^2 + i\delta)\,dp} \qquad (12.68)$$

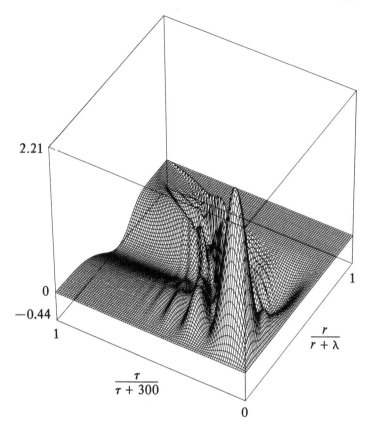

and the total cross section is given by

$$\sigma(k) = \frac{4\pi}{k^2} \sum_l (2l+1) T_l(k) T_l^*(k) \qquad (12.69)$$

There will be a resonance in the lth partial wave if the real part of the denominator of $T_l(k)$ has a zero for $E = E_R > 0$. Each partial wave can have, moreover, at most one boundstate for $g < 0$ at an energy $E_l^{bs} < 0$ determined by

$$1 + \frac{g}{2\pi^2} \int_0^\infty \frac{p^2 v_l(p) v_l^*(p)\, dp}{|E_l^{bs}| - p^2/2m} = 0 \qquad (12.70)$$

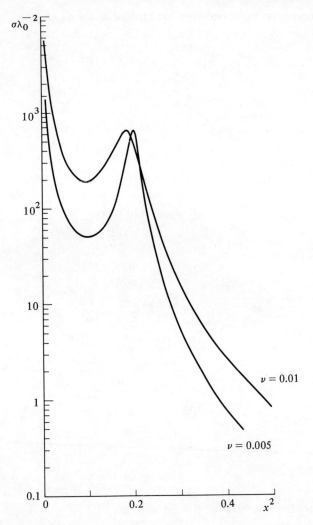

Fig. 12.12. Normalized scattering cross section for two potentials like eqn (12.73), both showing a resonance.

with a boundstate wave function

$$\psi_l(p) = \frac{N_l}{E_l^{bs} - p^2/2m} v_l(p) \qquad (12.71)$$

where

$$N_l = \left[\frac{1}{(2\pi)^3} \int_0^\infty \frac{p^2 v_l(p) v_l^*(p)\, dp}{(E_l^{bs} - p^2/2m)^2} \right]^{-\frac{1}{2}} \qquad (12.72)$$

is the normalization constant.

In particular, for a potential‡

$$V_{\mathbf{k}\mathbf{k}'} = \frac{g}{(2\pi)^3} \frac{1}{k^2 + \gamma^2} \frac{1}{k'^2 + \gamma^2} \qquad (12.73)$$

acting in s waves only, we obtain

$$E_0^{bs} = -\left[\frac{\gamma}{\sqrt{2m}} - \left(\frac{|g|}{8\pi\gamma}\right)^{\frac{1}{2}} \right]^2 \qquad (12.74)$$

if

$$\left(\frac{-g}{\gamma 8\pi}\right)^{\frac{1}{2}} - \gamma\sqrt{2m} > 0 \qquad (12.75)$$

For $g > 2^5 \gamma^3/m$, a resonance develops at an energy

$$E_R = \frac{g}{16\pi\gamma} - \frac{\gamma^2}{2m} + \left[\frac{g}{16\pi\gamma} \left(\frac{g}{16\pi\gamma} - \frac{2\gamma^2}{m} \right) \right]^{\frac{1}{2}} \qquad (12.76)$$

A numerical example of a cross section is given in Fig. 12.12 for the normalized variables introduced in eqn (12.31).

---

‡ This potential is referred to in nuclear physics as the Yamaguchi potential (Yamaguchi, 1954) but was already introduced by Wheeler (1936). See also McCarthy (1968).

Bibliography

ABRAGAM, A. (1961). *The Principles of Nuclear Magnetism.* Clarendon, Oxford.

ABRAHAM, D. B., BAROUCH, E., GALLAVOTTI, G., and MARTIN-LÖF, A. (1970). Thermalization of a magnetic impurity in the isotropic XY model. *Phys. Rev. Lett.* **25,** 1449–1450.

ABRAMOWITZ, M., and STEGUN, I. A. (1966). *Handbook of Mathematical Functions.* National Bureau of Standards, Washington, D.C.

AHLERS, G. (1974). Low temperature studies of the Rayleigh-Bénard instability and turbulence. *Phys. Rev. Lett.* **33,** 1185–1188.

―――― (1975). The Rayleigh-Bénard instability at helium temperatures. In *Fluctuations, Instabilities and Phase Transitions.* Ed. T. Riste, Plenum, New York.

ALDER, B. J. (1973). Computer dynamics. *Ann. Rev. Phys. Chem.* **24,** 325–337.

――――, and WAINWRIGHT, T. E. (1960). Studies in molecular dynamics. II. Behavior of a small number of elastic spheres. *J. Chem. Phys.* **33,** 1439–1451.

D'ANCONA, U. (1954). *The Struggle for Existence.* Brill, Leiden.

ANDRADE, E. N. DA C. (1934). A theory of the viscosity of liquids, Parts I and II. *Phil. Mag.* **17,** 497–511, 698–732.

ANSTIS, G. R., GREEN, H. S., and HOFFMAN, D. K. (1973). Kinetic theory of a one-dimensional model. *J. Math. Phys.* **14,** 1437–1443.

AVSEC, D. (1939). Ph.D. Thesis, University of Paris.

BALESCU, R. (1963). *Statistical Mechanics of Charged Particles.* Interscience, London.

―――― (1975). *Equilibrium and Non-equilibrium Statistical Mechanics.* Wiley, New York.

BARKER, J. A., and HENDERSON, D. (1976). What is "liquid"? Understanding the states of matter. *Rev. Mod. Phys.* **48,** 587–671.

BAXTER, R. J. (1971). Distribution functions. In *Physical Chemistry, An Advanced Treatise*, vol. VIII A/*Liquid State*, pp. 267–334. Ed. D. Henderson, Academic, New York.

BÉNARD, H. (1900). Les tourbillons cellulaires dans une nappe liquide. *Revue Générale des Sciences Pures et Appliquées* **11**, 1261–1271 and 1309–1328.

BERNARD, W., and CALLEN, H. B. (1959). Irreversible thermodynamics of nonlinear processes and noise in driven systems. *Rev. Mod. Phys.* **31**, 1017–1044.

BERTRAND, J. L. F. (1887). *Thermodynamique*. Paris.

BLATT, J. M. (1959). An alternative approach to the ergodic problem. *Progr. Theor. Phys.* **22**, 745–756.

——, and OPIE, A. H. (1974a). A new derivation of the Boltzmann transport equation. *J. Phys.* **A7**, L113–115.

—— and —— (1974b). Non-equilibrium statistical mechanics I. The Boltzmann transport equation. *J. Phys.* **A7**, 1895–1906.

BLOCH, C., and DOMINICIS, C. DE. (1958). Un développement du potentiel de Gibbs d'un système quantique composé d'un grand nombre de particules. *Nucl. Phys.* **7**, 459–479.

BLOCK, M. J. (1956). Surface tension as the cause of Bénard cells and surface deformation in a liquid film. *Nature* **178**, 650–651.

BOCCHIERI, P., and LOINGER, A. (1957). Quantum recurrence theorem. *Phys. Rev.* **107**, 337–338.

BOGOLYUBOV, N. N. (1946). Problems of a dynamical theory in statistical physics. *J. Phys.* (U.S.S.R.) **10**, 256–265. Transl. in *Studies in Statistical Mechanics*, vol. 1. Eds. J. de Boer and G. E. Uhlenbeck, North-Holland, Amsterdam, 1962.

—— (1970). Lectures on Quantum Statistics. I: Quantum Statistics. II: Quasi-averages. Gordon and Breach, New York.

BOHR, N. (1932). Faraday Lecture: Chemistry and the quantum theory of atomic constitution. *J. Chem. Soc.*, 349–384.

BOLTZMANN, L. (1872). Weitere Studien über das Wärmegleichgewicht unter Gasmolekülen. *Wien. Ber.* **66**, 275–370.

—— (1874). Zur Theorie der elastischen Nachwirkung. *Sitzungsbericht der Math Naturw. Kl. Kaiserl. Akad. Wien* **70** (2), 175–306.

—— (1877). Bemerkungen über einige Probleme der mechanischen Wärmetheorie. *Wien. Ber.* **75**, 62–100.

—— (1887). Zur Theorie der thermoelektrischen Erscheinungen. *Wien. Ber.* **96**, 1258–1297.

—— (1895). On certain questions of the theory of gases. *Nature*, **51**, 413–415.

—— (1895). Erwiderung an Culverwell. *Nature* **51**, 581.

—— (1896). Entgegnung auf die wärmetheoretischen Betrachtungen des Hrn. E. Zermelo. *Wied. Ann.* **57**, 773–784.

────── (1897). Zu Hrn. Zermelos Abhandlung 'über die mechanischen Erklärung irreversibler Vorgänge.' *Wied. Ann.* **60,** 392–398.

BOLTZMANN, L. (1897). Über irreversible Strahlungsvorgänge. *Sitzungsberichte der Preuss. Akad. der Wissensch. zu Berlin,* 660–662.

────── (1897). Über irreversible Strahlungsvorgänge. Zweite Mittheilung. *Sitzungsberichte der Preuss. Akad. der Wissensch. zu Berlin,* 1016–1018.

────── (1898). Über vermeintlich irreversibele Strahlungsvorgänge. *Sitzungsberichte der Preuss. Akad. der Wissensch. zu Berlin,* 182–187.

────── (1909). Wissenschaftliche Abhandlungen. Ed. F. Hasenöhrl, Barth, Leipzig. Reproduced by Chelsea Publications, New York, 1968.

BORCKMANS, P., and WALGRAEF, D. (1967). Irreversibility in paramagnetic spin systems. *Physica,* **35,** 80–96.

BORN, M., and GREEN, H. S. (1946). A general kinetic theory of liquids. I. The molecular distribution functions. *Proc. Roy. Soc. London* **A188,** 10–18.

────── and ────── (1947). A general kinetic theory of liquids. III. Dynamical properties. *Proc. Roy. Soc. London* **A190,** 455–474.

BOUSSINESQ, J. (1903). Théorie analytique de la chaleur. Gauthier-Villars, Paris, **2,** 172.

BRAUN, W., and HEPP, K. (1977). The Vlasov dynamics and its fluctuations in the 1/N limit of interacting classical particles. *Commun. Math. Phys.* **56,** 101–113.

BRITTIN, W. E., and CHAPPELL, W. R. (1962). The Wigner distribution function and second quantization in phase space. *Rev. Mod. Phys.* **34,** 620–627.

BROWN, A. L., and PAGE, A. (1970). *Elements of Functional Analysis.* Van Nostrand-Reinhold, London.

BURNETT, D. (1935). The distribution of molecular velocities and the mean motion in a non-uniform gas. *Proc. Lond. Math. Soc.* **40,** 382–435.

BUSSE, F. H. (1962). Inaugural Dissertation. Ludwig-Max. University, Munich. Transl. S. H. Davis, Rand Corp. LT 66–19.

──────, and WHITEHEAD, J. A. (1971). Instabilities of convection rolls in a high Prandtl number fluid. *J. Fluid Mech.* **47,** 305–320.

────── and ────── (1974). Oscillatory and collective instabilities in large Prandtl number convection. *J. Fluid Mech.* **66,** 67–79.

CALLEN, H. B. (1960). *Thermodynamics.* Wiley, New York.

──────, and GREENE. R. F. (1952). On a theorem of irreversible thermodynamics. *Phys. Rev.* **86,** 702–710.

──────, and ────── (1952). On a theorem of irreversible thermodynamics II. *Phys. Rev.* **88,** 1387–1391.

———, and WELTON, T. A. (1951). Irreversibility and generalized noise. *Phys. Rev.* **83,** 34–40.

CASIMIR, H. B. G. (1945). On Onsager's principle of microscopic reversibility. *Rev. Mod. Phys.* **17,** 343–350.

CERCIGNANI, C. (1975). *Theory and Application of the Boltzmann Equation.* Scottish Academic Press, Edinburgh.

CHANDRASEKHAR, S. (1954). On characteristic value problems in high order differential equations which arise in studies on hydrodynamic and hydromagnetic stability. *Amer. Math. Monthly,* **61,** 32–45.

CHANDRASEKHAR, S. (1961). *Hydrodynamic and Hydromagnetic Stability.* Clarendon, Oxford.

CHAPMAN, S., and COWLING, T. G. (1939). *Mathematical Theory of Non-uniform Gases.* Cambridge University Press, Cambridge. (Revised 1951.)

CHARLSON, G. S. and SANI, R. L. (1970). Thermoconvective instability in a bounded cylindrical fluid layer. *Intern. J. Heat Mass Trans.* **13,** 1479–1496.

CHEN, S. H. (1971). Structure in liquids. In *Physical Chemistry: An Advanced Treatise,* vol. VIIIa, pp. 85–156. Eds. H. Eyring, D. Henderson, and W. Jost, Academic, New York.

CHESTER, G. V., and THELLUNG, A. (1959). On the electrical conductivity of metals. *Proc. Phys. Soc.* **73,** 745–766.

———, and ——— (1961). The law of Wiedemann and Franz. *Proc. Phys. Soc.* **77,** 1005–1013.

CHOH, S. T. (1958). Ph.D. Thesis, University of Chicago.

———, and UHLENBECK, G. E. (1958). The kinetic theory of dense gases. University of Michigan report.

CHRISTENSEN, R. M. (1971). *Theory of Viscoelasticity, An Introduction.* Academic, New York.

CHRISTOPHERSON, D. G. (1940). Note on the vibration of membranes. *Quart. J. Math.* (Oxford Series), **11,** 63–65.

CLARKE, B. L. (1980). Stability of complex reaction networks. *Adv. Chem. Phys.* **43,** 1–215.

———, and RICE, S. A. (1970). Generalization of hydrodynamics to include single particle modes and fluctuations. *Physics of Fluids* **13,** 271–290.

CLEVER, R. M., and BUSSE, F. H. (1974). Transition to time-dependent convection. *J. Fluid Mech.* **65,** 625–645.

CLUSIUS, K., KÖLSCH, W., and WALDMANN, L. (1941). Isotopentrennung und Verbrennungsmechanismus in aufsteigenden H_2-D_2-Flammen. *Z. Physikal. Chem.* **A189,** 131–162.

COHEN, E. G. D. (1966). On the statistical mechanics of moderately dense gases not in equilibrium. In Lectures in Theoretical Physics,

VIII A, pp. 145–181. Ed. W. E. Brittin, University of Colorado Press, Boulder.

COLEMAN, B. D. (1970). On the stability of equilibrium states of general fluids. *Arch. Ration. Mech. Anal.* **36,** 1–32.

———, and TRUESDELL, C. (1960). On the reciprocal relations of Onsager. *J. Chem. Phys.* **33,** 28–31.

COOPERSMITH, M., and MANDEVILLE, G. (1974). Irreversible behavior of interacting systems. I. The approach to equilibrium. *J. Stat. Phys.* **10,** 391–403.

DAVIDSON, R., and KOZAK, J. J. (1975). On the relaxation to quantum-statistical equilibrium of the Wigner-Weisskopf atom in a one-dimensional radiation field VII. *J. Math. Phys.* **16,** 1013–1022.

DAVIES, R. O. (1952). Transformation properties of the Onsager relations. *Physica* **18,** 182.

DAVIS, S. H. (1967). Convection in a box: Linear theory. *Fluid Mech.* **30,** 465–478.

———, and SEGEL, L. A. (1968). Effects of surface curvature and property variation on cellular convection. *Phys. Fluids* **11,** 470–476.

DEBYE, P., and HÜCKEL, E. (1923). Zur Theorie der Elektrolyte I. Gefrierpunktserniedrigung und verwandte Erscheinungen. *Physik. Z.* **24,** 185–206.

DEGN, H. (1972). Oscillating chemical reactions in homogeneous phase. *J. Chem. Educ.* **49,** 301–307.

DIRICHLET, G. L. (1846). Über die Stabilität des Gleichgewichts. *J. Reine Angew. Math.* **32,** 85–88.

DOMENICALI, C. A. (1954). Irreversible thermodynamics of thermoelectricity. *Rev. Mod. Phys.* **26,** 237–275.

DORFMAN, J. R., and COHEN, E. G. D. (1965). On the density expansion of the pair distribution function for a dense gas not in equilibrium. *Phys. Lett.* **16,** 124–125.

ECKART, C. (1940). The thermodynamics of irreversible processes. I. The simple fluid. *Phys. Rev.* **58,** 267–269.

——— (1940). The thermodynamics of irreversible processes. II. Fluid mixtures. (Correction in *Phys. Rev.* **58,** 924.) *Phys. Rev.* **58,** 269–275.

——— (1940). The thermodynamics of irreversible processes. III. Relativistic theory of the simple fluid. *Phys. Rev.* **58,** 919–924.

EDDINGTON, A. S. (1929). *The Nature of the Physical World.* Cambridge University Press, Cambridge.

EDELSON, D., FIELD, R. J., and NOYES, R. M. (1975). Mechanistic details of the Belousov-Zhabotinskii oscillations. *Int. J. Chem. Kinet.* **7,** 417–432.

EGELSTAFF, P. (1967). *An Introduction to the Liquid State.* Academic, New York.

EHRENFEST, P., and EHRENFEST, T. (1911). Begriffliche Grundlagen der statistischen Auffassung in der Mechanik. *Encykl. Math. Wiss.* **4**(32). Trans. in *The Conceptual Foundations of the Statistical Approach in Mechanics*, Cornell University Press, Ithaca, 1959.

EIGEN, M. (1971). Selforganization of matter and the evolution of biological macromolecules. *Die Naturwissenschaften* **58**, 465–523.

EINSTEIN, A. (1909). Über die Entwicklung unserer Anschauungen über das Wesen und die Konstitution der Strahlung. *Phys. Z.* **10**, 817–826.

EINSTEIN, A. (1910). Theorie der Opaleszenz von homogenen Flüssigkeiten und Flüssigkeitsgemischen in der Nähe des kritischen Zustandes *Ann. d. Physik* **33**, 1275–1298.

ELIASHBERG, G. M. (1961). Transport equation for a degenerate system of Fermi particles. *JETP* (U.S.S.R.) **41**, 1241. (Also in *Sov. Phys. JETP* **14**, 886–892, 1962.)

ELTON, C. (1942). *Voles, Mice and Lemmings*. Oxford University Press, Oxford.

ENDERBY, J. E. (1968). Neutron scattering studies of liquids. In *Physics of Simple Fluids*. Eds. H. N. V. Temperley, J. S. Rowlinson, and G. S. Rushbrooke, North-Holland, Amsterdam.

ENSKOG, D. (1917). Kinetic theory of processes in moderately dense gases. Dissertation, Uppsala University.

———— (1929). Über die Entropie der Gase bei irreversiblen Prozessen. *Z. Phys.* **54**, 498–504.

ERNST, M. H., HAINES, L. K., and DORFMAN, J. R. (1969). Theory of transport coefficients for moderately dense gases. *Rev. Mod. Phys.* **41**, 296–316.

EYRING, H. (1936). Viscosity, plasticity, and diffusion as examples of absolute reaction rates. *J. Chem. Phys.* **4**, 283–291.

———— (1962). The transmission coefficient in reaction rate theory. *Rev. Mod. Phys.* **34**, 616–619.

————, and JHON, M. S. (1969). *Significant Liquid Structures*. Wiley, New York.

FEYNMAN, R. P. (1955). Application of quantum mechanics to liquid helium. In *Progress in Low-Temperature Physics*, vol. I. Ed. C. J. Gorter, North-Holland, Amsterdam.

FIELD, R. J., KÖRÖS, E., and NOYES, R. J. (1972). Oscillations in chemical systems. II. Thorough analysis of temporal oscillations in the bromate-cerium-malonic acid system. *J. Amer. Chem. Soc.* **94**, 8649–8664.

————, and NOYES, R. M. (1974). Oscillations in chemical systems. IV. Limit cycle behavior in a model of a real chemical reaction. *J. Chem. Phys.* **60**, 1877–1884.

FLEMING, P. D., and COHEN, C. (1976). Hydrodynamics of solids. *Phys. Rev.* **13**, 500–516.

FLÜGGE, W. (1967). *Visco-elasticity.* Blaisdell, Waltham, N.Y.

FORSTER, D. (1975). *Hydrodynamic Fluctuations, Broken Symmetry and Correlation Functions.* W. A. Benjamin, Reading.

———, LUBENSKY, T. C., MARTIN, P. C., SWIFT, J., and PERSHAM, P. S. (1971). Hydrodynamics of liquid crystals. *Phys. Rev. Lett.* **26**, 1016–1019.

FORSTER, D., MARTIN, P. C., and YIP, S. (1968). Moments of the momentum density correlation functions in simple liquids. *Phys. Rev.* **170**, 155–159.

———, ———, and ——— (1968). Moment method approximation for the viscosity of simple liquids: Application to Argon. *Phys. Rev.* **170**, 160–163.

FRASER, J. T., HABER, F. C., and MULLER, G. H. (eds.) (1972). *The Study of Time I.* Springer-Verlag, New York.

——— and LAWRENCE, N. (Eds.) (1975). *The Study of Time II.* Springer-Verlag, New York.

———, ———, and PARK, D. (eds.) (1978). *The Study of Time III.* Springer-Verlag, New York.

FRENKEL, J. (1946). *Kinetic Theory of Liquids.* Clarendon, Oxford.

FRÖHLICH, H. (1967). Microscopic derivation of the equations of hydrodynamics. *Physica* **37**, 215–226.

——— (1968). Long-range coherence and energy storage in biological systems. *Int. Journ. Quantum Chem.* **2**, 641.

——— (1969). The macroscopic wave equations of superfluids. *Phys. Kondens. Materie* **9**, 350–358.

——— (1969). General remarks on the connection of the laws of micro- and macrophysics. *J. Phys. Soc. Japan* **26**, Suppl. 189–195.

——— (1973). The connection between macro- and microphysics. *Rivista del Nuovo Cimento* **3**, 490–534.

——— (1974). Theory of superconductivity without solution of many body problems. *Collective Phenomena* **1**, 173–184.

FUJITA, S. (1962). Theory of transport coefficients. II. Viscosity coefficients of quantum gases obeying the Boltzmann statistics. *J. Math. Phys.* **3**, 359–367.

GALASIEWICZ, Z. M. (1970). *Superconductivity and Quantum Fluids.* Pergamon, New York.

GILL, J. C. (1975). The establishment of thermal equilibrium in paramagnetic crystals. *Rep. Prog. Phys.* **38**, 91–150.

GLANSDORFF, P., and PRIGOGINE, I. (1954). Sur les propriétés différentielles de la production d'entropie. *Physica* **20**, 773–780.

——— and ——— (1971). *Thermodynamic Theory of Structure, Stability and Fluctuations.* Wiley, New York.

GOEL, N. S., MAITRA, S. C., and MONTROLL, E. W. (1971). On the Volterra and other nonlinear models of interacting populations. *Rev. Mod. Phys.* **43,** 231–276.

GOLD, T. (ed.) (1967). *The Nature of Time.* Cornell University Press, Ithaca.

GOLLUB, J. P., and SWINNEY, H. L. (1975). Onset of turbulence in a rotating fluid. *Phys. Rev. Lett.* **35,** 927–930.

GRAD, H. (1949). On the kinetic theory of rarefied gases. *Comm. Pure Appl. Math.* **2,** 331–407.

——— (1958). Principles of the kinetic theory of gases. In *Encyclopedia of Physics*, vol. XII, *Thermodynamics of Gases*. Ed. S. Flügge, Springer, Berlin.

——— (1963). Asymptotic theory of the Boltzmann equation. *Phys. Fluids* **6,** 147–181.

GRAHAM, A. (1934). VIII. Shear patterns in an unstable layer of air. *Phil. Trans. Roy. Soc. London* **232A,** 285–296.

GRECOS, A. P. (1977). Lectures on dissipative processes in dynamical systems. In *Proceedings of the International Conference on Frontiers of Theoretical Physics*. Ed. F. C. Auluck, Indian National Science Academy, New Delhi.

———, and MARESCHAL, M. (1976). Evolution of observables in a Laurent hamiltonian system. *Physica* **83A,** 419–424.

———, and PRIGOGINE, I. (1972). Kinetic and ergodic properties of quantum systems—the Friedrichs model. *Physica* **59,** 77–96.

———, and ——— (1978). On the dynamical theory of irreversible processes and the microscopic interpretation of non-equilibrium entropy. In *Proceedings of the 13th IUPAP Conference on Statistical Physics*, Haifa, 1977, Eds. D. Cabib, C. G. Kuper, and I. Riess. (Also in *Ann. Israel Phys. Soc.* **2**(1), 83–97.)

GREEN, H. S. (1952). *The Molecular Theory of Fluids.* North-Holland, Amsterdam. (Also published by Dover, New York, 1969).

——— (1971). Self-consistent kinetic equations. In *Kinetic Equations*. Eds. R. L. Liboff and N. Rostoker, Gordon and Breach, New York.

GREEN, M. S. (1952). Markoff random processes and the statistical mechanics of time-dependent phenomena. *J. Chem. Phys.* **20,** 1281–1295.

——— (1954). Markoff random processes and the statistical mechanics of time-dependent phenomena. II. Irreversible processes in Fluids. *J. Chem. Phys.* **22,** 398–413.

GRODZKA, P. G., and BANNISTER, T. C. (1972). Heat flow and convection demonstration experiments aboard Apollo 14. *Science* **176,** 506–508.

DE GROOT, S. R. and MAZUR, P. (1969). *Non-equilibrium Thermodynamics.* North-Holland, Amsterdam.

GROSS, M. J. (1967). Laboratory analogies for convection problems. In *Mantles of the Earth and Terrestrial Planets.* Ed. S. K. Runkorn, Wiley, New York.

———, and PORTER, J. E. (1966). Electrically induced convection in dielectric liquids. *Nature* **212,** 1343–1345.

GUGGENHEIM, E. A. (1939). Grand partition functions and so-called "thermodynamic probability." *J. Chem. Phys.* **7,** 103–107.

GUTII, E., and MAYERHÖFER, J. (1940). On the deviations from Ohm's law at high current densities. *Phys. Rev.* **57,** 908–915.

GYARMATI, I. (1970). *Non-equilibrium Thermodynamics: Field Theory and Variational Principles.* Springer, Berlin.

HAAR, D. TER (1961). Theory and applications of the density matrix. *Rep. Prog. Phys.* **24,** 304–362.

HAASE, R. (1969). *Thermodynamics of Irreversible Processes.* Addison-Wesley, Reading, Mass.

HADAMARD, J. (1923). *Lectures on Cauchy's Problem in Linear Partial Differential Equations.* Yale University Press, New Haven. (Reprinted by Dover, New York, 1952.)

HAHN, E. L. (1950). Spin echoes. *Phys. Rev.* **80,** 580–594.

HAKEN, H. (1975). Cooperative phenomena in systems far from thermal equilibrium and in non-physical systems. *Rev. Mod. Phys.* **47,** 67–121.

HARMAN, T. C., and HONIG, J. M. (1967). *Thermoelectric and Thermodynamic Effects and Applications.* McGraw-Hill, New York.

HARTREE, D. R. (1928). The wave mechanics of an atom with a non-Coulomb central field. Part I. Theory and methods. *Proc. Cambridge Phil. Soc.* **24,** 89–110.

HASHITSUMÈ, N. (1952). A statistical theory of linear dissipative systems *Progr. Theoret. Phys.* **8,** 461–478.

HEIKES, R. R., and URE, R. W., JR. (1961) *Thermoelectricity, Science and Engineering.* Interscience, New York.

HEMMER, P. C., MAXIMON, L. C., and WERGELAND, H. (1958). Recurrence time of a dynamical system. *Phys. Rev.* **111,** 689–694.

HERRING, C., and NICHOLS, M. H. (1949). Thermionic emission. *Rev. Mod. Phys.* **21,** 185–270.

HILBERT, D. (1924). *Grundzüge einer allgemeinen Theorie der Linearen Integralgleichungen.* Teubner, Vienna.

HIRSCHFELDER, J. O., CURTISS, C. F., and BIRD, R. B. (1954, 1966). *Molecular Theory of Gases and Liquids.* Wiley, New York.

HOARD, C. Q., ROBERTSON, C. R., and ACRIVOS, A. (1970). Experiments on the cellular structure in Bénard convection. *Int. J. Heat Mass Trans.* **13,** 849–856.

HOBSON, A. (1966). Irreversibility in simple systems. *Am. J. Phys.* **34,** 411–416.

―――― (1968). The status of irreversibility in statistical mechanics. *Phys. Lett.* **26A,** 649–650.

――――, and LOOMIS, D. N. (1968). Exact classical nonequilibrium statistical-mechanical analysis of the finite ideal gas. *Phys. Rev.* **173,** 285–295.

HOFELICH, F. (1969). On the definition of entropy for non-equilibrium states. *Z. Physik* **226,** 395–408.

HOHENBERG, P. C., and MARTIN, P. C. (1965). Microscopic theory of superfluid helium. *Ann. Phys.* **34,** 291–359.

HOOYMAN, G. J., DE GROOT, S. R., and MAZUR, P. (1955). Transformation properties of the Onsager relations. *Physica* **21,** 362–366.

VAN HOVE, L. (1955). Quantum mechanical perturbations giving rise to a statistical transport equation. *Physica* **21,** 517–540.

―――― (1957). The approach to equilibrium in quantum statistics. A perturbation treatment to general order. *Physica* **23,** 441–480.

―――― (1959). The ergodic behavior of quantum many-body systems. *Physica* **25,** 268–276.

―――― (1961). The problem of master equations. In *Proceedings of the International School of Physics.* Ed. P. Caldirola, Academic, New York.

――――, and VERBOVEN, E. (1961). The generalized transport equation for an electron: A solution in a simple case. *Physica* **27,** 418–432.

HUANG, K., and YANG, C. N. (1957). Quantum-mechanical many-body problem with hard-sphere interaction. *Phys. Rev.* **105,** 767–775.

HUNTINGTON, H. B. (1958). The elastic constants of crystals. In *Solid State Physics*, vol. 7. Eds. F. Seitz and D. Turnbull, Academic, New York.

HUSIMI, K. (1940). Some formal properties of the density matrix. *Proc. Phys. Math. Soc. Japan* **22,** 264.

ISRAEL, W., and STEWART, J. M. (1979). Progress in relativistic thermodynamics and electrodynamics of continuous media. In *Einstein Centenary Volume.* Ed. A. Held, Plenum, New York.

JANCEL, R. (1969). *Foundations of Classical and Quantum Statistical Mechanics.* Pergamon, Oxford.

JANNER, A., VAN HOVE, L., and VERBOVEN, E. (1962). The evolution to quantum statistical equilibrium for a simple model: The strong-coupling limit. *Physica* **28,** 1341–1360.

JANSSEN, H. K. (1974). Stochastisches Reaktionsmodell für einen Nichtgleichgewichtsphasenübergang. *Z. Physik* **270,** 67–73.

JEANS, J. (1954). *The Dynamical Theory of Gases*, 4th ed. Dover, New York.

JEFFREYS, H, (1951). The surface elevation in cellular convection. *Quart. J. Mech. Appl. Math.* **4,** 283–288.

JEPSEN, D. W. (1965). Dynamics of a simple many-body system of hard rods. *J. Math. Phys.* **6**, 405–413.

KADANOFF, L. P., and MARTIN, P. C. (1963). Hydrodynamic equations and correlation functions. *Ann. Phys.* **24**, 419–469.

VAN KAMPEN, N. G. (1954). Quantum statistics of irreversible processes. *Physica* **20**, 603–622.

―――― (1956). Grundlagen der Statistischen Mechanik der Irreversiblen Prozesse. *Fortschr. Physik* **4**, 405 437.

―――― (1961). A power series expansion of the master equation. *Can. J. Phys.* **39**, 551–567.

―――― (1965). Fluctuations in nonlinear systems. In *Fluctuation Phenomena in Solids.* Ed. R. E. Burgess, Academic, New York.

―――― (1969). Thermal fluctuations in nonlinear systems. *Adv. Chem. Phys.* **15**, 65–77.

―――― (1971). The case against linear response theory. *Physica Norvegica* **5**, 279–284.

KAN Y., and DORFMAN, J. R. (1977). Logarithmic terms in the density expansion of transport coefficients of moderately dense gases. *Phys. Rev.* **A16**, 2447–1469.

KESTIN, J., and DORFMAN, J. R. (1971). *A Course in Statistical Thermodynamics.* Academic, New York.

KIRKWOOD, J. G. (1946). The statistical mechanical theory of transport processes. *J. Chem. Phys.* **14**, 180–201.

KITTEL, C. (1973). On the nonexistence of temperature fluctuations in small systems. *Am. J. Phys.* **41**, 1211–1212.

―――― (1976). *Introduction to Solid State Physics*, 5th ed. Wiley, New York.

KLEIN, F., and SOMMERFELD, A. (1897). *Über die Theorie des Kreisels.* Teubner, Leipzig.

KLEIN, O. (1931). Zur quantemechanischen Begründung des zweiten Hauptsatzes der Wärmelehre. *Z. Physik* **72**, 767–775.

KLIMONTOVICH, YU. L. (1958). On the method of "second quantization" in phase space. *JETP* **6**, 753–760.

KNOPS, R. J., and WILKES, E. W. (1973). Theory of elastic stability. In *Handbuch der Physik, VIa 3: Mechanics of Solids III.* Ed. C. Truesdell, Springer, Berlin.

KNUDSEN, M. (1950). *Kinetic Theory of Gases: Some Modern Aspects;* 3rd ed. Methuen, London.

KOSCHMIEDER, E. L. (1966). On convection on a uniformly heated plane. *Beitr. Phys. Atmos.* **39**, 1–11.

―――― (1974). Bénard convection. *Adv. Chem. Phys.* **26**, 177–212.

KRAMERS, H. A. (1940). Brownian motion in a field of force and the diffusion model of chemical reactions. *Physica* **7**, 284–304.

KREUZER, H. J. (1975). On the microscopic derivation of balance equations in hydrodynamics. *Physica* **80A,** 585–594.

―――― (1976). Equilibrium properties of a mixture of reacting gases: A soluble model. *Am. J. Phys.* **44,** 970–973.

―――― (1978). Microscopic hydrodynamics in a nonlocal potential. *Nuovo Cimento* **45B,** 169–200.

――――, and HIOB, E. (1976). Time evolution of large multicomponent systems: Numerical studies in a solvable model. *Phys. Rev.* **14A,** 2321–2328.

――――, and KURIHARA, Y. (1977). Time evolution in a two-component system with catalytic reactions. *Physica* **87A,** 94–116.

――――, and NAKAMURA, K. (1974). On the time evolution of large systems: A soluble model. *Physica* **78,** 131–142.

――――, and TESHIMA, R. (1977). Response of an ideal gas in an external potential. *Physica,* **87A,** 453–472.

――――, and ―――― (1977). Time evolution and thermalization of an ideal gas in a box. *Can. J. Phys.* **55,** 189–193.

――――, and ZASADA, C. S. (1976). Microscopic derivation of hydrodynamic balance equations in the presence of an electromagnetic field. *Physica* **83A,** 573–583.

KRISHNAMURTI, R. (1968a). Finite amplitude convection with changing mean temperature. Part 1: Theory. *J. Fluid Mech.* **33,** 445–455.

―――― (1968b). Finite amplitude convection with changing mean temperature. Part 2: An experimental test of the theory. *J. Fluid Mech.* **33,** 457–463.

―――― (1970). On the transition to turbulent convection. Part 1: The transition from two- to three-dimensional flow. *J. Fluid Mech.* **42,** 295–307.

―――― (1970). On the transition to turbulent convection. Part 2: The transition to time-dependent flow. *J. Fluid Mech.* **42,** 309–320.

KUBO, R. (1957). Statistical mechanical theory of irreversible processes. I. General theory and simple applications to magnetic and conduction problems. *J. Phys. Soc. Japan* **12,** 570–586.

――――, YOKOTA, M. and NAKAJIMA, S. (1957). Statistical-mechanical theory of irreversible processes. II. Response to thermal disturbance. *J. Phys. Soc. Japan* **12,** 1203–1211.

LAGRANGE, J. L. (1853). *Mecanique Analytique I,* 3rd ed. Mallet-Bachelier, Paris.

LANDAU, L. D., and LIFSHITZ, E. M. (1958). *Statistical Physics.* Pergamon, London.

――――, and ―――― (1959). *Fluid Mechanics.* Pergamon, London.

LANDAUER, R. (1975). Stability and entropy production in electrical circuits. *J. Stat. Phys.* **13,** 1–16.

LANFORD, O. E., III (1975). Time evolution of large classical systems. In *Lecture Notes in Physics*, vol. 38, *Dynamical Systems, Theory and Applications*, Battelle 1974 Rencontres. Ed. J. Moser, Springer, Berlin.

LASALLE, J. P., and LEFSCHETZ, S. (1961). *Stability by Liapounov's Direct Method.* Academic, New York.

VON LAUE, M. (1917). Temperatur- und Dichteschwankungen. *Phys. Z.* **18,** 542–544.

LEE, C. T. (1974). Exact analysis of a spatially nonequilibrium Knudsen gas. *Can. J. Phys.* **52,** 1139–1143.

LENARD, A. (ed.) (1973). Transport phenomena. In *Lecture Notes in Physics*, vol. 31, Battelle 1971 Rencontres. Springer, Berlin.

LIAPOUNOFF, A. M. (1949). Probléme general de la stabilité du mouvement, Société Mathématique de Kharkov. *Ann. Math.* **17.** Princeton University Press.

LOSCHMIDT, J. (1876). Über den Zustand des Wärmegleichgewichtes eines Systems von Körpern mit Rücksicht auf die Schwerkraft. *Wien. Ber.* **73,** 135, 366.

LODGE, A. (1964). *Elastic Liquids.* Academic, London.

LOTKA, A. J. (1910). Contribution to the theory of periodic reactions. *J. Phys. Chem.* **14,** 271–274.

——— (1920). Analytical note on certain rhythmic relations in organic systems. *Proc. Natl. Acad. Sci. (U.S.)* **6,** 410–415.

——— (1956). *Elements of Mathematical Biology.* Dover, New York.

LUGIATO, L., and MILANI, M. (1976). On the structure of the generalized master equation for open systems. *Physica* **85A,** 1–17.

LUTTINGER, J. M. (1964). Theory of thermal transport coefficients. *Phys. Rev.* **135A,** 1505–1514.

MACDONALD, D. K. C. (1962). *Thermoelectricity: An Introduction to the Principles.* Wiley, New York.

MANDEVILLE, G., and COOPERSMITH, M. (1974). Irreversible behavior of interacting systems. II. Fluctuations in equilibrium. *J. Stat. Phys.* **10,** 405–420.

MARTIN, P. C. (1968). Measurements and correlation functions. In *Many-Body Physics*, pp. 37–136. Eds. C. de Witt and R. Balian, Gordon and Breach, New York.

———, PARODI, O., and PERSHAN, P. S. (1972). Unified hydrodynamic theory for crystals, liquid crystals, and normal fluids. *Phys. Rev.* **6,** 2401–2420.

MATHESON, I., WALLS, D. F., and GARDINER, C. W. (1975). Stochastic models of first-order nonequilibrium phase transitions in chemical reactions. *J. Stat. Phys.* **12,** 21–34.

Mazur, P., and Montroll, E. (1960). Poincaré cycles, ergodicity and irreversibility in assembles of coupled harmonic oscillators. *J. Math. Phys.* **1,** 70–84.

McCarthy, I. E. (1968). *Introduction to Nuclear Theory.* Wiley, New York.

McFee, R. (1973). On fluctuations of temperature in small systems. *Am. J. Phys.* **41,** 230–234.

McIntyre, D., and Sengers, J. V. (1968). Study of fluids by light scattering. In *Physics of Simple Fluids*, pp. 447–505. Eds. H. N. V. Temperly, J. S. Rowlinson, and G. S. Rushbrooke, North-Holland, Amsterdam.

McLennan, J. A., Jr., and Swenson, R. J. (1963). Theory of transport coefficients in low-density gases. *J. Math. Phys.* **4,** 1527–1536.

Meissner, W., and Meissner, G. (1939). Die gaskinetischen Vorgänge in einem Expansionszylinder. *Ann. Phys.* (Leipzig) **36,** 303–318.

Meixner, J. (1941). Zur Thermodynamik der Thermodiffusion. *Ann. Phys.* **39,** 333–356.

―――― (1943). Zur Thermodynamik der irreversiblen Prozesse. *Z. Physikal. Chemie* **B53,** 235–263.

―――― (1943). Zur Thermodynamik der irreversiblen Prozesse in Gasen mit chemisch reagierenden, dissoziierenden und anregbaren Komponenten. *Ann. Phys.* (Leipzig) **43,** 244–270.

―――― (1969). Thermodynamik der Vorgänge in einfachen fluiden Medien und die Charakterisierung der Thermodynamik irreversibler Prozesse. *Z. Phys.* **219,** 79–104.

―――― (1970). On the foundations of thermodynamics of processes. In *A Critical Review of Thermodynamics.* Eds. E. B. Stuart, B. Gal-Or, and A. J. Bainard, Mono Book, Baltimore.

―――― (1973). Consistency of the Onsager-Casimir reciprocal relations. *Adv. Molec. Relax. Phenom.* **5,** 319–331.

――――, and Reik, H. G. (1959). Thermodynamik der irreversiblen Prozesse. In *Handbuch der Physik*, III/2. Ed. S. Flügge, Springer, Berlin.

Metiu, H., Kitahara, K., and Ross, J. (1976). Stochastic theory of the kinetics of phase transitions. *J. Chem. Phys.* **64,** 292–299.

Miller, D. G. (1960). Thermodynamics of irreversible processes: The experimental verification of the Onsager reciprocal relations. *Chem. Rev.* **60,** 15–37.

―――― (1974). The Onsager relation: Experimental evidence. In *Foundations of Continuum Thermodynamics.* Eds. J. J. Delgado Domingos, M. N. R. Nina, and J. H. Whitelaw, MacMillan, London.

Montroll, E. W. (1962). Some remarks on the integral equations of statistical mechanics. In *Fundamental Problems in Statistical*

Mechanics, pp. 230–249. Ed. E. G. D. Cohen, North-Holland, Amsterdam.
MORI, H. (1958). Time correlation functions in statistical mechanics of transport processes. *Phys. Rev.* **111**, 694–706.
MÜNSTER, A. (1959). Prinzipien der statistischen Mechanik. In *Handbuch der Physik*, III/2. Ed. S. Flügge, Springer, Berlin.
NAKAJIMA, S. (1958). On quantum theory of transport phenomena. *Prog. Theor. Phys.* **20**, 948–959.
NAKANO, H. (1956). A method of calculation of electrical conductivity. *Prog. Theor. Phys.* **15**, 77–79.
VON NEUMANN, J. (1929). Beweis des Ergodensatzes and des H-Theorems in der neuen Mechanik. *Z. Physik* **57**, 30–70.
NEWTON, R. G. (1966). *Scattering Theory of Waves and Particles.* McGraw-Hill, New York.
NICOLIS, G., and PORTNOV, J. (1973). Chemical oscillations. *Chem. Rev.* **73**, 365–384.
———, and PRIGOGINE, I. (1977). *Self-organization in Nonequilibrium Systems: From Dissipative Structures to Order through Fluctuations.* Wiley, New York.
ONSAGER, L. (1931). Reciprocal relations in irreversible processes. I. *Phys. Rev.* **37**, 405–426.
——— (1931). Reciprocal relations in irreversible processes. II. *Phys. Rev.* **38**, 2265–2279.
———, and MACHLUP, S. (1953). Fluctuations and irreversible processes. *Phys. Rev.* **91**, 1505–1512.
ONSAGER, L., and MACHLUP, S. (1953). Fluctuations and irreversible process. II. Systems with kinetic energy. *Phys. Rev.* **91**, 1512–1515.
OPPENHEIM, I., SHULER, K. E., and WEISS, G. H. (1977). *Stochastic Processes in Chemical Physics: The Master Equation.* MIT Press, Cambridge, Mass.
ORBAN, J., and BELLEMANS, A. (1967). Velocity-inversion and irreversibility in a dilute gas of hard disks. *Phys. Lett.* **24A**, 620–621.
PALM, E., ELLINGSEN, T., and GJEVIK, B. (1967). On the occurrence of cellular motion in Bénard convection. *J. Fluid Mech.* **30**, 651–661.
PAULI, W., (1928). Über das H-Theorem vom Anwachsen der Entropie vom Standpunkt der neuen Quantenmechanik. In *Probleme der modernen Physik*, Arnold Sommerfeld zum 60. Geburtstage gewidmet von seiner Schülern, Verlag, Leipzig.
——— (1964). *Collected Scientific Papers.* Eds. R. Kronig and V. F. Weisskopf, Interscience, New York.
———, and FIERZ, M. (1937). Über das H-Theorem in der Quantenmechanik. *Z. Physik* **106**, 572–587.

PEGG, D. T. (1975). Absorber theory of radiation. *Rep. Prog. Phys.* **38**, 1339–1383.

PEIERLS, R. (1934). Über die statistischen Grundlagen der Elektronentheorie der Metalle. *Helv. Phys. Acta* **7** (Suppl. 2), 24–30.

────── (1965). *Quantum Theory of Solids*. Oxford University Press, Oxford.

PELLEW, A., and SOUTHWELL, R. V. (1940). On maintained convective motion in a fluid heated from below. *Proc. Roy. Soc.* (London) **176A**, 312–343.

PENROSE, O., and ONSAGER, L. (1956). Bose–Einstein condensation and liquid helium. *Phys. Rev.* **104**, 576–584.

PIČMAN, L. (1967). Bethe-Salpeter equation and the transport coefficient of simple fluids. *Ann. Acad. Sci. Fennicae* A VI, **235**, 1–26.

PLANCK, M. (1896). Über elektrische Schwingungen, welche durch Resonanz erregt und durch Strahlung gedämpft werden. Sitzungsberichte der Kgl. Preuss. Akademie der Wissenschaften, 151–170.

────── (1897). Über irreversible Strahlungsvorgänge. Erste Mittheilung. Sitzungsberichte der Kgl. Preuss. Akademie der Wissenschaften, 57–68.

────── (1898). Über irreversible Strahlungsvorgänge. Zweite Mittheilung. Sitzungsberichte der Kgl. Preuss. Akademie der Wissenschaften, 715–717.

────── (1897). Über irreversible Strahlungsvorgänge. Dritte Mittheilung. Sitzungsberichte der Kgl. Preuss. Akademie der Wissenschaften, 1122–1145.

────── (1898). Über irreversible Strahlungsvorgänge. Vierte Mittheilung. Sitzungsberichte der Kgl. Preuss. Akademie der Wissenschaften, 449–476.

────── (1899). Über irreversible Strahlungsvorgänge. Fünfte Mittheilung (Schluss). Sitzungsberichte der Kgl. Preuss. Akademie der Wissenschaften, 440–480.

────── (1901). Über das Gesetz der Energieverteilung im Normalspectrum. *Ann. Physik* **4**, 553–563.

────── (1901). Über irreversible Strahlungsvorgänge (Nachtrag). Sitzungsberichte der Kgl. Preuss. Akademie der Wissenschaften, 544–555.

────── (1945). *Treatise on Thermodynamics*, 3rd ed. (transl. from the 7th German ed.). Dover, New York.

────── (1960). *Survey of Physical Theory*. Dover, New York.

POINCARÉ, H. (1890). Sure le problème des trois corps et les équations de la dynamique. *Acta Math.* **13**, 1–271.

POKROWSKY, L. A., and SERGEEV, M. V. (1973). Thermodynamics of a crystal lattice on the basis of a local-equilibrium ensemble. *Physica* **70**, 62–82.

POPPER, K. R. (1956). The arrow of time. *Nature* **177,** 538.
——— (1957). Irreversible processes in physical theory. *Nature* **179,** 297.
PRIGOGINE, I. (1949). Le domaine de validité de la thermodynamique des phenoménes irreversibles. *Physica* **15,** 272–284.
——— (1962). *Nonequilibrium Statistical Mechanics.* Wiley-Interscience, New York.
——— (1968). Introduction to nonequilibrium statistical physics. In *Topics In Nonlinear Physics,* pp. 216–350. Ed. N. J. Zabusky, Springer, New York.
——— (1969), *Introduction to Thermodynamics of Irreversible Processes,* 3rd ed. Interscience, New York.
——— (1973). Time, irreversibility, and structure. In *The Physicist's Conception of Nature.* Ed. J. Mehra, Reidel, Dordrecht.
———, GEORGE, C., HENIN, F., and ROSENFELD, L. (1973). A unified formulation of dynamics and thermodynamics. *Chemica Scripta* **4,** 5–32.
———, and RÉSIBOIS, P. (1961. On the kinetics of the approach to equilibrium. *Physica* **27,** 629–646.
———, and RICE, S. A. (eds.) (1975). *Advances in Chemical Physics* Wiley, New York. In particular vol. 32.
RAYLEIGH, L. (1916). On convective currents in a horizontal layer of fluid when the higher temperature is on the underside. *Phil. Mag.* **32,** 529–546.
RAZAVY, M. (1975). Remarks concerning the derivation and the expansion of the master equation. *Physica* **84A,** 591–602.
——— (1977). On the quantization of dissipative structures. *Z. Phys.* **B26,** 201–206.
——— (1978). Hamilton's principal function for the Brownian motion of a particle and the Schrödinger-Langevin equation. *Can. J. Phys.* **56,** 311–320.
——— (1979). *Quantum-Mechanical Irreversible Motion of an Infinite Chain. Can. J. Phys.* **57,** 1731–1737.
RÉSIBOIS, P. (1970). On linearized hydrodynamic modes in statistical physics. *J. Stat. Phys.* **2,** 21–51.
——— (1972). Hydrodynamical concepts in statistical physics. In *Irreversibility in the Many Body Problem.* Eds. J. Biel and J. Rae, Plenum Press, New York.
———, and DE LEENER, M. (1977). *Classical Kinetic Theory of Fluids.* Wiley, New York.
———, and MARESCHAL, M. (1978). Kinetic equations, initial conditions and time-reversal: A solvable one-dimensional model revisited. *Physica* **94A,** 211–253.
RHIM, W. K., PINES, A., and WAUGH, J. S. (1971). Time-reversal experiments in dipolar-coupled spin systems. *Phys. Rev.* **B3,** 684–696.

RICE, S. A., BOON, J. P., and DAVIS, H. T. (1968). Comments on the experimental and theoretical study of transport phenomena. In *Simple Dense Fluids*. Eds. H. L. Frisch and Z. W. Salsburg. Academic, New York.

RISTE, T. (ed.) (1975). *Fluctuations, Instabilities, and Phase Transitions*. Plenum, New York.

ROBERTS, P. H. (1969). Electrohydrodynamic convection. *Q. J. Mech. Appl. Math.* **22**, 211–220.

ROSENFELD, L. (1961). Questions of irreversibility and ergodicity. In Rendiconti Scuola Int. Fisica, Corso XIV, Varenna, 1960. Ed. P. Caldirola, Academic, New York.

RUELLE, D., and TAKENS, F. (1971). On the nature of turbulence. *Commun. Math. Phys.* **20**, 167–192.

SCHECHTER, R. S., VELARDE, M. G., and PLATTEN, J. K. (1974). The two-component Bénard problem. *Adv. Chem. Phys.* **26**, 265–301.

SCHLÖGL, F. (1971). Fluctuations in thermodynamic non-equilibrium states. *Z. Physik* **244**, 199–205.

——— (1971). On thermodynamics near a steady state. *Z. Physik* **248**, 446–458.

——— (1972). Chemical reaction models for nonequilibrium phase transitions. *Z. Physik* **253**, 147–161.

SCHMIDT, R. J., and MILVERTON, S. W. (1935). On the instability of a fluid when heated from below. *Proc. Roy. Soc.* (London) **152A**, 586–594.

SCHRÖDINGER, E. (1914). Zur Dynamik elastisch gekoppelter Punktsystems. *Ann. Physik* **44**, 916–934.

——— (1960). *Statistical Thermodynamics*. Cambridge University Press, Cambridge.

SEGEL, L. A. (1966). Non-linear hydrodynamic stability theory and its applications to thermal convection and curved flows. In *Non-Equilibrium Thermodynamics, Variational Techniques and Stability*, pp. 165–197. Eds. R. J. Donnelly, R. Herman, and I. Prigogine, University of Chicago Press, Chicago.

SIGGIA, E. D. (1977). Origin of intermittency in fully developed turbulence. *Phys. Rev.* **A15**, 1730–1750.

SILVESTON, P. L. (1958). Wärmedurchgang in waagerechten Flüssigkeitsschichten. *Forsch. Ing. Wes.* **24**, 29–32, 59–69.

SNEDDON, I. N. (1951). *Fourier Transforms*. McGraw-Hill, New York.

SOMERSCALES, E. F. C., and DOUGHERTY, T. S. (1970). Observed flow patterns at the initiation of convection in a horizontal liquid layer heated from below. *J. Fluid Mech.* **42**, 755–768.

STEPHENSON, J. (1975). A study of fluid argon via the constant volume specific heat, the isothermal compressibility, and the speed of sound and their extremum properties along isotherms. *Can. J. Phys.* **53**, 1367-1384.

STORK, K., and MÜLLER, U. (1972). Convection in boxes: Experiments. *J. Fluid Mech.* **54**, 599-611.

SWENSON, R. J. (1962). Derivation of generalized master equations. *J. Math. Phys.* **3**, 1017-1022.

TAKASHIMA, M., and ALDRIDGE, K. D. (1976). The stability of a horizontal layer of dielectric fluid under the simultaneous action of a vertical DC electric field and a vertical temperature gradient. *Q. J. Mech. Appl. Math.* **29**, 71-87.

TERREAUX, C. (1969). Macroscopic equations of motion for a two-component fluid. *Physica* **44**, 301-317.

THOMSON, W. (1882). Experimental researches in thermo-electricity. *Proc. Roy. Soc.* (Edinburgh), May 1854. (Also in *Collected Papers I*, pp. 460-468, Cambridge University Press, Cambridge.)

TIPPLESKIRCH, H. V. (1956). *Beitr. Phys. Atmos.* **29**, 219-233.

TRITTON, D. J. (1977). *Physical Fluid Dynamics.* Van Nostrand Reinhold, New York.

TROFIMENKOFF, P. N., and KREUZER, H. J. (1973). Novel solution of the Boltzmann equation for a constant relaxation time. *Am. J. Phys.* **41**, 292-293.

TRUESDELL, C., and NOLL., W. (1965). The Non-linear field theories of mechanics. In *Handbuch der Physik* III/3. Ed. S. Flügge, Springer, Berlin.

────── (1969). A precise upper limit for the correctness of the Navier-Stokes theory with respect to the kinetic theory. *J. Stat. Phys.* **1**, 313-318.

────── (1969). *Rational Thermodynamics* (*A Course of Lectures on Selected Topics*). McGraw-Hill, New York.

TURING, A. M. (1952). The chemical basis of morphogenesis. *Phil. Trans. Royal Soc. Lond.* **B237**, 37-72.

VIGFUSSON, J. O. (1976). Ergodic properties of the linear chain with arbitrary masses and force constants. I. Correlation functions. II. Ergodic properties. *Physics* **85A**, 211-236, 237-260.

VERSCHAFFELT, J. E. (1951). *Bull. Clin. Sci., Acad. Roy. de Belg.* **37**, 853.

VLASOV, A. A. (1938). *JETP* **8**, 291

VOLTERRA, V. (1927). Variazoni e fluttuazioni del numero d'individui in specie animali conviventi. Memorie del R. Comitato talasso-grafico italiano CXXXI. Reprinted in Opere Matematische di Vito Volterra, vol. 5, Accademia Nazionale dei Lincei, Rome, 1962.

────── (1931). Lecon sur la theorie mathématique de la lutte pour la vie. Gauthier-Villars, Paris.

——— (1937). Principes de biologie mathématique. *Acta Biotheoret.* **3**, 1–36.
WALDMANN, L. (1958). Transporterscheinungen in Gasen von mittlerem Druck. In *Handbuch der Physik*, vol. 12, *Thermodynamics of Gases*. Ed. S. Flügge, Springer, Berlin.
WALGRAEF, D., and BORCKMANS, P. (1969). Irreversibility in paramagnetic spin systems. I. Kinetic equations for reduced density matrix. *Phys. Rev.* **187**, 421–429.
———, and ——— (1969). Irreversibility in paramagnetic Spin Systems. II. Non-Markovian processes and spin-spin relaxation. *Phys. Rev.* **187**, 430–441.
——— (1974). Quantum statistics of a monomode-laser model. *Physica* **72**, 578–596.
WALLS, D. F. (1970). Higher order effects in the master equation for coupled systems. *Z. Physik* **234**, 231–241.
WATANABE, S. (1935). Le deuxième théorème de la thermodynamique et la mecanique ondulatoire. *Act. Sci. Ind.* No. 308, Hermann, Paris.
WEISSKOPF, V., and WIGNER, E. (1930). Berechnung der natürlichen Linienbreite auf Grund der Diracschen Lichttheorie. *Z. Physik* **63**, 54–73. (Transl. by W. R. Hindmarsh in Calculation of the natural line width on the basis of Dirac's theory of light. In *Atomic Spectra*, p. 304, Pergamon, 1967.)
———, and ——— (1930). Über die natürliche Linienbreite in der Strahlung des harmonischen Oszillators. *Z. Physik* **65**, 18–29.
WHEELER, J. A. (1936). The dependence of nuclear forces on velocity. *Phys. Rev.* **50**, 643–649.
———, and FEYNMAN, R. P. (1945). Interaction with the absorber as the mechanism of radiation. *Rev. Mod. Phys.* **17**, 157–181.
WHITEHEAD, J. A., JR. (1975). A survey of hydrodynamic instabilities. In *Fluctuations, Instabilities and Phase Transitions*. Ed. T. Riste, Plenum, New York.
WIGNER, E. (1932). On the quantum correction for thermodynamic equilibrium. *Phys. Rev.* **40**, 749–759.
WIGNER, E. P. (1954). Derivations of Onsager's reciprocal relations. *J. Chem. Phys.* **22**, 1912–1915.
WOODS, L. C. (1975). *The Thermodynamics of Fluid Systems*. Clarendon, Oxford.
YAMAGUCHI, Y. (1954). Two-nucleon problem when the potential is non-local but separable. I and II. *Phys. Rev.* **95**, 1628–1634, 1635–1643.
YANG, C. N. (1962). Concept of off-diagonal long-range order and the quantum phases of liquid He and of superconductors. *Rev. Mod. Phys.* **34**, 694–704.

YARNELL, J. L., KATZ, M. J., WENZEL, R. G., and KOENIG, S. H. (1973). Structure factor and radial distribution function for liquid argon at 85 K. *Phys. Rev.* **A7,** 2130–2144.

YIH, C. S. (1968). Fluid motion induced by surface tension variation. *Phys. Fluids* **11,** 477–480.

YVON, J. (1937). Recherches sur la théorie cinétique des liquides. *Act. Sci. Ind.* No. 542 and 543, Hermann, Paris.

ZERMELO, E. (1896a). Über einen Satz der Dynamik und die mechanische Wärmetheorie. *Wied. Ann.* **57,** 485.

——— (1896b). Über die mechanische Erklärung irreversibler Vorgänge. Eine Antwort auf Hrn. Boltzmann's 'Entgegnung.' *Wied. Ann.* **59,** 793–801.

ZIMAN, J. M. (1969). *Principles of the Theory of Solids.* Cambridge University Press, Cambridge.

ZUBAREV, D. N. (1974). *Nonequilibrium Statistical Thermodynamics.* Consultants Bureau, New York.

ZWANZIG, R. J. (1960). Ensemble methods in the theory of irreversibility. *J. Chem. Phys.* **33,** 1338–1341.

——— (1964). On the identity of three generalized master equations. *Physica* **30,** 1109–1123.

——— (1965). Time correlation function and transport coefficients in statistical mechanics. *Ann. Rev. Phys. Chem.* **16,** 67–102.

Author Index

Abragam, A., 331*
Abraham, D.B., 349
Abramowitz, M., 285
Acrivos, A., 128
Ahlers, G., 131
Alder, B.J., 190, 331
Aldridge, K.D., 112
d'Ancona, U., 99
Andrade, E.N. da C., 173
Anstis, G.R., 190, 359
Avsec, D., 127

Balescu, R., 179, 312, 321
Bannister, T.C., 133
Barker, J.A., 267, 271
Barouch, E., 349
Baxter, R.J., 159
Bellemans, A., 330, 332
Bénard, H., 111
Bernard, W., 237*
Bertrand, J.L.F., 2
Bird, R.B., 174
Blatt, J.M., 189, 340, 359, 361
Bloch, C., 282
Block, M.J., 131
Bocchieri, P., 337*
Bogolyubov, N.N., 150, 154, 177, 189, 211, 226, 230
Bohr, N., 11
Boltzmann, L., 28, 151, 177, 193, 326, 327, 330, 335–40
Boon, J.P., 190
Borckmans, P., 316

Born, M., 154, 177
Boussinesq, J., 114
Braun, W., 179
Brittin, W.E., 223
Brown, A.L., 73
Burnett, D., 206
Busse, F.H., 86, 134, 139–43

Callen, H.B., 7, 76, 237
Casimir, H.B.G., 41, 45, 47, 50, 55, 60
Cercignani, C., 205
Chandrasekhar, S., 112, 116, 122, 129, 138, 145, 146, 148
Chapman, S., 198, 202
Chappell, W.R., 223
Charlson, G.S., 131
Chen, S.H., 263
Chester, G.V., 239, 286–88, 291, 292
Choh, S.T., 190
Christensen, R.M., 28
Christopherson, D.G., 128
Clarke, B.L., 161
Clever, R.M., 86, 139–43
Clusius, K., 204
Cohen, C., 38
Cohen, E.G.D., 190
Coleman, B.D., 58, 88
Coopersmith, M., 328
Cowling, T.G., 202
Curtiss, C.F., 174

Davidson, R., 351
Davies, R.O., 58

*The work by these authors listed in the bibliography is relevant to the discussion on the page indicated, but not explicitly quoted there.

Davis, H.T., 190
Davis, S.H., 126, 131
Debye, P., 178
Degn, H., 107-9
Dirichlet, G.L., 72, 83
Domenicali, C.A., 67
Dominicis, C. De., 282
Dorfman, J.R., 12, 14, 190, 280
Dougherty, T.S., 133, 135, 140

Eckart, C., 2
Eddington, A.S., 328
Edelson, S., 107
Egelstaff, P., 30, 262, 263, 270, 284, 318
Ehrenfest, P., 365
Ehrenfest, T., 365
Eigen, M., 110
Einstein, A., 9, 48, 326
Eliashberg, G.M., 290
Ellingsen, T., 134
Elton, C., 99
Enderby, J.E., 263
Enskog, D., 3, 198, 205
Ernst, M.H., 190, 280
Eyring, H., 173, 174, 180

Feynman, R.P., 227, 324
Field, R.J., 107, 109
Fierz, M., 300
Fleming, III. P.D., 38
Flügge, W., 31
Forster, D., 37, 237, 278, 280
Fraser, J.T., 328, 340
Frenkel, J., 28
Fröhlich, H., 211, 214, 221
Fujita, S., 237, 271, 280

Galasiewicz, Z.M., 230, 271
Gallavotti, G., 349
Gardiner, C.W., 98
George, C., 317, 369
Gill, J.C., 334
Gjevik, B., 134
Glansdorff, P., 18, 69, 76, 82, 85-87, 90, 109, 143, 145, 148
Goel, N.S., 101, 103, 104, 106, 107
Gold, T., 328
Gollub, J.P., 142
Grad, H., 156, 185, 188, 197, 198, 231
Graham, A., 136
Grecos, A., 316, 317, 369
Green, H.S., 152, 154, 173, 177, 180, 182, 189, 190, 359

Green, M.S., 55, 237
Greene, R.F., 237*
Grodzka, P.G., 133
de Groot, S.R., 2, 18, 41, 58
Gross, M.J., 112
Guggenheim, E.A., 11
Guth, E., 208, 236
Gyarmati, I., 67

Haar, D. Ter, 211
Haase, R., 18
Hadamard, J., 73
Hahn, E.L., 331
Haines, L.K., 190, 280
Haken, H., 2, 97, 351
Harman, T.C., 67
Hartree, D.R., 178
Hashitsume, N., 55
Heikes, R.R., 63*
Hemmer, P.C., 336, 342
Henderson, D., 267, 271
Henin, F., 317, 369
Hepp, K., 179
Herring, C., 208
Hilbert, D., 197
Hiob, E., 355, 357, 358
Hirschfelder, J.O., 174
Hoard, C.Q., 128*
Hobson, A., 359, 361
Hofelich, F., 33
Hoffman, D.K., 190, 359
Hohenberg, P.C., 37
Honig, J.M., 67
Hooyman, G.J., 58
van Hove, L., 289, 297, 306, 308, 310-12, 326, 385
Huang, K., 263
Hückel, E., 178
Hungtington, H.B., 25
Husimi, K., 211

Israel, W., viii

Jancel, R., 300, 358, 366
Janner, A., 312
Janssen, H.K., 98
Jeans, J., 62
Jeffreys, H., 133
Jepsen, D.W., 359
Jhon, M.S., 174

Kandanoff, L.P., 271
van Kampen, N.G., 294, 297, 300, 304

AUTHOR INDEX

Kan, Y., 190
Katz, M.J., 271
Kestin, J., 12, 14
Kirkwood, J.G., 55, 154, 177, 187
Kitahara, K., 98
Kittel, C., 10, 293
Klein, F., 72, 83
Klein, O., 366
Knops, R.J., 72, 83, 86, 88
Knudsen, M., 62
Koenig, S.H., 271
Kölsch, W., 204
Körös, E., 109
Koschmieder, E.L., 112, 131–35
Kozak, J.J., 351
Klimontovich, Yu.L., 223
Kramers, H.A., 98
Kreuzer, H.J., 207, 211, 222, 232, 348, 349, 355, 357–59, 361, 371, 386
Krishnamurti, R., 128, 130, 133, 134, 140, 144
Kubo, R., 237, 246, 254, 259, 261, 271
Kurihara, Y., 386

Lagrange, J.L., 72
Landau, L.D., 11, 12, 142
Landauer, R., 33, 88
Lanford, O.E., III, 156, 188
Lasalle, J.P., 72, 75
von Laue, M., 11, 12
Lee, C.T., 359
de Leener, M., 316, 317, 321
Lefschetz, S., 72, 75
Liapounoff, A.M., 72
Lifshitz, E.M., 11, 12, 142
Lodge, A., 28
Loinger, A., 337
Loomis, D.N., 359
Loschmidt, J., 330
Lotka, A.J., 99
Lubensky, T.C., 37*
Lugiato, L., 316
Luttinger, J.M., 271

McCarthy, I.E., 400
MacDonald, D.K.C., 63*
McFee, R., 11
Machlup, S., 55
McIntyre, D., 263
McLennan, J.A. Jr., 237, 280
Maitra, S.C., 101, 103, 104, 106, 107
Mandeville, G., 328

Mareschal, M., 316
Martin, P.C., 37, 237, 263, 271, 280
Martin-Löf, A., 349
Matheson, I., 98
Maximon, L.C., 336, 342
Mayerhöfer, J., 208, 236
Mazur, P., 2, 18, 41, 58, 337
Meissner, G., 205
Meissner, W., 205
Meixner, J., 2, 3, 18, 33, 35, 58, 205
Metiu, H., 98
Milani, M., 316
Miller, D.G., 56, 66, 67
Milverton, S.W., 130
Montroll, E.W., 101, 103, 104, 106, 107, 297, 337
Mori, H., 229, 237, 271, 280
Müller, U., 131
Münster, A., 10, 12

Nakajima, S., 271, 297
Nakamura, K., 348, 358
Nakano, H., 287, 291
von Neumann, J., 300
Newton, R.G., 263, 396
Nicolis, G., 107
Nichols, M.H., 208
Noll, W., 25, 32
Noyes, R.M., 107, 109

Onsager, L., 2, 41, 45, 51, 55, 67, 221, 227
Opie, A.H., 189
Oppenheim, I., 297
Orban, J., 330, 332

Page, A., 73
Palm, E., 134
Parodi, O., 37
Pauli, W., 289, 297, 298, 300, 367
Pegg, D.T., 324
Peierls, R., 291
Pellew, A., 119
Penrose, O., 221
Persham, P.S., 37
Pičman, L., 271
Pines, A., 331, 333–35
Planck, M., 121, 322, 325–27, 336, 385, 386
Platten, J.K., 112*
Poincaré, H., 336
Pokrovsky, L.A., 38, 227
Popper, K.R., 326
Porter, J.E., 112

Portnov, J., 107
Prigogine, I., 2, 18, 67, 69, 76, 82, 85–87, 90, 107, 109, 112, 143, 145, 148, 205, 297, 312, 315–17, 321, 325, 341, 369, 385, 386

Rayleigh, Lord, 115, 119, 122
Razavy, M., 297, 345, 347, 348
Reik, H.G., 2, 18, 35
Resibois, P., 297, 315–17, 321
Rhim, W.K., 331, 333–35
Rice, S.A., 112, 161, 190
Riste, T., 112
Roberts, P.H., 112
Robertson, C.R., 128*
Rosenfeld, L., 11, 317, 369
Ross, J., 98
Ruelle, D., 142

Sani, R.L., 131
Schechter, R.S., 112*
Schlögl, F., 88, 95–97
Schmidt, R.J., 130
Schrödinger, E., 11, 343
Segel, L.A., 112, 126, 134
Sengers, J.V., 263
Sergeev, M.V., 38, 227
Shuler, K.E., 297
Siggia, E.D., 142
Silveston, P.L., 130, 131, 133, 135
Sneddon, I.N., 355
Somerscales, E.F.C., 133, 135, 140
Sommerfeld, A., 72, 83
Southwell, R.V., 119
Stegun, I.A., 285
Stephenson, J., 16
Stewart, J.M., viii
Stork, K., 131
Swenson, R.J., 237, 280, 297
Swift, J., 37*
Swinney, H.L., 142,

Takashima, M., 112
Takens, F., 142
Terreaux, C., 221
Teshima, R., 359, 361, 371
Thellung, A., 239, 286–88, 291, 292

Thomson, W., 2, 66
Tippelskirch, H.V., 136
Tritton, D.J., 112*
Trofimenkoff, P.N., 207
Truesdell, C., 25, 32, 58, 206
Turing, A.M., 107

Uhlenbeck, G.E., 190
Ure, Jr., R.W., 63*

Verlarde, M.G., 112*
Verboven, E., 312
Verschaffelt, J.E., 58
Vigfusson, J.O., 341
Vlasov, A.A., 178
Volterra, V., 99

Wainwright, T.E., 331
Waldmann, L., 185, 202–4, 206
Walgraef, D., 316
Walls, D.F., 98, 364
Watanabe, S., 300
Waugh, J.S., 331, 333–35
Weiss, G.H., 297
Weisskopf, V., 351
Welton, T.A., 237
Wenzel, R.G., 271
Wergeland, H., 336, 342
Wheeler, J.A., 324, 372, 400
Whitehead, J.A., 112, 140, 141
Wigner, E.P., 52, 55, 223, 351
Wilkes, E.W., 72, 83, 86, 88
Woods, L.C., 18

Yamaguchi, Y., 400
Yang, C.N., 221, 263
Yarnell, J.L., 271
Yih, C.S., 133*
Yip, S., 280
Yokota, M., 271
Yvon, J., 154

Zasada, C.S., 222
Zermélo, E., 336
Ziman, J.M., 290
Zubarev, D.N., 229
Zwanzig, R.J., 237, 271, 280, 297, 314

Subject Index

Absorber theory of radiation, 324 n
Adiabatic switching, 239, 246
Admittance, 254
Affinity
 and chemical potential, 93
 definition, 35, 92
 for ideal gas, 94
Angular momentum density, 24–27
 balance equation for, 24–25
 internal, 25–27. See also
  Spin
 relaxation of, 26–27
Annihilation operator, 281, 349, 371
 and field operator, 281, 371
 for quasi-particles, 353
 time evolution of, 372–75
Anticommutation relations, 210, 213, 222, 281
Antiferromagnet
 spin wave in, 37
Antikinetic evolution, 331–35
Antisymmetric part of pressure tensor, 26
Approach to equilibrium, 1, 327–69, 389
Argon, 13, 16
 isothermal compressibility of, 16
 specific heat of, 16
 temperature fluctuation in, 16
Assumption of local equilibrium. See
  Local equilibrium
Asymptotic stability, 76. See also
  Stability
 of Volterra-Lotka model, 100
Autocatalytic reactions, 109
Autocorrelation function, 267
Avogadro's number, 13

Balance equations, 2, 6, 17, 40, 79, 84, 112–14, 118, 200, 272, 273. See
  also Boussinesq approximation
 from classical statistical mechanics, 158–69
 in Euler coordinate system, 19
 Fröhlich's derivation of, 214–21
 in Lagrange coordinate system, 21
 for mechanical quantities, 18–27
 from operator hierarchy, 226–34
 from quantum statistical mechanics, 209–34
 for solids, 36–39
 for tensor operators, 221–26
Balance equation for angular momentum, 24–25
Balance equation for entropy, 32–36, 168–169, 323
 from Boltzmann equation, 195, 205, 329
 from classical statistical mechanics, 168–69
 See also Balance equation for
  $\mathcal{H}$-quantity; Irreversibility; Second
  law of thermodynamics
Balance equation for internal energy, 24, 113, 174, 192, 232 n, 272–73
 from Boltzmann equation, 192
 from classical statistical mechanics, 164–68
Balance equation for kinetic energy, 22
 from classical statistical mechanics, 164
 from quantum statistical mechanics, 231
Balance equation for kinetic pressure, 232–34

Balance equation for mass density, 19, 112, 272, 351
  from Boltzmann equation, 192
  from classical statistical mechanics, 159–60
  from quantum statistical mechanics, 216, 230
Balance equation for momentum, 21, 112, 272, 273, 351
  from Boltzmann equation, 192
  from classical statistical mechanics, 160–64
  from quantum mechanics, 217, 231
Balance equation for potential energy, 23
Balance equation for strain tensor in solids, 38
Balance equation for total energy, 22–24, 232 n
Balance equation for viscoelastic stress tensor, 28–32
Barycentric coordinate system, 19
Barycentric derivative, 20
Barycentric velocity, 22
BBGKY hierarchy, 154, 155–58, 159, 166, 177, 178, 180, 186, 189, 191, 294, 296, 327, 328, 335
Belousov-Zhabotinskii reaction, 98, 107–9
Bénard convection, 85, 111–48,
  boundary conditions for, 116–18
  Boussinesq approximation for, 114–16
  experiments, 128–36, 140–43
  linear stability analysis, 116–28
  nonlinear stability analysis, 135–43
Bénard instability. See Bénard convection
Bénard problem. See Bénard convection
Biological clocks, 110 n
Blackbody radiation, 327
Boltzmann's constant, k_B, 7
Boltzmann distribution function. See Distribution function
Boltzmann equation, 152, 154, 158, 169, 173, 177, 178, 186, 195, 206, 231 n, 280, 286, 290, 291, 294, 296, 328–31, 335–36, 370
  Chapman-Enskog theory of, 197–206
  derivation of, 179–90
  discussion of, 186–90
  Hilbert expansion of, 197–98
  relaxation time approximation, 206
Born approximation, 264, 385
Bose-Einstein statistics, 349, 354, 358, 376
Bosons, 210, 222, 281, 371
Boundary conditions for Bénard convection, 116–18
Bound states, 372, 386–90
Boussinesq approximation for balance equations, 114–16, 121 n, 124, 134, 144, 145
Breit-Wigner form of the T-matrix, 392
Broken symmetry, 37, 247 n
Brownian motion, 55 n
Brusselator, 109
Brussels formalism of NESM, 314
Brussels school, 312, 369
Bulk modulus, 30, 234
Bulk viscosity, 26–27, 113, 173, 221. See also Viscosity coefficients
Buoyancy-driven convection. See Bénard convection
Buoyant force, 111, 114
Burnett and super-Burnett equations, 206

Canonical ensemble, 253, 287, 317, 371
Canonical statistical operator, 212, 258
Canonical variables, 240
Cell patterns for Bénard convection, 123–28
Center-of-mass coordinate system, 19
Center-of-mass velocity, 20
Chapman-Enskog solution of the Boltzmann equation, 3, 13, 155, 205, 235, 236
Chemical oscillations, 107–10
Chemical potential, 3, 34, 92, 93
  for ideal gas, 94, 354
Chemical reactions, 20, 91–110, 180 n
  triangle, 45–47
Circulation, 25
Classical response theory, 239–43
Classical statistical mechanics and kinetic theory, 149–208
Clausius-Carnot theorem, 36
Clausius inequality, 33
Coarse-graining, 6, 187, 188, 300, 302, 305, 365–66
Coefficient of heat conduction. See Thermal conductivity

SUBJECT INDEX 431

Coherent cross section, 267
Collision frequency, 12. *See also* Collision time
Collision integral, 178, 182, 185 n, 188, 192
Collision time, 151, 188, 196, 290, 294, 370
  for various gases, 13
Commutation relations, 210, 213, 222, 251, 281
Complementarity, 11 n, 11
Compressibility (isothermal), 9
  argon, 16
  mechanical stability, 79
Compression, 22, 114
Conduction process, 21
Canonical statistical operator, 211–12
Conservation laws. *See* Balance equations
Constitutive laws
  from Boltzmann equation, 196–206
  derivation, 169–77
  from linear response theory, 239
  See also Linear phenomenological laws
Constitutive relations. *See* Linear phenomenological laws
Continuity equation. *See* Balance equation for mass density
Convection due to surface tension, 133
Convection polygons, 127
Convection rolls, 126–27, 138–42
  oscillating, 143
Convective current, 21
Conversion
  internal to kinetic energy, 23
  kinetic to potential energy, 23
  potential to internal energy, 23
Correlations, 182, 219 n, 233
Correlation function, 253, 262, 318
  current-current, 259
  density-density, 261–69, 283–84
  See also Autocorrelation function
Correlation length, 15
Correspondence principle, 230
Couette flow, 142 n
Creation operator, 281, 314, 349, 371.
  See also Annihilation operator; Destruction operator; Field operator
Creep, 29
Criterion of local equilibrium. *See* Local equilibrium

Criterion for local equilibrium in a gas, 13
Criterion for local equilibrium in a liquid, 15
Criterion for local equilibrium in a solid, 15
Critical point, 6, 247 n
Cross-roll instability in Bénard convection, 140
Curie's principle, 43
Current response, 258–61
Cyclic invariance, 255

Dawson's integral, 285
de Broglie wavelength, 151, 262
Debye-Hückle screening, 179
Debye-Hückle theory of electrolytes, 178
Debye solid, 337
Debye temperature, 293
Degenerate quantum gas, 283
Density-density response function, 253
Density evolution in an ideal gas, 376–401
Density martrix, 151, 209, 211, 228, 299, 313–14, 351, 367–69. *See also* Density operator; Statistical operator
Density of thermodynamic function per unit mass, 18
Density operator, 229, 240, 244, 371. *See also* Density matrix; Statistical operator
Destruction operator, 314
Detailed balance, 46, 49, 53, 54, 369
Diffusion, 2, 20, 61, 229
Diffusion lifetime, 274
Diffusion-reaction model, 98
Dipole moment. *See* Electric dipole moment
Director in liquid crystal, 37
Dispersion relations, 238, 239, 247–50
Dissipation of energy in linear response, 254–56
Dissipative processes, 84, 89, 100
Dissipativity condition, 316
Distribution function, 155, 222–23
  Maxwell-Boltzmann, 163, 360
  radial, 267, 271
  reduced, 155, 156 n
  Wigner, 223, 386
  See also BBGKY hierarchy; Reduced density matrix

Dyadic, 21
Dynamical system, 72, 74

Effusion, 62
Einstein's fluctuation formula, 9, 11, 48, 52, 84. *See also Fluctuations*
Elastic constants, 29
Electrical conduction, 2, 65, 280, 286–95, 312
Electrical conductivity tensor, 287, 290
  static, 288
Electric current, 63–67, 208, 287–95
  microscopic definition of, 64, 286
Electric dipole moment, 64, 240, 244, 259
Electrodiffusion, 45
Electrokinetic phenomena, 2
Electron gas, 240, 243
Energy balance. *See* Balance equation for energy
Energy conservation. *See* Balance equation for energy
Energy convection coefficient in thermodiffusion, 61
Energy density, 23. *See also* Balance equation for energy
Energy dissipation, 256
  principle of least-, 67 n
Energy-flux tensor, 230
Energy representation of thermodynamics, 10
Ensemble probability function, 153, 241
Ensemble theory of limitations, 11
Entropy, 3, 33
  instantaneous, 7
  See also $\mathcal{H}$-quantity; Second law of thermodynamics
Entropy balance. *See* Balance equation for entropy
Entropy current, 33, 169, 194–95
Entropy density, 33, 35, 169
  microscopic expression for, 194
Entropy of an ideal gas, 60
Entropy production, 1, 2, 35, 40, 68, 92
  in Volterra-Lotka model, 101
  See also Minimum entropy production
Entropy representation of thermodynamics, 7, 10

Equation of motion. *See* Hamilton's equations; Heisenberg's equation; Schrödinger's equation; Balance equation for momentum
Equation of state
  of ideal gas, 13
  local, 3, 6, 165
  van der Waals (virial), 96
Equilibrium thermodynamics
  fluctuations, 7–12
  fundamental differential form of, 3, 34
  Tisza's postulational approach, 7
Ergodic problem, 49, 341
Error function, 375
Euler coordinate system, 19
Evolution criterion due to Glansdorff and Prigogine, 69, 85–90
Evolution of biological macromolecules, 110 n
Excess balance equations, 84
  linearized for Bénard convection, 116–18, 139
Excess entropy production, 100
Exchange of stability, 121
Exclusion principle, 15
Expansion of a fluid, 27
Eyring's theory of rate processes, 174

Fading memory hypothesis, 32
Fermi-Dirac gas, 15, 293, 295, 357, 378, 387
  statistics, 207, 354, 376
Fermi energy, 207, 291
Fermi surface, 293
Fermions, 210, 222, 281, 371
Fermi's golden rule, 265
Ferroelectrics, 25 n
Fick's law of diffusion, 43
Field operator in second quantization, 210, 244
First law of thermodynamics, 24
Flow terms in balance equations, 18–19
Fluctuation-dissipation theorem, 239, 251, 253–58
Fluctuations
  Einstein's formula, 9, 48
  equilibrium, 7–12
  in extensive variables, 8–10
  in intensive variables, 10–12
  time-reversal invariance, 4, 47
  See also Local equilibrium
Fokker-Planck equation, 297 n, 348

Fourier's law of heat conduction, 2, 14, 42, 64, 68, 113, 130, 169, 174, 177, 272
 from Boltzmann equation, 201–2
 microscopic derivation, 168, 192, 174–77
Free surface in Bénard convection, 117–18, 123
Fröhlich's derivation of hydrodynamic equations, 214–21
Fundamental differential form of thermodynamics, 3, 34, 37
Fundamental relation of equilibrium thermodynamics, 34
Fusion curve, 16 n, 16

Gauge symmetry, broken, 37
Gaussian distribution for second-order fluctuation moments, 9
Gauss' theorem, 183, 193
Gels, 28
General evolution criterion (Glansdorff and Prigogine), 85–90
Generalized barycentric derivative, 186
Generalized Hooke's law, 29
Gibbs free energy, 100
 activation energy, 174
 ensembles, 49
 equality, 35
 free energy of ideal gas, 46
 relation, 34
 relation for a solid, 37
Gibbs-Duhem theory of stability, 76
Global stability criterion for nonequilibrium, 84
Grad limit for hard-sphere gas, 188–89, 286
Grad's thirteen moment equations, 231 n

Hamiltonian density, 231, 232 n
Hamiltonian operator, 209, 228, 243
Hamilton's equations, 49, 153–54, 183, 323–24, 328
Hard sphere gas
 grad limit of, 188
Harmonic chain, 336–37, 341–48, 356
Hartree approximation, 178 n
Heat conduction, 15, 61, 88, 89, 114. *See also* Fourier's law
Heat current, 36, 63–65, 113, 192. *See also* Heat flux; Fourier's law

Heat flux, 24, 113
Heat per unit mass, 24
Heat reservoir, 8, 10, 361
Heisenberg operator, 242
Heisenberg uncertainty relation, 284, 300, 301, 305
Heisenberg's equation of motion, 210, 213, 222, 251, 259, 260, 323, 350
 for two-component system, 353
Hilbert expansion, 197–98
History of nonequilibrium thermodynamics, 2, 2 n
History of stability theory, 72 n
Hooke's law, generalized, 29
$\mathcal{H}$ quantity, Boltzmann's, 192–96, 329
 coarse-grained, 365
 Gibbs' fine-grained, 361, 364
$\mathcal{H}$ theorem, Boltzmann's, 194, 324, 327, 331, 358
 coarse-grained, 365, 369
 generalized, 364, 369
Hydrodynamic approximation to Boltzmann equation, 198–206
Hydrodynamic derivative, 20, 35. *See* Barycentric, material, or substantial derivative
Hydrodynamic modes, 318–20
Hydrodynamic regime, 150–51, 197, 235
Hydrostatic pressure, 22, 24, 38, 113, 169. *See also* Pressure
Hysteresis effects in Bénard convection, 140 n

Ideal gas
 correlations, 280–86
 equation of state, 13
 in the grad limit, 188
Incoherent cross section, 267
Infinitesimally slow process, 5
Information theory, 327, 361–64
Instantaneous entropy, 7, 9
Intensive thermodynamic variables
 fluctuations, 10–12
 as Lagrange multipliers, 10–11
Interacting populations, 99 n, 101 n
Interaction picture, 244
Internal body torques, 26
Internal energy, 3, 4, 10
 fluctuations, 8
 instantaneous, 7–8
 See also Balance equation for internal energy

Internal energy convection, 168
Internal energy density, 23
  microscopic definition, 158
Internal energy flux. See Heat flux
Internal quantum numbers, 211
Irreversible process, 1, 323
Irreversibility, 152, 182, 187, 322–69, 370, 374
  in the Boltzmann equation, 327–40
  conservative, 325, 327, 363
  consumptive, 325, 327, 363
  in an ensemble, 358–64
  in systems of coupled oscillators, 341–58
  due to thermodynamic limit, 358

Jacobi identity, 222 n

Kelvin model for viscoelastic solids, 31
Kelvin's stationary phase, 355
Kinematic viscosity, 115, 273, 278, 320
Kinetic energy, 22
  microscopic definition, 158
  See Balance equation for kinetic energy
Kinetic energy-stress tensor, 230, 352
Kinetic equations, 151, 154, 158, 177, 183, 187, 205, 296
  See Boltzmann equation; Master equations; Vlasov equation
Kinetic pressure tensor, 161, 165, 224, 231
  See Balance equation for kinetic energy
Kinetic regime, 150–51, 370
Kinetic theory
  density expansion, 190
Klein's lemma, 366–67
Knudsen's formula for thermodiffusion, 62
Kramers-Kronig dispersion relations for response functions, 238

Lagrange coordinate system, 19
Landau condition in superfluidity, 37
Langevin equation, 50, 51, 351 n
Laplace equation, 73
Legendre transform, 8
Lennard-Jones potential, 151, 174
Liapounoff function, 75, 76, 82, 84, 85, 90, 102
  in Bénard convection, 145
  for linear stability, 83–85
  for Volterra-Lotka model, 102
  See also Stability
Liapounoff stability, 73–76. See also Stability
Liapounoff's second method, 75. See also Stability
Lie algebra, 222 n
Limitation of ensemble theory, 11
Limit cycle, 107–10
Linear constitutive laws. See Linear phenomenological laws
Linear phenomenological laws, 40–70, 94, 152, 194
  from Boltzmann equation, 196–206
  microscopic theory, 169–77
  See also Fourier's law; Viscosity
Linear response theory, 152, 235–95
Linearized hydrodynamics, 269–276
Liouville's equation, 154, 156, 157, 168, 240, 296, 313, 328, 359 360, 361
Liouville's operator, 241, 242, 317
Liouville's theorem, 53, 181, 191
Liouville-von Neumann equation, 211, 244, 313, 315
Liquid crystals, 37
Local equilibrium
  from Boltzmann equation, 3, 204
  criterion, 2–7
  density matrix, 228–29
  deviations from, 205 n
  in a gas, 12–15
  in a liquid, 15–16
  in a solid, 15
  strict, 194 n, 229
Local potential in stability analysis, 90
Long range order, 221
Lorentz force, 277
Lorenz number, 291
Loschmidt demon, 334
Loschmidt paradox, 327, 330–36

Macroscopic operators, 300–2
Macrovariables, 149–50
Markov process, 288
Markoffian rate equation, 288, 297
Mass current, 19
Mass density, 18
  microscopic definition, 158, 215, 230
  See also Balance equation for mass density
Master equation, 289, 291, 292, 296–321, 366, 369, 370

Brussels, 312–21
Pauli, 297–306, 312
Prigogine-Resibois, 312–21
van Hove, 289, 306–12
Material derivative, 20, 35. *See also* Hydrodynamics, barycentric, or substantial derivative
Maxwell-Boltzmann statistics, 283, 338, 354, 376, 378
Maxwell demon, 334
Maxwell distribution function. *See* Distribution functions
Maxwell gas, 202–3, 384
Maxwell model for viscoelastic fluids, 31, 32
Maxwell's equations, 326
Mean field equation, 179
Mean-free path, 3, 13–15, 62, 170, 183, 187–88, 196, 204
  for various gases, 13
Memory effects, 29
Microscopic reversibility, 47, 50
  and Loschmidt paradox, 330–36
  and macroscopic irreversibility, 323–27, 366
  and Zermélo paradox, 336–40
Minimum entropy production, 69, 71, 109
  theorem of, 67–70, 87, 90
Modulus
  bulk, 30
  shear, 30
Molecular dynamics, 330
Mole number fluctuations, 10. *See also* Fluctuations; Local equilibrium
Moment of inertia per unit mass, 25
Momentum balance. *See* Balance equation for momentum
Momentum density, 29. *See also* Balance equation for momentum
Multicomponent system, 19–20, 210, 348–58

Navier-Stokes equation, 27, 206, 219, 221, 233
  Fröhlich's derivation, 214–21
Nematic liquid crystal, 37
Neutron diffraction, 271
Neutron scattering, 263
Newtonian fluid, 30, 113, 272
Newton's law of friction, 43
Nonequilibrium phase transitions, 2, 123, n

  in chemical reactions, 95–98
  of first order, 97
Nonlinear theory of Bénard convection, 136–43
Nonsymmetric pressure tensor, 25–27
Normal mode analysis for Bénard convection, 118–23
Normal solution of Boltzmann equation, 203. *See also* Chapman-Enskog solution; Hydrodynamic solution
Nusselt number, 130

Ohm's law of electrical conduction, 3, 42, 236 n, 294
  derivation from Boltzmann equation, 206–8
  deviations from, 208 n
Onsager reciprocity relations, 41, 44–60, 261
  and detailed balance, 46–56
  experimental verification, 66–67
  from linear response theory, 258–61
  in thermodiffusion, 60–62
  in thermoelectricity, 64–67
  theorem, 57
  and transformation properties of forces and fluxes, 56–60
Onset of turbulence, 142 n
Operator balance equations, 221–26
Oscillating chemical reactions, 107–10
Oscillating convection rolls, 143
Oscillations in Volterra-Lotka model, 104–6
Oscillatory instability in Bénard convection, 140
Overstability, 121

Parity transformation, 249
  and symmetry of response functions, 249–50
Particle in a separable potential, 396–401
Peltier effect, 66
Perturbation theory, 236–37, 298, 385–86
Phase coherence, 227 n
Phase space
  Γ space, 52, 153, 241, 300, 302
  Υ space, 156 n
Phase transitions, fluctuations near, 5. *See also* Nonequilibrium phase transitions

Phonons
  density of states, 15
  mean free path, 15
Plasma
  Vlasov equation, 179
Poincaré cycle. *See* Poincaré recurrence
Poincaré recurrence, 336–43, 361
Poincaré recurrence time, 336, 339–41, 364
  for harmonic systems, 337
Poisson bracket, 154, 242, 313
Polar elastic materials, 25
Polymer rheology, 28
Potential energy
  external, 23
  internal, 158
  See also Balance equation for potential energy
Power, 22
Prandtl number, 115, 119, 121, 137
  for air, 142
  from Boltzmann equation, 203
  for water, 140
Pressure (hydrostatic), 3, 24
  fluctuations, 12, 16
  microscopic expression, 163, 173, 192
Pressure tensor, 21–22
  microscopic expression of, 162, 173, 218, 220–21
  nonsymmetric, 25–27
  and stress tensor, 29
  viscous, 26
  See also Kinetic pressure tensor
Prigogine's approach to nonequilibrium statistical mechanics, 312–21
Principle of the least dissipation of energy, 67
Probability function for fluctuations, 7, 9, 48. *See also* Fluctuations
Projection operator, 314

Quantum hydrodynamics, 209
Quantum-mechanical response theory, 243–46
Quasiparticle, 310
Quasistatic process, 4–5

Radial distribution function. *See* Distribution functions
Radiation damping, 324, 325 n, 326
Random phase approximation, 299, 306, 312

Rate equations, 94
Rate of strain tensor, 29
Rayleigh number, 115–16, 119, 121–24, 134, 137, 140
  critical, 116, 119, 121–24, 136, 146
  measured, 128–31
Reaction rate, 20, 91
Reduced density matrix, 209–15, 219, 221–22, 226
Reduced distribution function. *See* Distribution functions
Regression of fluctuations, 51–52, 54, 237–38
Regression of time. *See* Time scales
Relaxation function, 246, 257
Relaxation time. *See* Time scales
Relaxation time approximation in Boltzmann equation, 206
Resolvent of evolution operator, 313
Resonances, 372, 390–96
Response functions, 242–43, 246
  analyticity, 247–50
  and correlation functions, 253–54
  Fourier transforms, 247–50
  for ideal gas, 284
  symmetries, 247–50
Reversible process, 1, 4, 5, 33
Rigid body rotation, 27
Rigid surface in Bénard convection, 117, 123
Rotational viscosity, 26. *See also* Viscosity coefficients

Scattering cross section, 186, 188, 264, 266–67, 401
Scattering length, 264
Schrödinger annihilation operator, 281
Schrödinger equation, 49 n, 263, 345–47, 397
Schrödinger picture, 244
Second law of thermodynamics, 32–36, 77, 339
Second quantization, 201, 244, 371
Seebeck effect, 65, 66
Self-organization of matter, 110 n
Shear modulus, 30, 234
Shear stress, 117
Shear viscosity, 26–27, 113, 173, 203, 221. *See also* Kinematic viscosity; Viscosity coefficients
Solenoidal velocity field, 137
Sound mode, 320
Sound velocity, v_s, 4, 275

Sources in balance equations, 18–19
Specific heat at constant pressure, 9
Specific heat per mole at constant volume, 8, 275
Spin, 25–27
Spin-echo experiment, 331–36
Spin system, 209, 316. See also Antiferromagnet
Spin wave in antiferromagnet, 37
Stability
  asymptotic, 75, 100
  of chemical reactions, 79, 92
  of diffusion, 79
  general theory, 71–76
  hydrodynamic, 85, 145
  Liapounoff's second method, 75–76
  linear, 183
  marginal, 121, 123, 140, 146
  mechanical, 79
  practical, 75
  thermal, 79, 88–90
  thermodynamic, 85, 143, 145
Stability and fluctuations, 71–90
Stability and the evolution criterion of Glansdorff and Prigogine 85–90, 143–48
Stability diagram for Bénard covection, 135–36
Stability of equilibrium states, 76–82
Stability of nonequilibrium states, 82–85
Statistical operator, 211–12, 281. See also Density matrix; Density operator
Stochastic reaction model, 98
Stoichiometric coefficient, 20, 71
Stokes' flow, 54 n
Stosszahlansatz, 182–83, 187, 189, 237, 305, 324, 327, 335, 339
Strain flux tensor, 38
Strain tensor, 29
  balance equation for, 38
  local, 38
  rate of, 29
Streamlines, 126–27
Stress tensor, 29–30
  balance equation for, 30
  viscoelastic, 30
Strict local equilibrium, 194, 194 n, 228
Subdynamics, 317, 369
Substantial derivative, 20. See also Barycentric, hydrodynamics, or material derivative
Sum rules from linear response theory, 238, 251–53
  for particle density, 25
Superconductors, 209, 221
Superfluid, 31
  hydrodynamics, 209, 221
  velocity, 37
Superoperator, 313
Superposition principle, 185, 187
Surface tension-driven convection, 133
Surface traction, 29

Temperature fluctuations, 10–16. See also Fluctuations; Local equilibrium
Tensor operators, 221–26
  balance equations, 221
Tensor product (dyadic), 21
Theory of subdynamics, 317, 369
Thermal conduction. See Fourier's law; Heat conduction
Thermal conductivity, 14, 65, 68, 292–93
  from Boltzmann equation, 201–2
Thermal diffusivity, 115, 276, 320
Thermal expansion, 9, 113, 115
Thermal time, 151 n
Thermocouple, 63, 66
Thermodiffusion, 60–62
Thermodynamic flux, 36, 40–42
Thermodynamic force, 36, 40–42
Thermodynamic limit, 338, 358
Thermoelectric current, 62
Thermoelectricity, 63–67
Thomson relation, second, 66–67
Time evolution. See Transient regime; Kinetic regime; Hydrodynamic regime
Time evolution of an ideal gas, 370–401
Time evolution of a two-component gas, 348–48
Time parity, 41 n, 42, 248, 250
Time reversal (inversion), 41–42, 52, 248, 328
  See also Loschmidt paradox; Microscopic reversibility
Time scales, 4–7, 12–13, 150–51. See also Transient regime; Kinetic regime; Hydrodynamic regime
Tisza's approach to thermodynamics, 7

T-matrix
  Breit-Wigner form, 391
  partial wave, 398
Torque, 25–27
  internal body torque and symmetry of pressure tensor, 26
Transformation properties of Onsager relations, 56–60
Transient effects in an ideal gas, 370–401
Transient regime, 150–51, 235, 370
Transition probabilities, 297, 299, 304, 308, 310, 367–69
  Markoffian, 298
Translation invariance, 254
Transpiration of a gas through a porous plug, 62
Transport coefficients, 42
  from Boltzmann equation, 196–206
  in linear response theory, 238, 260–61, 269–86
  microscopic theory, 169–77
  See also Fourier's law; Onsager reciprocity relations; Viscosity
Transport laws. *See* Linear phenomenological laws
Triple collisions, 180, 182, 190
  in gas reactions, 180 n
Two-body interaction, 210, 217, 220
Two-body scattering
  summational invariants, 191, 198
Two-particle correlation length, 15

Umklapp processes, 15
Uncertainty principle and time scales, 151 n
Uncertainty relation, 284, 305
Unit tensor, 22

van der Waals forces, 179
van der Waals gas, 96, 98
van der Waals potential, 174. *See also* Lennard-Jones potential
van Kampen's criticism of linear response theory, 294–95
Variational principle, 148, 229 n

Velocity, 19
  angular, 25
  center-of-mass, 20
  fluctuations, 81, 144–46
  microscopic definition of, 158, 216, 226–27
Velocity field
  solenoidal, 137
  irrotational, 227
Velocity inversion experiment, 330
Velocity operator, 216
Virial equation of state, 96
Viscoelastic fluid, 29–31
Viscoelastic solid, 29, 31–32
Viscoelastic stress tensor, 30
  balance equation for, 28–32
Viscoelasticity, 28–32, 222, 234
Viscosity coefficients, 26–27
  Andrade's theory, 173
  from Boltzmann equation, 202–3
  in ideal gas, 286
  from linear response theory, 278–79
  from master equation, 320–21
  microscopic theory, 170–73, 221
  See also Bulk viscosity; Navier-Stokes equation; Rotational viscosity; Shear viscosity
Viscous pressure tensor, 26–27
Vlasov equation, 154, 158, 188
  derivation, 177–79
Volterra-Lotka model, 98–107
  of predator-prey interactions, 99
Volume fluctuations, 9. *See also* Fluctuations; Local equilibrium
Vorticity, 116–18

Wiedemann and Franz law, 286
Wigner distribution function. *See* Distribution functions
Wigner-Weisskopf model, 351 n

Zermélo paradox, 327, 336–40
Zig-zag instability in Bénard convection, 140
Zubarev's variational principle, 229

This book on nonequilibrium physics provides a concise exposition of the basic theoretical ideas of all schools of thought with applications to explicit physical systems. Where possible, the experimental findings are also discussed. The author begins with a concise description of nonequilibrium thermodynamics encompassing an analysis of stability. He then applies these ideas to chemical and biological reaction networks and to Benard convection. The foundations of nonequilibrium thermodynamics are laid in a classical as well as a quantum-mechanical theory. Linear response theory, master equations, and irreversibility and the approach to equilibrium are then discussed. The book concludes with a look at transient effects in the time evolution of large systems based on exactly soluble models.

H. J. Kreuzer is Killam Research Professor in the Department of Physics, Dalhousie University, Nova Scotia.

From reviews of the hardback edition:

... the author has woven the parts together in a pleasing way and provides what is certainly one of the best available introductions to the subject ... *Physics Bulletin*

There is probably no more formidable subject than that of nonequilibrium statistical physics ... It is clear throughout Professor Kreuzer's monograph that he has spared no effort in coming to grips with this problem. *Contemporary Physics*

... the author successfully gives us a quick bird's-eye view of this rapidly growing field. *Mathematical Review*

ALSO PUBLISHED BY OXFORD UNIVERSITY PRESS

Thermodynamics and statistical mechanics
P. T. Landsberg, 1983

OXFORD UNIVERSITY PRESS

£15.00 net in UK
Also available in hardback

ISBN 0 19 851375 5